Sharon Sead
Chicago
1988

OPTICS

OPTICS

K. D. Möller

Professor of Physics
Fairleigh Dickinson University
Teaneck, New Jersey

University Science Books
Mill Valley, California

University Science Books
20 Edgehill Road, Mill Valley, CA 94941

Library of Congress Catalog Card Number 86-050343
ISBN 0-935702-145-8

PRINTED IN THE UNITED STATES OF AMERICA
10 9 8 7 6 5 4 3 2 1

Production Credits

Book Production: Greg Hubit Bookworks
Copyediting: Janet Greenblatt
Text Design: Robert Ishi
Jacket and Color Insert Design: Nancy Benedict
Jacket Sculpture: Susan Luery Franchi
Jacket Photo: Chris Armstrong

Frequent reference to the following publications is made throughout this text in abbreviated form.
Full publication data follows. Grateful acknowledgment is given for permission to reproduce
materials from these books as well as from other sources cited in the text.

Hecht, Eugene, and Alfred Zajac. *Optics*. Reading, Mass.: Addison-Wesley Publishing Company,
1979.
Cagnet, Michel, Maurice Françon, and Jean Claude Thrierr. *Atlas of Optical Phenomena*. Berlin:
Springer-Verlag, 1962.
Pohl, R. W. *Einführung in die Optik*. Berlin: Springer-Verlag, 1976.

To
Robert Wichard Pohl
1884–1976
Professor of Experimental Physics
University of Göttingen

Preface

This book describes optics first by using the principles of geometrical optics, inter-ference, and diffraction and then by using Maxwell's theory. The book may be used for a one-semester introductory junior-level course for students in physics and electrical engineering. It is assumed that the students have some knowledge of calculus and have had some introduction to optics in a general physics course. The table below suggests which sections of Chapters 1 through 7 may be used for an introductory course. For a one-semester advanced junior or senior course, Chapters 1 to 7 may be covered in their entirety.

If a two-semester course is desired, the instructor may select from the remaining chapters of the book those sections most relevant to the students' needs. It is also possible to teach Chapters 8 through 18 as a one-semester course in applied optics.

Chapters	Introductory Optics (1 Semester) Sections	Intermediate Optics (1 Semester) Sections	Optics for Electrical Engineering (2 Semesters) Sections
1	1–5, 8	all	1–5, 8
2	all	all	all
3	1, 2	all	1, 2
4	1–3	all	1–3
5	1–3	all	1–3
6	1–4	all	1–4
7	1–3	all	1–3
8		all	all
9		all	all
10			1–4, 6
11			1–5
12			1–2
13			—
14			all
15			all
16			—
17			all
18			all

The problems are often formulated as working problems including hints for the solutions. The *Solutions Manual* contains not only solutions to all problems but also some derivations not included in the text. The availability of solved problems may help applications-oriented readers find solutions to their particular problems more quickly by analogy.

I would like to thank V. P. Tomaselli, R. Bell, C. Belorgeot, A. Schadowitz, and N. B. Bai for reviewing the manuscript and providing many helpful suggestions. For assistance in reading proofs, I also thank E. Angelidis, F. Fahnert, O. Haase, B. Kremer, H. S. Lakaraju, C. Miller, B. C. Moeller, P. Petrone III, A. Pluchino, M. K. Poutous, N. G. Ugras, K. B. Soldano, P. Walsh, R. G. Zoeller, and Yu-Faye Chao. Finally, I thank my wife for her continuous support and for always keeping me in good spirit.

K. D. Möller

Contents

Contents

Geometrical Optics | 1

1. INTRODUCTION

In this chapter, we will develop and discuss the principles of geometrical optics. We will find that a large number of optical image-forming phenomena can be described by a few simple laws. These laws may be deduced from **Fermat's principle** and, ultimately, from **Maxwell's theory of electromagnetic waves**.

It is assumed that light travels in straight lines and is partially refracted and reflected at an interface. Interfaces are considered to be either flat or spherical. Also, light rays are assumed to travel close to the axis of the system, that is, in accordance with paraxial theory. Deviations from this assumption will be discussed in Chapter 13 (Aberrations). The formulas and principles developed can be applied to thin and thick lenses and lens systems, as well as to mirrors and mirror systems.

2. LIGHT: ITS TRAVEL IN STRAIGHT LINES AND ITS SPEED IN A VACUUM

The formation of shadows by light is a simple demonstration of its travel in straight lines. But if very accurate measurements were taken along the edge of a shadow, we would not find a sharp contrast between light and shadow. This phenomenon is discussed in Chapter 3. In this chapter, however, we will assume that a discontinuous change of intensity exists.

A. Image Formation and the Pinhole Camera

A topic of particular interest in optics is image formation. The pinhole camera is essentially a box with a small hole (or pinhole) in the front. If we place an illuminated object in front of this pinhole, we will observe an inverted image on the back of the box, which serves as an observation screen. As seen in Figure 1.1, the object, pinhole, and image together form two similar triangles, one inside the

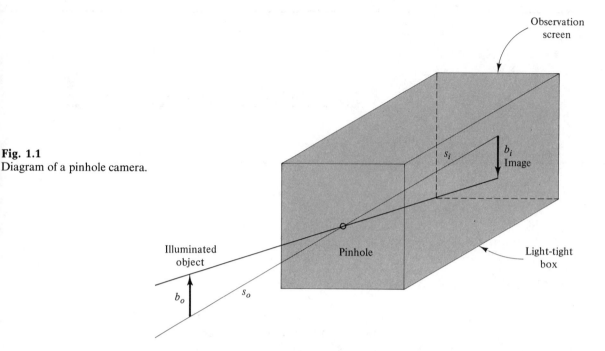

Fig. 1.1
Diagram of a pinhole camera.

box and one outside the box. Using the properties of similar triangles, we then get the relationship given in equation 1.1, where b_o is the height of the object, b_i is the height of the image, s_i is the distance between the base of the image and the pinhole, and s_o is the distance between the base of the object and the pinhole.

$$\frac{b_i}{s_i} = \frac{b_o}{s_o} \quad \text{or} \quad \frac{b_i}{b_o} = \frac{s_i}{s_o} \tag{1.1}$$

An image is formed when rays of light extend from each point on the object through the pinhole to a specific image point. Although the image produced is perfect, it is very faint because the pinhole does not admit much light into the camera. The magnification of the image, b_i/b_o, is equal to the ratio of the distances, s_i/s_o.

The formation of an image by a pinhole can be demonstrated in a simple experiment. Place a piece of paper with a pinhole close to your eye and hold it at varying distances from a page of text. You will find that you can read the page at distances shorter than the naked eye can as long as there is sufficient light. In fact, at very short distances, magnification is observed. In this experiment, the refraction at the eye is not used: the pinhole and the sensitive area of the retina together form a pinhole camera (Figure 1.2). We then see writing on the page upright, whereas according to Figure 1.1, the image is upside down. Our brain produces an inversion of the observed image.

B. The Speed of Light

The speed at which light travels in a vacuum, c, is a universal constant, and its magnitude can be demonstrated by modern techniques. For example, laser light

1. Geometrical Optics

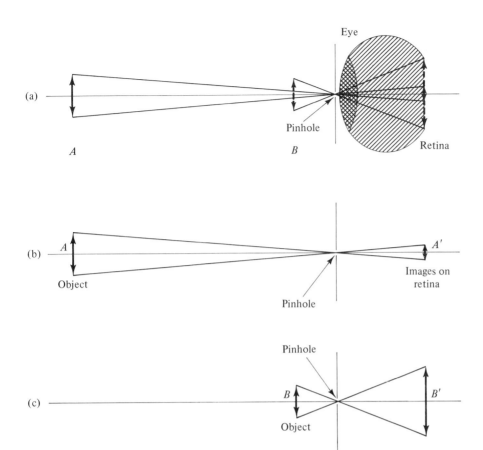

Fig. 1.2
Pinhole camera consisting of a pinhole and the retina of the eye. Since only the center of the eye is used, it has no image-forming effect. The image A' of the object A is reduced, whereas the image B' of the object B is magnified.

can be reflected from the moon via a reflector placed there by astronauts in 1969. The earth-moon distance is known, and the time it takes for emission and return of the laser flash can be measured. In a vacuum, the ratio of the light path distance to the travel time results in a value of 3×10^{10} cm/s. This value (more accurately 2.997924562×10^8 m/s) represents the rate at which light is propagated in a vacuum. In air, the speed of light is slightly slower. Furthermore, in other materials, it might be considerably slower, as we will see in almost all chapters, but particularly in Chapter 7.

3. REFRACTION AT PLANE SURFACES

A. The Law of Refraction

You are probably familiar with such optical instruments as a magnifying glass, binoculars, a telescope, and a microscope, and you may be wearing glasses or contact lenses. Also, our eyes provide us with an independent imaging system. We can understand how all these optical devices form images if we consider the image-forming properties of a lens made of glass or some other transparent material.

The bending of a light ray on passing through an interface (say, an air-glass interface) is called **refraction**. When we observe a rock at the bottom of a pond, the rock appears in a different position than it really is because the light rays are refracted; that is, the light does not travel in a straight line (Figure 1.3).

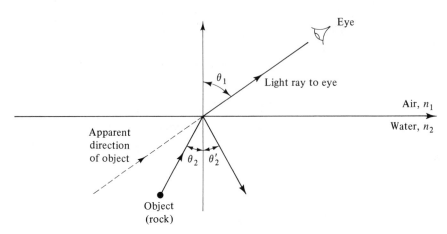

Fig. 1.3
Diagram showing how light rays are reflected and refracted at a water-air interface. The object, a rock in this case, is illuminated by stray light. Light coming from the object is partly refracted at the interface and traveling to the eye, and partly reflected. The direction of the apparent object is as indicated. The incident, refracted, and reflected rays are in one plane.

The quantitative relationship between angles θ_1 and θ_2 of Figure 1.3 is called **Snell's law of refraction**, which is given by $n_1 \sin \theta_1 = n_2 \sin \theta_2$. The constants n_1 and n_2, which are called the **indices of refraction** (or **refractive indices**), depend on the materials used. The constant n is related to the velocity v of the light passing through a material by $n = c/v$.

For reflected light, the law of reflection holds; that is, $\theta_2 = \theta_2'$ (Figure 1.3). Since transparent media absorb little light, the progression of light in such media can be described by considering Snell's law and the law of reflection only. In most of the discussions on refraction, however, we do not pay attention to the reflected light. Reflection at mirrors is discussed at the end of this chapter.

B. Fermat's Principle

We may ask, Is there a single principle that can be used to deduce all the properties of light? And can these properties then be utilized to analyze any observation? Maxwell's equations of electromagnetic theory do, in fact, constitute such a principle. However, the laws of reflection and refraction can be derived from Fermat's principle. This principle is very useful for the analysis of geometrical optical phenomena. Furthermore, it states that while light travels in straight lines in each medium, it always chooses the path requiring the shortest *total time*.

1. Geometrical Optics

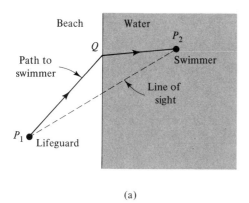

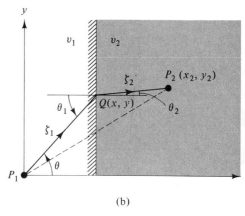

(a)   (b)

Fig. 1.4
The application of Fermat's principle to a lifeguard rescuing a swimmer. Figure (a) identifies the
essential features, and (b) establishes the coordinates.

Fermat's principle can be illustrated by the following example (Figure 1.4). A
lifeguard (analogous to the light beam) on the beach spots a swimmer in distress
in the water. The goal is clear: Reach the swimmer in the minimum time. The
lifeguard can run on the beach at a speed v_1 and can swim in the water at a speed
v_2, where v_1 is greater than v_2, since the lifeguard can run on the beach faster
than he can swim in water. Thus, the problem is to determine the path the
lifeguard should take to minimize the *total time*.

    Let $Q(x, y)$ be the point at which the lifeguard enters the water. Now,
Q may vary in y, but not in x. The times of travel on the beach and in
the water are $t_1 = \zeta_1/v_1$ and $t_2 = \zeta_2/v_2$, respectively, where $\zeta_1^2 = x^2 + y^2$ and
$\zeta_2^2 = (x_2 - x)^2 + (y_2 - y)^2$. Minimizing the time yields

$$\frac{d}{dt}(t_1 + t_2) = 0 \quad \text{implying that} \quad \frac{d}{dy}\left(\frac{\zeta_1}{v_1} + \frac{\zeta_2}{v_2}\right) = 0 \qquad (1.2)$$

Substituting for ζ_1 and ζ_2 and taking the derivatives with respect to y while
keeping v_1 and v_2 constant, we get

$$\frac{1}{2}\left(\frac{2y}{v_1\sqrt{x^2 + y^2}}\right) + \frac{1}{2}\left(\frac{2(y_2 - y)(-1)}{v_2\sqrt{(x_2 - x)^2 + (y_2 - y)^2}}\right) = 0 \qquad (1.3)$$

From Figure 1.4,

$$\sin\theta_1 = \frac{y}{\sqrt{x^2 + y^2}}$$

and

$$\sin\theta_2 = \frac{y_2 - y}{\sqrt{(x_2 - x)^2 + (y_2 - y)^2}}$$

Thus, substituting in equation 1.3 yields

$$\frac{\sin\theta_1}{v_1} = \frac{\sin\theta_2}{v_2}$$

3. Refraction at Plane Surfaces

By recalling that $n_1 = c/v_1$ and $n_2 = c/v_2$, we can apply this result to the analogy of an optical system to get Snell's law:

$$n_1 \sin \theta_1 = n_2 \sin \theta_2 \qquad (1.4)$$

Since $v_1 > v_2$ and $\sin \theta_1 / \sin \theta_2 = v_1/v_2$, then $\theta_1 > \theta_2$ for a minimum time of travel. Note that although the time required to traverse the path $P_1 Q P_2$ is a minimum, the distance is clearly not.

The law of reflection can be obtained as well from Fermat's principle.

In equation 1.2, it was shown that $t_1 + t_2$ is a minimum if $\zeta_1/v_1 + \zeta_2/v_2$ is also a minimum. Therefore, according to the definition of the index of refraction, $n_1\zeta_1/c + n_2\zeta_2/c$ (or $n_1\zeta_1 + n_2\zeta_2$) is also a minimum. The quantity $n\zeta$ is called the **optical path length**, and Fermat's principle states that the total optical path length is a minimum, regardless of the media being traversed.

C. Two Refracting Surfaces

Two plane parallel interfaces. When a light beam is incident on two parallel interfaces forming a plane parallel plate of refractive index n surrounded by a vacuum (or, approximately, air), the law of refraction is applied twice, first when the beam of light passes from the medium with index $n = 1$ to the medium with index n, and then again when the beam passes from the medium with index n to the medium with index $n = 1$. Figure 1.5 shows that although the emerging ray is parallel to the incident ray, it is parallel **displaced**.

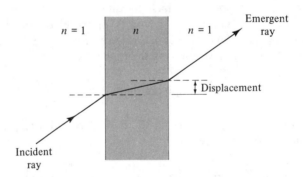

Fig. 1.5
A light beam is refracted twice on passing through a plane parallel plate.

Two interfaces at an angle. Two refracting surfaces at an angle A form a refracting prism. The propagation of light through such a prism is more complicated than the propagation of light through a plane parallel plate. Figure 1.6 shows the angles of incidence and refraction for the two refraction processes. The angle of incidence θ_1 and the angle of refraction θ_2 are involved in the first refraction process. The angle of incidence θ_2' and the angle of refraction θ_1' are involved in the second refraction process.

We can express the angle of deviation δ as a function of the angle of incidence θ_1 and the apex angle A by considering the following relations:

1. Geometrical Optics

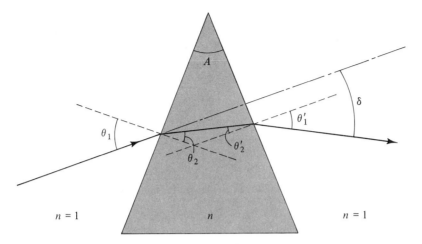

Fig. 1.6
Refraction at a prism with an apex angle A and refractive index n. θ_1 is the angle of incidence and θ_2 is the angle of refraction for the first process. For the second process, θ_2' is the angle of incidence and θ_1' is the angle of refraction. The angle of deviation is δ.

$$\delta = \theta_1 - \theta_2 + \theta_1' - \theta_2'$$
$$A = \theta_2 + \theta_2'$$

Then, applying the law of refraction

$$\sin\theta_1 = n\sin\theta_2$$
$$n\sin\theta_2' = \sin\theta_1'$$

yields

$$\delta = \theta_1 + \sin^{-1}[(n^2 - \sin^2\theta_1)^{1/2}\sin A - \sin\theta_1\cos A] - A$$

If the path through the prism is symmetric, the value of δ is a **minimum** (see Problem 4 in Chapter 7) and is expressed as

$$\delta_m = 2\sin^{-1}\left(n\sin\frac{A}{2}\right) - A$$

The refractive index n of a material can be determined from the angle of minimum deviation measured in a prism with a known apex angle. Although this simple method is an old one, it is still used for accurate measurement of the index of refraction.

D. Combination of Prisms

Two similar prisms can be combined to produce different effects. Figure 1.7 shows two such combinations and the path of light through each. In (a), the arrangement is that of a plane parallel plate. The setup in (b), on the other hand, is a prototype of image formation, since two symmetric rays from one point on the axis meet at another point on the axis after traversal of the prism.

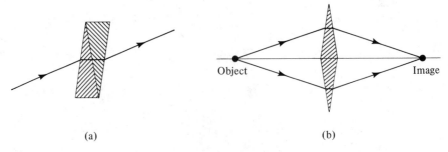

(a)   (b)

Fig. 1.7
Combination of two prisms, showing the path of the light rays passing through each. (a) Plane parallel plate arrangement. (b) Arrangement for image formation.

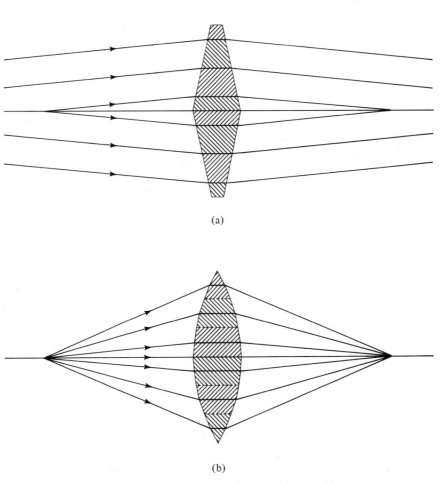

(a)

(b)

Fig. 1.8
Image formation with prisms. (a) Symmetric pass of light through two prisms. (b) Symmetric pass of light through a set of prisms with different apex angles.

Image formation with a set of prisms. If light passes symmetrically through a prism, then parallel rays of light incident on the prism will emerge parallel. For two prisms, light passes as shown in Figure 1.8a. To produce an image from an on-axis point, the apex angle of each section of the prism can be changed as shown in Figure 1.8b. Here, the prism surfaces can be thought of as approximating tangents to a circle. In fact, as the height of the prism is made smaller and smaller, a circle is actually approached. If rotatory symmetry around the axis of the system is also considered, then a spherical surface is approached; that is, we obtain the characteristic structure of lenses.

E. Fresnel Lenses

Figure 1.9a shows a set of prisms symmetrically arranged. The cross sections approximate the cross section of a lens. If the middle part of each prism is cut out, the wedges in Figure 1.9b are produced. These wedges have the same refraction properties as the prisms. A notably symmetric arrangement of the prisms is called a **Fresnel lens**. A. J. Fresnel (1788–1827) was a French physicist best known for his work on diffraction. Fresnel lenses are usually made with rotary symmetric grooves on one surface only for convenience of production (Figure 1.9c). The main advantages of Fresnel lenses are that they are lightweight and that they absorb less light compared to a standard lens if made of light-absorbing material. In Chapter 7, we will see that the refractive index depends in general on the color (wavelength) of thc light. Therefore, Fresnel lenses are achromatic; that is, they show different image points of the same object point for different colors (wavelengths) of light.

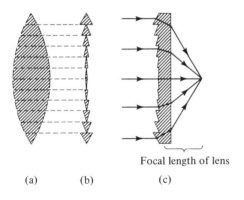

(a)   (b)   (c)

Focal length of lens

Fig. 1.9
(a) A set of prisms refracts light from one point on the axis to an image point, as in Figure 1.8b. (b) The wedges produced when the central part of each prism is cut out. (c) Cross section of a rotary symmetric arrangement of the grooves of a Fresnel lens.

4. REFRACTION AT SPHERICAL SURFACES

A. Refraction at One Surface

A study of the optical properties of spherically curved surfaces is of particular relevance because such a study can be extended to explain the refraction and imaging properties of lenses. Figure 1.10 shows the path of a light ray refracted

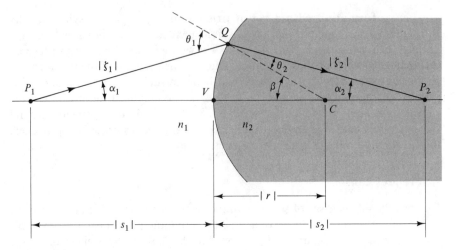

Fig. 1.10
Optical diagram for refraction at a spherically curved surface with semiinfinite media on each side. All angles and distances are positive.

at a spherically curved surface. The light ray travels from an object at P_1 in a medium of refractive index n_1 to a point Q on the interface, and then to the image point P_2 in a medium of refractive index n_2. C is the center of curvature, V is the vertex, and $|r|$ is the radius of curvature. The object and image distances are denoted by $|s_1|$ and $|s_2|$, respectively. All distances are positive, as indicated by the absolute sign, and all angles are associated with positive numbers.

We will now derive the equation for image formation at a spherical interface. Considering only those light rays having small angles α with the optical axis $P_1 P_2$, we are using the **paraxial approximation**. First we apply the sine-theorem to get

$$\frac{|s_1| + |r|}{\sin(180 - \theta_1)} = \frac{|\zeta_1|}{\sin \beta} \quad \text{and} \quad \frac{|s_2| - |r|}{\sin \theta_2} = \frac{|\zeta_2|}{\sin(180 - \beta)} \tag{1.5}$$

From trigonometry, $\sin(180 - \alpha) = \sin \alpha$, and rearranging the terms in equation 1.5 yields

$$\frac{|s_1| + |r|}{|\zeta_1|} = \frac{\sin \theta_1}{\sin \beta} \quad \text{and} \quad \frac{|s_2| - |r|}{|\zeta_2|} = \frac{\sin \theta_2}{\sin \beta} \tag{1.6}$$

Substituting Snell's law of refraction, $\sin \theta_1 = (n_2/n_1)\sin \theta_2$, and combining equations 1.6, we obtain

$$\frac{|s_1| + |r|}{|\zeta_1|} = \frac{n_2}{n_1}\left(\frac{|s_2| - |r|}{|\zeta_2|}\right) \tag{1.7}$$

In the paraxial approximation, $|s_1| \simeq |\zeta_1|$ and $|s_2| \simeq |\zeta_2|$, so from equation 1.7,

$$1 + \frac{|r|}{|s_1|} = \frac{n_2}{n_1}\left(1 - \frac{|r|}{|s_2|}\right)$$

or rearranging,

1. Geometrical Optics

$$\frac{n_1}{|s_1|} + \frac{n_2}{|s_2|} = \frac{n_2 - n_1}{|r|} \qquad (1.8)$$

To investigate whether these equations are valid for all values of $|s_1|$, $|s_2|$, and $|r|$, consider the following two examples.

Example 1. If $n_1 = 1$, $n_2 = 1.5$, $|s_1| = 4$ m, and $|r| = 1$ m, then $|s_2| = 6$ m, as shown in Figure 1.11.

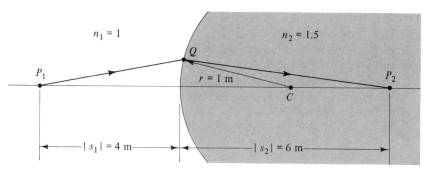

Fig. 1.11
Example 1 for refraction at a spherical (convex) surface. The image point P_2 is to the right of the refracting surface.

Example 2. If, however, $|s_1| = 0.5$ m and $|r| = 1$ m, then we formally obtain $|s_2| = -1$ from equation 1.8. Since we cannot accept a negative value as the result of this calculation, we must reformulate equation 1.8.

We introduce coordinates for the description of object and image distance from the refracting surface. Coordinates may carry positive and negative numbers. If we interpret $|s_2|$ as image distance to the left of the refracting surface and call distances to the left negative, we have a way out of our problem (Figure 1.12). But then we also must give a negative value to $|s_1|$ on the left.

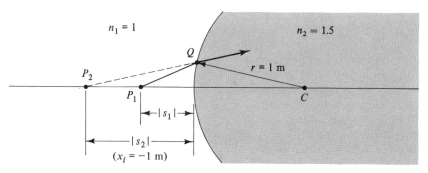

Fig. 1.12
Example 2 for refraction at a spherical (convex) surface. The object point is to the left of the surface, as it is in Example 1. But now the image point is also to the left.

4. Refraction at Spherical Surfaces

B. Sign Convention

Object and image distances, as well as the radius of curvature, can be considered variables with positive or negative values. For this purpose, we will use the following symbols:

object distance, x_o object size, y_o

image distance, x_i image size, y_i

radius of curvature, r

All these variables have values corresponding to distances measured from the vertex V and signs as indicated in Figure 1.13.

If we now assume that light is always incident from the left, then the refracting surface is **convex** if r is positive (from V to C) and **concave** if r is negative (from V to C'). This convention results in a negative value for the object distance, and, consequently, a negative sign for the first term in equation 1.8. Equation 1.8 can then be rewritten as

$$\frac{n_1}{-x_o} + \frac{n_2}{x_i} = \frac{n_2 - n_1}{r} \tag{1.9}$$

where x_o is the object distance, x_i is the image distance, and r is the radius of curvature.

Applying equation 1.9 to the examples shown in Figures 1.11 and 1.12, where $n_1 = 1$ and $n_2 = 1.5$, we obtain the following:

Example 1: $r = 1$ m $x_o = -4$ m $x_i = 6$ m

Example 2: $r = 1$ m $x_o = -.5$ m $x_i = -1$ m

The negative value for x_i in example 2 indicates that the distance 1 m is to the left of V in Figure 1.13.

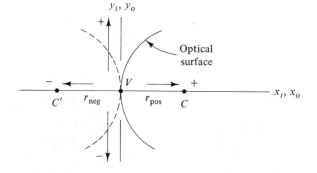

Fig. 1.13
Numerical sign convention for refraction at a curved optical surface. Light is assumed to be incident from the left.

When describing angles, the convention here is that angles are positive if the axis or normal is rotated through an acute angle in the mathematically positive sense—that is, *counterclockwise*. For example, for those angles defined in Figure 1.10, α_1, θ_1, and θ_2 are positive angular displacements, while α_2 and β are negative angular displacements (Figure 1.14).

1. Geometrical Optics

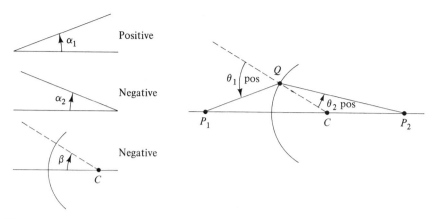

Fig. 1.14
Partial reproductions of Figure 1.10 with emphasis on the angles involved. Both angles θ_1 and θ_2 are positive.

C. Virtual Images and Focal Points

We will now illustrate the use of equation 1.9 and the relationship between the object and image distances for a positive (convex) refracting surface. For $n_1 = 1$ and $n_2 = 1.5$, equation 1.9 becomes:

$$\frac{2}{-x_o} + \frac{3}{x_i} = \frac{1}{r}$$

For various values of x_o expressed in terms of r, we arrive at the values of x_i presented in Table 1.1 and Figure 1.15.

Table 1.1
The relationship between x_o and x_i for a single refracting surface (*convex, r positive*), as computed from equation 1.9 (for $r = 10$)

x_o	x_i	Figure 1.15
$-\infty$	$3r$ $3(10)$	(a)
$-6r$ $-6(10)$	$4.5r$ $4.5(10)$	(b)
$-2r$ $-2(10)$	∞	(c)
$-r$ -10	$-3r$ $-3(10)$	(d)

A similar treatment can be applied when the refracting surface is concave, that is, when r is negative. For example, when $r = -10$, equation 1.9 becomes

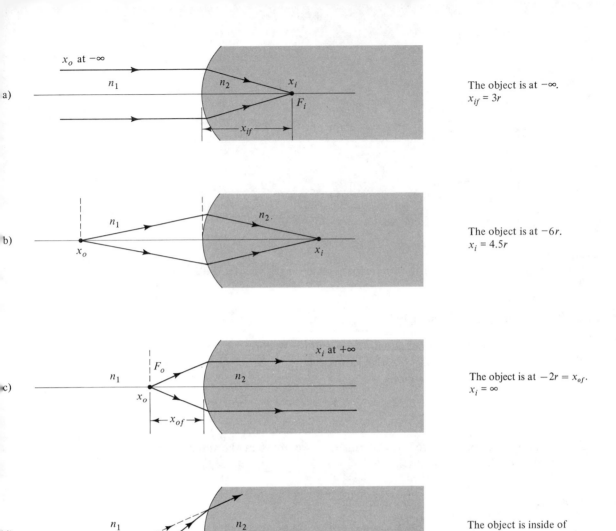

Fig. 1.15

Optical diagrams for single refracting surfaces (convex, r positive) with the object of various positions on the left. The focal points F_i and F_o and their coordinates x_{if} and x_{of} are illustrated here and defined later in the text.

$$\frac{2}{-x_o} + \frac{3}{x_i} = -\frac{1}{10}$$

Values for x_o and x_i are shown in Table 1.2 and Figure 1.16.

Table 1.2
The relationship between x_o and x_i for a
single refracting surface (concave, r
negative), as computed from equation
1.9 (for $r = -10$)

x_o	x_i	Figure 1.16
$-\infty$	$3r$ $3(-10)$	(a)
$6r$ $6(-10)$	$2.25r$ $2.25(-10)$	(b)
$2r$ $2(-10)$	$1.5r$ $1.5(-10)$	(c)
r (-10)	r (-10)	(d)
$-2r$ $-2(-10)$	∞	(e)

In Figures 1.15d and 1.16a–d, the image point is not on the right-hand side of the interface, as in Figure 1.15a and b. In these cases, the light does not converge to a point, but rather diverges. This diverging light can be traced back to a point on the left-hand side of the interface. For an observer on the right-hand side of the interface, the light appears to originate from this point, which is therefore called a **virtual image point**.

When the incident light is parallel to the axis (that is, the object is at $-\infty$) the rays meet at a point on the right-hand side of a convex refracting surface as shown in Figure 1.15a. This point is called the **image focal point** and is designated F_i. Similarly, if the light emerging from the refracting surface is parallel to the axis, as shown in Figure 1.15c, it originates from a single point. This point is called the **object focal point**, designated F_o. The coordinate of the image focal point F_i is called x_{if}. We obtain x_{if} by setting $x_o \to -\infty$ in equation 1.9. Thus, for x_{if},

$$\lim_{\substack{x_o \to -\infty \\ (x_i \to x_{if})}} \left(\frac{n_1}{-x_o} + \frac{n_2}{x_i} = \frac{n_2 - n_1}{r} \right)$$

or

$$\frac{1}{x_{if}} = \frac{n_2 - n_1}{n_2 r} \tag{1.10}$$

For F_o, the coordinate is called x_{of}, and we have

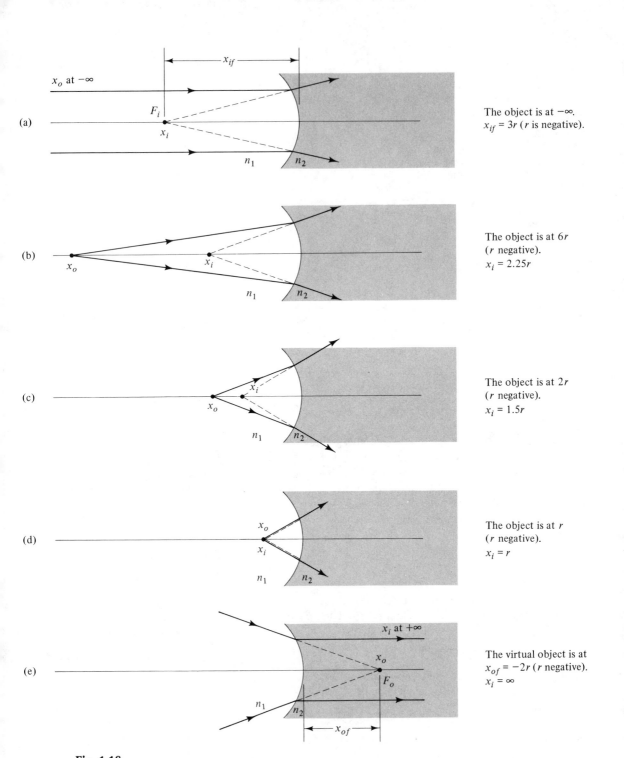

The object is at $-\infty$.
$x_{if} = 3r$ (r is negative).

The object is at $6r$
(r negative).
$x_i = 2.25r$

The object is at $2r$
(r negative).
$x_i = 1.5r$

The object is at r
(r negative).
$x_i = r$

The virtual object is at
$x_{of} = -2r$ (r negative).
$x_i = \infty$

Fig. 1.16
Optical diagrams for single refracting surfaces (concave, r negative) with the object at various positions on the left. The focal points F_i and F_o and their coordinates x_{if} and x_{of} are illustrated here and defined later in the text.

$$\lim_{\substack{x_i \to \infty \\ (x_o \to x_{of})}} \left(\frac{n_1}{-x_o} + \frac{n_2}{x_i} = \frac{n_2 - n_1}{r} \right)$$

or

$$\frac{1}{-x_{of}} = \frac{n_2 - n_1}{n_1 r} \tag{1.11}$$

Equations 1.10 and 1.11 lead to the following relation:

$$\frac{n_2}{x_{if}} = \frac{n_1}{-x_{of}} = \frac{n_2 - n_1}{r} \tag{1.12}$$

For a convex refracting surface (r positive) where $n_1 = 1$ and $n_2 = 1.5$, equation 1.12 yields

$$x_{if} = \frac{n_2 r}{n_2 - n_1} = 3r \quad \text{and} \quad x_{of} = -2r$$

These values are shown in Figure 1.15a and c. Since equation 1.12 also holds for negative values of r, the same result is obtained for a concave refracting surface; that is,

$$x_{if} = 3r \quad \text{and} \quad x_{of} = -2r$$

These values are shown in Figure 1.16a and e.

For a concave (refracting) surface, x_{of} represents the virtual object focal distance. This is shown in Figure 1.16e, where the incident converging light leaves the surface after refraction as parallel light. In this case, the incident light can be traced to converge at F_o (but it does not actually emerge from this point). For magnification on a single surface, see Section 5 of Chapter 9.

D. The Thin-Lens Equation

We can derive the thin-lens equation by considering the image formation for two spherical surfaces. Figure 1.17 shows the process of image formation when three

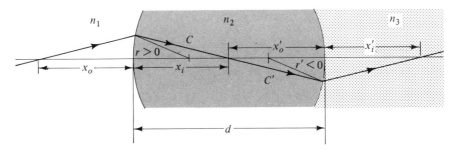

Fig. 1.17
Tracing an optical ray through two refracting surfaces derived from a thick lens of refractive index n_2 with a medium of refractive index n_1 on the object side and a medium of refractive index n_3 on the image side. The distance between the vertices of the two refracting surfaces is d. For a thin lens, d is made very small such that $d \ll r$ and $|r'|$.

different media are involved. Note that the image point of the first imaging process serves as the object point for the final image. Applying the general sign convention to both surfaces, the object distance is associated with a negative number and the image distance with a positive number. Thus,

$$-\frac{n_1}{x_o} + \frac{n_2}{x_i} = \frac{n_2 - n_1}{r} \tag{1.13}$$

and

$$-\frac{n_2}{x_o'} + \frac{n_3}{x_i'} = \frac{n_3 - n_2}{r'} \tag{1.14}$$

The distance d between the two refracting surfaces is expressed as

$$d = x_i + |x_o'| \tag{1.15}$$

where x_o' has a negative value because it is to the left of the second surface. Substituting in equation 1.13 then yields

$$-\frac{n_1}{x_o} + \frac{n_2}{d + x_o'} = \frac{n_2 - n_1}{r} \tag{1.16}$$

To derive the thin-lens equation, we assume d to be small compared to all the other distances involved, including r and $|r'|$. This is best visualized by considering the special case where the first surface produces a virtual image (Figure 1.18). Setting $d = 0$ and substituting the expression for n_2/x_o' from equation 1.16 into equation 1.14 yields

$$-\frac{n_1}{x_o} - \frac{n_3 - n_2}{r'} + \frac{n_3}{x_i'} = \frac{n_2 - n_1}{r} \tag{1.17}$$

Rearranging the terms, we get

$$-\frac{n_1}{x_o} + \frac{n_3}{x_i'} = \frac{n_2 - n_1}{r} + \frac{n_3 - n_2}{r'} \tag{1.18}$$

We assume that $n_1 = n_3 = 1$ (i.e., in air) and $n_2 = n$, and we obtain

$$-\frac{1}{x_o} + \frac{1}{x_i'} = \frac{n - 1}{r} + \frac{1 - n}{r'} \tag{1.19}$$

Defining

$$\frac{n - 1}{r} + \frac{1 - n}{r'} = \frac{1}{f} \tag{1.20}$$

we have (calling x_i' now x_i)

$$-\frac{1}{x_o} + \frac{1}{x_i} = \frac{1}{f} \tag{1.21}$$

1. Geometrical Optics

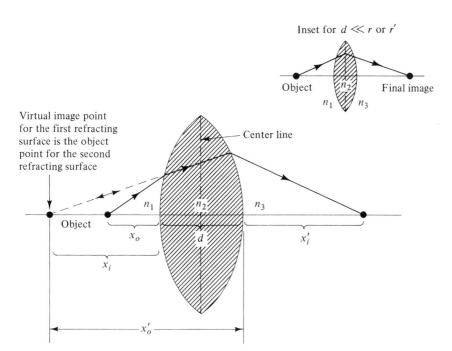

Fig. 1.18

Optical diagram of a thick lens soon to be a thin lens as d is made very small compared to the radii of the surfaces. The virtual image for the convex first surface ($x_i < 0$) becomes the object ($x_o' < 0$) of the concave second surface. The final image is at x_i' from the second surface. For the thin-lens approximation, where $d \ll r$ and $|r'|$, all distances are measured from the center line of the lens and *not* from the vertices of the two curved surfaces (see inset).

Equation 1.21 is called the **thin-lens equation**, and f of equation 1.20 is the focal length for the thin lens. For a convex first surface, $r > 0$; for a concave second surface, $r' < 0$. The same conventions are applied as defined for the single refracting surface (see Section 4C).

If we consider a symmetric **biconvex** lens, r is positive, and $|r| = |r'|$, where r' is negative; that is, we have from equation 1.20

$$\frac{1}{f} = 2\left(\frac{n-1}{r}\right) \qquad (1.22)$$

E. Positive and Negative Lenses

The quantity f depends on the refractive index n of the lens and the radii of curvature r and r'. It is a **characteristic number** of the lens. If f is positive, the lens is called a positive or convergent lens; if negative, a negative or divergent lens. The terms *convergent* and *divergent* describe the path of the light on the right of the lens when the light is incident from the left as parallel light with x_o at $-\infty$.

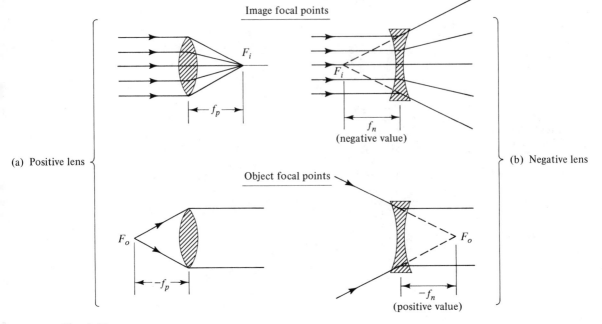

Fig. 1.19
Image and object focal points. (a) For a positive lens, the image focal point is on the right, and the object focal point is on the left. (b) For a negative lens, the image focal point is on the left, and the object focal point is on the right.

F. Object and Image Focal Points

For a positive lens, parallel light incident from the left converges on the right to one point. This is called the image focal point F_i. On the left, at the same distance, is the object focal point, F_o (Figure 1.19a).

For a negative lens, parallel light incident from the left diverges on the right side. The focal point on the left is called the image focal point; the focal point on the right, the object focal point (Figure 1.19b).

We also use f for the coordinate of the focal points F_o and F_i. We may indicate a positive lens with the subscript p and a negative lens with the subscript n. Then, for a positive lens, $F_o \rightarrow -f_p$ and $F_i \rightarrow f_p$, and for a negative lens, $F_o \rightarrow -f_n$ (which is a positive number) and $F_i \rightarrow f_n$ (which is a negative number).

The absolute value of the coordinates of the focal points is often called the focal length. This focal length is the same on the left and right of the lens. This is different from the refraction at one single surface (see Section 4C). Often, the distinction between object and image focal point is ignored.

G. Magnification

Consider a positive lens of focal length $f = 10$ cm. In Figure 1.20a and b, we show the information obtained from the thin-lens equation. For a given lens, if we know the object distance, we can find the image distance.

If we place an arrow at the object position, we expect an arrow at the image

1. Geometrical Optics

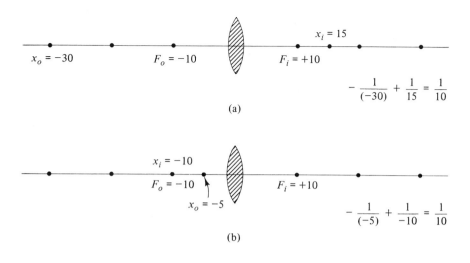

$$-\frac{1}{(-30)} + \frac{1}{15} = \frac{1}{10}$$

(a)

$$-\frac{1}{(-5)} + \frac{1}{-10} = \frac{1}{10}$$

(b)

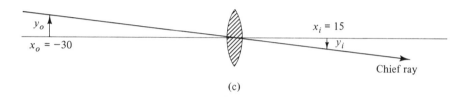

(c)

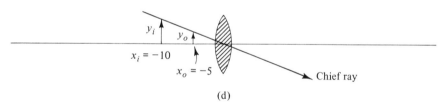

(d)

Fig. 1.20
(a, b) Image position calculated for two different object positions using the thin-lens equation with $f = 10$ cm. (c, d) If the object has length y_o, then the image has length y_i, determined by a chief ray.

position. Its height is determined by the ray passing undeviated from the object through the center of the lens. This ray is called a chief ray (see Figure 1.20c and d). We call y_o and y_i the object and image size, respectively. If y points up, it is positive; if down, negative. From similar triangles (see Figure 1.20d), we have

$$\frac{y_i}{x_i} = \frac{y_o}{x_o} \tag{1.23}$$

The **magnification** m is defined as

$$m = \frac{y_i}{y_o} = \frac{x_i}{x_o} \tag{1.24}$$

4. Refraction at Spherical Surfaces

We have read off these relations from Figure 1.20d, where the y's and x's are all on the same side of the lens and therefore pose no sign problems; but they hold in general. For the example in Figure 1.20c,

$$m = \frac{y_i}{y_o} = \frac{15}{-30} = -\frac{1}{2} \quad \text{or} \quad y_i = -\frac{1}{2}y_o \tag{1.25}$$

The minus sign for m tells us that the image points from the axis in the opposite direction of the object. The image is inverted. For the example in Figure 1.20d, we have

$$m = \frac{y_i}{y_o} = \frac{-10}{-5} = 2 \quad \text{or} \quad y_i = 2y_o \tag{1.26}$$

The image is twice as large and points into the same direction as the object. The image is erect and m is positive.

H. Real and Virtual Images for a Real Object at a Positive Lens

Comparing the image position in Figure 1.20c and d, we observe that in Figure 1.20c, the object is on the left side of the lens and the image is on the right. A real object forms a real image. In Figure 1.20d, the object and image are both on the same side of the lens. Unlike the case in Figure 1.20c, the image comes first, then the object. The image is called a **virtual image**. The real image in Figure 1.20c could be detected with a photographic plate placed at the image position; but in Figure 1.20d, we need an imaging system, such as another lens or our eye, to form a real image on a photographic plate or on the retina.

I. Graphical Method

Consider a special ray passing through the end of the object parallel to the axis. It will pass through the image focal point F_i after refraction by the lens

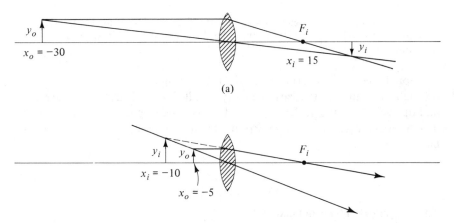

(a)

Fig. 1.21
Graphical construction of image position using a chief ray and the ray that is parallel to the axis and through the focal point.

(Figure 1.21). In Figure 1.21a, the ray through the focal point F_i and the chief ray **converge** toward the image. In Figure 1.21b, the same rays **diverge** on the right side of the lens. But tracing them back on the left side makes them meet at the virtual image.

The graphical method, as done with graph paper, can fairly accurately determine the position and size of the image. We use two rays through the end of the object. One ray goes through the center of the lens; the other is parallel to the axis on one side and goes through the focal point on the other side. Converging in a forward direction (to the right) will lead to the real image. If these rays diverge, they will converge in backward direction and meet at the virtual image.

J. Positive Lenses

We now consider the thin-lens equation for a positive lens. There are four possible combinations for the signs of x_o and x_i. However, the case where x_o is positive and x_i is negative cannot be realized. We would obtain a negative number on the left side of equation 1.21.

In Figure 1.22, we consider the object at various positions from the far left of the lens to the far right. In Figure 1.22a, we cannot use the graphical method because the size of the image is zero. When such a situation occurs in a lens system, either we approximate by using a large finite distance on the left side or we place an arrow of exaggerated size at the focal point (see Figure 1.23). Figure 1.22b has already been discussed. Figure 1.22c presents the simple case

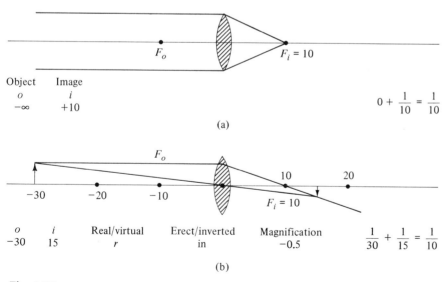

Object	Image		
o	i		
$-\infty$	$+10$		$0 + \dfrac{1}{10} = \dfrac{1}{10}$

(a)

| o | i | Real/virtual | Erect/inverted | Magnification | |
| -30 | 15 | r | in | -0.5 | $\dfrac{1}{30} + \dfrac{1}{15} = \dfrac{1}{10}$ |

(b)

Fig. 1.22
Calculated image positions for a positive lens of $f = 10$ cm when the object is at various assumed positions from the far left to the far right. The image position and size are determined with the graphical method. o = object position, i = image position, r = real, vi = virtual, e = erect, in = inverted, and m = magnification.

4. Refraction at Spherical Surfaces

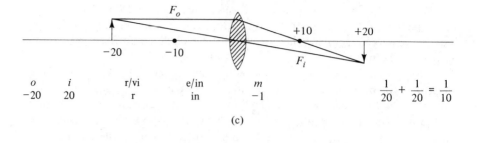

o	i	r/vi	e/in	m
-20	20	r	in	-1

$$\frac{1}{20} + \frac{1}{20} = \frac{1}{10}$$

(c)

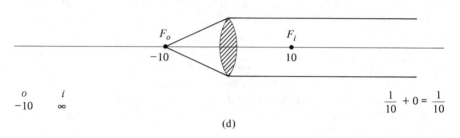

o	i
-10	∞

$$\frac{1}{10} + 0 = \frac{1}{10}$$

(d)

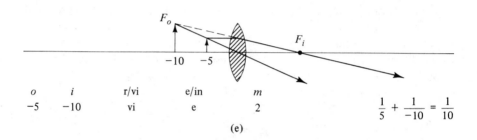

o	i	r/vi	e/in	m
-5	-10	vi	e	2

$$\frac{1}{5} + \frac{1}{-10} = \frac{1}{10}$$

(e)

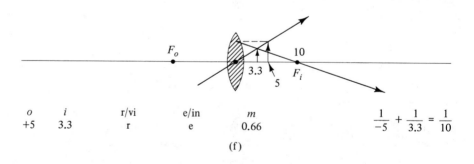

o	i	r/vi	e/in	m
$+5$	3.3	r	e	0.66

$$\frac{1}{-5} + \frac{1}{3.3} = \frac{1}{10}$$

(f)

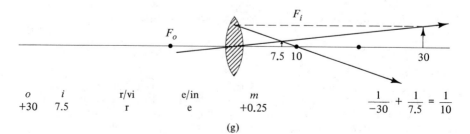

Fig. 1.22 (cont.)

o	i	r/vi	e/in	m
$+30$	7.5	r	e	$+0.25$

$$\frac{1}{-30} + \frac{1}{7.5} = \frac{1}{10}$$

(g)

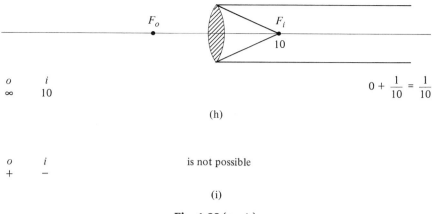

$$0 + \frac{1}{10} = \frac{1}{10}$$

o	i
∞	10

(h)

o	i	is not possible
$+$	$-$	

(i)

Fig. 1.22 (cont.)

where image and object distance are equally large. Figure 1.22d is the reverse of Figure 1.22a, and Figure 1.22e has already been discussed.

Figure 1.22f presents a new situation. But now the object is assumed to be produced by another positive lens on the left. It is a real image of a real object. The calculation shows that the image position is closer to the lens. Coming from the left, we have first the image, then the object. The object is called a virtual object. The image construction using the graphical method is based on the idea that the parallel ray that forms the object in the first process will be bent when the lens is inserted. It will pass through the focal point F_i. To do the construction, we trace from the object parallel to the axis back to the lens and then to F_i. The chief ray is also used, and the image is obtained at their intersection.

The case of Figure 1.22g is similar to that of Figure 1.22f. When the object is moved far enough to the right, Figure 1.22h is obtained from Figure 1.22g. In all constructions of Figure 1.22, we use the image focal point F_i. At each figure, we indicate the values for x_o and x_i, and we see that the combination x_o positive and x_i negative does not appear. We also indicate for each case in Figure 1.22 if the image is real or virtual, erect or inverted, and give the value of m.

In Figure 1.23, we show how the graphical method is used if the object is larger than the diameter of the lens and if the object is at infinity.

K. Application of the Graphical Method to Image Formation by Several Lenses

If the graphical method is applied to two lenses, each lens is considered separately. The image formed by the first lens serves as the object for the second lens. While going through the second image-forming process, the first lens and all rays used in the first process are ignored. For a system consisting of many lenses, we proceed in this way in a step-by-step fashion.

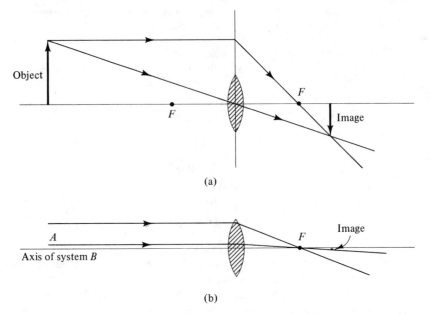

(a)

(b)

Fig. 1.23
(a) Image formation of an object larger than the lens diameter. The plane of the lens is used. (b) Image formation for an object at infinity. The axis B of the system is the ray from the object through the center of the lens. It will be replaced by ray A, which comes from the object if the object is at a very large finite distance.

L. Negative Lenses

For a negative lens, we could go through a similar number of figures, moving the object from the left to the right of the lens. However, considering the combinations of signs for x_o and x_i, we see from Figure 1.22 that for the positive lens, there are only three principal different cases: Figure 1.22b and c; Figure 1.22e; and Figure 1.22f and g. We exclude the cases involving an infinite distance. We also see that the case where x_o is negative and x_i is positive is not possible for a negative lens. In Figure 1.24, we present examples for the remaining three combinations of positive and negative values for x_o and x_i for a negative lens.

Figure 1.24a and b show similar cases (same sign combinations). The graphical construction makes use of the parallel ray, which diverges after passing through the lens and therefore is traced back to F_i. The other ray used is the chief ray.

In Figure 1.24c, the object is on the right side as a virtual object. The graphical construction proceeds as before, going parallel to the lens, then to F_i and extending forward. The chief ray then meets at the image. In Figure 1.24d, a similar construction is used, but the image is now obtained on the other side of the lens, and it is inverted.

For all constructions, F_i is again used. But comparing with the positive lens, F_i is now "on the other side." For each case in Figure 1.24, we indicate the signs of x_o and x_i, if the image is real or virtual, erect or inverted, and the value of m.

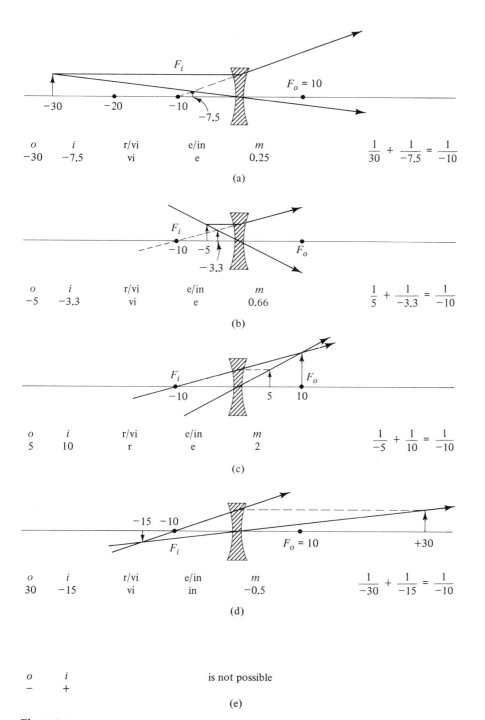

Fig. 1.24
Four important image-object positions for a negative lens with $f = -10$ cm. The image position and size are determined using the graphical method. o = object position, i = image position, r = real, vi = virtual, e = erect, in = inverted, and m = magnification.

M. Comparison of Cases for Positive and Negative Lenses

We will now describe some of the cases shown in Figure 1.22 by having the object:

	Figure	Abbreviation
Far away to the left of lens	1.22b, c	Far L
Within focal length of lens, left	1.22e	In foc. length, L
On the right side	1.22f, g	R

We will compare these three cases with the following cases in Figure 1.24.

	Figure	Abbreviation
Far away to the right of lens	1.24d	Far R
Within focal length, right	1.24c	In foc. length, R
On the left side	1.24a, b	L

The three cases chosen have different combinations for positive and negative values of x_o and x_i. We list these six cases in Table 1.3 and see the correspondence between the positive and negative lenses. If we interchange + and − (for f, x_o and x_i), r and vi, and left and right, one part of the table is transformed into the other. We see that e/in and m remain the same. Constructing such a table helps us to consider all important cases for the positive and negative lens. Its apparent symmetry ensures us that simple errors like mistakes in signs have been eliminated.

Table 1.3

$f = 10$ $f(+)$	o	i	r/vi	e/in	m	left/right
			Positive Lens (Figure 1.22)			
1.22b	−30	15	r	in	−0.5	Far L
1.22e	−5	−10	vi	e	2	In foc. lens, L
1.22g	30	7.5	r	e	0.25	R

$f = -10$ $f(-)$	o	i	r/vi	e/in	m	left/right
			Negative Lens (Figure 1.24)			
1.24d	30	−15	vi	in	−0.5	Far R
1.24c	5	10	r	e	2	In foc. lens, R
1.24a	−30	−7.5	vi	e	0.25	L

All real images have positive signs at column i, all virtual images negative signs.

o object position	e erect
i image position	in inverted
r real	m magnification
vi virtual	

We see from the table that all erect images have positive signs for m, all inverted images negative signs. All real images have positive signs in column i, all virtual images negative signs.

N. Lenses in a Different Medium

So far, we have assumed that the lens with index n is in air, with index $n' = 1$. If we have a lens in a medium of index n', we obtain in analogy to equations 1.18 and 1.20

$$\frac{1}{f} = \frac{1}{n'}\left(\frac{n-n'}{r} + \frac{n'-n}{r'}\right) \qquad (1.27)$$

Since r' is negative, $1/f$ changes sign if n' changes from smaller n to larger n. But all we need to do is calculate the value of f and apply the thin-lens equation for either the positive or negative lens, whatever is appropriate. For example, a biconvex lens-shaped air inclusion in plastic acts as a negative lens, a biconcave inclusion as a positive lens (Figure 1.25).

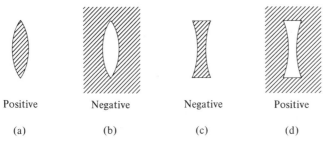

Positive	Negative	Negative	Positive
(a)	(b)	(c)	(d)

Fig. 1.25
(a) A biconvex lens of optical dense material acts as a positive lens. (b) If the lens is made of less dense material than the surrounding material, a biconvex lens acts as a negative lens. (c, d) Analogous cases for a biconcave lens.

5. THE USE OF LENSES AND LENS SYSTEMS

Having described the imaging properties of a single lens, we will now discuss some practical applications of lenses and lens systems.

A. The Camera and the Eye

The photographic camera uses a lens (or a lens system) to form a real image of an object on the plane of the photographic plate. Using a lens system improves the quality of the image. Since large angles are employed (Figure 1.26), the paraxial theory can only be used approximately. The subject of aberration, that is, nonparaxial theory of image formation, is discussed in Chapter 13.

We consider here our eye in an oversimplified way using a single but flexible

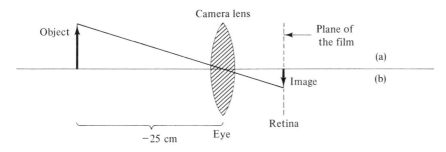

Fig. 1.26
Optical diagram of (a) an elementary camera or (b) the eye.

lens to form a real image of an object on the retina. The camera and the eye have some similarity from the point of view of image formation (see Figure 1.26). The eye lens can change its focal length. This is done by the ciliary muscles and is called accommodation. For a normal eye, the shortest focal length corresponds to the distance from the lens to the retina, and the eye focuses an object at "infinity" onto the retina. The closest object distance for which the eye lens can accommodate is called the **near point**, which is taken here as 25 cm (see Figure 1.26). In this situation, the focal length of the eye lens is smaller than the distance from the lens to the retina. An eye that is not functioning normally does not focus an infinitely distant object on the retina: **A nearsighted** eye focuses objects between the lens and retina while a **farsighted** eye focuses objects behind the retina.

Figure 1.27 shows schematically how additional lenses can be used to correct defects in vision. Note that the addition of a suitable lens (glasses or contact lenses) changes the effective focal length of the eye in the desired direction. The distance between the glasses and the eye lens may be neglected. For contact lenses, it is zero. As we will see later (in Section 6E), if two thin lenses of focal length f_1 and f_2 are placed in contact, the resulting focal length f is given by

$$\frac{1}{f} = \frac{1}{f_1} + \frac{1}{f_2}$$

This equation can be used to estimate the focal length of the corrective lenses, since the distance between the corrective lens and the eye may be neglected. Although the image formed on the retina is an inverted image, the brain **compensates** so that we perceive the image as right side up.

B. Magnifiers

The size of the image on the retina is increased and kept in focus through a change of the eye lens curvature as the object distance decreases from infinity to the **near point** (25 cm). If the lens of the eye could accommodate a shorter object distance, the size of the image would increase further. In practice, this is made possible by placing a biconvex lens before the eye. The effect of this auxiliary lens, called a **magnifier**, is to enlarge the image on the retina. When the object is at the near point, the image is magnified by

$$m = \frac{y_i}{y_o} = \frac{x_i}{x_o} = \frac{A}{-25}$$

where A is the distance from the eye lens to the retina (Figure 1.28). When the object is closer than the near point, a magnifier is placed at distance D before the lens (Figure 1.29). The magnification of the system in Figure 1.29 is $M = y_i'/y_o$. Multiplying by y_i/y_i yields

$$M = \left(\frac{y_i'}{y_i}\right)\left(\frac{y_i}{y_o}\right)$$

Since $y_i = y_o'$ we can substitute to get

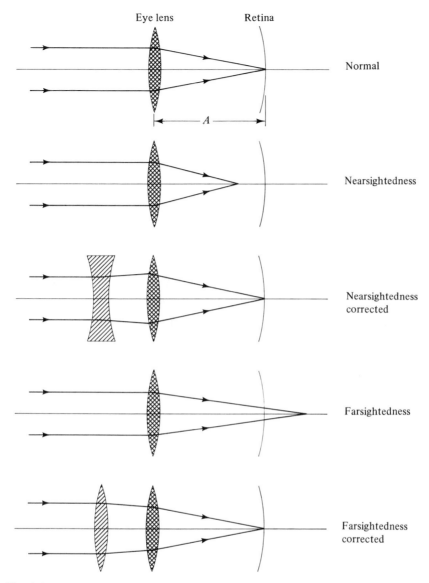

Fig. 1.27
Optical diagrams to show how normal, nearsighted, and farsighted eyes focus objects. Nearsightedness is corrected by a diverging lens, and farsightedness is corrected by a converging lens. The distance from the eye lens to the retina is noted by A.

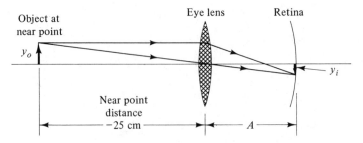

Fig. 1.28
Optical diagram of an object-eye-retina system when the object is at the near point (about 25 cm from the normal eye's lens). This example also illustrates the parameters involved in estimating the magnification: $m = y_i/y_o = A/-25$ cm for the real, inverted image as indicated when m is negative.

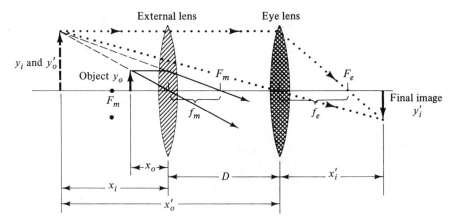

Fig. 1.29
Optical diagram of an external lens (magnifier) and eye lens acting as a magnifier system. Parameters x and y without primes refer to the external lens. Parameters bearing primes refer to the eye lens. F_m and f_m refer to magnifier, F_e and f_e to eye. The distance between the two lenses is D.

$$M = \left(\frac{y_i'}{y_o'}\right)\left(\frac{y_i}{y_o}\right)$$

According to the figure, $m_m = y_i/y_o$ and $m_e = y_i'/y_o'$, where m_m stands for magnification due to the magnifier lens and m_e represents magnification due to the eye lens. The total magnification M for the system is therefore

$$M = m_m m_e \tag{1.28}$$

This magnification must be compared with the magnification due to the eye alone, m_e, so that only m_m needs to be considered.

We can determine m_m by placing the object so that the virtual image y_i is at the *near point* (-25 cm). For the magnifier alone, we have

$$m_m = \frac{y_i}{y_o} = \frac{x_i}{x_o}$$

Then, by eliminating x_o using the thin-lens equation 1.21,

$$m_m = \frac{x_i}{x_o} = 1 - \frac{x_i}{f_m} \tag{1.29}$$

where f_m (positive) is the focal length of the magnifier lens. At the near point with the magnifier close to the eye (that is, $|x_i| > |x_o| \gg D$) $x_i + D$ is approximately equal to $x_i = -25$ cm, so that

$$m_m = 1 + \frac{25}{f_m} \tag{1.30}$$

In a slightly different case, the normal eye looks at the object with relaxed muscles; that is, the final image is on the retina, which is at the eye lens's focal point $x_i' = f_e$, the image produced by the magnifier and thus the object for the eye lens is a virtual image at *negative infinity*, the original object must be at F_m to the left of the magnifier, as shown in Figure 1.29. For this case, $x_i = -\infty$ and $x_o = -f_m$. For x_o' we have $x_i + D$, and the magnification in this case is

$$m_m = -\frac{x_i}{f_m} \quad \text{and} \quad m_e = \frac{f_e}{x_i + D} \tag{1.31}$$

For m_m to be meaningful, it is necessary to assume that x_i (negative) is large but not $-\infty$. Since x_i and $x_i + D$ have approximately the same value, they can cancel each other in the product $M = m_m m_e$, so that finally

$$M = -\frac{f_e}{f_m} \tag{1.32}$$

This case is of special interest when a magnifying glass is used as the ocular in optical instruments.

C. Angular Magnification

Another measure of magnification is the **angular magnification**, or **magnifying power**, given by

$$MP = \frac{\beta}{\alpha} \tag{1.33}$$

where α is the angular size seen on the retina *without* the magnifier lens (Figure 1.30) and β is the angular size seen on the retina *with* the magnifier lens (Figure 1.31). We obtain the largest value for α by placing the object at the near point, $x_o = -25$ cm. For small angles, this results in

$$\alpha = \frac{y_o}{x_o} = \frac{y_o}{-25} \tag{1.34}$$

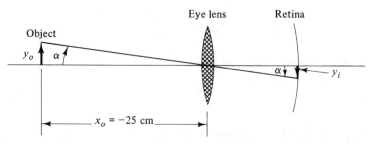

Fig. 1.30
Angular size α of an object placed in front of the naked eye. The optical axis and the chief ray define the maximum angular size of the object with respect to the center of the eye lens. In this example, the distance of the object is equal to the distance of the near point.

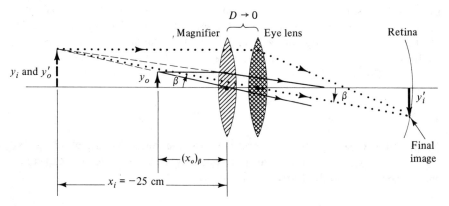

Fig. 1.31
Angular size β of a virtual object (the virtual image of the magnifier lens) found in front of the eye lens. The optical axis and the chief ray through the center of the eye lens define the maximum angular size of the object. In this example, $D \to 0$, and the distance of the object is approximately equal to the distance of the near point.

as shown in Figure 1.30, where α is negative in agreement with our sign convention.

Let us now consider two examples of angular magnification. Recall that the magnifier lens is usually placed close to the eye, so that the distance D between the magnifier lens and the eye lens is negligible.

In the first case the object is placed at a distance $(x_o)_\beta$, so that the virtual image is at the near point; that is, $x_i = -25$ cm, as shown in Figure 1.31. From Figure 1.31 and equation 1.21, we get

$$\beta = \frac{y_i}{x_i} = \frac{y_o}{(x_o)_\beta} = y_o \left(\frac{1}{x_i} - \frac{1}{f_m} \right) \tag{1.35}$$

where f_m (positive) is the focal length of the magnifier lens. For the angular magnification MP,

1. Geometrical Optics

$$MP = \frac{\beta}{\alpha} = \frac{y_o(1/x_i - 1/f_m)}{y_o/-25} = (-25)\left(\frac{1}{x_i} - \frac{1}{f_m}\right) \qquad (1.36)$$

The distance x_i (negative) is at the near point, giving

$$MP = 1 + \frac{25}{f_m} \qquad (1.37)$$

In the second case, the virtual image is at $x_i \rightarrow \infty$; then the eye is accommodated to infinity (relaxed eye), and equations 1.33 to 1.35 yield

$$MP = \frac{y_o(1/x_i - 1/f_m)}{y_o/x_o} = x_o\left(\frac{1}{x_i} - \frac{1}{f_m}\right) \rightarrow \frac{-x_o}{f_m} = \frac{25}{f_m} \qquad (1.38)$$

On optical instruments (e.g., microscope objectives), $MP = 25$ cm$/f_m$ is marked as the numerical ratio 25 cm$/f_m$ times $\times$. For example, for $f_m = 5$ cm, we have $5\times$; for $f_m = 12.5$ cm, we have $2\times$.

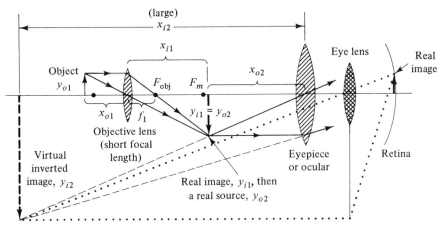

Fig. 1.32
Optical diagram of a simple microscope. A real object, when placed in front of a lens with a short focal length (the *objective lens*), produces a real inverted image, which becomes the object for the magnifier lens (the *eyepiece*, or *ocular*). The eyepiece produces a large virtual inverted image at x_{i2}, and that image is the object for the eye lens, which focuses the real erect image on the retina. (But we "see" the object upside down.)

D. Microscopes

The **objective lens** of a microscope makes a real image of the object. The **ocular**, or magnifier is applied to this real image, as shown in Figure 1.32. The objective lens magnification is

$$m_1 = \frac{y_{i1}}{y_{o1}} = \frac{x_{i1}}{x_{o1}}$$

When rearranged via lens equation 1.21 (f_1 positive), the result is

5. The Use of Lenses and Lens Systems   35

$$m_1 = 1 - \frac{x_{i1}}{f_1}$$

Combining m_1 (the objective lens magnification) and $m_2 = m_m$ (the ocular magnification), we get

$$M = m_1 m_m = \left(1 - \frac{x_{i1}}{f_1}\right)\left(1 - \frac{x_{i2}}{f_m}\right) \tag{1.39}$$

where, using equation 1.29, we treat the objective lens like the magnifier.

In most microscopes, x_{i1} is fixed at 16 cm. If the virtual image produced by the ocular is placed at the near point ($x_{i2} = -25$ cm), the result is

$$M = \left(1 - \frac{16}{f_1}\right)\left(1 + \frac{25}{f_m}\right) \tag{1.40}$$

When the object is close to the focal point of the objective lens, a real image is formed at x_{i1}, and x_{i1} is large compared to f_1. The first term in equation 1.39 may therefore be approximated by $-x_{i1}/f_1$. The large virtual image formed by the eyepiece is at x_{i2}, where $|x_{i2}| \gg f_m$. Thus, the term $(1 - x_{i2}/f_m)$ in equation 1.39 approaches $-x_{i2}/f_m$. The overall approximation for M is now

$$M \simeq \left(-\frac{16}{f_1}\right)\left(\frac{25}{f_m}\right)$$

The minus sign indicates that the virtual image of the microscope is inverted. The eye produces a real, upright image on the retina, but we "see" the object upside down because the brain produces another inversion.

E. Telescopes

The telescope, like the microscope, has two convex (positive) lenses; but the objective lens of a telescope forms an image of a faraway object onto its focal plane. The ocular (magnifier) makes a virtual image of this real first image at infinity, and it is this virtual image that is seen by the eye (Figure 1.33).

The magnification of the objective lens is

$$m_{ob} = \frac{y_{i1}}{y_{o1}} = \frac{x_{i1}}{x_{o1}} \simeq \frac{f_1}{x_{o1}}$$

The magnification of the ocular (equation 1.31) is

$$m_m \simeq \frac{x_{i2}}{x_{o2}} = -\frac{x_{i2}}{f_m}$$

Here we have used the characteristic numbers f_1 of the first lens and f_m of the magnifier. Since

$$x_{o1} \simeq x_{i2}$$

the magnification M of the system is

$$M = m_{ob} m_m = -\frac{f_1}{f_m} \tag{1.41}$$

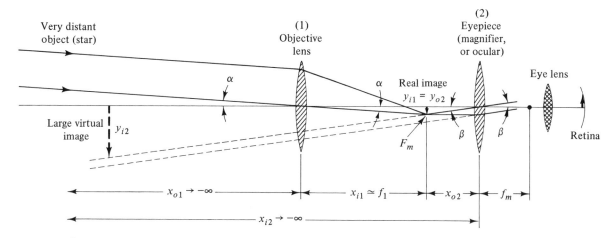

Fig. 1.33
Optical diagram of a simple telescope. The objective lens forms a real, inverted image in front of the eyepiece; $x_{i1} \simeq f_1$. The image y_{i1} is the object for the eyepiece at a distance $x_{o2} \simeq -f_m$. The virtual image is at a large distance, x_{i2}, and is the object for the eye. The eye lens then forms a real, erect image on the retina, but the object is still "seen" as upside down. The distance $f_1 + f_m$ is approximately the length of the telescope.

For the calculation of MP, we note that because of the large distance to the object, the angle through which the unaided eye sees the object is the same as the angle α the object makes with the objective lens, where

$$\alpha = \frac{y_{o1}}{x_{o1}} = \frac{y_{i1}}{f_1}$$

The angle β under which the magnifier looks at the first image is $-y_{o2}/f_m$ or $-y_{i1}/f_m$; thus

$$MP = \frac{\beta}{\alpha} = -\frac{y_{i1}/f_m}{y_{i1}/f_1} = -\frac{f_1}{f_m} \tag{1.42}$$

Notice that the result obtained for MP (angular magnification) is the same as that for M because of the small-angle approximation.

The telescope discussed here is the simplest form of an astronomical telescope. For terrestrial telescopes, we must have an inverted, real image on the retina so that images will be "seen" right side up. This can be accomplished by adding another lens. The Galilean telescope uses a negative lens for the ocular to accomplish the same thing (Chapter 12).

Magnification is not the only interesting property of these optical instruments. The **resolution** of such instruments (i.e., the smallest objects that can be separated) will be considered in subsequent chapters. We will also discuss the light throughput of these instruments, demonstrated with the example of a wrongly assembled projector. Aberration produced by the lenses of these instruments will be considered in Chapter 13.

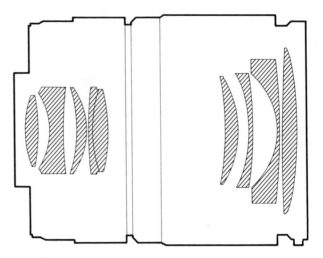

Fig. 1.34
Zoom lens made of nine lenses, some of which have
different refractive indices.

6. MATRIX FORMULATION FOR THE THIN LENS AND
SYSTEMS OF THIN LENSES

So far, we have discussed the imaging properties of the thin lens and thin-lens
systems. Our model utilized the refraction and reflection properties of light
on spherical surfaces. Our most important result was the thin-lens equation
(equation 1.21). The paraxial theory was assumed in the derivation of the thin-
lens equation. We will now reformulate the mathematical formulas for image
formation. We do this by describing mathematically the refraction process on
one interface. Refraction on two interfaces will turn out to be the application of
a similar mathematical operation on each of them. In this way, complicated lens
systems (Figure 1.34) may be treated by using the same process repeatedly.

Figure 1.34 shows a zoom lens made of nine lenses, some of which may have
different refractive indices. Such a lens system will improve the final image
quality. The way in which this is accomplished will be discussed in the chapter
on aberration (Chapter 13). To get the final image of such a lens system, the
calculation process can be broken down into a sequence of steps associated
with each surface and the distance between the surfaces. The ray passes surface
1, is translated to surface 2, passes surface 2, and so on. Each step can be
represented by a certain matrix, and multiplying these matrices will determine
the final transmission process. The steps are similar to those taken in a ray-
tracing procedure. This concept will make the image-forming process easier to
understand for complicated systems and more applicable for computer use. It
will also show how *thick* lenses are used and describe the image-forming process
using an equation similar to the thin-lens equation (equation 1.21). The matrix
method also can be used for systems of thin and thick lenses and even for mirror
systems.

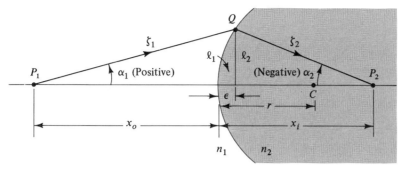

Fig. 1.35
Optical diagram for the passage of a light ray from P_1, through a medium with index of refraction n_1, through a curved surface with radius r (at Q), into a material with index of refraction n_2, and on to point P_2.

A. Matrix Formulation for a Single Surface

The equation for refraction at a single surface, equation 1.9, can be rewritten using variables that describe the imaging but lend themselves to a symmetric formation. Each ray is given a height and an angle, as shown in Figure 1.35. For paraxial rays (α_1 and α_2 are small), the distance ε on the axis (called the sagitta) is very small (i.e., $\varepsilon \simeq 0$) and the chord and segment at the vertex are very nearly equal. Taking $\alpha_1 = \ell_1/-x_o$ and $\alpha_2 = -\ell_2/x_i$, where counterclockwise angles are positive and clockwise angles are negative, equation 1.9 can be rewritten as

$$\frac{n_1 \alpha_1}{\ell_1} + n_2\left(-\frac{\alpha_2}{\ell_2}\right) = \frac{n_2 - n_1}{r} \tag{1.43}$$

Also, since $\ell_1 = \ell_2$, two simultaneous equations can be written in the variables α_j and ℓ_j ($j = 1, 2$) after solving equation 1.43 for α_2. These two equations are

$$\ell_2 = \ell_1 \tag{1.44}$$

and

$$\alpha_2 = \frac{n_1}{n_2}\alpha_1 - \left(\frac{n_2 - n_1}{n_2 r}\right)\ell_1 \tag{1.45}$$

Equations 1.44 and 1.45 can then be rewritten in **matrix form** as

$$\begin{pmatrix} \ell_2 \\ \alpha_2 \end{pmatrix} = \begin{pmatrix} 1 & 0 \\ -\dfrac{1}{r}\left(\dfrac{n_2 - n_1}{n_2}\right) & \dfrac{n_1}{n_2} \end{pmatrix}\begin{pmatrix} \ell_1 \\ \alpha_1 \end{pmatrix} \tag{1.46}$$

where the terms are defined as in Figure 1.35.

A general example of matrix multiplication of 2 times 2 matrices is

$$\begin{pmatrix} A & B \\ C & D \end{pmatrix}\begin{pmatrix} a & b \\ c & d \end{pmatrix} = \begin{pmatrix} Aa + Bc & Ab + Bd \\ Ca + Dc & Cb + Dd \end{pmatrix}$$

$$= \begin{pmatrix} (1,1) & (1,2) \\ (2,1) & (2,2) \end{pmatrix} \tag{1.47}$$

6. Matrix Formulation for the Thin Lens and Systems of Thin Lenses

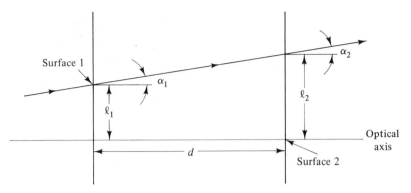

Fig. 1.36
Partial optical diagram showing a ray being translated a distance d in some homogeneous medium.

where in the last matrix the elements are numbered (i, k) with $i = 1, 2$ and $k = 1, 2$. If b and d are zero, we have the multiplication of a matrix and a vector

$$\begin{pmatrix} A & B \\ C & D \end{pmatrix} \begin{pmatrix} a \\ c \end{pmatrix} = \begin{pmatrix} Aa + Bc \\ Ca + Dc \end{pmatrix} \tag{1.48}$$

In equation 1.46, the image described by the parameters (ℓ_1, α_1) in medium 1 is transformed into an image described by (ℓ_2, α_2) in medium 2. The transformation is done by the refraction matrix

$$R_{12} = \begin{pmatrix} 1 & 0 \\ -\dfrac{1}{r}\left(\dfrac{n_2 - n_1}{n_2}\right) & \dfrac{n_1}{n_2} \end{pmatrix} \tag{1.49}$$

The column vectors

$$I_1 = \begin{pmatrix} \ell_1 \\ \alpha_1 \end{pmatrix} \quad \text{and} \quad I_2 = \begin{pmatrix} \ell_2 \\ \alpha_2 \end{pmatrix} \tag{1.50}$$

represent the *initial* and *final* images, respectively. Then in matrix notation, equation 1.46 is

$$I_2 = R_{12}I_1 \tag{1.51}$$

Therefore, the mechanics of the imaging process are that R operating I_1 gives I_2. If the refracting surface is concave rather than convex, r must be replaced by $-r$. For a **planar refracting surface**, where $r \to \infty$, the result is

$$R = \begin{pmatrix} 1 & 0 \\ 0 & \dfrac{n_1}{n_2} \end{pmatrix} \tag{1.52}$$

We also need to represent the (parallel) **translation** of a ray a linear distance d in some homogeneous medium (Figure 1.36). We have $\alpha_1 = \alpha_2$, and for small angles,

$$\ell_2 = \ell_1 + \alpha_1 d \tag{1.53}$$

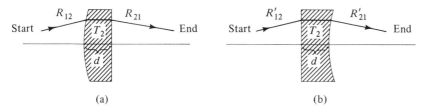

Fig. 1.37
General optical diagrams for thick lenses, showing the sequences of refraction, translation, and refraction for planoconvex lenses. (a) The spherical surface is first; (b) the plane surface is first.

When written in matrix form, equation 1.53 becomes

$$\begin{pmatrix} \ell_2 \\ \alpha_2 \end{pmatrix} = \begin{pmatrix} 1 & d \\ 0 & 1 \end{pmatrix} \begin{pmatrix} \ell_1 \\ \alpha_1 \end{pmatrix} \tag{1.54}$$

or

$$I_2 = T_2 I_1 \tag{1.55}$$

where T_2 is the **translation matrix**:

$$T_2 = \begin{pmatrix} 1 & d \\ 0 & 1 \end{pmatrix} \tag{1.56}$$

For example, consider the lenses shown in Figure 1.37. Light incident from the left is refracted at the first surface, transmitted through the glass, and refracted at the second surface. For the lens in Figure 1.37a, the final ray parameters I_2 are related to the initial ray parameters I_1 by the equation

$$I_2 = R_{21} T_2 R_{12} I_1 \tag{1.57}$$

where R_{12} is the refraction matrix at the first surface (air-to-glass), T_2 is the translation matrix in the second medium (glass), and R_{21} is the refraction matrix at the second (glass-to-air) surface. When read from left to right, the order of the matrices is opposite to the progression of the rays through the system. This convention was developed as a natural ordering of the way each matrix acts on the initial image.

The lens in Figure 1.37b was generated by interchanging the two refracting surfaces of the lens in Figure 1.37a. These two lenses refract light differently. Indeed, the image parameters are now

$$I_2' = R_{21}' T_2 R_{12}' I_1$$

where $R_{12} \neq R_{12}'$ and $R_{21} \neq R_{21}'$. In general, matrix multiplication is noncommutative; that is, for any two matrices A and B, $AB \neq BA$. The exception to this rule is if either A or B is a diagonal matrix. For example, in a 2×2 diagonal matrix, only elements $(1, 1)$ and $(2, 2)$ are not equal to zero. Diagonal matrices exist for plane refracting surfaces. The point here, however, is that $I_2 \neq I_2'$ because the refraction matrices are different, not because the matrices are noncommutative.

6. Matrix Formulation for the Thin Lens and Systems of Thin Lenses

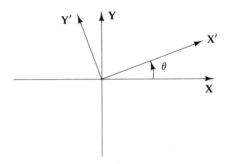

Fig. 1.38
Vectors **X′** and **Y′** rotated by the angle θ (positive) with respect to the vectors **X** and **Y**, respectively.

B. Matrix Multiplication Demonstrated with the Rotation Matrix

Matrix multiplication may be demonstrated using an example in which the matrix represents rotation in an X, Y coordinate system. If one matrix represents the rotation by the angle θ, then the product of two such matrices would represent the rotation by 2θ. The components of a vector in a coordinate system rotated about the origin can be obtained in this way, and the multiplication of two rotation matrices can also be demonstrated.

The diagram in Figure 1.38 shows the rotation of the coordinate system X', Y' relative to the system X, Y by the angle θ (positive). Any vector with components X', Y' may be expressed in the X, Y coordinate system as

$$X'\mathbf{i}' + Y'\mathbf{j}' = X\mathbf{i} + Y\mathbf{j}$$

where **i**, **j** and **i′**, **j′** are unit vectors. Multiplication with the dot product of **i** and **j** yields two equations:

$$X = X'(\mathbf{i}' \cdot \mathbf{i}) + Y'(\mathbf{j}' \cdot \mathbf{i}) \tag{1.58}$$

$$Y = X'(\mathbf{i}' \cdot \mathbf{j}) + Y'(\mathbf{j}' \cdot \mathbf{j}) \tag{1.59}$$

The dot products can easily be calculated, giving

$$X = X' \cos \theta - Y' \sin \theta \tag{1.60}$$

$$Y = X' \sin \theta + Y' \cos \theta \tag{1.61}$$

In matrix form, this becomes

$$\begin{pmatrix} X \\ Y \end{pmatrix} = \begin{pmatrix} \cos \theta & -\sin \theta \\ \sin \theta & \cos \theta \end{pmatrix} \begin{pmatrix} X' \\ Y' \end{pmatrix} \tag{1.62}$$

This means that the system X', Y' is rotated by $-\theta$ into the system X, Y; that is, the rotation is clockwise. Two successive rotations by the angle $-\theta$ results in

$$\begin{pmatrix} \cos \theta & -\sin \theta \\ \sin \theta & \cos \theta \end{pmatrix} \begin{pmatrix} \cos \theta & -\sin \theta \\ \sin \theta & \cos \theta \end{pmatrix} = \begin{pmatrix} (\cos^2 \theta - \sin^2 \theta) & (-2 \cos \theta \sin \theta) \\ (2 \cos \theta \sin \theta) & (\cos^2 \theta - \sin^2 \theta) \end{pmatrix}$$

$$= \begin{pmatrix} \cos 2\theta & -\sin 2\theta \\ \sin 2\theta & \cos 2\theta \end{pmatrix} \tag{1.63}$$

The last matrix represents the rotation of angle -2θ.

1. Geometrical Optics

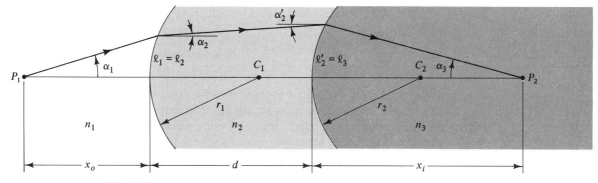

Fig. 1.39
Optical diagram showing refraction at a convex surface, translation through the material of index of refraction n_2, and refraction at a second convex surface. The thickness of this double-convex lens is d.

This simple example for two identical matrices serves well to demonstrate matrix multiplication. We will make use of the rotation matrix in later chapters.

C. Matrix Formulation for a Pair of Surfaces

In an optical system composed of two curved refracting surfaces separating three different media, let us assume that both surfaces have positive radii of curvature. To find the final image for the system shown in Figure 1.39, we use equations 1.46 and 1.54, respectively, to get

$$\begin{pmatrix} \ell_3 \\ \alpha_3 \end{pmatrix} = \begin{pmatrix} 1 & 0 \\ -\dfrac{1}{r_2}\left(\dfrac{n_3 - n_2}{n_3}\right) & \dfrac{n_2}{n_3} \end{pmatrix} \begin{pmatrix} 1 & d \\ 0 & 1 \end{pmatrix} \begin{pmatrix} 1 & 0 \\ -\dfrac{1}{r_1}\left(\dfrac{n_2 - n_1}{n_2}\right) & \dfrac{n_1}{n_2} \end{pmatrix} \begin{pmatrix} \ell_1 \\ \alpha_1 \end{pmatrix} \quad (1.64)$$

or, symbolically,

$$I_3 = R_{23} T_2 R_{12} I_1 \quad (1.65)$$

Compare the order of the matrices in equation 1.65 with the path of the ray through the optical system. The notation R_{12} signifies refraction from medium 1 to medium 2, R_{23} signifies refraction from medium 2 to medium 3, and T_2 denotes translation through medium 2. This labeling scheme will be used later in the text when we study multi-element systems.

It is convenient to make the following substitutions in the refraction matrices of equation 1.64:

$$P_{12} = -\frac{1}{r_1}\left(\frac{n_2 - n_1}{n_2}\right) \quad \text{and} \quad P_{23} = -\frac{1}{r_2}\left(\frac{n_3 - n_2}{n_3}\right) \quad (1.66)$$

where subscript 12 refers to the movement from medium 1 to medium 2 and where subscript 23 signifies the movement from medium 2 to medium 3. The symbol P is used because the terms are associated with **refracting power** of the surfaces. When the multiplication is carried out in equation 1.64 using the abbreviations of equation 1.66, the result is

6. Matrix Formulation for the Thin Lens and Systems of Thin Lenses

$$\begin{pmatrix} \ell_3 \\ \alpha_3 \end{pmatrix} = \begin{pmatrix} 1 + dP_{12} & d\dfrac{n_1}{n_2} \\ P_{23} + dP_{12}P_{23} + \dfrac{n_2}{n_3}P_{12} & d\dfrac{n_1}{n_2}P_{23} + \dfrac{n_1}{n_3} \end{pmatrix} \begin{pmatrix} \ell_1 \\ \alpha_1 \end{pmatrix} \qquad (1.67)$$

We call the 2×2 matrix in equation 1.67—that is, $R_{23}T_2R_{12}$—the **surface-pair matrix**. Using this matrix, we can describe thin lenses by setting $d = 0$. We can also describe thin-lens systems by using translation matrices between the surface pair matrices. Moreover, thick lenses can be considered if we specify that $d \neq 0$, and thick-lens systems can be described in conjunction with translation matrices.

D. The Thin-Lens Matrix

Many optical systems are composed of thin lenses. We obtain the **thin-lens matrix** from equation 1.67 by setting $d = 0$ and substituting for P_{12} and P_{23} from equation 1.66:

$$\begin{pmatrix} \ell_3 \\ \alpha_3 \end{pmatrix} = \begin{pmatrix} 1 & 0 \\ -\dfrac{1}{r_2}\left(\dfrac{n_3 - n_2}{n_3}\right) - \dfrac{1}{r_1}\left(\dfrac{n_2 - n_1}{n_2}\right)\left(\dfrac{n_2}{n_3}\right) & \dfrac{n_1}{n_3} \end{pmatrix} \begin{pmatrix} \ell_1 \\ \alpha_1 \end{pmatrix} \qquad (1.68)$$

Multiplying out the terms gives

$$\ell_3 = \ell_1 \qquad (1.69)$$

and

$$\alpha_3 = \left[-\dfrac{1}{r_2}\left(\dfrac{n_3 - n_2}{n_3}\right) - \dfrac{1}{r_1}\left(\dfrac{n_2 - n_1}{n_3}\right) \right]\ell_1 + \dfrac{n_1}{n_3}\alpha_1 \qquad (1.70)$$

Equation 1.69 simply reproduces the thin-lens assumption. We will now show that equation 1.70 is equivalent to the thin-lens equation derived earlier.

Let us use the following abbreviation:

$$\frac{1}{f} = \frac{1}{r_2}\left(\frac{n_3 - n_2}{n_3}\right) + \frac{1}{r_1}\left(\frac{n_2 - n_1}{n_3}\right) \qquad (1.71)$$

Later we will see that f is not just a collection of terms, but the focal length of the thin lens. Equation 1.70 then becomes

$$\alpha_3 = -\frac{\ell_1}{f} + \frac{n_1}{n_3}\alpha_1$$

When the object in Figure 1.39 is to the left, x_o is negative and the counterclockwise angle α is positive, and the result is

$$\alpha_1 = \frac{\ell_1}{-x_o}$$

When the image is to the right and the angle is clockwise, we have

$$\alpha_3 = -\frac{\ell_3}{x_i}$$

Substituting for α_1 and α_3, using the thin-lens approximation that $\ell_1 = \ell_2 = \ell_3$, and substituting in equation 1.70 gives

$$-\frac{1}{x_i} = -\frac{1}{f} + \frac{n_1}{n_3}\left(\frac{1}{-x_o}\right) \quad \text{or} \quad \frac{1}{f} = -\frac{1}{x_o}\left(\frac{n_1}{n_3}\right) + \frac{1}{x_i} \tag{1.72}$$

We have identified f correctly as the **focal length**, because for $n_1 = n_3 = 1$, we once again get the thin-lens equation of equation 1.21,

$$-\frac{1}{x_o} + \frac{1}{x_i} = \frac{1}{f} \tag{1.73}$$

The term n_1/n_3 in equation 1.72 generalizes the thin-lens equation so that the lens can be in contact with materials other than air so that different indices of refraction can be used on either side.

Using equation 1.71, the refractive matrix for the thin lens (equation 1.68) can be written in terms of the focal length as

$$\begin{pmatrix} 1 & 0 \\ -\dfrac{1}{f} & \dfrac{n_1}{n_3} \end{pmatrix} \quad \text{or, for } n_1 = n_3, \quad \begin{pmatrix} 1 & 0 \\ -\dfrac{1}{f} & 1 \end{pmatrix} \tag{1.74}$$

E. Thin-Lens Systems and the Principal Planes

To use the thin-lens matrix, we obtain from equations 1.68 and 1.71 the value for f and proceed as usual with equation 1.73 to determine x_i if x_o is given. So far, we have not obtained a simplification, but matrices are useful both for lens systems and for thick lenses.

For example, consider two thin lenses having refractive indices n_1 and n_2 that are a distance a apart in air. The refraction matrices for the two lenses are given in Figure 1.40, as well as the translation matrix (equation 1.56). The sequence in which the three matrices must be multiplied is

$$\begin{pmatrix} 1 & 0 \\ -\dfrac{1}{f_2} & 1 \end{pmatrix}\begin{pmatrix} 1 & a \\ 0 & 1 \end{pmatrix}\begin{pmatrix} 1 & 0 \\ -\dfrac{1}{f_1} & 1 \end{pmatrix} = \begin{pmatrix} 1 - \dfrac{a}{f_1} & a \\ -\dfrac{1}{f_2}\left(1 - \dfrac{a}{f_1}\right) - \dfrac{1}{f_1} & 1 - \dfrac{a}{f_2} \end{pmatrix} \tag{1.75}$$

For the special case of $a = 0$ we obtain

$$\begin{pmatrix} 1 & 0 \\ -\dfrac{1}{f_2} - \dfrac{1}{f_1} & 1 \end{pmatrix} \tag{1.76}$$

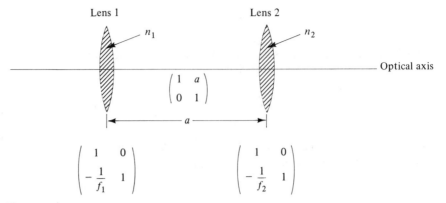

Fig. 1.40
Diagram of the three matrices used when an optical ray passes through (thin) lens 1 (which has a refractive index of n_1), is translated from lens 1 to lens 2, and then passes through (thin) lens 2 (which has a refractive index of n_2). The distance between the lenses is a.

Proceeding as before and using $-1/f = -1/f_1 - 1/f_2$, we get the correct focal length of a system of two lenses with zero distance between them.

If we write out the matrix-like equation 1.68, where we describe the incoming ray with (ℓ_1 and α_1) and the outgoing ray with (ℓ_3 and α_3), we get (for $a \neq 0$)

$$\begin{pmatrix} \ell_3 \\ \alpha_3 \end{pmatrix} = \begin{pmatrix} 1 - \dfrac{a}{f_1} & a \\ -\dfrac{1}{f_2}\left(1 - \dfrac{a}{f_1}\right) - \dfrac{1}{f_1} & 1 - \dfrac{a}{f_2} \end{pmatrix} \begin{pmatrix} \ell_1 \\ \alpha_1 \end{pmatrix} \tag{1.77}$$

or, breaking the equations down,

$$\ell_3 = \left(1 - \frac{a}{f_1}\right)\ell_1 + a\alpha_1$$

$$\alpha_3 = \left[-\frac{1}{f_2}\left(1 - \frac{a}{f_1}\right) - \frac{1}{f_1}\right]\ell_1 + \left(1 - \frac{a}{f_2}\right)\alpha_1 \tag{1.78}$$

In contrast to the case of the single-thin-lens matrix, equation 1.69, ℓ_1 is not equal to ℓ_3 because a is not zero. It is therefore not possible to extract a simple equation relating focal length, object and image distance, from equation 1.77. The difficulty could be remedied if the element in the (1, 1) position of the 2×2 matrix in equation 1.77 were unity and the (1, 2) element vanished (see also equation 1.74). Then, we would have $\ell_3 = \ell_1$ and we could proceed as in the thin-lens case. This modification can be achieved by a transformation obtaining a set of parameters that will realize the above-mentioned conditions.

Figure 1.41 depicts two new planes, H and H', which are located at distances h and h', respectively, from the first and second refracting surfaces. The distances are chosen such that $\ell_3' = \ell_1'$. Translation matrices (see equation 1.56) can be used to account for the coordinate shifts h and h'. The new parameters (ℓ_3', α_3')

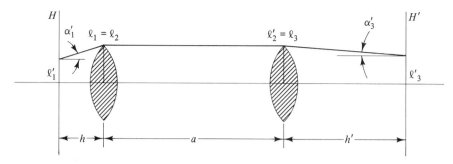

Fig. 1.41
The principal planes H and H' and their locations at distances h and h' from the refracting surfaces of their respective lenses. (For thin lenses at the center.)

are given by

$$\begin{pmatrix} \ell'_3 \\ \alpha'_3 \end{pmatrix} = \begin{pmatrix} 1 & h' \\ 0 & 1 \end{pmatrix} \begin{pmatrix} 1 - \dfrac{a}{f_1} & a \\ P & 1 - \dfrac{a}{f_2} \end{pmatrix} \begin{pmatrix} 1 & h \\ 0 & 1 \end{pmatrix} \begin{pmatrix} \ell'_1 \\ \alpha'_1 \end{pmatrix} \tag{1.79}$$

where

$$P = -\frac{1}{f_2}\left(1 - \frac{a}{f_1}\right) - \frac{1}{f_1} \tag{1.80}$$

Multiplication of the three 2×2 matrices gives

$$\begin{pmatrix} Ph' + 1 - \dfrac{a}{f_1} & h\left(1 - \dfrac{a}{f_1}\right) + h'\left(1 - \dfrac{a}{f_2}\right) + a + Phh' \\ P & Ph + 1 - \dfrac{a}{f_2} \end{pmatrix} \tag{1.81}$$

The $(2, 1)$ element P of the resulting matrix remains invariant. Considering equations 1.68 and 1.74, where $n_1 = n_3 = 1$, we can speculate that $P = -1/f$ where f is the focal length of the two-lens system. This would be in analogy to the thin-lens case, where we found $-1/f$ for the $(2, 1)$ element and f was the focal length of the thin lens. Proceeding with the analogy to the one-thin-lens matrix (see equation 1.75), we want to have the $(1, 1)$ and $(2, 2)$ elements equal to 1. Then we would obtain expressions for h' and h:

$$h' = \frac{a}{Pf_1} \tag{1.82}$$

and

$$h = \frac{a}{Pf_2} \tag{1.83}$$

Thus, the $(1, 1)$, $(2, 1)$, and $(2, 2)$ elements are 1, $P = -1/f$, and 1, respectively, in accordance with the matrix of equation 1.74 and the requirements of equations 1.79, 1.82, and 1.83.

6. Matrix Formulation for the Thin Lens and Systems of Thin Lenses

If we substitute these equations in the (1, 2) element and obtain zero, we would have the desired form for the two-thin-lens matrix. The (1, 2) matrix element is

$$h\left(1 - \frac{a}{f_1}\right) + h'\left(1 - \frac{a}{f_2}\right) + a + Phh'$$

and substituting for h and h' from equations 1.82 and 1.83 yields

$$\frac{a}{Pf_2}\left(1 - \frac{a}{f_1}\right) + \frac{a}{Pf_1}\left(1 - \frac{a}{f_2}\right) + a + P\left(\frac{a^2}{P^2 f_1 f_2}\right) \tag{1.84}$$

If a/P is factored out of the matrix, we have

$$\frac{a}{P}\left(\frac{1}{f_2} - \frac{a}{f_1 f_2} + \frac{1}{f_1} - \frac{a}{f_1 f_2} + P + \frac{a}{f_1 f_2}\right) \tag{1.85}$$

Substituting for P inside the parentheses, we find

$$\frac{a}{P}\left(\frac{1}{f_2} - \frac{a}{f_1 f_2} + \frac{1}{f_1} - \frac{a}{f_1 f_2} - \frac{1}{f_2} + \frac{a}{f_1 f_2} - \frac{1}{f_1} + \frac{a}{f_1 f_2}\right) \tag{1.86}$$

All these terms cancel out, and the matrix element is indeed zero, as desired. That is, the values of h and h', determined by having $(1, 1) = 1$ and $(2, 2) = 1$, cause the $(1, 2)$ element to vanish. The planes H and H' are called the **principal planes** of the system. The goal of introducing principal planes is to have thin-lens-type equations for the total system of thin lenses. Object and image distances are measured from the new principal planes and not from any center of the lenses involved.

To determine the significance of the signs for h and h', refer to Figure 1.42. If the vertex V_1 of the first lens is to the right of H, then h is positive; if it is to the left, then h is negative (Figure 1.42a). If V_2 is to the left of H', then h' is positive; if it is to the right, then h' is negative (Figure 1.42b).

To summarize, using h and h' in equations 1.82 and 1.83, the refraction matrix for the lens system once again has the simple form

$$\begin{pmatrix} 1 & 0 \\ -\dfrac{1}{f} & 1 \end{pmatrix} \tag{1.87}$$

where $-1/f$ is obtained from equation 1.80 and represents the refracting power of the total system. The imaging process is the same as it was for the thin lens. The thin-lens equation $-1/x_o + 1/x_i = 1/f$ can be used to find the image if the object distance x_o is measured from the H-plane and the image distance x_i is measured from the H' plane. Both planes are generally displaced from the surface of the lenses. The final steps of obtaining the image parameters makes no use of the vectors

$$\begin{pmatrix} \ell_1 \\ \alpha_1 \end{pmatrix}, \begin{pmatrix} \ell_3 \\ \alpha_3 \end{pmatrix} \quad \text{and} \quad \begin{pmatrix} \ell_1' \\ \alpha_1' \end{pmatrix}, \begin{pmatrix} \ell_3' \\ \alpha_3' \end{pmatrix}$$

used initially in equations 1.77 and 1.79.

1. Geometrical Optics

(a) First lens surface

(b) Second lens surface

Fig. 1.42
Signs of h and h' with respect to H and V_1 and H' and V_2, respectively.
(a) If V_1 is to the right of H, then h is positive; if to the left, h is
negative. (b) If V_2 is to the left of H', then h' is positive; if to the right,
h' is negative.

F. Two Thin Lenses: Comparison of Methods

Consider a lens system consisting of two thin lenses with focal lengths $f_1 = 10$
cm and $f_2 = 10$ cm. First we will calculate the final image position by using the
thin-lens equation twice. The object distance is $x_{o1} = -30$ cm, and the distance
between the two lenses is $a = 40$ cm. This is shown in Figure 1.43 and equation
1.88. We then apply the matrix method and find the final image location directly.

Using thin-lens equation 1.21 and our sign convention first on the first lens
and then on the second lens, we get

$$a = 40 \text{ cm}$$

First Lens   *Second Lens*

$x_{o1} = -30$ $x_{o2} = -25$

$$-\left(\frac{1}{-30}\right) + \frac{1}{x_{i1}} = \frac{1}{10} \qquad -\left(\frac{1}{-25}\right) + \frac{1}{x_{i2}} = \frac{1}{10} \qquad (1.88)$$

$$\frac{1}{x_{i1}} = \frac{3}{30} - \frac{1}{30} = \frac{1}{15} \qquad \frac{1}{x_{i2}} = \frac{2.5}{25} - \frac{1}{25} = \frac{1.5}{25}$$

$$x_{i1} = 15 \qquad\qquad\qquad x_{i2} = 16.7$$

Using the definitions involved in the matrix method from equation 1.80 yields

$$P = -\frac{1}{f} = -\frac{1}{f_2}\left(1 - \frac{a}{f_1}\right) - \frac{1}{f_1} = -\frac{1}{10}\left(1 - \frac{40}{10}\right) - \frac{1}{10} = +\frac{2}{10}\left(\frac{1}{\text{cm}}\right)$$

6. Matrix Formulation for the Thin Lens and Systems of Thin Lenses

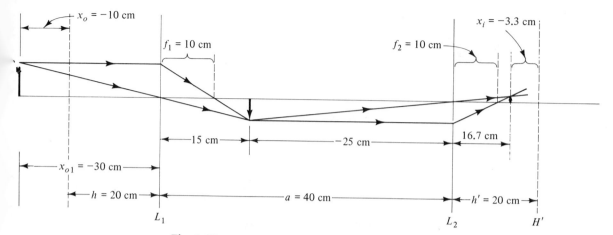

Fig. 1.43
Illustration of the use of principal planes for two lenses.

for the refractive power P. Using equation 1.82 gives

$$h' = \frac{a}{Pf_1} = \frac{40}{(2/10)10} = 20$$

and

$$h = \frac{a}{Pf_2} = \frac{40}{(2/10)10} = 20$$

The thin-lens equation

$$\frac{1}{f} = \frac{1}{-x_o} + \frac{1}{x_i}$$

is now used with the understanding that the focal length of this two-thin-lens system is $f = -5$ cm and that the object is located at $x_o = -10$ cm measured from the principal plane H. The image distance is measured from the H' principal plane to the image at x_i. (For the sign convention of h and h', see Figure 1.44a and b.) Substituting for f and x_o yields

$$\frac{1}{-5} = -\left(\frac{1}{-10}\right) + \frac{1}{x_i}$$

This makes $x_i = -3.3$ cm, which places the image on the left of H' toward the second lens (see Figure 1.43). The position of the final image, measured for comparison from the second lens, is 16.7 cm. This is the same distance obtained previously by applying the thin-lens equation twice.

7. MATRIX FORMULATION FOR THICK LENSES AND THICK-LENS SYSTEMS

A. Thick Lenses and Principal Planes

To consider image formation by thick lenses ($d \neq 0$), we return to the surface pair matrix of equation 1.67. Writing out the two equations gives

$$\ell_3 = (1 + dP_{12})\ell_1 + d\frac{n_1}{n_2}\alpha_1 \qquad (1.89)$$

and

$$\alpha_3 = \left(P_{23} + dP_{12}P_{23} + \frac{n_2}{n_3}P_{12}\right)\ell_1 + \left(d\frac{n_1}{n_2}P_{23} + \frac{n_1}{n_3}\right)\alpha_1 \qquad (1.90)$$

Figure 1.44 shows this thick-lens case.

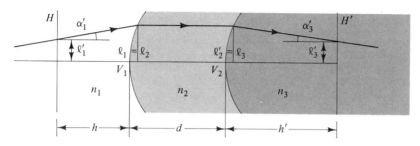

Fig. 1.44
The principal planes for a thick lens.

Equation 1.89 shows that $\ell_1 \neq \ell_3$, unlike the case for thin lenses (see equation 1.69). As with the thin-lens system, it is not possible to extract from equations 1.89 and 1.90 a simple equation relating focal length, object distance, and image distance. This difficulty could again be remedied if the (1, 1) and (2, 2) elements were unity and the (1, 2) element vanished (equation 1.74). Then, of course, we would have $\ell_3 = \ell_1$ and it would be possible to proceed as in the single-thin-lens case. Again, this modification can be obtained by introducing principal planes located at distances h and h', respectively, from the first and second refracting surfaces.

One has similarly to equation 1.79

$$\begin{pmatrix}\ell_3' \\ \alpha_3'\end{pmatrix} = \begin{pmatrix} 1 & h' \\ 0 & 1 \end{pmatrix}\begin{pmatrix} 1 + dP_{12} & d\dfrac{n_1}{n_2} \\ P & \dfrac{n_1}{n_3} + dP_{23}\dfrac{n_1}{n_2} \end{pmatrix}\begin{pmatrix} 1 & h \\ 0 & 1 \end{pmatrix}\begin{pmatrix}\ell_1' \\ \alpha_1'\end{pmatrix} \qquad (1.91)$$

where

$$P = P_{23} + dP_{12}P_{23} + \frac{n_2}{n_3}P_{12}$$

Multiplying out the 2 × 2 matrices in equation 1.91 gives

$$\begin{pmatrix} 1 + dP_{12} + h'P & h(1 + dP_{12}) + d\dfrac{n_1}{n_2} + h'\left(Ph + \dfrac{n_1}{n_3} + dP_{23}\dfrac{n_1}{n_2}\right) \\ P & Ph + \dfrac{n_1}{n_3} + dP_{23}\dfrac{n_1}{n_2} \end{pmatrix} \qquad (1.92)$$

7. Matrix Formulation for Thick Lenses and Thick-Lens Systems

The (2, 1) element P of the matrix again remains invariant. If we look back to equations 1.68 and 1.71, we are led to speculate that P is the negative inverse of the focal length of the thick lens (as with the thin-lens case).

The (1, 1) and (2, 2) elements of equation 1.92 should be unity; therefore, we have two equations from which h and h' can be determined:

$$1 + dP_{12} + h'P = 1 \tag{1.93}$$

and

$$Ph + \frac{n_1}{n_3} + dP_{23}\frac{n_1}{n_2} = 1 \tag{1.94}$$

It follows that

$$h' = -\frac{P_{12}}{P}d \tag{1.95}$$

To determine h, we set $n_1 = n_3$, where n_1 and n_3 are not necessarily equal to unity and get with equation 1.94

$$Ph + 1 + dP_{23}\frac{n_1}{n_2} = 1$$

and have

$$h = -\frac{P_{23}}{P}\left(\frac{n_1}{n_2}\right)d \tag{1.96}$$

Substituting for h and h' into the (1, 2) element of matrix (1.92) gives us ($n_1 = n_3$)

$$-\frac{P_{23}}{P}\left(\frac{n_1}{n_2}\right)d(1 + dP_{12}) + d\frac{n_1}{n_2} + \left(-\frac{P_{12}}{P}d\right)\left(\left(-P_{23}\frac{n_1}{n_2}\right)d + 1 + dP_{23}\frac{n_1}{n_2}\right)$$

By factoring out n_1/n_2P, substituting for P, and canceling, this matrix element is equal to zero. That is, the values found for h and h' do indeed cause the (1, 2) element of equation 1.92 to vanish. The planes H and H' are called the principal planes of the thick lens.

To summarize, with the choice of h and h' given previously, the refraction matrix for the thick lens now has the simple form

$$\begin{pmatrix} 1 & 0 \\ -\dfrac{1}{f} & 1 \end{pmatrix}, \tag{1.97}$$

such that for the thick lens,

$$\begin{pmatrix} \ell'_3 \\ \alpha'_3 \end{pmatrix} = \begin{pmatrix} 0 & h' \\ 0 & 1 \end{pmatrix} \begin{pmatrix} 1 & 0 \\ -\dfrac{1}{f} & 1 \end{pmatrix} \begin{pmatrix} 1 & h \\ 0 & 1 \end{pmatrix} \begin{pmatrix} \ell'_1 \\ \alpha'_1 \end{pmatrix} \tag{1.98}$$

(see Figure 1.44). Here again we have set $P = -1/f$. We can now use the thin-lens equation $-1/x_o + 1/x_i = 1/f$ to find the image distance x_i measured from the H'

plane if we have measured the object distance x_o from the H-plane. Both planes are displaced from the surfaces of the lenses.

We now collect the relevant equations for the thick-lens problem (with $n_1 = n_3$).

From equations 1.95 and 1.96,

$$h = -d\left(\frac{P_{23}}{P}\right)\left(\frac{n_1}{n_2}\right) \quad \text{and} \quad h' = -d\left(\frac{P_{12}}{P}\right) \tag{1.99}$$

and from the text after equation 1.91,

$$P = -\frac{1}{f} = P_{23} + dP_{12}P_{23} + \frac{n_2}{n_1}P_{12} \tag{1.100}$$

where (see equation 1.66)

$$P_{12} = -\frac{1}{r_1}\left(\frac{n_2 - n_1}{n_2}\right) \quad \text{and} \quad P_{23} = -\frac{1}{r_2}\left(\frac{n_1 - n_2}{n_1}\right) \tag{1.101}$$

From equations 1.100 and 1.101, $-1/f$ can be calculated. And since

$$-\frac{1}{x_o} + \frac{1}{x_i} = \frac{1}{f} \tag{1.102}$$

the image position can be determined if the object position is given using equation 1.99.

B. Glass Hemisphere and Comparison of Methods

We will now consider a glass hemisphere with refractive index of $n = n_2 = 1.5$ in air ($n_1 = n_3 = 1$) and set $r_1 = r$ and $r_2 = \infty$; because it is a hemisphere, $d = r$. Then

$$P_{12} = -\frac{1}{r}\left(\frac{n - 1}{n}\right) = -\frac{1}{r}\left(\frac{1.5 - 1}{1.5}\right) = -\frac{1}{3r}$$

$$P_{23} = 0$$

$$P = nP_{12} = -\frac{n - 1}{r} = -\frac{1}{2r}, \quad \text{implying } f = 2r, \tag{1.103}$$

$$h = 0,$$

$$\text{and} \quad h' = -r\left(-\frac{1}{3r}\right)(-2r) = -\frac{2}{3}r < 0$$

In Figure 1.45, H is at V_1, whereas H' is at distance $2r/3$ to the left of V_2. If $x_o = -2f = -4r$, then equation 1.102 gives $x_i = 4r = 2f$.

For comparison, we repeat the calculation, this time using the single-thin-lens approach; that is, we use $d = 0$ in the previous calculations. The corresponding quantities are then $P_{12} = -1/3r$, $P_{23} = 0$ from equation 1.100, and $P = nP_{12} = -1/2r$, resulting in $f = 2r$, and $h = h' = 0$ from equation 1.99. The same value is obtained for f, and both h and h' vanish. Since d was assumed to be zero,

7. Matrix Formulation for Thick Lenses and Thick-Lens Systems

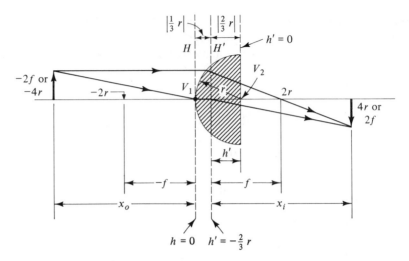

Fig. 1.45
Hemispherical lens treated as a thick lens.

both H and H' for the single thin lens coincide with the center of the thick lens. For the thin-lens case, all distances are measured from the center as shown in Figure 1.46. For $x_o = -2f = -4r$, the thin-lens equation gives $x_i = 2f$ as before.

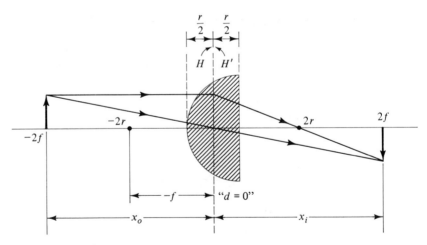

Fig. 1.46
Hemispherical lens with the thin-lens approximation.

To summarize, the numerical results for x_o and x_i are the same for either case, but the positions of the image and object with respect to the lens are different. Compare Figures 1.45 and 1.46 (Table 1.4). Only the value of h' differs. A reasonable question at this point might be: What is the difference between the thick-lens and thin-lens treatments? The answer is that for accurate positioning

1. Geometrical Optics

Table 1.4
Comparison of precise thick-lens parameters with a thin-lens approximation

	x_o	x_i	f	h	h'
Thick lens	$-2f$	$2f$	$2r$	0	$-2/3r$
Thin lens	$-2f$	$2f$	$2r$	0	0

of the object and image points and for accurate design of multielement lens systems, the correct thick-lens approach must be used. The thin-lens approach is only a special case of the thick lens, but for many applications the difference is small and can be neglected.

C. Application to Glass Spheres

Consider the case of a glass sphere ($n_2 = 1.5$) in air ($n_1 = n_3 = 1$). The particular choice of n_2 leads to an interesting result. Using $r_1 = r$, $r_2 = -r$, $d = 2r$ for spherical surfaces, and equations 1.99–1.101 give us

$$P_{12} = -\frac{1}{3r} \quad \text{and} \quad P_{23} = -\frac{1}{2r} \tag{1.104}$$

and then

$$P = -\frac{1}{2r} + 2r\left(-\frac{1}{2r}\right)\left(-\frac{1}{3r}\right) + \frac{3}{2}\left(-\frac{1}{3r}\right) = -\frac{2}{3r} = -\frac{1}{f}$$

which implies $f = 1.5r$, and

$$h = -2r\frac{-1/2r}{(-2/3r)(3/2)} = -r \quad \text{and} \quad h' = -2r\frac{-1/3r}{-2/3r} = -r \tag{1.105}$$

In Figure 1.47, we have arbitrarily assumed equal object and image distances. Although the sphere is taken as a thick lens ($d = 2r$), the principal planes are at

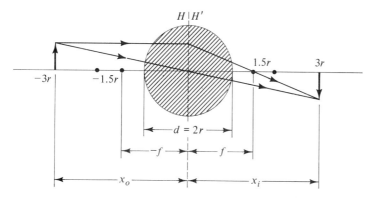

Fig. 1.47
Optical diagram of a spherical (two-hemisphere) lens using the thick-lens approach.

7. Matrix Formulation for Thick Lenses and Thick-Lens Systems

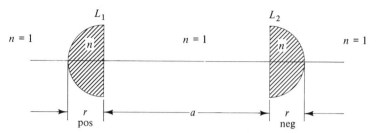

Fig. 1.48
Two hemispherical thick lenses separated by a distance a.

the geometrical center of the lens. For this special case, the results are identical to those obtained with the single-thin-lens equation.

D. Thick-Lens Systems

To introduce multi-lens systems, we consider the case of two hemispherical glass lenses (L_1 and L_2) with radii r separated by a distance a (Figure 1.48). For lens L_1, the parameters f, h, and h' have already been found (see Figure 1.44), and the same can be done for lens L_2. However, the purpose here is to find h, h', and the focal length of the total system. To begin, we write down the sequence of matrices that transform an object point from the left of L_1 into an image point to the right of L_1. For L_1, we use equations 1.49, 1.52, and 1.56 (also see equations 1.64 and 1.65) to get ($n_1 = n_3 = 1$, $n_2 = n$)

$$L_1 = \begin{pmatrix} 1 & 0 \\ 0 & n \end{pmatrix} \begin{pmatrix} 1 & d \\ 0 & 1 \end{pmatrix} \begin{pmatrix} 1 & 0 \\ P_{12}^{(1)} & 1/n \end{pmatrix} \tag{1.106}$$

refraction translation refraction at
at second   through   first curved
(flat) L_1 surface of L_1
surface of
L_1 with
$r_2 = \infty$

where equation 1.66 with $r_2 = \infty$ gives

$$P_{12}^{(1)} = -\frac{1}{r}\left(\frac{n-1}{n}\right) \quad \text{and} \quad P_{23}^{(1)} = 0$$

The superscript (1) in $P_{12}^{(1)}$ identifies the lens in question. Translation through the air gap is represented by

$$\begin{pmatrix} 1 & a \\ 0 & 1 \end{pmatrix}$$

The three matrices for L_2 are similarly obtained (although the sequence of matrices for the plane and curved surfaces is different). The resulting matrix for the system is

$$
\begin{pmatrix} 1 & 0 \\ P_{23}^{(2)} & n \end{pmatrix}\begin{pmatrix} 1 & d \\ 0 & 1 \end{pmatrix}\begin{pmatrix} 1 & 0 \\ 0 & 1/n \end{pmatrix}\begin{pmatrix} 1 & a \\ 0 & 1 \end{pmatrix}\begin{pmatrix} 1 & 0 \\ 0 & n \end{pmatrix}\begin{pmatrix} 1 & d \\ 0 & 1 \end{pmatrix}\begin{pmatrix} 1 & 0 \\ P_{12}^{(1)} & 1/n \end{pmatrix} \quad (1.107)
$$

where $P_{23}^{(2)} = (1 - n)/r$ (see equation 1.66). Matrix multiplication yields

$$
\begin{pmatrix}
1 + (2d + a)P_{12}^{(1)} & 2\dfrac{d}{n} + a \\[2ex]
[1 + (2d + a)P_{12}^{(1)}]P_{23}^{(2)} + nP_{12}^{(1)} & \left(2\dfrac{d}{n} + a\right)P_{23}^{(2)} + 1
\end{pmatrix} \quad (1.108)
$$

which is called the matrix M with elements M_{ij} (at the ith row and jth column). For example, $M_{12} = 2d/n + a$.

To find the principal planes, we proceed as in equation 1.79, forming the matrix product

$$
\begin{pmatrix} 1 & h' \\ 0 & 1 \end{pmatrix}\begin{pmatrix} M_{11} & M_{12} \\ M_{21} & M_{22} \end{pmatrix}\begin{pmatrix} 1 & h \\ 0 & 1 \end{pmatrix} \quad (1.109)
$$

Matrix multiplication yields

$$
\begin{pmatrix} M_{11} + h'M_{21} & (M_{11} + h'M_{21})h + M_{12} + h'M_{22} \\ M_{21} & M_{21}h + M_{22} \end{pmatrix} \quad (1.110)
$$

as the general transformation matrix. Following the procedure used previously, M_{21} is again identified with $-1/f$, and h and h' are obtained from the requirement that the resulting matrix be

$$
\begin{pmatrix} 1 & 0 \\ -\dfrac{1}{f} & 1 \end{pmatrix}
$$

Comparing matrix elements yields

$$
M_{11} + h'M_{21} = 1 \quad (1.111)
$$

and

$$
M_{21}h + M_{22} = 1 \quad \text{and} \quad M_{21} = -\frac{1}{f} \quad (1.112)
$$

which results in

$$
h' = \frac{1 - M_{11}}{M_{21}} = \frac{1 - [1 + (2d + a)P_{12}^{(1)}]}{-1/f} = f(2d + a)P_{12}^{(1)} \quad (1.113)
$$

and

$$
h = \frac{1 - M_{22}}{M_{21}} = \frac{1 - [2(d/n) + a]P_{23}^{(2)} - 1}{-1/f} = f\left(2\frac{d}{n} + a\right)P_{23}^{(2)} \quad (1.114)
$$

These values of h and h' will reduce the $(1, 2)$ element of equation (1.110) to zero.

7. Matrix Formulation for Thick Lenses and Thick-Lens Systems

The terms in equations 1.113 and 1.114 cannot easily be recognized by inspection. Therefore, to check the results, we will use equations 1.113 and 1.114 to treat the familiar case of the glass sphere (Figure 1.47) in which $a = 0$ and $n = 1.5$. Referring to equations 1.106 and 1.107, we see that d is the thickness of each hemispherical lens and that d is employed in each of the two halves that make up the spherical lens (equation 1.107). (Note that $d = r$ and not $2r$.) Substituting into the matrix element $(2, 1)$ of equation 1.108 with

$$P_{12}^{(1)} = -\frac{1}{r_1}\frac{n-1}{n} \quad \text{and} \quad P_{23}^{(2)} = -\frac{1-n}{(-r)},$$

we get

$$-\frac{1}{f} = M_{21} = \left[1 + 2r\left(-\frac{n-1}{rn}\right)\right]\left(\frac{1-n}{r}\right) + n\left(-\frac{n-1}{rn}\right) = -\frac{2}{3r}$$

which implies

$$f = 1.5r \tag{1.115}$$

Substituting for f and using $n = 1.5$, equation 1.114 gives us

$$h = 1.5r\left(2\frac{r}{n} + 0\right)\left(\frac{1-n}{r}\right) = -r \tag{1.116}$$

and equation 1.113 gives us

$$h' = 1.5r(2r)\left[-\frac{1}{r}\left(\frac{n-1}{n}\right)\right] = -r \tag{1.117}$$

All these values agree with those obtained previously when the glass sphere was treated as a single thick lens. Compare equations 1.104 and 1.105 with equations 1.115–1.117.

The validity of equations 1.111 and 1.112 can be proved in general. For any lens system, we have only to set up the matrix M. From the elements of matrix M, we can find h, h', and $-1/f$ for the total system. Then image and object distances measured from the principal planes can be obtained using the thin-lens equation $-1/x_o + 1/x_i = 1/f$.

8. MIRRORS AND MIRROR EQUATIONS

A. Flat Surface Mirrors and Virtual Images

A surface that reflects most of the incident light is called a **mirror**. A typical aluminized front surface mirror will have a normal incident reflectance of $\simeq 95\%$ at a wavelength of 6000 Å. Reflecting surfaces may be planar, spherical, elliptical, paraboloidal, toroidal, or any other shape. A mirror is used to redirect (i.e., collect and focus) light. In regions of the electromagnetic spectrum where optically transparent lens material is not available, optical systems use mirrors to direct the light rays. For example, in the infrared region (wavelengths of $1-50$ μm), there is no equivalent of "glass." That is, the available infrared transmitting materials (such as sodium chloride, quartz or germanium) either have only window regions

1. Geometrical Optics

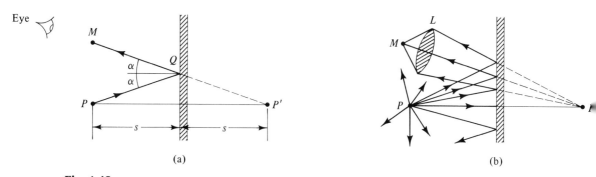

Fig. 1.49
Reflection by a plane mirror. (a) One ray emanating from the object. (b) Divergent rays emanating from the object.

(selected regions free of absorption), are very expensive, or are soluble in water and will degrade if used in a humid environment. For these reasons, infrared instruments normally use aluminized mirrors to direct and focus the optical beams. Silver-coated first-surface mirrors tarnish too quickly to be widely used.

In any discussion of mirrors, we must consider the virtual image formed by a plane mirror. Figure 1.49a shows a diagram of the virtual image point P' of the object point P. The virtual image point has the same perpendicular distance from the mirror as the object point has. The virtual image point is usually detected by the eye. Because distances can only be seen with a cone of light, at least two rays are needed with a finite angle (Figure 1.49b). The eye detects the diverging cone of light from the object point, which seems to come from the virtual image point and which converges on the retina. A converging lens is used to form a real image from a virtual object.

A similar situation exists for refracting surfaces. Figure 1.3 demonstrated the case of the eye viewing an object through water. The positon of the object appears closer to the surface than it actually is. The position can be determined by considering the refraction of a cone of light emerging from the object. The refracted rays are traced back along straight lines to form the virtual image (Figure 1.50). Again, the eye lens detects the apparent object. The exact location of the apparent object will be discussed in detail in Chapter 12.

B. Deduction of the Law of Reflection from the Law of Refraction

Fermat's principle can be applied to reflection. In the law of reflection, a light beam travels from one position to another in a minimum amount of time.

The conditions for image formation, however, can be obtained using a mathematical trick. Using the sign convention that positive angles are represented by counterclockwise rotations and negative angles by clockwise rotations, we obtain for the law of reflection (Figure 1.51):

$$\theta_1 = -\theta_2 \tag{1.118}$$

This equation can also be obtained from the law of refraction (equation 1.4) by substituting $n_1 = -n_2$. Since this substitution produces the correct law of reflec-

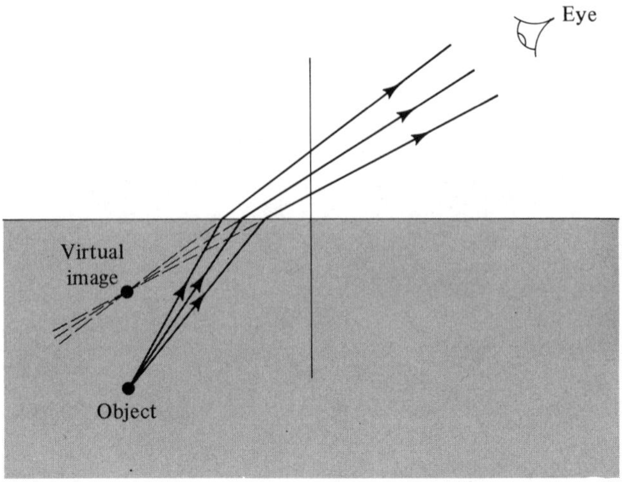

Fig. 1.50
Location of a virtual image seen by the eye on a refracting surface.

Eye

Virtual image

Object

Fig. 1.51
The angles in the law of reflection.

θ Positive | θ Negative

tion, we continue with our derivation of laws of reflection. We substitute $n_2 = -n_1$ into equation 1.9, the image-forming equation on one refracting surface, and have

$$-\frac{n_1}{x_o} + \frac{-n_1}{x_i} = \frac{-n_1 - n_1}{r} \tag{1.119}$$

or

$$\frac{1}{x_o} + \frac{1}{x_i} = \frac{2}{r} = \frac{1}{f} \tag{1.120}$$

We obtain the equation for image formation on a spherical reflecting surface. These considerations are, of course, limited to the paraxial theory. The same sign convention that was introduced for refracting surfaces can be utilized here. Note that the first term in equation 1.120 has no negative sign in contrast to the thin-lens equation 1.21.

C. Sign Convention for Spherical Mirrors

In spherical mirrors, the light is incident from the left. For concave mirrors, the radius of curvature r is negative; the radius of curvature is positive for convex mirrors. The object distance x_o is negative because it is to the left of the mirror. The image distance x_i can vary, however, depending on whether the image is real or virtual. The sign of x_i is negative for real images (left of the mirror) and positive for virtual images (right of the mirror).

The imaging process for spherical mirrors is similar to the imaging process for the thin lens. Figure 1.52 shows image formation for on-axis points. The magnification is defined as $m = y_i/y_o = -x_i/x_o$. As a result, a real, inverted image will have a negative magnification (see also Section 4G). For virtual images, the magnification is positive.

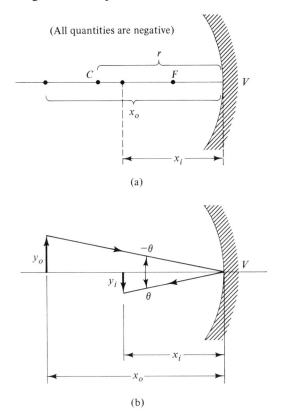

(a)

Fig. 1.52
(a) Image formation for on-axis points for a concave mirror.
(b) Image sizes and magnification for a concave mirror.

(b)

D. Graphical Method and Summary of Results

Figure 1.53 shows two object positions for a concave mirror. In Figure 1.53a, a real image is formed, and in Figure 1.53b, a virtual image is formed.

The graphical construction for mirrors is similar to graphical construction for lenses (see Section 4I).

1. The incident ray parallel to the optical axis is reflected back through the focal point F.

2. The ray incident at angle $-\theta$ at the vertex is reflected back at angle θ.

3. The ray incident through the center of curvature C is reflected back onto itself.

Only the first two conditions are necessary for image construction.

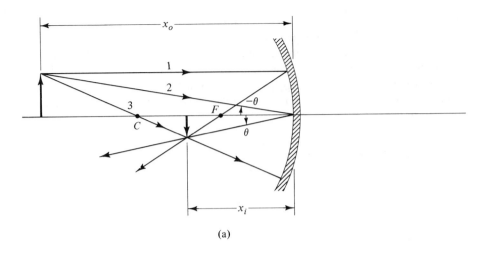

(a)

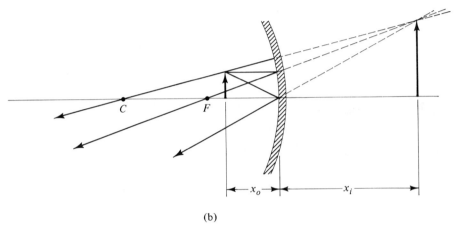

(b)

Fig. 1.53
Image formation by concave spherical mirrors. (a) Formation of a real image of a faraway object. (b) Formation of an enlarged virtual image of an object close to the mirror.

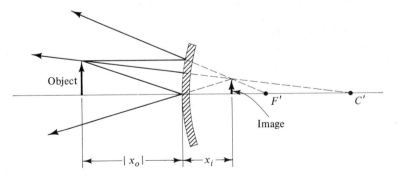

Fig. 1.54
Image formation with a convex mirror ($r > 0$, $f > 0$), any $|x_o| = 0$: upright, virtual, reduced image.

1. Geometrical Optics

Figure 1.54 shows image formation with a convex mirror. Any position of the object results in a virtual image. Table 1.5 summarizes the image-forming parameters for concave and convex mirrors.

Table 1.5
Image-forming parameters

	Concave $(r < 0)$	Concave $(r < 0)$	Concave $(r < 0)$	Concave $(r < 0)$	Convex $(r > 0)$
Object distance	$|x_o| > |r|$	$|x_o| = 2|f|$ $= |r|$	$|x_o| = |f|$	$|x_o| < |f|$	Any
Image distance	x_i negative	$|x_o| = |x_i|$	$x_i = -\infty$	x_i positive	x_i positive
Magnification	Negative	Negative	—	Positive	Positive
Character	Real, inverted, reduced	Real, inverted, same size	—	Virtual, upright, enlarged	Virtual, upright, enlarged

All that has been said about virtual images for lenses applies to mirrors as well. A virtual object can also appear in mirror systems and mirror-lens systems. It is treated here analogous to the discussion for the thin lens (see Figure 1.22f and Table 1.3). Large telescopes use a combination of mirrors and lenses. The resolution of the mirror-lens system is important and depends on the diameter of the mirror. The resolution properties of optical instruments will be discussed in Chapter 9.

Mirror systems are also important from another point of view. An arrangement of two flat plane parallel mirrors forms a resonant cavity. Light generated between the planes and traveling along the direction normal to the mirrors will be reflected back and forth. Such resonant cavities are used in lasers.

9. MATRIX FORMULATION FOR MIRRORS AND MIRROR SYSTEMS

A. Matrices for Mirror Reflection and Translation

We have now seen that substituting $n_2 = -n_1$ into the image-forming equation for refraction gives us the corresponding equation for a spherical mirror. In this section, we will treat mirror systems using the matrix method, finding the appropriate matrix representation for reflections and translations. For **reflection** at a curved surface,

$$\begin{pmatrix} 1 & 0 \\ -\dfrac{1}{r}\left(\dfrac{n_2 - n_1}{n_2}\right) & \dfrac{n_1}{n_2} \end{pmatrix} \xrightarrow{\text{for } n_2 = -n_1} \begin{pmatrix} 1 & 0 \\ -\dfrac{2}{r} & -1 \end{pmatrix} \tag{1.121}$$

Refraction at a
curved surface
(from equation 1.49)

Reflection at
a curved
surface

For a **plane mirror**, $r \to \infty$,

$$\begin{pmatrix} 1 & 0 \\ 0 & -1 \end{pmatrix}$$

(1.122)

For translation through a distance d,

$$\begin{pmatrix} 1 & d \\ 0 & 1 \end{pmatrix}$$

(1.123)

We will now discuss reflections from mirror systems, such as optical cavities.

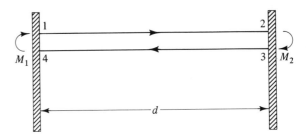

Fig. 1.55
Optical diagram of a plane mirror cavity, showing a multireflecting optical ray at positions 1 through 4.

B. Reflecting Cavities and the Eigenvalue Problem

Figure 1.55 shows a plane mirror cavity. No light should escape the mirror cavity, even after many traversals back and forth. Therefore, the light ray must travel in the same direction after each **cycle** is completed. The sequence of the matrices is as follows:

$$\begin{pmatrix} 1 & 0 \\ 0 & -1 \end{pmatrix}\begin{pmatrix} 1 & -d \\ 0 & 1 \end{pmatrix}\begin{pmatrix} 1 & 0 \\ 0 & -1 \end{pmatrix}\begin{pmatrix} 1 & d \\ 0 & 1 \end{pmatrix} = \begin{pmatrix} 1 & 2d \\ 0 & 1 \end{pmatrix} = M \qquad (1.124)$$
$$ 4 \to 1 3 \to 4 2 \to 3 1 \to 2$$

where $1 \to 2$ is a translation through a distance d, $2 \to 3$ is a reflection at M_2, $3 \to 4$ is a translation through the distance d, and $4 \to 1$ is a reflection at M_1.

The matrix M is apparently equal to a unit matrix. Figure 1.55 shows that after two traversals of length d, the ray is back at its original position and direction. The matrix M represents a traversal by the length $2d$. To find out if the action of M on the vector (ℓ_1, α_1) is equivalent to the action of a unit matrix, we have

$$\begin{pmatrix} 1 & 2d \\ 0 & 1 \end{pmatrix}\begin{pmatrix} \ell_1 \\ \alpha_1 \end{pmatrix} = \begin{pmatrix} \lambda & 0 \\ 0 & \lambda \end{pmatrix}\begin{pmatrix} \ell_1 \\ \alpha_1 \end{pmatrix}$$

(1.125)

where λ must be determined. If λ turns out to be 1, M acts as a unit matrix. Equation 1.125 can be written as a system of two linear, homogeneous equations:

$$(1 - \lambda)\ell_1 + 2d\alpha_1 = 0$$
$$(1 - \lambda)\alpha_1 = 0$$

(1.126)

To get nonzero solutions of such a system, the determinant must vanish

$$\begin{vmatrix} 1 - \lambda & 2d \\ 0 & 1 - \lambda \end{vmatrix} = (1 - \lambda)^2 = 0 \rightarrow \lambda = 1 \tag{1.127}$$

and we have the result that λ is equal to 1. The matrix M is indeed equivalent to the unit matrix. In mathematical terms, we state that the eigenvalues of the matrix M are all 1. For the two plane parallel mirrors, we find the result that the light ray travels back and forth without escaping. The two mirrors form a resonant cavity.

C. Confocal Cavities

Consider a cavity consisting of a convex and a concave spherical mirror, both having the same radius of curvature $|r|$ and separated by a distance $d = |r|$. The eight matrices in the proper sequence are (see Figure 1.56)

$$\begin{pmatrix} 1 & 0 \\ -\dfrac{2}{r} & -1 \end{pmatrix} \begin{pmatrix} 1 & -d \\ 0 & 1 \end{pmatrix} \begin{pmatrix} 1 & 0 \\ \dfrac{2}{r} & -1 \end{pmatrix} \begin{pmatrix} 1 & d \\ 0 & 1 \end{pmatrix}$$

$$\underbrace{\overbrace{\underset{\text{Convex}}{4} \quad \underset{}{3}}^{} \quad \overbrace{\underset{\text{Concave}}{2} \quad \underset{}{1}}^{}}_{M_A}$$

$$\begin{pmatrix} 1 & 0 \\ -\dfrac{2}{r} & -1 \end{pmatrix} \begin{pmatrix} 1 & -d \\ 0 & 1 \end{pmatrix} \begin{pmatrix} 1 & 0 \\ \dfrac{2}{r} & -1 \end{pmatrix} \begin{pmatrix} 1 & d \\ 0 & 1 \end{pmatrix} \tag{1.128}$$

$$\underbrace{\underset{\text{Convex}}{8} \quad 7 \quad \underset{\text{Concave}}{6} \quad 5}_{M_B}$$

We have $M_A = M_B$. The matrices are computed in pairs, giving (e.g., for M_A)

$$\begin{pmatrix} 1 & 0 \\ \dfrac{2}{r} & -1 \end{pmatrix} \begin{pmatrix} 1 & d \\ 0 & 1 \end{pmatrix} = \begin{pmatrix} 1 & d \\ \dfrac{2}{r} & \dfrac{2d}{r} - 1 \end{pmatrix}$$

$$\begin{pmatrix} 1 & 0 \\ -\dfrac{2}{r} & -1 \end{pmatrix} \begin{pmatrix} 1 & -d \\ 0 & 1 \end{pmatrix} = \begin{pmatrix} 1 & -d \\ -\dfrac{2}{r} & \dfrac{2d}{r} - 1 \end{pmatrix} \tag{1.129}$$

and by multiplication

$$\begin{pmatrix} 1 - \dfrac{2d}{r} & 2d\left(1 - \dfrac{d}{r}\right) \\ -\dfrac{4}{r}\left(1 - \dfrac{d}{r}\right) & \left(\dfrac{2d}{r} - 1\right)^2 - \dfrac{2d}{r} \end{pmatrix} \tag{1.130}$$

Introducing $d = r$ gives

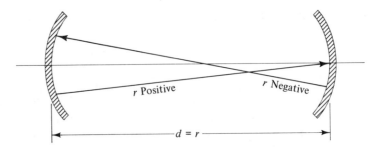

Fig. 1.56
Cavity formed by convex and concave mirrors of the same radius of curvature and at distance $d = r$.

$$M_A = M_B = \begin{pmatrix} -1 & 0 \\ 0 & -1 \end{pmatrix} \qquad (1.131)$$

Multiplying M_A and M_B yields

$$M = \begin{pmatrix} 1 & 0 \\ 0 & 1 \end{pmatrix}$$

For the special case where $d = r$, the matrix M is the unit matrix. After four traversals, the ray travels in the original direction. The mirrors are forming a resonance cavity for $d = r$. The light path is shown in Figure 1.57.

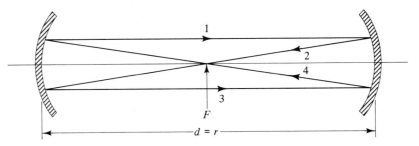

Fig. 1.57
A light ray traveling in a mirror cavity. The two spherical mirrors have radii of curvature r and are placed at a distance $d = r$. After one "round trip" (1 to 4), the ray is again traveling in the original direction 1.

In this chapter we saw how the laws of reflection and refraction may be applied to image formation. Later chapters will examine the roles played by other phenomena in image formation. We will also see how the simple laws used in this chapter can be derived from Maxwell's theory. It is amazing just how many optical phenomena are governed by a few simple laws. Although basic algebra is used in the mathematical formulations, the expressions for more complicated systems do get lengthy. For such systems, the matrix formulation shows us more clearly what is going on at different sections of the system.

Problems

1. *Prism.* Consider a prism of apex angle A and refractive index n. Calculate the dependence of the angle of deviation δ as a function of n and the angle of incidence θ_1.

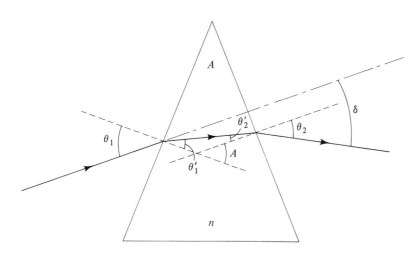

Ans. $\delta = \theta_1 + \sin^{-1}[(\sin A)\sqrt{n^2 - \sin^2 \theta_1} - \cos A \sin \theta_1] - A$

2. *Image formation of a real object for a single thin biconvex lens.*
   a. Fill in the following table, assuming the object is placed as indicated in column x_o.

x_o	x_i	Magnifi- cation m	Image Real	Virtual	Image Enlarged	Reduced
(A) $-\infty$						
(B) $-3f$						
(C) $-1.5f$						
(D) $-f$						
(E) $-0.5f$						

   b. Make sketches for the five cases.

3. *Negative lens.* Consider a negative lens with $f = -5$ cm. The object is at $x_o = -10$ cm.
   a. Find x_i.
   b. Make a sketch.
   Ans. $x_i = -3.3$ cm.

4. *System of two thin lenses.* The first lens produces a real image, and the second lens is placed so far away that the real image serves as a real object.
   a. Make a sketch.

b. Calculate the final image position for the case where $x_{o1} = -30$ cm, $f_1 = 10$ cm, $f_2 = 15$ cm. Assume the distance between the lenses to be $d = 45$ cm.

c. Find the magnification.

Ans. b. $x_{i2} = 30$ cm; c. $m_{total} = 1/2$

5. *Two thin lenses in the magnifier configuration.* The first lens produces a virtual image. This image serves as the real object for the second lens (on account of its position). The focal length and distance from the object with respect to the second lens are chosen in such a way that the final image is real.

a. Make a sketch and calculate the final image location for $x_{o1} = -6$ cm, $f_1 = 12$ cm, $f_2 = 10$ cm, and a distance between the lenses of 8 cm.

b. Find the final magnification $m_{total} = m_1 \cdot m_2$.

Ans. a. $x_{i2} = 20$ cm; b. $m_{total} = -2$

6. *Two thin lenses in the "near-point configuration" of the magnifier.* If the second lens is the eye, there is a limit as to how small x_{o2} may be chosen. This distance is called the near point and is about 25 cm.

a. Make a sketch.

b. If $d = 8$ cm, $f_1 = 12$ cm, and $x_{o2} = -25$ cm, find x_{o1}.

c. If $x_{i2} = 2$ cm, what value for f_2 must we choose?

d. For this case, the angular magnification MP is given as $MP = 25/f_m + 1$. Calculate MP and give it as MP_x (see Section 5C).

Ans. b. $x_{o1} = -7.03$ cm; c. $f_2 = 1.85$ cm; d. $MX = 2.08 \times$

7. *Two thin lenses, magnifier and eye, with the first image at ∞.* Consider the eye looking at a virtual image of the first lens (magnifier) at $-\infty$. Assume $f_1 = 12$ cm, $d = 8$ cm, and $x_{i2} = 2$ cm.

a. Where is the object to be placed; that is, find x_{o1}.

b. Make a sketch.

c. What is the value of f_2?

d. Find the angular magnification MP.

Ans. a. $x_{o1} = -12$ cm; c. $f_2 = 2$ cm; d. $MP = 2.08$

8. *Two-lens system with the real image of the first lens located within the focal length of the second.* If the real image of the real object of the first lens is located within the focal length of the second lens, it produces a final virtual image.

a. If $x_{o1} = -20$ cm, $f_1 = 10$ cm, $f_2 = 15$ cm, and $d = 30$ cm, make a sketch and calculate x_{i1} and x_{i2}.

b. If $x_{o1} = -20$ cm, $f_1 = 10$ cm, $f_2 = 15$ cm, and $d = 10$ cm, make a sketch and find x_{i2}.

Ans. a. $x_{i1} = 20$ cm, $x_{i2} = -30$ cm; b. $x_{i2} = 6$ cm

9. *Laser beam expander.* A laser beam of diameter 2 mm should be expanded to a beam of 20 mm.

a. A biconvex and a biconcave lens should be used. A biconvex lens L_2 of diameter 30 mm and focal length $f_2 = 50$ mm is available. What is the focal length of the biconcave lens L_1? How far to the left of the biconvex lens is the biconcave lens to be placed?

b. Using two biconvex lenses, one with $f_2 = 50$ mm, the other $f_1 = 5$ mm, what is the distance between the two lenses?

Ans. a. $f_1 = -5$ mm, distance L_1 to L_2 is 45 mm; b. distance L_1 to L_2 is 55 mm

10. *Three-lens system; microscope.* In the microscope, the objective lens forms a magnified real image. Then a magnifier configuration is applied and forms a real final image on the retina.
   a. Find the final image location x_{i3} for $x_{o1} = -2.5$ cm, $f_1 = 2$ cm, $f_m = 6$ cm, and $f_3 = 2$ cm. Assume that the distance between the first and second lens is 16 cm and that the distance between the second and third lens is 1 cm. Give x_{i1}, x_{i2}, and x_{i3}, and make a sketch (see Section 5D).
   b. The magnification M of the system is given as $M = -(16/f_1)(25/f_2)$. Calculate M.
   Ans. a. $x_{i1} = 10$ cm, $x_{i2} = -\infty$, $x_{i3} = 2$ cm; b. $M = -33.3$

11. *Telescope.* The telescope is similar to the microscope. But the objective lens forms a real image from a faraway object. This object is looked at with a magnifier configuration. Assume that for the magnifier configuration, the image of this object is at $-\infty$.
   a. For $f_1 = 30$ cm, $f_2 = 6$ cm, $f_3 = 2$ cm, $d_{23} = 2$ cm, $x_{o1} = -\infty$, and $x_{i2} = -\infty$, give x_{i1}, x_{o2}, and x_{i3}, and make a sketch.
   b. What is the distance between lenses 1 and 2?
   c. The angular magnification is defined as $MP = f_1/f_2$. Give the value of MP.
   Ans. a. $x_{i1} = 30$ cm, $x_{o2} = -6$ cm, $x_{i3} = 2$ cm; b. $d_{12} = 36$ cm; c. $MP = 5$.

12. *Refraction and translation matrices.* We considered two glass lenses formed by interchanging the order of the refracting surfaces.

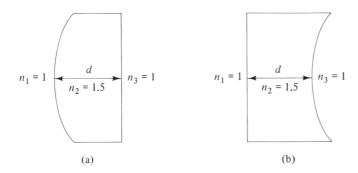

(a)   (b)

   a. For the lens with the convex surface to the left, obtain the matrices describing refraction at both surfaces and translation through the lens. Take the refractive index of the glass to be 1.5. Evaluate the product of the matrices and show that the result is
   $$\begin{pmatrix} 1 - \dfrac{d}{3r} & \dfrac{2}{3}d \\[2mm] -\dfrac{1}{2r} & 1 \end{pmatrix}$$

   b. Repeat the calculation for the lens with the plane surface to the left. Show that the result is now
   $$\begin{pmatrix} 1 & \dfrac{2}{3}d \\[2mm] \dfrac{1}{2r} & 1 + \dfrac{d}{3r} \end{pmatrix}$$

13. *Comparison of a thick and thin lens.* Consider a thick, biconvex lens in air with radii of curvature $r_1 = 20$ cm and $r_2 = -10$ cm, index of refraction 1.5, and thickness 5 cm. An object is located 8 cm to the left of the first surface.

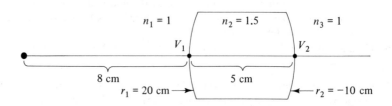

a. Calculate the image distance from the second surface by successively using the single-surface refraction equation for each surface.
b. Apply the thin-lens equation to this lens. First calculate the focal length f, using

$$\frac{1}{f} = (n_2 - 1)\left(\frac{1}{r_1} - \frac{1}{r_2}\right)$$

and then determine the image distance from the second surface. (Now all distances should be measured from the center of the lens.)
c. Treat the original thick-lens using the matrix method.
Ans. a. $x_{i1} = -15$ cm, $x_{i2} = -40$ cm; b. $f = 13.33$ cm, $x_i = -49.41$ cm, x_i from $V_2 = -52.5$ cm; c. $f = 14.1$ cm, $h' = -1.18$ cm, $x_i = -38.8$ cm from H'

14. *Biconcave lens.* Consider a biconcave lens in air with radii $r_1 = -10$ cm and $r_2 = 5$ cm and a thickness at the vertex of 3 cm. An object is 5 cm to the left of the first vertex.

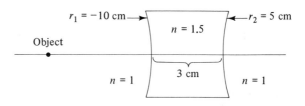

a. Using the equation for image formation

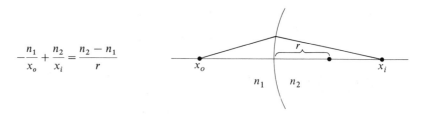

$$-\frac{n_1}{x_o} + \frac{n_2}{x_i} = \frac{n_2 - n_1}{r}$$

at each surface of the lens, find the final image.
b. Use the thin-lens equation to solve this problem, taking the center of the lens as a reference point.
c. Treat the thick, biconcave lens using the matrix method.

Ans. a. $x_{i2} = -3.75$ to left of V_2; b. $f = -6.67$ cm, $x_i = -3.29$ cm, and it is -4.79 to the left of V_2; c. $P = -6.25$ cm; $x_i = -3.125$ cm from H', -3.75 cm to left of V_2.

15. *Short focal length biconvex lens.* A quartz lens should be used for concentrating parallel light on a detector. In the wavelength region considered, quartz has $n = 2$. A biconvex lens of thickness 4 mm should be used, and the detector should be placed 2 mm to the right of the second surface. We want to find the radii of curvature.
    a. Express x_{i1} by the radius r_1 from the first surface.
    b. Express x_{o2} with x_{i1} and d, and write the imaging equation for the second surface.
    c. Assume $r_1 = -r_2$ and solve the quadratic equation for r_1.
       Make the following checks:
    d. Calculate x_{i1} from (a), using the value of r_1, and convert to x_{o2}.
    e. Calculate x_{i2} from (b) (should be 2 mm).
    Ans. a. $x_{i1} = 2r_1$; b. $x_{i2} = 2$ mm; c. $r_1 = 5.23$ mm; d. $x_{o2} = 6.46$ mm; e. $x_{i2} = 1.996$ mm.

16. *Short-focal-length biconvex lens as thick lens.* In problem 15, we saw that a quartz lens of index of refraction $n = 2$ and radii of curvature $-r_1 = +r_2 = -5.23$ mm concentrates parallel light from the left onto a detector at a distance 2 mm from the second surface to the right. Check the validity of all numbers by application of the thick-lens formulas ($d = 4$ mm).
    Ans. $x_i = f = -1/P = 3.233$ mm, 1.998 mm to right of V_2

17. *Short-focal length planoconvex lens.* In problems 15 and 16, we considered a biconvex lens of thickness 4 mm and radii of curvature $r_1 = -r_2 = -5.23$ mm. The light was incident parallel from the left and the image position was sought for a lens with $n = 2$. Now we want to find the image of a planoconvex lens of thickness $d = 4$ mm.
    a. Calculate the image position from the second plane surface.
    b. Make the lens so thick that the image is on the second surface. What is the thickness d' of the lens?
    c. Check the calculation of (b) with the matrix formulation of the thick lens.
    Ans. a. $x_{i2} = 3.23$ mm; b. $d' = 10.46$ mm; c. $x_i = f = 5.23$ mm

18. *Glass sphere.* Consider a glass sphere ($n = 1$) of radius 10 cm. An object is located 120 cm to the left of the sphere (measured from the first refracting surface).

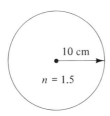

10 cm

$n = 1.5$

    a. Determine the image point of the object, using the single-surface refraction equation at each surface.
    b. Assume that the thin-lens equation can be applied. Solve for the final image point position, using the center of the sphere as a reference.
    c. Use the matrix method to solve this problem and apply the formulas derived in Section 7C.
    d. Determine the parameters for equal object and image distances for the sphere,

using the center as a reference point and $f = nr/2(n - 1)$. Make a sketch and express all distances in terms of the diameter of the sphere.

19. *Corner mirror.* Show that a corner mirror (90°) will reflect incident light through an angle of 180° regardless of the angle α between incident ray and mirror. ($\mathbf{n}_1$ is a normal.)

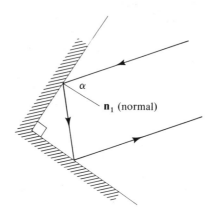

20. *Concave mirror.* Consider a concave mirror with $r = -12$ cm. The object is at $x_o = -20$ cm.
   a. Find x_i.
   b. Make a sketch.
   Ans. $x_i = -8.6$ cm

21. *Convex mirror.* Consider a convex mirror with $r = 12$ cm. The object is at $x_o = -20$ cm.
   a. Find x_i.
   b. Make a sketch.
   Ans. $x_i = 4.6$ cm

22. *Concave spherical mirror, analytical solution.* Use $r = -10$ cm and find x_i for the following:
   a. $x_o = -\infty$
   b. $|x_o| > |r|$, $x_o = -15$ cm
   c. $|x_o| = |r|$, $x_o = -10$ cm
   d. $|r| \geq |x_o| > |f|$, $x_o = -8$ cm
   e. $|x_o| < |f|$, $x_o = -3$ cm
   Ans. a. $x_i = f - 5$ cm; b. $x_i = -7.5$ cm; c. $x_i = -10$ cm; d. $x_i = -13.3$ cm; e. $x_i = 7.5$ cm

23. *Concave spherical mirror, graphical solution.* Using the graphical method, find the image produced by a concave spherical mirror (r negative) for the following values of object distances:
   a. $x_o = -\infty$
   b. $|x_o| > |r|$
   c. $|x_o| = |r|$
   d. $|r| \geq |x_o| > |f|$
   e. $|x_o| < |f|$

1. Geometrical Optics

24. *Convex spherical mirror, analytical solution.* Use $r = 5$ cm and find x_i for the following:

  a. $x_o = -\infty$
  b. $|x_o| > r$, $x_o = -7$ cm
  c. $|x_o| = r$, $x_o = -5$ cm
  d. $r \geq |x_o| > f$, $x_o = -4$ cm
  e. $|x_o| < f$, $x_o = -1$ cm
  Ans. a. $x_i = f = 2.5$ cm; b. $x_i = 1.84$ cm; c. $x_i = 1.67$ cm; d. $x_i = 1.54$ cm; e. $x_i = 0.714$ cm

25. *Convex spherical mirror, graphical solution.* Using the graphical method, find the image produced by a convex spherical mirror (r positive) for the following values of the object (to the left of the mirror):

  a. $x_o = -\infty$
  b. $|x_o| > r$
  c. $|x_o| = r$
  d. $r \geq |x_o| > f$
  e. $|x_o| < f$

26. *Hemispherical cavity.* Consider the optical cavity consisting of a plane mirror (M_1) and a spherical mirror (M_2), separated by a distance d. The path of the ray, indicated in the figure shown here, retraces itself after eight steps. For clarity, the eight steps are shown in the figure as $1, 2, 3, \ldots, 8$.

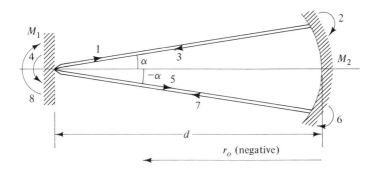

  a. Write the eight matrices in the correct sequence and show that the resulting product is

$$M = \begin{pmatrix} \left(1 - \dfrac{2d}{r_o}\right)^2 - \dfrac{4d}{r_o}\left(1 - \dfrac{d}{r_o}\right) & 4d\left(1 - \dfrac{d}{r_o}\right)\left(1 - \dfrac{2d}{r_o}\right) \\ -\dfrac{4}{r_o}\left(1 - \dfrac{2d}{r_o}\right) & \left(1 - \dfrac{2d}{r_o}\right)^2 - \dfrac{4d}{r_o}\left(1 - \dfrac{d}{r_o}\right) \end{pmatrix}$$

  b. The matrix is of the form $\begin{pmatrix} a & b \\ c & a \end{pmatrix}$; that is, we expect a double root from the determinant. Show that the eigenvalue of this problem may be expressed as $\lambda = a \pm \sqrt{bc}$ and give λ as a function of d/r.
  c. For what value of d/r do we get $\lambda = 1$, and what does it mean?
  Ans. c. $\lambda = 1$

Interference of Light | 2

1. INTRODUCTION

In Chapter 1, we saw that light propagates in straight lines and obeys the laws of reflection and refraction. In this chapter, we will discuss experiments whose results can only be explained if we assume that light has all the properties of a wave. Again, light waves travel in a vacuum with speed c and in matter with speed $v = c/n$ (where c is the speed of light and n is the refractive index).

The wave property of light is complementary to what we discussed in Chapter 1. Light waves obey the laws of reflection and refraction but also show features that result from their **wave nature**. The most striking evidence of the wave property is the production of fringes when two light waves are superimposed. The intensity of two superimposed waves may be zero at certain positions. The wavelike character of light follows from electromagnetic theory and is discussed in Chapter 4. In this chapter, we discuss light as a simple harmonic wave depending on one space coordinate and time.

We will consider the different ways to superimpose light, leading to the observation of interference patterns. With such experimental setups, we will study the phase relations between superimposed waves for laser light, atomic emission, blackbody radiation, and secondary waves produced by **Huygens' principle**. In certain experiments, light may produce fringes, whereas in other, very similar experiments, no fringes are produced. Through a detailed analysis of this subject, we will learn when light is considered coherent or incoherent.

Examples of fringe patterns are obtained by using monochromatic light, such as laser light, or a light source with a narrow band filter. While this can be demonstrated in the laboratory, we are much more familiar with naturally occurring interference phenomena, such as the colored spectrum of a soap bubble. Such phenomena are more complicated because they involve waves with many different wavelengths. The analysis considers the superposition of waves for each wavelength separately and then adds the intensity patterns.

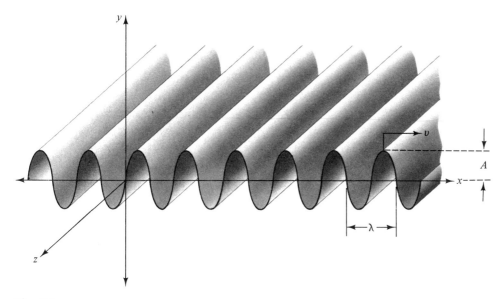

Fig. 2.1
A wave oscillating in the y direction, propagating in the x direction, and extending to infinity in the z direction. The wave has amplitude A, wavelength λ, and propagation velocity v.

A. The Superposition Principle and the Assumption About Light as Waves

Light is one realization of electromagnetic waves. Historically, this was concluded when the speed of light was found to be the propagation velocity of electromagnetic waves with much longer wavelengths. The description of light using Maxwell's theory will be discussed in Chapter 4.

The **superposition principle** follows from Maxwell's theory. It tells us that the superposition of two waves, each having the wavelength λ, results again in a wave with the same wavelength λ. This principle is the backbone of the discussion in this chapter. The main procedure is to count the **path difference** between two waves to be superimposed. The path difference must then be converted into a **phase difference**. This can be given as a fraction of 2π or as a multiple or fraction of the wavelength.

For the mathematical description, we represent light by harmonic waves. This description fits laser light especially well. We can imagine these waves as oscillating in the y direction, propagating in the x direction, and extending to infinity in the z direction (Figure 2.1). In describing such a wave, we use only the part indicated by the solid line in Figure 2.1, that is, a one-dimensional oscillation depending on x and time t. We assume the following for the description of light as waves:

1. Light is described by a transverse harmonic wave traveling in a vacuum with speed c, and having wavelength λ, time period T, frequency $v = 1/T$ and amplitude A. In matter, the speed is c/n, and the wavelength is λ/n.

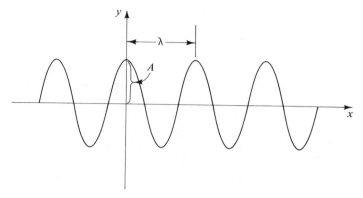

Fig. 2.2
Plot of a harmonic wave of peak amplitude A and wavelength λ.

2. The intensity of light is proportional to the square of the amplitude of the wave.

3. A phase shift of π occurs if a wave is reflected at a medium with higher index of refraction or at a metal surface. No shift of π occurs on reflection at a medium with lower index of refraction.*

2. LIGHT AS A HARMONIC WAVE DEPENDING ON SPACE AND TIME

A. Superposition of Waves Depending on Space Coordinates Only

The mathematical description of a harmonic wave (extending from $-\infty$ to $+\infty$) is

$$y = A \cos\left(2\pi \frac{x}{\lambda}\right) \qquad (2.1)$$

where y is the amplitude, A is the maximum amplitude, and λ is the wavelength in the x direction (see Figure 2.2). The argument $2\pi(x/\lambda)$ is called the phase. It gives the fraction x/λ of the total angle corresponding to 2π. If x exceeds the value of λ, we can write $x = m\lambda + x'$, where m is an integer and x' is the excess x value beyond integer multiples of λ. The phase is now $2\pi m + 2\pi(x'/\lambda)$. Since the cosine function is periodic in 2π, the phase is reduced to $2\pi(x'/\lambda)$, giving the same value for y. Similarly, a sin function can be used.

Let us now consider the superposition of two waves that have the same amplitude and wavelength. In Figure 2.3, the points (1) and (2) of the two waves are separated by a distance δ. According to equation 2.1,

---

*A mechanical wave is reflected with a phase shift of π at an interface if the velocity is smaller on the other side; if larger, no phase shift occurs.

2. Interference of Light

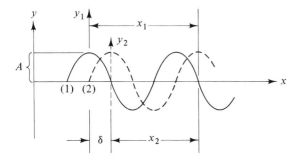

Fig. 2.3
Superposition of waves shifted a distance δ from each other. Both waves have the same amplitude and wavelength.

$$y_1 = A \cos\left(2\pi\frac{x_1}{\lambda}\right) \quad \text{and} \quad y_2 = A \cos\left(2\pi\frac{x_2}{\lambda}\right)$$

The superposition is expressed by the addition of the amplitudes:

$$y = y_1 + y_2 = A \cos\left(2\pi\frac{x_1}{\lambda}\right) + A \cos\left(2\pi\frac{x_2}{\lambda}\right) \tag{2.2}$$

If we relate x_1 and x_2 by $x_2 + \delta = x_1$, where δ is the shift between the two waves, we can plot the two waves in a common coordinate system involving only x_1 and δ, as shown in Figure 2.4. We then have

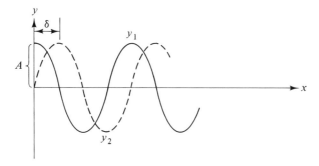

Fig. 2.4
Two waves with amplitude A and wavelength λ, with $y_1 = A$ for $x_1 = 0$ and $y_2 = A$ for $x_2 = x_1 - \delta = 0$.

$$y = A \cos\left(2\pi\frac{x_1}{\lambda}\right) + A \cos\left(2\pi\frac{x_1 - \delta}{\lambda}\right) \tag{2.3}$$

We can rewrite equation 2.3 using the formula for the addition of two cosine functions:

$$\cos\phi_1 + \cos\phi_2 = 2\cos\left(\frac{\phi_1 + \phi_2}{2}\right)\cos\left(\frac{\phi_1 - \phi_2}{2}\right) \tag{2.4}$$

Letting $\phi_1 = 2\pi x_1/\lambda$ and $\phi_2 = 2\pi x_2/\lambda = 2\pi(x_1 - \delta)/\lambda$ and applying equation 2.4,

we have

$$y = y_1 + y_2 = A \left[\cos 2\pi \frac{x_1}{\lambda} + \cos \left(2\pi \frac{x_1 - \delta}{\lambda} \right) \right]$$

$$= 2A \left[\cos \frac{2\pi(2x_1/\lambda) - 2\pi(\delta/\lambda)}{2} \right] \cos \left(\frac{2\pi(\delta/\lambda)}{2} \right) \tag{2.5}$$

$$= 2A \left(\cos 2\pi \frac{\delta/2}{\lambda} \right) \cos \left(2\pi \frac{x_1}{\lambda} - 2\pi \frac{\delta/2}{\lambda} \right)$$

| Amplitude factor (independent of x) | Phase factor | Phase constant |

The superimposed waves result in a new wave with the same wavelength but a modified amplitude. The new wave is shifted by a phase constant depending on δ, and the new amplitude also depends on δ. We call δ the optical path difference and $2\pi(\delta/\lambda)$ the phase difference of the two waves (see Figure 2.4).

Independent of the choice of δ, the cosine wave remains a cosine wave depending on x_1 as a variable. On the other hand, the amplitude factor depends not on x_1 but on δ and can take on all values between 0 and $\pm 2A$.

The amplitude factor is the one we are most interested in. If the absolute value of this factor is a maximum, we have **constructive interference**. If this factor is zero, we have **destructive interference**.

Constructive interference:

The case where $\delta = 0, \lambda, 2\lambda, \ldots \left(\text{i.e.,} \left| 2A \cos 2\pi \frac{\delta/2}{\lambda} \right| = 2A \right)$

Destructive interference:

The case where $\delta = \frac{\lambda}{2}, \frac{3\lambda}{2}, \frac{5\lambda}{2}, \ldots \left(\text{i.e., } 2A \cos 2\pi \frac{\delta/2}{\lambda} = 0 \right)$

We see that destructive interference is independent of the point x_1 we choose for observation. The amplitude factor is zero, as is the resulting wave. This is not so in the case of constructive interference. There we obtain $2A$ only for specific choices of x_1, that is, when the phase factor is 1.

According to assumption 2 in Section 1A, the intensity is proportional to the square of the amplitude. Thus,

$$y^2 = \left[4A^2 \cos^2 2\pi \frac{\delta/2}{\lambda} \right] \left[\cos \left(2\pi \frac{x_1}{\lambda} - 2\pi \frac{\delta/2}{\lambda} \right) \right]^2 \tag{2.6}$$

B. Superposition of Waves Depending on Space and Time Coordinates

Light waves, like water waves, travel from their point of origin. The waves discussed so far have depended only on space coordinates. But to describe traveling waves, we must take the time dependence into account. We can consider the time-independent wave as a snapshot of a traveling wave. A sequence of such snapshots depends on time as a parameter. Let us start with the wave having the amplitude $y_1 = A \cos 2\pi(x/\lambda)$ depending on x. We can consider this the first

2. Interference of Light

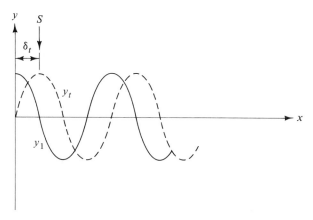

Fig. 2.5
Amplitude of a wave y_1 at $t = 0$ and position x. At time t, the wave peak at $x = 0$ has moved to position S.

snapshot and then draw a second wave y_t that would be seen if we took the next snapshot (Figure 2.5). The amplitude y_t at time t may be expressed as

$$y_t = A \cos 2\pi \left(\frac{x - \delta_t}{\lambda} \right)$$

with the shift δ_t depending on time. At time $t = 0$, the maximum of y_1 is at $x = 0$; at a later time t, the maximum is at $x = \delta_t$. If we express δ_t as $\delta_t = vt$, where v is the velocity of the wave, we have

$$y_t = y(x, t) = A \cos 2\pi \left(\frac{x - vt}{\lambda} \right) \tag{2.7}$$

We can express the velocity as $v = \lambda/T$, where during a time of one period T, the wave has moved a distance of one wavelength λ. Introducing this into equation 2.7 yields

$$y(x, t) = A \cos 2\pi \left(\frac{x}{\lambda} - \frac{t}{T} \right) \tag{2.8}$$

We can see the symmetric dependence of $y(x, t)$ on x and t. The final phase is determined by the sum of the fractions x/λ of 2π and t/T of 2π. Any additional phase, such as ψ in

$$y(x, t) = A \cos 2\pi \left(\frac{x}{\lambda} - \frac{t}{T} + \psi \right) \tag{2.9}$$

may be added to the space fraction or the time fraction.

We will now add two time-dependent waves in the same way we added the time-independent waves. We have

$$y_1 = A \cos 2\pi \left(\frac{x_1}{\lambda} - \frac{t}{T} \right) \quad \text{and} \quad y_2 = A \cos 2\pi \left(\frac{x_2}{\lambda} - \frac{t}{T} \right) \tag{2.10}$$

and as before, $x_2 = x_1 - \delta$. For the superposition, we get

$$y = y_1 + y_2 = 2A \cos\left(2\pi \frac{\delta/2}{\lambda}\right) \cos\left[2\pi\left(\frac{x_1}{\lambda} - \frac{t}{T}\right) - 2\pi\frac{\delta/2}{\lambda}\right] \qquad (2.11)$$

Amplitude factor (independent of x and t) — Phase factor — Phase constant

The superimposed waves result in a time-dependent wave with the same wavelength and frequency that y_1 and y_2 had. The amplitude factor and the phase constant are the same as in the time-independent case.

The conditions for constructive and destructive interference are the same as in the time-independent case. For constructive interference,

$$\delta = 0, \lambda, 2\lambda, \ldots$$

where the amplitude y varies between $+2A$ and $-2A$, depending on the values of x_1 and t. For destructive interference,

$$\delta = \frac{1}{2}\lambda, \frac{3}{2}\lambda, \frac{5}{2}\lambda, \ldots$$

where the amplitude y is zero regardless of the values of x_1 and t.

The intensity is proportional to the square of the amplitude:

$$y^2 = \left[2A \cos\left(2\pi\frac{\delta/2}{\lambda}\right)\right]^2 \cos^2\left[2\pi\left(\frac{x_1}{\lambda} - \frac{t}{T}\right) - 2\pi\frac{\delta/2}{\lambda}\right] \qquad (2.12)$$

Amplitude factor — Wave factor

We see that the intensity is zero for destructive interference, independent of the phase factor.

We can now apply this mathematical formulation of the superposition of two traveling waves to the description of interference of light. For constructive interference, the second factor in equation 2.11 oscillates with a very high frequency ($v = 1/T$). Using the formula

$$\cos^2 \alpha = \tfrac{1}{2}(1 + \cos 2\alpha)$$

we see from equation 2.12 that this frequency is twice the frequency of light (where α is in terms of $v = 1/T$). The human eye and indeed most detectors will not respond to oscillations in this frequency range. Only the **time average** is recorded. (See problem 1.) Therefore, the maxima we observe are proportional to the time average of the second factor in equation 2.12. Taking this time average as a_v^2, the intensity is

$$y^2 = 4A^2 a_v^2 \cos^2\left(2\pi\frac{\delta/2}{\lambda}\right) \qquad (2.13)$$

As a result of this study, we see that the amplitude factor is only important in finding the minima and maxima of interference phenomena.

2. Interference of Light

We can ignore the factor a_v^2 if we call the intensity the square of the *amplitude factor*. In many experiments, we will make relative intensity measurements of a pattern and introduce a reference intensity, perhaps at the center of the pattern. If we consider

$$I = I_0 \cos^2 \left(2\pi \frac{\delta}{2\lambda} \right)$$

then I_0 is the value of I when δ is such that

$$\cos^2 \left(2\pi \frac{\delta}{2\lambda} \right) = 1$$

The time average factor a_v^2 does not play a role in that case. We will obtain the quantitative relation between the amplitude of an electromagnetic wave and its power in Chapter 4.

3. TWO-BEAM INTERFERENCE PROBLEMS

A. Spatially Separated Sources

When we observe a fringe pattern, we can usually see the maxima and minima next to each other. Such a pattern is produced when two waves are simultaneously generated at two points in space. In Figure 2.6, if $\theta = 90°$, we have the case discussed previously. For other angles, the two waves will superimpose in the direction of arrows 1 and 2, shown in Figure 2.6. Consider the superposition of circular water waves generated at two points by oscillating studs (Figure 2.7). The generated waves are considered "in phase" if their phase difference is constant. In the case shown in Figure 2.6, the phase difference is zero for the direction $\theta = 0$.

For a description of the interference phenomenon, let us consider waves traveling along the lines from S_1 and S_2 to the observation point P in Figure 2.8a. They do not propagate in the same direction, and for water waves this must be taken into account quantitatively. For light waves, we will observe the

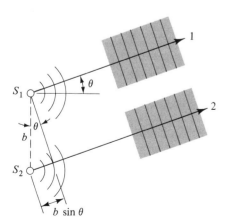

Fig. 2.6
Two sources emitting waves simultaneously.

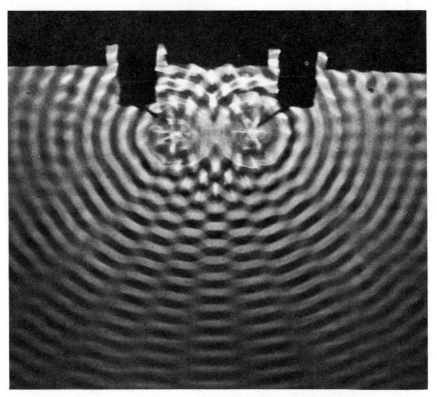

Fig. 2.7
The pattern caused by the interference of water waves emitted simultaneously from two sources separated by a distance b. (From *PSSC Physics*, 6th ed., D. C. Heath and Company, Lexington, Mass., 1986. Reprinted by permission of D. C. Heath and Company.)

emission of spherical waves from point sources, of cylindrical waves from line sources. We assume that the two emitted light waves are in phase. How this can be accomplished experimentally will be discussed later.

The wavelength and the distance b between the sources is small compared to the distance X of the observation screen. Waves arriving at the observation point are well approximated by plane waves. The interference phenomena are therefore described by the superposition of two traveling waves having the angles θ and θ' in Figure 2.8a. The difference between these angles is negligible because of the magnitude of X. As a result, we can treat the waves as if they were traveling along parallel lines (Figure 2.8b). We can also employ a lens and observe the fringe pattern in the focal plane of the lens for a shorter distance X' than X, as shown in Figure 2.8c.

For the intensity of the fringe pattern at the observation point, we have with equation 2.12 and $b \sin \theta = \delta$

$$y^2 = \left[2A \cos \left(\pi \frac{b \sin \theta}{\lambda} \right) \right]^2 \cos^2 \left[2\pi \left(\frac{x}{\lambda} - \frac{t}{T} \right) - \pi \frac{b \sin \theta}{\lambda} \right] \qquad (2.14)$$

For the intensity as the square of the amplitude factor, we have

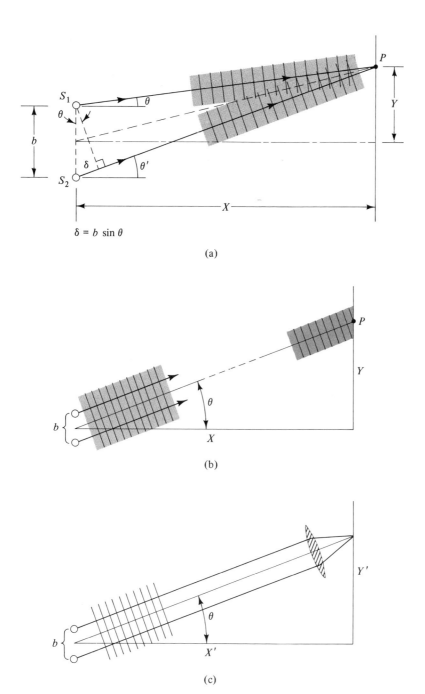

$\delta = b \sin \theta$

(a)

(b)

(c)

Fig. 2.8
Observation of interference patterns generated by two sources. The generated waves
have a constant phase relation. (a) Geometry for the experiment with water waves (see
Figure 2.7). (b) Observation at a screen far away from the source. (c) Using a lens to
reduce the distance X in (b) to distance X'. In this case, only parallel rays meet in the
focal plane of the lens at one point.

$$y^2 = 4A^2 \cos^2 \left(\pi \frac{b \sin \theta}{\lambda} \right) \tag{2.15}$$

We observe constructive interference for

$$b \sin \theta = \delta = 0, \lambda, 2\lambda, 3\lambda, \ldots \tag{2.15a}$$

or

$$\sin \theta = 0, \frac{\lambda}{b}, 2\frac{\lambda}{b}, 3\frac{\lambda}{b}, \ldots \tag{2.15b}$$

and destructive interference for

$$b \sin \theta = \delta = \frac{\lambda}{2}, \frac{3}{2}\lambda, \frac{5}{2}\lambda, \ldots \tag{2.16a}$$

or

$$\sin \theta = \frac{1}{2}\frac{\lambda}{b}, \frac{3}{2}\frac{\lambda}{b}, \frac{5}{2}\frac{\lambda}{b}, \ldots \tag{2.16b}$$

For water waves, we see in Figure 2.7 that strips of waves are separated by gray lines. In the direction of these lines we have destructive interference. The amplitude factor is zero and no oscillating waves are present. In the direction of the strips of waves we have constructive interference. If we actually performed this water wave interference experiment, we would see the waves moving.

For light waves, the picture is similar. Maxima now appear as the time averages of moving waves; minima as "no light."

A numerical example is given in Figure 2.9, where we see the angular dependence of the maxima and minima approaching the position $\theta = 0$. In principle, the pattern can be obtained with two lasers. In practice, we use one laser with two openings at one end (Figure 2.10).

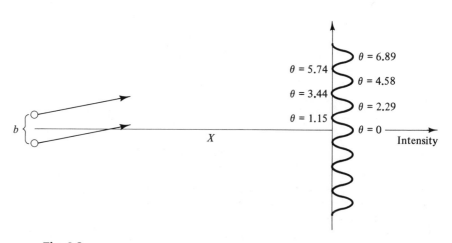

Fig. 2.9
Maxima and minima obtained for $b = 12.5~\mu m$, $X = 4$ m, and $\lambda = 0.5~\mu m$.

2. Interference of Light

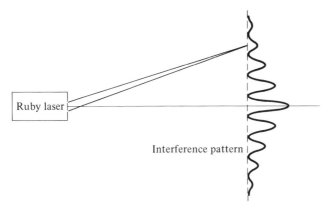

Fig. 2.10
The light leaving the two openings of a ruby laser have the same phase and produce an
interference pattern with a central maximum.

B. Interference Phenomena with Light Waves from Sources of Different Types: The One-Source Approach

In the preceding section, we assumed that the two light waves were emitted in
phase. The superposition of those waves resulted in an interference pattern, an
observation that could be verified by using a laser. Two laser sources will also
produce an interference pattern. There is a fixed phase difference between the
two emitted waves, similar to the situation for water waves.

If we use two incandescent lamps, each with a narrow-band frequency filter,
or two spectral lamps of atomic emission, each emitting one spectral line of the
same wavelength, we will not observe an interference pattern. The reason is that
light from the incandescent lamp or atomic emission is composed of a large
number of waves emitted randomly in space and time. The phase relation
between two individual waves, both from one source, varies. Pairs of individual
waves, each from one source, also have no constant phase relation. A pair of
waves, each from one of the two sources, may produce a maximum; the next may
produce a minimum; and the third may give something in between. On the
average, no fringe pattern can be observed.

This situation is found for all sources where random processes are used for
the production of light, such as raising the temperature or producing collisions
of electrons and atoms. However, interference fringes can be observed using just
one such source. An experimental setup is chosen whereby each emitted wave is
divided into two parts. Interference is then produced by superposition of these
two parts. The splitting of the wave is accomplished in most cases by partial
reflection and transmission.

The experiment by Thomas Young uses a slightly different approach. Each
wave from the "random" source passes through a very small hole. Then it travels
to a screen with two holes, where it is divided into two waves. Passing through
the first hole deforms the wave in such a way that there is a fixed phase relation
between the two parts passing through the two holes on the second plane. This

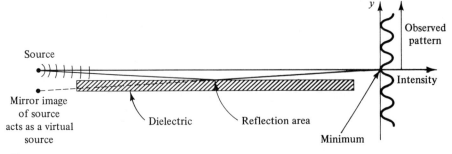

Fig. 2.11
Optical diagram for the Lloyd's mirror experiment. Each wave from the source is partly reflected and partly traverses straight to the observation point. The optical path difference is called δ. For $\delta = 0$, one has a minimum.

experiment uses **Huygens' principle** for its description, and we will discuss it later in this chapter.

In many experiments (see Figure 2.8), the angle θ is small. We can, then, approximate

$$\sin \theta \simeq \theta \simeq \tan \theta = \frac{Y}{X}$$

and, using equation 2.15, get the **small-angle approximation** for the amplitude factor

$$y^2 = 4A^2 \cos^2 \pi \frac{b}{\lambda}\left(\frac{Y}{X}\right) \tag{2.17}$$

C. Lloyd's Mirror

In the **Lloyd's mirror** experiment, named after Humphrey Lloyd, a nineteenth-century Irish physicist, light from a source is reflected under a large angle to the normal at a dielectric plate or mirror (see Figure 2.11). The reflected part of the wave may be considered as originating at the mirror image of the source point. Taking real and virtual source points as emitting points, we have a fixed phase relation between them and are back to what was discussed in Section 3A. We expect the same mathematical description and fringe pattern. However, there is an important difference. According to assumption 3 in Section 1A, we have a phase shift of π when a wave is reflected, regardless of whether we use a dielectric or a metal mirror. This phase shift of π is translated into an additional path difference of $\lambda/2$. We see this in the superposition of two waves:

$$A \cos 2\pi \left(\frac{x_1}{\lambda} - \frac{t}{T}\right) \quad \text{and} \quad A \cos 2\pi \left(\frac{x_2}{\lambda} - \frac{t}{T}\right)$$

If the second wave undergoes a phase shift of π, we can write

$$A \cos \left[2\pi \left(\frac{x_2}{\lambda} - \frac{t}{T}\right) + \pi\right] = A \cos 2\pi \left(\frac{x_2 + \lambda/2}{\lambda} - \frac{t}{T}\right) \tag{2.18}$$

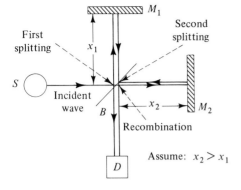

M_1

First splitting

x_1

Second splitting

S

Incident wave

B

x_2 M_2

Recombination

D

Fig. 2.12
The Michelson interferometer, with arms of unequal lengths x_1 and x_2 in the same medium.

Assume: $x_2 > x_1$

In the superposition formula (equation 2.11), this phase shift appears as an addition (or subtraction) of $\lambda/2$ to δ (we now have $x_2 = x_1 - \delta - \lambda/2$):

$$2A \cos\left(2\pi \frac{\delta + \lambda/2}{2\lambda}\right) \cos\left[2\pi\left(\frac{x_1}{\lambda} - \frac{t}{T}\right) - 2\pi \frac{\delta + \lambda/2}{2\lambda}\right]$$

or

$$2A \cos\left(2\pi \frac{\delta/2}{\lambda} + \frac{\pi}{2}\right) \cos\left[2\pi\left(\frac{x_1}{\lambda} - \frac{t}{T}\right) - 2\pi \frac{\delta/2}{\lambda} - \frac{\pi}{2}\right] \qquad (2.19)$$

For the square of the amplitude factor, we get

$$y^2 = 4A^2 \cos^2\left(2\pi \frac{\delta/2}{\lambda} + \frac{\pi}{2}\right) \qquad (2.20)$$

Comparing the pattern produced by two laser sources (Figures 2.9 and 2.10) with the pattern of Lloyd's mirror (Figure 2.11), we find that they are complementary. The center maximum in Figure 2.9 for zero path difference between the two waves is now replaced by a minimum in Figure 2.11 (for $\delta = 0$). While we have no optical path difference, we have a shift of $\lambda/2$ from reflection.

So far, we have considered only source points in a small volume. How small this volume must be will be discussed in the section on spatial coherence.

D. The Michelson Interferometer

In this section, we will study the superposition of light waves using a **Michelson interferometer**. This device, invented by Albert Michelson in 1880, is one of the most useful measuring devices in optics.

A simplified diagram of the Michelson interferometer is shown in Figure 2.12. The incident wave is directed toward a **beam splitter** B, which (ideally) transmits and reflects each 50 percent of the incident wave without introducing any additional phase shifts. The reflected and transmitted beams travel to the plane mirrors M_1 and M_2. After reflection from these mirrors, the beams (shown as being displaced for graphic purposes) recombine at the beam splitter into a single beam and travel toward detector D (the eye). Any light continuing back toward the source S is of no interest and may be disregarded.

3. Two-beam Interference Problems 87

We assume laser light to be incident at B. After respective reflections at M_1 and M_2 and recombination at B, the waves have traveled different distances of $2x_2$ and $2x_1$. If we assume that x_2 is greater than x_1, then the beam incident toward and reflected by M_2 has traversed an optical path that is $\delta = 2(x_2 - x_1)$ further than that reflected from M_1.

The intensity (square of the amplitude factor) of the two superimposed waves is described using equation 2.12:

$$y^2 = 4A^2 \cos^2\left(2\pi\frac{\delta/2}{\lambda}\right) \tag{2.21}$$

We observe constructive interference for

$$\delta = 2(x_2 - x_1) = m\lambda \tag{2.22}$$

and destructive interference for

$$\delta = 2(x_2 - x_1) = \left(m + \frac{1}{2}\right)\lambda \qquad \text{where } m = 0, 1, 2, 3 \tag{2.23}$$

Since the difference in length of the two arms of the Michelson interferometer is directly expressed in multiples of λ, the Michelson interferometer can be used for very exact length measurements. Indeed, Michelson had originally invented the interferometer to test the ether hypothesis and to study atomic spectra, the latter earning him the Nobel Prize in 1907.

Can we also use atomic emission from spectral lamps or an incandescent lamp to obtain fringes with the Michelson interferometer? The situation is similar to that discussed for Lloyd's mirror. The wave from one source point is divided by the beam splitter, and after recombination, the phase difference depends only on the path difference. Each emitted wave contributes to the intensity of the resulting interference pattern. If we have an extended source, the individual waves will arrive at the beam splitter at different angles, and a ring pattern results on the observation screen (Figure 2.13). The details are analogous to the ring pattern observed at a plane parallel plate and will be discussed later in detail.

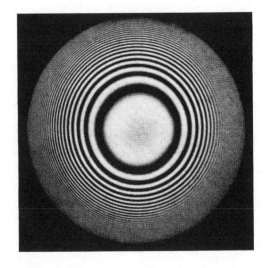

Fig. 2.13
Ring pattern observed with a Michelson interferometer using an extended source. (From Cagnet, Françon, and Thrierr, *Atlas of Optical Phenomena*.)

2. Interference of Light

E. The Wedge-Shaped Gap

We can place two dielectric plates (microscope slides) in such a way that a wedge-shaped air gap is produced (Figure 2.14a). The incident light is partly reflected and transmitted at the front and back surfaces of the first plate. The light traveling to the second plate is also partly reflected and transmitted at both surfaces. All the reflected light may be superimposed, and an interference pattern will result, depending on the optical path difference of the individual component waves. The same is true for the transmitted light.

The interference pattern obtained from one plane parallel plate of constant thickness will be discussed in Section 4C. The general problem of two inclined plates is very complicated. Therefore, here we will restrict our discussion to a qualitative description of the interference pattern resulting from the wedge-shaped air gap. To obtain the conditions for constructive and destructive inter-

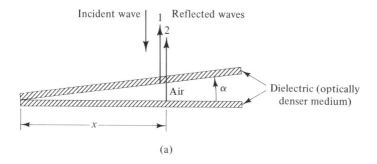

(a)

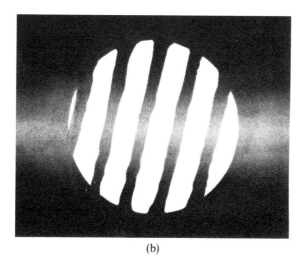

(b)

Fig. 2.14
(a) Optical diagram for a wedge-shaped gap of air between two dielectric slabs (glass microscope slides). The gap angle α is assumed small. As x increases, the optical path difference between waves 1 and 2 also increases, resulting in an interference pattern. (b) Fringe pattern observed with $1/10 \lambda$ optical flat plates. (Courtesy of Republic Lens Company, Inc., Englewood, N.J.)

ference, we consider only the two waves 1 and 2 reflected at the interfaces of the wedge (Figure 2.14a). We assume that their amplitudes are the same (which is not exactly true). We must then find the path difference and phase shift in order to describe the interference phenomena.

The gap difference at distance x is $x \tan \alpha$, where α is the wedge angle. Wave 2 travels twice the distance of wave 1, that is, the path difference δ is $\delta = 2x \tan \alpha$. In addition, we have a phase shift of π when wave 2 is reflected at the interface air-dense material. This phase shift is equivalent to $\lambda/2$ and we have for the condition of constructive interference

$$2x \tan \alpha + \frac{\lambda}{2} = 0, \lambda, 2\lambda, \ldots, m\lambda \tag{2.24}$$

and for destructive interference

$$2x \tan \alpha + \frac{\lambda}{2} = \frac{1}{2}\lambda, \frac{3}{2}\lambda, \ldots, (m + 1/2)\lambda \tag{2.25}$$

For the distance x from the vertex where these fringes occur, we have, for constructive interference,

$$x = \frac{(m - \frac{1}{2})\lambda}{2 \tan \alpha} \qquad \text{where } m = 1, 2, 3, \ldots \tag{2.26}$$

and for destructive interference,

$$x = \frac{m\lambda}{2 \tan \alpha} \qquad \text{where } m = 0, 1, 2, 3, \ldots$$

Is the first fringe dark or bright? For $x = 0$, we have $m = 0$, so we have a dark fringe. (This is the same situation seen with Lloyd's mirror.)

By observing fringes in such a wedge-shaped arrangement, we can measure small thicknesses, such as the diameter of a human hair (see problem 2). If placed at a distance x, the number of fringes will determine α, and the value of the thickness is obtained from $x \tan \alpha$. This experiment can also be used in the observation of Newton's rings, which is our next topic of discussion.

F. Newton's Rings

Fringes obtained with a wedge depend on the thickness of the gap. If we consider nonparallel plates, we would obtain fringes along the lines where the gap width is constant (fringes of equal thickness). This is demonstrated with **Newton's rings**, the pattern formed when a spherical surface and a flat surface are put together (Figure 2.15 and problem 3). Light is incident from the top. The fringes now extend in circles at gaps of equal thickness. For reflected light, we will see a dark spot in the middle; for transmitted light, a bright spot (Figure 2.16).

The qualitative description of the interference pattern is similar to that given for the wedge-shaped gap. For reflected light, a phase shift of π corresponding to a path difference of $\lambda/2$ is obtained at the center, resulting in destructive interference. For transmitted light, there is no phase difference, and constructive interference is observed.

2. Interference of Light

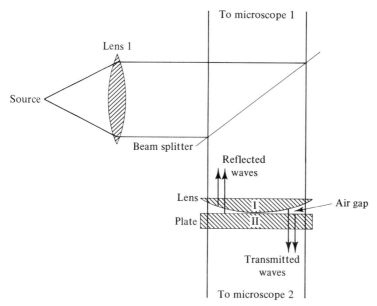

Fig. 2.15
Experimental setup for the observation of Newton's rings. Light is made parallel by lens 1 and is reflected by the beam splitter to the lens-plate assembly. Reflected light from surfaces I and II travels to microscope 1 for observation of fringes in reflection. Transmitted light from surfaces I and II travels to microscope 2 for observation of transmission fringes.

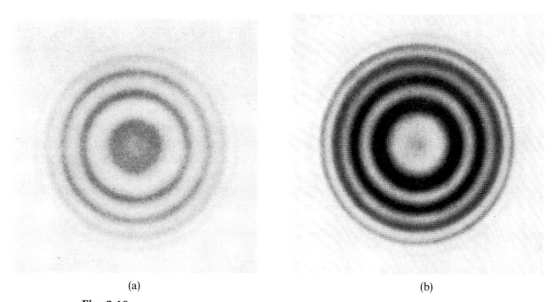

(a)   (b)

Fig. 2.16
Newton's rings observed (a) using reflected light and (b) using transmitted light. (From Cagnet, Françon, and Thrierr, *Atlas of Optical Phenomena*.)

3. Two-beam Interference Problems   91

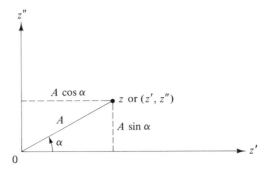

Fig. 2.17
Complex numbers are plotted with the real part z' as the abscissa and the imaginary part z'' as the ordinate. We can also plot complex numbers as a vector z of length A at an angle α; its z' component is $A \cos \alpha$, and its z'' component is $A \sin \alpha$.

As we discussed for the Michelson interferometer, we are not limited to laser light for the observation of fringes. Each emitted wave contributes individually to the intensity of the resulting interference pattern, allowing the use of atomic emission or an incandescent lamp.

G. The Use of the Complex Plane

To get the amplitude factor for the superposition of two waves, we would like to develop an elegant mathematical description. This method is also applicable to more than two waves. It becomes more powerful when there are a number of waves to be superimposed, compared to the use of trigonometric formulas. We will accomplish this with the aid of complex numbers.

Consider the wave $y = \cos \alpha$, where α stands for the total phase expression. The **complex number** $z = Ae^{i\alpha}$ is defined by

$$z = Ae^{i\alpha} = A(\cos \alpha + i \sin \alpha) = z' + iz'' \tag{2.27}$$

The real part of z is z'; the imaginary part of z is z''; A is the absolute value of z; and α is the phase angle. The cosine wave of equation 2.1 is represented by the real part. In addition, we have as the imaginary part a sine wave with the same phase angle α. The graphical representation is given in Figure 2.17. The complex conjugate of a complex number $z = z' + iz''$ is defined by $z^* = z' - iz''$, that is, substituting $-i$ for i. Therefore, the complex conjugate of $z = Ae^{i\alpha}$ is $z^* = Ae^{-i\alpha}$.

The **complex conjugate** may be used to split more complicated complex numbers into their real and imaginary parts. For example,

$$\begin{aligned} z &= \frac{a + ib}{c + id} = \frac{(a + ib)(c - id)}{(c + id)(c - id)} = \frac{ac + bd + i(bc - ad)}{c^2 + d^2} \\ &= \frac{ac + bd}{c^2 + d^2} + i\left(\frac{bc - ad}{c^2 + d^2}\right) = z' + iz'' \end{aligned} \tag{2.28}$$

We can also use the complex conjugate to obtain the absolute value of a complex number. The absolute value of z is the distance A from 0 to z in Figure

2. Interference of Light

2.17. We can find the absolute value by multiplying z by its complex conjugate $z^* = Ae^{-i\alpha}$ and then taking the square root:

$$\sqrt{zz^*} = \sqrt{Ae^{i\alpha}Ae^{-i\alpha}} = \sqrt{A^2} = A \qquad (2.29)$$

Let us use complex numbers to calculate the **addition** of two waves $y_1 = A\cos\alpha_1$ and $y_2 = A\cos\alpha_2$, where $\alpha_2 = \alpha_1 - \phi$. We complement y_1 and y_2 by a sine wave to obtain complex numbers:

$$z_1 = A(\cos\alpha_1 + i\sin\alpha_1) = Ae^{i\alpha_1}$$
$$z_2 = A(\cos\alpha_2 + i\sin\alpha_2) = Ae^{i\alpha_2} = Ae^{i(\alpha_1-\phi)} \qquad (2.30)$$

For the addition $z_1 + z_2$, we get

$$z = Ae^{i\alpha_1} + Ae^{i(\alpha_1-\phi)}$$
$$= Ae^{i\alpha_1}(1 + e^{-i\phi})$$
$$= Ae^{i\alpha_1}(e^{i(\phi/2)} + e^{-i(\phi/2)})e^{-i(\phi/2)} \quad \left(\text{Note: } \frac{e^{ix} + e^{-ix}}{2} = \cos x\right) \qquad (2.31)$$
$$= A\left(2\cos\frac{\phi}{2}\right)e^{i(\alpha_1-(\phi/2))}$$

(See also problem 4.) Following the procedure for obtaining the absolute value, we have

$$\sqrt{zz^*} = \sqrt{2A\left(\cos\frac{\phi}{2}\right)e^{i(\alpha_1-(\phi/2))}\,2A\left(\cos\frac{\phi}{2}\right)e^{-i(\alpha_1-(\phi/2))}}$$
$$= 2A\cos\frac{\phi}{2} \qquad (2.32)$$

If we set

$$\frac{\phi}{2} = 2\pi\left(\frac{\delta/2}{\lambda}\right)$$

we have obtained the amplitude factor of equation 2.11. See also equations 2.4 and 2.5.

We can also get the entire expression of equation 2.11 by first rewriting equation 2.31 as

$$z = 2A\cos\frac{\phi}{2}\left[\cos\left(\alpha_1 - \frac{\phi}{2}\right) + i\sin\left(\alpha_1 - \frac{\phi}{2}\right)\right] \qquad (2.33)$$

Then we take the real part of z; that is,

$$z' = y = 2A\cos\frac{\phi}{2}\cos\left(\alpha_1 - \frac{\phi}{2}\right) \qquad (2.34)$$

Substitution for

$$\frac{\phi}{2} = 2\pi\left(\frac{\delta/2}{\lambda}\right) \quad \text{and} \quad \alpha_1 = 2\pi\left(\frac{x_1}{\lambda} - \frac{t}{T}\right)$$

results in equation 2.11.

The **intensity** is described as the square of the amplitude and again results in a wave. We have seen that the wave factor of the expression for the intensity (equation 2.12) is fast oscillating and counts only as a time average. We again call it a_v^2. Thus, for the intensity of the superimposed waves represented in equation 2.34, we have

$$(z')^2 = y^2 = 4A^2 a_v^2 \cos^2 \frac{\phi}{2} \qquad (2.35)$$

Let us compare this result with what we got from the treatment with complex numbers (equation 2.32),

$$zz^* = 4A^2 \cos^2 \frac{\phi}{2} \qquad (2.36)$$

We see that zz^* is the square of the amplitude factor (equation 2.12) and therefore proportional to the intensity. If we multiply equation 2.36 by a_v^2, we get the intensity obtained in equation 2.35.

Very often we are only interested in the square of the amplitude factor, since this factor determines constructive and destructive interference. In this case, we obtain all we need by calculating zz^*. The time dependence and unimportant phase factors appear as $e^{i\omega t}$ and $e^{i\chi}$, respectively (where ω is the angular frequency and χ the phase angle). These factors are automatically eliminated by calculation of zz^*. Therefore, it is not unusual to treat a problem without considering the time dependence. In the remainder of this chapter, when we talk about intensity, we mean the square of the amplitude factor.

4. MULTIPLE-BEAM INTERFERENCE PROBLEMS

A. An Interference Pattern Due to N Equally Spaced Sources

In Figure 2.18, N identical sources are indicated schematically. We assume that the emitted waves are all in phase. For water waves, we can realize this N-source arrangement by an extension of the two-source experiment of Section 3A. The N-studs are simultaneously moved up and down with respect to the water surface. Since we must assume that the N light waves are in phase, how can we accomplish this phase relation experimentally? Certainly, we cannot use N sources of hot wires or atomic emission. But we can use N laser sources. As we will see in Section 5B, N source points can emit N light waves in phase as an extension of Young's experiment and Huygens' principle.

In Figure 2.18, the distance from the αth source to the screen is $\rho_\alpha (\alpha = 1, 2, 3, \ldots, N)$, and all adjacent pairs of rays have the same optical path difference δ_1 for a particular direction of superposition (angle θ). As in the $N = 2$ case treated earlier, the screen is sufficiently far away so that the propagation direction of the waves is considered parallel. This is analogous to the case of the two waves discussed in Section 3A (see Figure 2.8). The waves will eventually superimpose at a point on a faraway screen, and a lens may be used to reduce the size of X and Y. Adding the wave amplitudes,

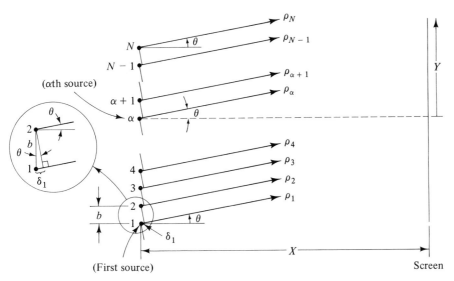

Fig. 2.18
Geometry for an N-source interference pattern.

$$u = A \cos\left[2\pi\left(\frac{\rho_1}{\lambda} - \frac{t}{T}\right)\right] + A \cos\left[2\pi\left(\frac{\rho_2}{\lambda} - \frac{t}{T}\right)\right] + \cdots \quad (2.37)$$

or in complex notation,

$$u = A e^{i(k\rho_1 - \omega t)} + A e^{i(k\rho_2 - \omega t)} + \cdots A e^{i(k\rho_N - \omega t)}$$

$$= A \sum_{\alpha=1}^{N} e^{i(k\rho_\alpha - \omega t)} \quad (2.38)$$

where we use the abbreviation $2\pi/\lambda = k$ and $2\pi/T = \omega$. The symbol k is the wave or propagation vector, and ω is the angular frequency. The path difference between any two adjacent sources is

$$\delta_1 = \rho_{\alpha+1} - \rho_\alpha \qquad \alpha = 1, 2, 3, \ldots, N-1 \quad (2.39)$$

and the path difference between any source and the first source is

$$\rho_{\alpha+1} - \rho_1 = \alpha\delta_1 \quad (2.40)$$

or

$$\rho_N - \rho_1 = (N-1)\delta_1 \quad (2.41)$$

The resulting amplitude at the screen is now

$$u = A e^{-i\omega t}(e^{ik\rho_1} + e^{ik\rho_2} + \cdots + e^{ik\rho_N})$$

or

$$u = A e^{-i\omega t} \sum_{\alpha=1}^{N} e^{ik\rho_\alpha}$$

4. Multiple-beam Interference Problems

and on factoring one term, $e^{ik\rho_1}$, we have

$$u = Ae^{i(k\rho_1 - \omega t)} \sum_{\beta=0}^{N-1} e^{ik(\beta\delta_1)} \tag{2.42}$$

where we have used equations 2.39–2.41, and where β replaces α as the index for summation. The factoring has left the sum in equation 2.42 dependent on the path difference (which equals δ_1) and not on each ρ_α. Note that there are only $N-1$ spaces between N sources. Using the formula

$$\sum_{\beta=0}^{N-1} x^\beta = \frac{1 - x^N}{1 - x} \tag{2.43}$$

where $x = e^{ik\delta_1}$, we have

$$u = Ae^{i(k\rho_1 - \omega t)} \frac{1 - e^{iNk\delta_1}}{1 - e^{ik\delta_1}}$$

Factoring $e^{iNk(\delta_1/2)}$ from the numerator and $e^{ik(\delta_1/2)}$ from the denominator, we get

$$u = Ae^{i(k\rho_1 - \omega t)} \left(\frac{e^{iNk(\delta_1/2)}}{e^{ik(\delta_1/2)}} \right) \left(\frac{e^{-iNk(\delta_1/2)} - e^{iNk(\delta_1/2)}}{e^{-ik(\delta_1/2)} - e^{ik(\delta_1/2)}} \right) \tag{2.44}$$

To obtain the intensity, we rearrange and multiply by $(1/2i)/(1/2i)$:

$$u = Ae^{i(k\rho_1 - \omega t + Nk(\delta_1/2) - k(\delta_1/2))} \frac{-(e^{iNk(\delta_1/2)} - e^{-iNk(\delta_1/2)})/2i}{-(e^{ik(\delta_1/2)} - e^{-ik(\delta_1/2)})/2i} \tag{2.45}$$

Now the first exponential factor disappears upon multiplication by its complex conjugate, and we rewrite the last factor with

$$\sin x = \frac{e^{ix} - e^{-ix}}{2i}$$

The result for uu^*, which is proportional to the intensity, is

$$uu^* = A^2 \frac{\sin^2[kN(\delta_1/2)]}{\sin^2[k(\delta_1/2)]} \tag{2.46}$$

From the inset in Figure 2.18, we have $\delta_1 = b\sin\theta$. Introduction into equation 2.46 with $k = 2\pi/\lambda$ gives us

$$uu^* = A^2 \frac{\sin^2[\pi N(b/\lambda)\sin\theta]}{\sin^2[\pi(b/\lambda)\sin\theta]} \tag{2.47}$$

For the small angle approximation, we have $\sin\theta \simeq \theta \approx \tan\theta = Y/X$, giving us

$$uu^* = A^2 \frac{\sin^2[\pi N(b/\lambda)(Y/X)]}{\sin^2[\pi(b/\lambda)(Y/X)]} \tag{2.48}$$

A plot extending to the first few principal maxima is shown in Figure 2.19. The essential features are the occurrence of the **principal maxima** (spaced a distance λ/b apart) and the **secondary maxima** and **minima**. To discuss the principal maxima (given by $b\sin\theta = \delta_1 = \rho_{\alpha+1} - \rho_\alpha = m\lambda$, where m is an integer), note

2. Interference of Light

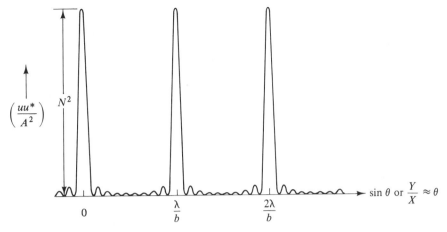

Fig. 2.19
Plot of uu^* versus $\sin \theta$ or for small angles Y/X for the N-source pattern for $N = 10$. There are $N - 1$ secondary minima between the principal maxima.

that equation 2.47 becomes indeterminate, that is,

$$\lim_{\delta_1 \to m\lambda} uu^* = \frac{0}{0}$$

L'Hôpital's rule can be used to evaluate uu^* as follows:

$$\lim_{\delta_1 \to m\lambda} \frac{\sin[N\pi(\delta_1/\lambda)]}{\sin[\pi(\delta_1/\lambda)]} = \lim_{\delta_1 \to m\lambda} \frac{(N\pi/\lambda)\cos[N\pi(\delta_1/\lambda)]}{(\pi/\lambda)\cos[\pi(\delta_1/\lambda)]} = N \qquad (2.49)$$

Then

$$uu^* = A^2 N^2 \qquad (2.50)$$

We see that the intensity is N^2 times the intensity due to a single wave arriving at the observation point, which is A^2. The light intensity has been redistributed on the observation screen.

The secondary minima are obtained for uu^* from equations 2.46–2.48 and are equal to zero with the additional condition that the denominator $\sin \pi(\delta_1/\lambda)$ does not equal zero. Therefore, the minima are determined from

$$\sin \frac{N\pi}{\lambda}\delta_1 = 0$$

that is,

$$\frac{N\delta_1}{\lambda} = 1, 2, \ldots, N - 1, N + 1, \ldots$$

The Nth term is omitted because

$$\frac{N\delta_1}{\lambda} = N \to \frac{\delta_1}{\lambda} = 1$$

which would make $\sin \pi(\delta_1/\lambda)$ equal to zero. The Nth-term corresponds to the principal maxima. There are $N - 1$ minima between two principal maxima.

The secondary maxima are positioned between the secondary minima. They are positioned at δ_1/λ values where uu^* has maxima. The exact location may be determined by setting the derivative of uu^* with respect to δ_1/λ equal to zero. However, the maxima are close to the maxima of $\sin[(N\pi/\lambda)\delta_1]$, and again we must assume that $\sin \pi(\delta_1/\lambda) \neq 0$. Therefore, the secondary maxima are close to

$$N\frac{\delta_1}{\lambda} = \frac{1}{2}, \frac{3}{2}, \frac{5}{2}, \ldots$$

The first secondary maximum is located within the wing of the principal maximum and is not observable. The same is true for the other secondary maxima located at the wings of principal maxima.

For the case $N = 2$, equation 2.47 yields the result obtained in Section 3A (equation 2.15):

$$uu^* = A^2 \frac{\sin^2[\pi 2(b/\lambda)\sin\theta]}{\sin^2[\pi(b/\lambda)\sin\theta]} = 4A^2\cos^2\left(\pi\frac{b}{\lambda}\sin\theta\right) \tag{2.51}$$

where the formula

$$\sin 2q = 2\sin q \cos q$$

has been used. For the small angle case, we can substitute $\sin\theta = Y/X$ to get

$$uu^* = 4A^2\cos^2\left[\pi\frac{b}{\lambda}\left(\frac{Y}{X}\right)\right]$$

We have constructive interference for

$$\frac{b}{\lambda}\sin\theta \quad \text{or} \quad \frac{b}{\lambda}\left(\frac{Y}{X}\right) = 0, 1, 2, 3, \ldots \tag{2.52a}$$

and destructive interference for

$$\frac{b}{\lambda}\sin\theta \quad \text{or} \quad \frac{b}{\lambda}\left(\frac{Y}{X}\right) = \frac{1}{2}, \frac{3}{2}, \frac{5}{2}, \ldots \tag{2.52b}$$

B. The Random Phase Problem

In the interference problem just considered, we assumed that (1) all waves were emitted from their respective sources simultaneously and (2) the spacing b between the N adjacent sources was fixed. In other words, we assumed equal phase differences in the spatial coordinates (ρ_α in Figure 2.18). We will now remove the second assumption and investigate the result on the interference pattern.

Let us consider a large number of sources randomly spaced along a line in the source plane such that the αth source is positioned anywhere between the first and Nth source. We will see that under these conditions, the interference pattern will collapse as N approaches infinity.

The starting point for this discussion is equation 2.40. Now, however, the path difference between the αth source and the first source is not α times the path

2. Interference of Light

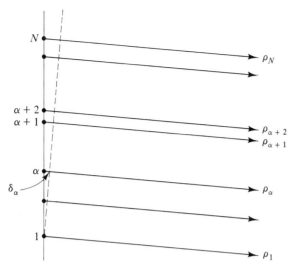

Fig. 2.20
N sources with random spacings between them in the "b"-direction. δ_α is the distance between the wave front (relative to the wave front from the first source) and the αth source.

difference between adjacent sources. That is, equations 2.40 and 2.41 are no longer true. Instead of equation 2.40, we have

$$\rho_{\alpha+1} - \rho_1 = \delta_\alpha \tag{2.53}$$

where $\alpha = 1, 2, \dots$. The amplitude at the screen is (equation 2.38)

$$u = A \sum_{\alpha=1}^{N} e^{i(k\rho_\alpha - \omega t)}$$

Substituting, we have

$$u = A \sum_{\alpha=1}^{N} e^{i[k(\rho_1 + \delta_{\alpha-1}) - \omega t]} \tag{2.54}$$

For convenience, let us call

$$\gamma_\alpha = \rho_1 + \delta_{\alpha-1}$$

where γ_α is now random because the sources are randomly spaced (see Figure 2.20 for the geometry). The intensity is given by

$$uu^* = A^2 \left(\sum_{\alpha=1}^{N} e^{ik\gamma_\alpha} \right) \left(\sum_{\beta=1}^{N} e^{ik\gamma_\beta} \right)^*$$

$$= A^2 \left(\underbrace{1 + 1 + \cdots + 1}_{N} + \sum_{\alpha \neq \beta} e^{ik(\gamma_\alpha - \gamma_\beta)} \right) \tag{2.55}$$

There are N times 1 as a result of our multiplying $e^{ik\gamma_\alpha}$ by $e^{-ik\gamma_\beta}$ for $\alpha = \beta$. The other explicit terms are for $\alpha \neq \beta$. We define

$$\gamma_{\alpha\beta} = \gamma_\alpha - \gamma_\beta$$

4. Multiple-beam Interference Problems 99

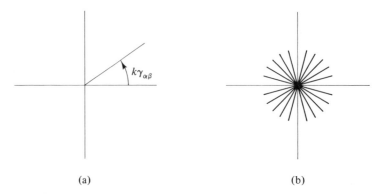

(a) (b)

Fig. 2.21
(a) Representation of $k\gamma_{\alpha\beta}$. (b) "Star" pattern generated by a random array of N-sources of equal amplitude or radius in the plot corresponding to random δ_α.

and have

$$uu^* = A^2\left(1 + 1 + 1 + \cdots + 1 + \sum_{\alpha\neq\beta} e^{ik\gamma_{\alpha\beta}}\right)$$

$$= A^2\left(N + \sum_{\alpha\neq\beta} e^{ik\gamma_{\alpha\beta}}\right) = A^2 N\left(1 + \frac{\sum\limits_{\alpha\neq\beta} e^{ik\gamma_{\alpha\beta}}}{N}\right) \tag{2.56}$$

Since γ_α and γ_β are random, we cannot evaluate the sum in equation 2.56 as we did in equations 2.42–2.48. The sum

$$\sum_{\alpha\neq\beta} e^{ik\gamma_{\alpha\beta}}$$

can be evaluated if we construct a **polar plot with N radial lines**, all of the same length. Each line is at a random angle representing a particular value of $k\gamma_{\alpha\beta}$, as in Figure 2.21a. Values of $k\gamma_{\alpha\beta}$ in excess of 2π are easily accounted for, since the exponential function is periodic, that is,

$$k\gamma_{\alpha\beta} \to k\gamma'_{\alpha\beta} + 2\pi m$$

where m is an integer. The evaluation of uu^* is thus reduced to a simple geometrical evaluation. For small N, a net value of the sum $\sum e^{ik\gamma_{\alpha\beta}}$ may exist; but for the interesting case of large N, the many values of $k\gamma_{\alpha\beta}$ in the polar plot lead to the "star" pattern shown in Figure 2.21b. For this latter case, phase factors differing by π (i.e., radii diametrically opposed) can always be found and cancel each other. That is,

$$\sum e^{ik\gamma_{\alpha\beta}} = 0$$

and

$$uu^* = A^2 N \tag{2.57}$$

Equation 2.57 tells us that no interference pattern exists; that is, removal of the periodicity requirement for the source spacing has destroyed the fringe pattern.

2. Interference of Light

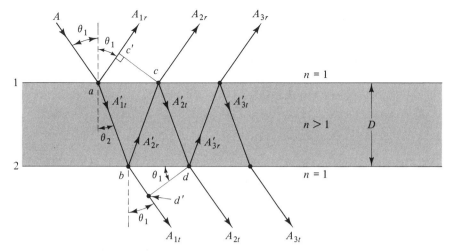

Fig. 2.22
Geometry for multiple-beam interference in a nonabsorbing plane parallel plate of index of refraction n and thickness D. The light in air is incident at an angle θ_1. Using Snell's law at the first surface and then at the second surface shows that the emerging angle is equal to the incident angle θ_1.

In equation 2.55, we see that the interference terms in the summation process cancel, leaving the series of terms $1 + 1 + 1 + 1 \cdots = N$, which yields no redistribution of the light. The random spacing translates into random phase differences (see Figure 2.20). Comparing equation 2.50 (for sources in phase) with equation 2.57 (for sources with random phases), we note that the intensity pattern for in-phase (coherent) sources is, on the average, N times stronger at the maximum than for random phase (incoherent) sources.

C. The Plane Parallel Plate

We will now consider the fringe pattern produced with a plane parallel plate of thickness D when illuminated by laser light. For atomic emission or radiation from an incandescent lamp, we have the same situation as discussed for the Michelson interferometer. In Figure 2.22, the incoming ray, which has amplitude A, strikes the plate at point a. At surface 1, part of the ray is reflected (A_{1r}) at an angle θ_1 and part is refracted (A'_{1t}) at an angle θ_2. It then travels to point b, where again reflection (A'_{2r}) and refraction (A_{1t}) take place. The successive reflections and refractions are as indicated in the sketch.

The plate of thickness D and index of refraction n is surrounded by a vacuum ($n = 1$). We calculate the phase difference between the two transmitted fractions A_{1t} and A_{2t}. The phase difference is the path difference in units of λ as a fraction of 2π, that is, $2\pi(\delta/\lambda)$. According to assumption 1 in Section 1A, the wavelength in the medium is reduced by n, that is, in terms of the phase

$$2\pi \frac{\delta}{\lambda/n} \quad \text{or} \quad 2\pi \frac{n\delta}{\lambda}$$

The product $n\delta$ is called the optical path length and was used in Chapter 1 for the formulation of Fermat's principle. We now obtain the final phase difference by calculating the optical path difference $n\delta$ and then multiplying by $k = 2\pi/\lambda$. For the optical path difference $n\delta$ between A_{1t} and A'_{2t}, we have

$$n\delta = n[bcd] - [bd'] = n\frac{2D}{\cos\theta_2} - [bd]\sin\theta_1$$

$$= \frac{2Dn}{\cos\theta_2} - 2[bc]\sin\theta_2\sin\theta_1 \tag{2.58}$$

$$= \frac{2Dn}{\cos\theta_2} - 2\frac{D}{\cos\theta_2}\sin\theta_2\sin\theta_1 = 2Dn\left(\frac{1}{\cos\theta_2} - \frac{\sin^2\theta_2}{\cos\theta_2}\right)$$

$$= 2nD\cos\theta_2$$

where brackets indicate points outlining a certain distance. Note that frequent reference is made to Figure 2.22, and the law of refraction, $n\sin\theta_2 = \sin\theta_1$, has been applied. According to assumption 3 in Section 1A, phase shifts upon reflection on the denser medium must be taken into account. For the transmitted waves A_{1t} and A_{2t}, we have no phase shifts at surfaces 1 and 2 and have for the final phase difference

$$\Delta = \frac{2\pi}{\lambda}n\delta = \frac{2\pi}{\lambda}2nD\cos\theta_2 \tag{2.59}$$

For the phase difference of A_{2r} and A_{3r}, we get the same result. The only reflection on the denser medium occurs for the incident wave. We represent the reflection coefficient (ratio of reflected to incident amplitude) by r for reflection on the denser medium and by r' for reflection on the less dense medium. Likewise, we represent the transmission coefficient (ratio of transmitted to incident amplitude) by t for transmission to the denser medium and by t' for transmission to the less dense medium. We can now list the various reflected and transmitted amplitudes, using for their calculation the ones preceding:

$$A_{1r} = rA$$
$$A'_{1t} = tA$$
$$A_{1t} = t'tA$$
$$A'_{2r} = r'tA$$
$$A_{2r} = t'r'tA$$
$$A'_{2t} = r'r'tA$$
$$A_{2t} = t'r'r'tA$$
$$A'_{3r} = r'r'r'tA$$
$$A_{3r} = t'r'r'r'tA$$
$$A'_{3t} = r'r'r'r'tA$$
$$A_{3t} = t'r'r'r'r'tA$$

and so on. All possible phase shifts on the denser medium are included in the definition of the reflection coefficient r. For the phase factor corresponding to

2. Interference of Light

the phase difference of two successive amplitudes on both sides of the plate, we use $e^{i\Delta}$ (see equation 2.59) and have

$$A_{1r} = rA \qquad\qquad A_{1t} = t'tA$$
$$A_{2r} = t'tr'Ae^{i\Delta} \qquad A_{2t} = t't(r')^2 Ae^{i\Delta}$$
$$A_{3r} = t't(r')^3 Ae^{i2\Delta} \qquad A_{3t} = t't(r')^4 Ae^{i2\Delta}$$
$$A_{4r} = t't(r')^5 Ae^{i3\Delta} \qquad A_{4t} = t't(r')^6 Ae^{i3\Delta}$$
$$A_{5r} = t't(r')^7 Ae^{i4\Delta} \qquad A_{5t} = t't(r')^8 Ae^{i4\Delta}$$

and so on. We now sum up all the reflected amplitudes:

$$
\begin{aligned}
A_r &= A_{r1} + A_{r2} + A_{r3} + A_{r4} + \cdots \\
&= A[r + t'tr'e^{i\Delta} + t't(r')^3 e^{i2\Delta} + t't(r')^5 e^{i3\Delta} + \cdots] \qquad (2.60) \\
&= A(r + t'tr'e^{i\Delta}[1 + (r')^2 e^{i\Delta} + (r')^4 e^{i2\Delta} + \cdots])
\end{aligned}
$$

Again, we can apply the sum formula (see equation 2.43); that is,

$$\sum_{\beta=0}^{N-1} x^\beta = \frac{1 - x^N}{1 - x}$$

and set $x = (r')^2 e^{i\Delta}$ and have

$$A_r = A\left[r + r'tt'e^{i\Delta}\left(\frac{1 - (r'^2 e^{i\Delta})^N}{1 - r'^2 e^{i\Delta}} \right) \right] \qquad (2.61)$$

We can assume that $r'^{2N} \to 0$, since $r' < 1$ and N is very large. So we approximate

$$A_r = A\left(r + \frac{r'tt'e^{i\Delta}}{1 - r'^2 e^{i\Delta}} \right) \qquad (2.62)$$

For the transmission using a similar calculation, we obtain

$$A_t = Att' \frac{1}{1 - r'^2 e^{i\Delta}} \qquad (2.63)$$

For further simplification of equations 2.62 and 2.63, we need relations between r, r', t, and t'. Quantitative formulas relating these quantities to the refractive indices of the media on both sides of the interface, as well as the angles of incidence and refraction, are called **Fresnel's formulas** and will be derived in Chapter 4. Here we will follow an idea of G. Stokes to obtain two useful relations between the four parameters just mentioned. Our description of interference makes use of waves that have all the properties of mechanical waves, although this is not true for light waves in reality. For a mechanical dynamic system, we can reverse the time dependence and let the system run backward. Let us apply this same principle to the reflection and refraction on an interface.

Figure 2.23a shows the incoming light being reflected and refracted. In Figure 2.23b and c, we reverse rA and tA, respectively, and obtain an additional refraction trA and an additional reflection $r'tA$. Using the reversibility (opposite direction in Figure 2.23a compared to Figure 2.23b and c), we have

$$A = r^2 A + t'tA \to r^2 + t't = 1 \qquad (2.64)$$

4. Multiple-beam Interference Problems

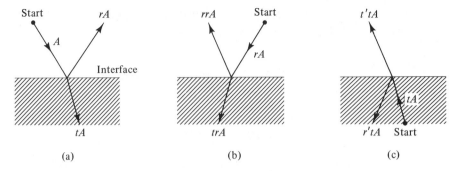

Fig. 2.23
According to Stokes' treatment, we follow rays through an interface going (a) first in the forward direction and (b, c) then in the reverse directions.

and for the additional refraction and reflection terms,

$$trA + r'tA = 0 \rightarrow r = -r' \tag{2.65}$$

These relations are also true for light and are in agreement with Fresnel's formulas. We take both energy and intensity proportional to the square of the amplitude. The flux of light through an interface will be discussed in Chapter 4.

Equation 2.65 tells us that the absolute value of reflection is the same, regardless of whether reflection occurs from the less dense medium or from the denser medium, but that these reflection coefficients differ by a minus sign. We know that the phase factor due to $\lambda/2$ or π results in a shift from a positive quantity to a negative quantity for a cosine or sine wave as well as for the exponential representation. This result of our reversibility consideration corresponds to that part of assumption 3 concerned with the phase shift of π on the denser medium.

Using the relations $tt' = 1 - r^2$ and $r' = -r$, we can eliminate the reflection and transmission coefficients from the less dense side; that is, we can eliminate t' and r'. Substituting into equations 2.62 and 2.63, we have

$$A_r = A\left(r - \frac{r(1 - r^2)e^{i\Delta}}{1 - r^2 e^{i\Delta}} \right) \tag{2.66a}$$

and

$$A_t = A\left(\frac{1 - r^2}{1 - r^2 e^{i\Delta}} \right) \tag{2.66b}$$

Multiplying equations 2.66a and 2.66b by the complex conjugates A_r^* and A_t^*, respectively, and then rearranging, results in expressions proportional to the intensity:

$$A_r A_r^* = A^2\left(\frac{2r^2(1 - \cos \Delta)}{1 + r^4 - 2r^2 \cos \Delta} \right) \tag{2.67}$$

and

$$A_t A_t^* = A^2 \left(\frac{(1 - r^2)^2}{1 + r^4 - 2r^2 \cos \Delta} \right) \tag{2.68}$$

Introducing the normalized transmitted and reflected intensities

$$I_t = \left| \frac{A_t}{A} \right|^2 \quad \text{and} \quad I_r = \left| \frac{A_r}{A} \right|^2$$

and the abbreviation

$$g^2 = \frac{4r^2}{(1 - r^2)^2}$$

and using the trigonometric formula

$$\cos \Delta = 1 - 2 \sin^2 \frac{\Delta}{2}$$

we have

$$I_r = \frac{g^2 \sin^2 \Delta/2}{1 + g^2 \sin^2 \Delta/2} \quad \text{and} \quad I_t = \frac{1}{1 + g^2 \sin^2 \Delta/2} \tag{2.69}$$

Adding equations 2.69, we find

$$I_r + I_t = 1 \tag{2.70}$$

telling us that we have taken care of all waves. (Note that equations 2.69 correspond to **energy conservation**.) From equation 2.59, the phase shifts $\Delta/2$ at all wavelengths and angles of refraction θ_2 are given by

$$\frac{\Delta}{2} = \frac{2\pi n D \cos \theta_2}{\lambda} \tag{2.71}$$

For perpendicularly incident light, one has $\cos \theta_2 = 1$. Peak transmission that is the case of constructive interference is then given by

$$\frac{\Delta}{2} = 0, \pi, 2\pi, \ldots, m\pi \quad \text{or} \quad D = 0, \frac{\lambda}{2n}, \frac{2\lambda}{2n}, \ldots, \frac{m\lambda}{2n} \tag{2.72}$$

For the case of destructive interference,

$$\frac{\Delta}{2} = \frac{\pi}{2}, \frac{3\pi}{2}, \frac{5\pi}{2}, \ldots, \left(m + \frac{1}{2} \right) \pi \tag{2.73}$$

or

$$D = \frac{\lambda}{4n}, \frac{3\lambda}{4n}, \frac{5\lambda}{4n}, \ldots, \frac{(m + \frac{1}{2})\lambda}{2n}$$

If the plate has the thickness of half the wavelength in the medium or a multiple of this, we obtain unity for the transmission. The reflection is zero regardless of the reflectivity r of the individual interfaces. The reflectivity of each plate may be almost 1, yet the transmission is still 1. This is the case of construc-

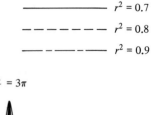

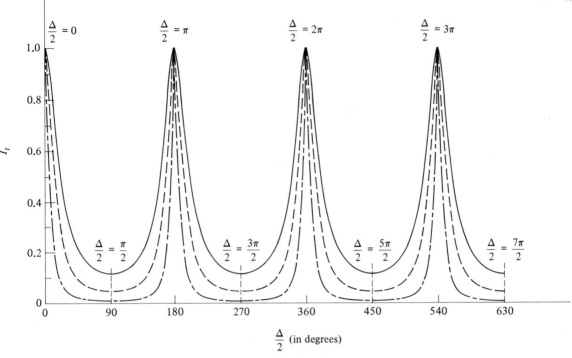

Fig. 2.24
Transmission through a dielectric plate as a function of phase shifts $\Delta/2$ and for different single-surface reflectivities. This type of pattern is also obtained with a Fabry-Perot etalon (see Chapter 8.)

tive interference, called the **resonance condition**. The two interfaces are forming a resonant cavity. The case of destructive interference results in a minimum transmission for $\sin^2(\Delta/2) = 1$, and the value of this transmission depends on the reflectivity r (Figure 2.24). So far we have considered an idealized case and neglected all possible losses.

In Figure 2.24, we see the maxima and minima as functions of $\Delta/2 = 2\pi n D/\lambda$ for normal incidence. The plot can be interpreted as depending on D for constant λ or depending on λ for constant D. If we vary D and restrict λ to a certain spectral region, our previous mathematical treatment applies to a **Fabry-Perot spectrometer**. If D is constant, the treatment applies to a **Fabry-Perot etalon** (see Chapter 8).

We now consider the case where $\cos \theta_2 \neq 1$, that is, oblique incidence of the light. For constructive interference, we have

$$D \cos \theta_2 = \frac{\lambda}{n}\left(\frac{m}{2}\right) \tag{2.74}$$

For destructive interference, we have

2. Interference of Light

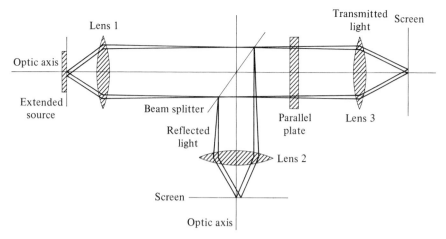

Fig. 2.25
Observation in reflection and transmission of fringes of a plane parallel plate. The emitted waves of an extended source form fringes at equal inclination in the focal plane of lenses 2 and 3. Ring patterns will result.

$$D \cos \theta_2 = \frac{\lambda}{n} \left(\frac{m + \frac{1}{2}}{2} \right) \tag{2.75}$$

We may wish to express θ_2, the angle of refraction in the medium, in terms of θ_1, the angle of incidence. We use Snell's law for the substitution $\sin \theta_1 = n \sin \theta_2$ along with the identity $\sin^2 \theta_2 + \cos^2 \theta_2 = 1$, and we get

$$\frac{\sin^2 \theta_1}{n^2} + \cos^2 \theta_2 = 1 \quad \text{or} \quad \cos \theta_2 = \pm \left(1 - \frac{\sin^2 \theta_1}{n^2} \right)^{\frac{1}{2}} \tag{2.76}$$

Interference in a plate illuminated at oblique incidence is obtained according to equations 2.74 and 2.75 for constructive interference

$$\sqrt{1 - \frac{\sin^2 \theta_1}{n^2}} = \frac{1}{D} \left(\frac{\lambda}{n} \right) \left(\frac{m}{2} \right) \tag{2.77}$$

and for destructive interference

$$\sqrt{1 - \frac{\sin^2 \theta_1}{n^2}} = \frac{1}{D} \left(\frac{\lambda}{n} \right) \left(\frac{m + \frac{1}{2}}{2} \right) \tag{2.78}$$

Since D is a constant, we see that the angles of incidence will determine whether we obtain bright or dark fringes. Fringes produced in such a way are called **fringes of equal inclination**. The observation of the interference pattern produced by reflected and transmitted light may be accomplished as shown in Figure 2.25.

Waves from different source points are usually incident on the plate at different angles. Only waves from points lying on a circle around the normal of the plate have the same angle of incidence on the plate and form fringes of equal inclination. The resulting fringe pattern is shown in Figure 2.26 for reflection and transmission.

4. Multiple-beam Interference Problems　　107

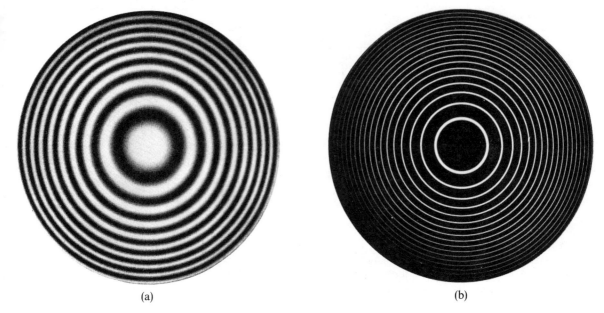

(a)   (b)

Fig. 2.26
Interference fringes observed in (a) reflection and (b) transmission at a plane parallel plate. The source is not "pointlike." (From Cagnet, Françon, and Thrierr, *Atlas of Optical Phenomena*.)

We see that the two patterns are complementary. For $\theta_1 = \theta_2 = 0$, we have a bright spot in reflection, a dark spot in transmission. But this situation depends on the particular value of D/λ used in the demonstration. For another value of D/λ, we could observe a dark spot in reflection and a bright spot in transmission. This is radically different from the observation of Newton's rings, which are fringes of equal thickness.

So far, we have discussed only the simple case for which the media on both sides of the plane parallel plate have the same refractive index, which we took to be 1. The general case of three different refractive indices is of course more complicated but can be calculated in a similar way.

5. COHERENCE

A. Temporal Coherence

Up to this point, we have assumed that we are superimposing two waves, each infinitely long and having a fixed phase difference between them. Most of our discussion has centered on the variation of the phase difference caused by changes in the path length.

We regarded laser light as an example of such infinitely long waves—but in reality, it is not. We will now consider the conditions under which interference may be observed for laser light, atomic emission, and blackbody emission. In Chapter 15, we will discuss how each type of light is produced and described.

Atomic emission is observed if an electron moves from a higher to a lower

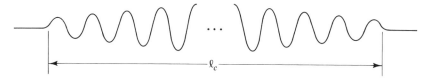

Fig. 2.27
A schematic of a wave train of finite length, ℓ_c. We refer to ℓ_c as the coherence length.

orbit in the atom. The energy $E_2 - E_1 = h\nu$ is emitted in the process taking a certain finite time. As a result, the emitted light wave might be considered a finite wave train with first increasing and then decreasing amplitude, as depicted in Figure 2.27.

The length of the wave train depends on the time duration of the emission process. The international standard of length uses the emission of krypton 86 (^{86}Kr) at $\lambda_{air} = 605.616$ nm (1 nm $= 10^{-9}$ m). The emission process takes about 10^{-9} s. The corresponding length of the wave train is about 1 m and is called the **coherence length** ℓ_c. If we consider a laser as a source, the wave train is much longer, and the coherence length is about 10^5 m.

Let us now examine finite wave trains of atomic emission in a Michelson interferometer. We will assume just one atom as a source. Its emitted finite wave train is split at the beam splitter (Figure 2.28a). The wave trains are reflected at M_1 and M_2 (Figure 2.28b), and a part of each is superimposed in the direction of the detector (Figure 2.28c). For finite wave trains, we can easily imagine that mirror M_2 of the Michelson interferometer is displaced so far that the two wave trains from M_1 and M_2 "miss" each other at recombination (Figure 2.28d). For ^{86}Kr, this distance $(x_2 - x_1)$ must be only 0.5 m and indeed, such a procedure is used to determine ℓ_c.

To study the superposition of finite wave trains, we need some knowledge of their mathematical representation. This is done by using Fourier theory, and

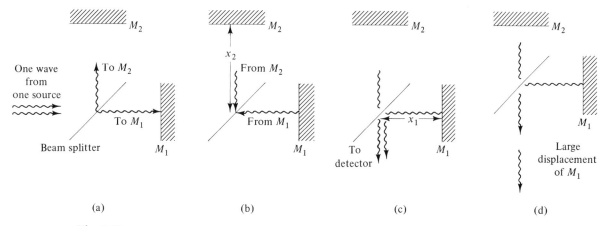

(a) (b) (c) (d)

Fig. 2.28
Splitting of the incident beam in a Michelson interferometer. Recombination depends on the displacement of M_1.

we will touch on this subject in another chapter. At the present, we only need to know that a finite wave train can be represented by a superposition of an infinite number of waves. The combination of the amplitudes of all the waves of different frequencies v determines the length of the wave train. A long wave train needs for its representation only a small band of frequencies and amplitudes of only a certain range of variation. The longer the wave train, the smaller the frequency interval necessary. For details, see Chapter 15. The dependence of ℓ_c on Δv may be expressed as

$$\ell_c \propto \frac{1}{\Delta v}$$

where Δv is the frequency range used in the superposition. Therefore, the frequency range Δv may be used for the characterization of the wave train (see Chapter 15).

In **atomic emission**, the frequency range, or **bandwidth**, is about 1000 MHz, whereas in laser emission, it can be of the order of 1000 Hz. Consequently, we can move the Michelson mirror much further before fringes disappear.

In **blackbody radiation** (incandescent lamp), the light is emitted from a hot object. Its radiation is continuous in frequency dependence, and its amplitude dependence over a small frequency interval may be considered constant. The energy of each wave is given by $E = hv$, and since the emission is a random process, there is no phase relation between the emitted waves. Using a filter, we can select a certain bandwidth Δv of the emitted light. Each wave with frequency within Δv has, on the average, a coherence length $\ell_c \propto 1/\Delta v$ and can be treated as before. As a consequence, we can use more and more narrow-band filters to select smaller and smaller bands of blackbody radiation, resulting in larger and larger coherence lengths. For our Michelson interferometer experiment, this means that we can move the Michelson mirror to larger and larger distances and still observe fringes. There is, of course, a limitation to this process. Smaller and smaller bandwidths Δv will result in smaller and smaller energies of this frequency interval. Eventually, the noise of the detection system will make the observation of the corresponding energy impossible.

A similar result would be obtained if one used a narrow filter with atomic emission.

B. Young's Experiment

In the experiment of Lloyd's mirror, we saw how a wave can be "folded" so that two parts of it can be superimposed to produce fringes. Another way to accomplish the superposition of two parts of a wave coming from one source is by performing **Young's two-slit interference experiment** (wave front division). See Figure 2.29.

In Young's experiment, a wave emitted at the source S_0 travels to the small aperture A_1. At the round opening, the incident wave generates a spherical wave in accordance with Huygens' principle. (Huygens' principle tells us that a wave generates new waves at an interface or opening. We will cover Huygens' principle in detail in Chapter 3.) The newly generated wave at A_1 travels a large distance and appears as a plane wave at the second aperture A_2. There it generates two

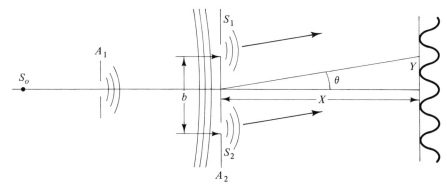

Fig. 2.29
Schematic of Young's experiment. Light is emitted by the source S_0. According to Huygens' principle, the first aperture A_1 produces spherical waves, which in turn generate spherical waves at S_1 and S_2 with a fixed phase relation to produce a fringe pattern. If S_0 is small and far away, aperture A_1 is not needed.

new waves at the small openings S_1 and S_2, each wave having a fixed phase relation to the incident wave. Consequently, there is a fixed phase relation between the two new generated waves. The two new waves at A_2 again travel a large distance X and produce an interference pattern on the observation screen. The two waves generated at S_1 and S_2 may be described in the same way we described the interference pattern of two light sources in Section 2B (equations 2.13 and 2.15).

For the intensity (square of the amplitude factor), we have

$$y^2 = 4A^2 \cos^2\left(\pi\frac{b\sin\theta}{\lambda}\right) \tag{2.79}$$

Since the distance X is very large compared to b and Y, we can use the small angle approximation:

$$y^2 = 4A^2 \cos^2 \pi\left(\frac{b}{\lambda}\right)\left(\frac{Y}{X}\right) \tag{2.80}$$

Although the mathematical expression is the same for the description of the interference pattern, we did not assume (as we did in Section 2B) that the light traveling from S_1 and S_2 has a fixed phase relation. Here we generate a phase relation between the two waves through Huygens' principle. The phase difference is zero for $\theta = 0$ because the wave incident on A_2 is like a plane wave.

Young's experiment can be performed with light from atomic emission and blackbody sources in a way analogous to the procedure for Lloyd's mirror (Section 3C). The appearance of fringes in Young's experiment is shown in Figure 2.30.

Young's experiment can also be done using X rays. When physicists first attempted to prove the wave character of X rays, they conducted diffraction experiments on a narrow slit. As we will see in the next chapter, the side loops of the diffraction pattern are less pronounced than the fringes of the interference pattern; therefore, these experiments were not convincing. The scattering on

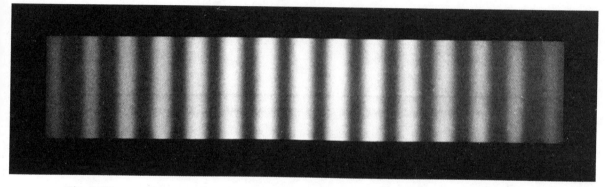

Fig. 2.30
Fringes resulting from Young's experiment. These fringes are observed with the interference of light generated at two openings according to Huygens' principle. For the decrease of the intensity at both ends, see Chapter 3, Section 2E and F. (From Cagnet, Françon, and Thrierr, *Atlas of Optical Phenomena*.)

atoms in crystals produced patterns similar to interference patterns. Max von Laue received the Nobel prize for this procedure in 1914.

Recently, X rays of $\lambda = 60$ Å have been used for obtaining interference fringes in an experiment very similar to Young's (Figure 2.31). The only difference in the experiment shown in Figure 2.31 is that on the second screen, a set of three openings is used instead of the one opening used in Young's experiment.

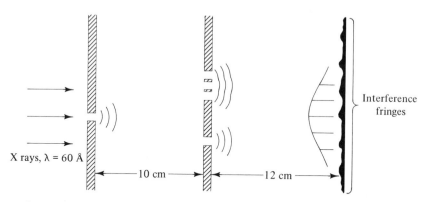

Fig. 2.31
Interference fringes obtained using *X* rays of 60 Å in a Young-type experiment.

We have used the expressions "small opening" and "large distance *X*." In Figure 2.32, we give the experimental dimensions for the observation of fringes with Young's experiment according to R. Pohl.

Young's experiment was historically very important, since it demonstrated for the first time the wavelike character of light some 150 years before laser light was available. To show the wave nature of light using nonlaser light he generated new waves at the apertures according to Huygens' principle. Through Huygens'

2. Interference of Light

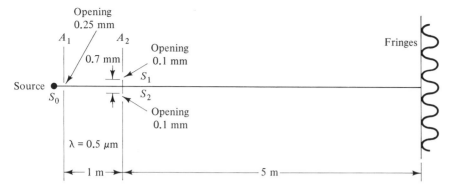

Fig. 2.32
Physical dimensions for the observation of fringes in Young's experiment according to R. Pohl.

principle, the phase relations of these two waves are guaranteed. These waves are regarded as **spatially coherent**.

C. Spatial Coherence

As we have seen, we can obtain fringes from two separate sources if the light from the two source points is in phase. If we use atomic emission at each point, we do not obtain fringes. To overcome this difficulty, we used only one source in Lloyd's experiment and produced a virtual source through the mirror (Section 3C). In Young's experiment, we used one source only but generated two new sources with fixed phase relation at the openings S_1 and S_2 (See Figure 2.29). The aperture A_1 was assumed small. The question of just how large we can make the opening will be discussed later.

Now we want to study the question of whether or not fringes can be observed with Young's experiment if we use two sources or apertures. What will be the permissible distance between them to observe fringes? In Figure 2.33, we use solid and broken lines to show the superposition of two Young's experiments. Our previous discussion holds for the case where light is coming from S (see Figure 2.29). For the case where light is coming from S', the wave arrives at the aperture A_2 as a plane wave, but at an angle ϕ (see the inset in Figure 2.33). The wave generates a new wave, first at S_2 and then at S_1. The time difference results in a path difference of $x = b \sin \phi$. This angle ϕ is also the angle for the direction of the central maximum of interference for the light from S'. For light from S, the central maximum appears in the direction of $\phi = 0$. If the condition $x = b \sin \phi \ll \lambda$ holds, the two *intensity patterns* are closely superimposed and fringes are observed (Figure 2.34). Fringes are observed despite the fact that the waves from S and S' have no phase relation with each other. If $x = b \sin \phi = \lambda/2$, then the two intensity patterns are superimposed in such a way that the position of the minima of one are at the maxima of the other, and no fringes can be seen. If x becomes larger, however, then fringes will reappear. It is customary to say that light from S and S' is *coherent* if a fringe pattern can be observed (e.g., with Young's experiment) and *incoherent* if no pattern can be observed.

5. Coherence   113

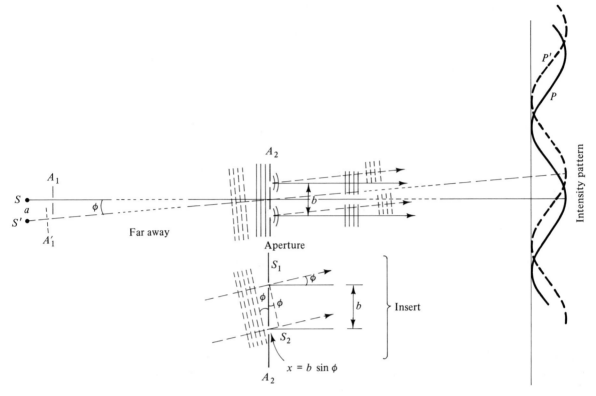

Fig. 2.33
Superposition of two Young's experiments using separate sources S and S' and apertures A_1 and A'_1. There is a distance a between the two sources, but both use the same aperture A_2. The inset shows ϕ for the light coming from S'.

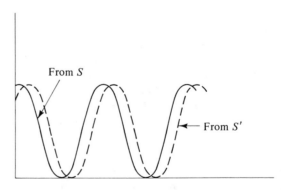

Fig. 2.34
Superposition of two intensity patterns in Young's experiment, each from one source without mutual phase relation. The distance $x = b \sin \phi$ (see Figure 2.33) is approximately $\frac{1}{10}(\lambda/2)$.

2. Interference of Light

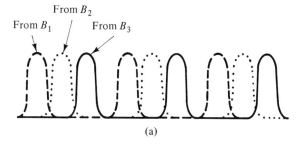

From B_2

From B_1 From B_3

(a)

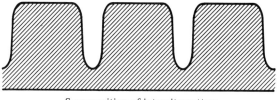

Superposition of intensity pattern

(b)

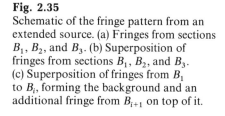

Fig. 2.35
Schematic of the fringe pattern from an extended source. (a) Fringes from sections B_1, B_2, and B_3. (b) Superposition of fringes from sections B_1, B_2, and B_3. (c) Superposition of fringes from B_1 to B_i, forming the background and an additional fringe from B_{i+1} on top of it.

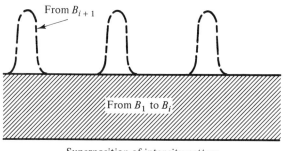

From B_{i+1}

From B_1 to B_i

Superposition of intensity pattern

(c)

D. Spatial Coherence and an Extended Source

We have seen that fringes can be observed with Young's experiment using two sources. Fringes can be observed if $b \sin \phi \ll \lambda/2$, where b is the separation of the two slits and ϕ is the angle from the two slits to the two sources S and S'. We will now consider source points between S and S'. The additional intensity pattern produced by a source point between S and S' will be superimposed on the fringe pattern produced by S and S' and will not obscure them. The question now is how large can we make the extension of the source.

We divide the source in sections of equal width B_1, B_2, ..., B_n between the source extension from $S^0 = S$ to $S^n = S'$. The fringes produced by each section will all superimpose (see Figure 2.35a and b for the first three). If the extension of the source is now made larger, say to S^i, we get to the point where we have a superposition of the sections B_1 to B_i ($i < n$), and no fringe pattern will be

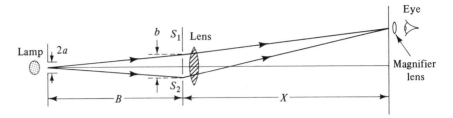

Fig. 2.36
Laboratory setup for the observation of fringes from an extended source. Experimental values: after R. Pohl $B = 20$ m; $X = 1$ m; width of S_1 and $S_2 = 0.4$ mm; $b = 6$ mm; $\lambda = 5.9 \times 10^{-7}$ m. At $2a = 2$ mm, the fringes disappear for the first time, and the upper limit of the experiment is at $2a = 4$ mm.

observable (see the background in Figure 2.35c). Enlarging the source further to S^{i+1} adds a new section B_{i+1}. The superposition of the interference patterns of sections B_1 to B_i results in the background, and B_{i+1} produces fringes (see Figure 2.35c). The fringes appear for a larger and larger size of the source on a larger and larger background.

Figure 2.36 shows an experiment by R. Pohl that demonstrates the observation of fringes with an extended source. It is customary to say that the light from the extended source is partially coherent if fringes are observed in this experimental setup. Note that the individual emitters of the extended light source have no phase relation with each other.

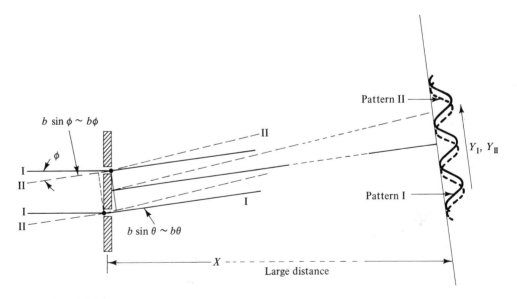

Fig. 2.37
Light waves from two stars, I and II, forming an angle ϕ when they arrive at a double slit of width b. The waves from each star produce a fringe pattern on a screen at distance X, described by the coordinates Y_I and Y_{II}.

2. Interference of Light

E. The Michelson Stellar Interferometer

The double-slit arrangement of Young's experiment was modified by Michelson so that he could measure the distance between two stars in a double-star system. To understand the modification needed for the **Michelson stellar interferometer**, consider the light coming from two stars incident on a double slit (Figure 2.37). The angle between the stars in ϕ. Each source produces a fringe pattern with the following intensities:

Source I: $u_I^2 = A^2 \cos^2\left(\pi \frac{b}{\lambda} \sin\theta\right) \rightarrow A^2 \cos^2 \pi \frac{b}{\lambda}\theta$ or $A^2 \cos^2 \pi \frac{b}{\lambda}\left(\frac{Y_I}{X}\right)$ \hfill (2.81)

Source II: $u_{II}^2 = A^2 \cos^2\left(\pi\frac{b\sin\theta - b\sin\phi}{\lambda}\right)$ or $A^2\cos^2\pi\frac{b\theta - b\phi}{\lambda}$ \hfill (2.82)

where the last expressions on the right are for the small-angle approximation.

The angle ϕ is what we wish to obtain. We consider the two special cases $\phi = \lambda/b$ and $\phi = \lambda/2b$ (Figure 2.38). For the case where $\phi = \lambda/b$, the two patterns

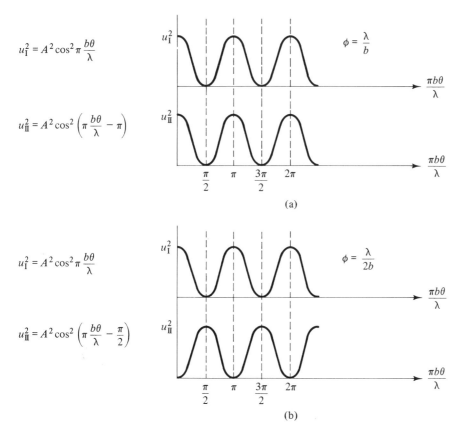

$u_I^2 = A^2\cos^2\pi\frac{b\theta}{\lambda}$

$u_{II}^2 = A^2\cos^2\left(\pi\frac{b\theta}{\lambda} - \pi\right)$

$u_I^2 = A^2\cos^2\pi\frac{b\theta}{\lambda}$

$u_{II}^2 = A^2\cos^2\left(\pi\frac{b\theta}{\lambda} - \frac{\pi}{2}\right)$

Fig. 2.38
Plot of u_I^2 and u_{II}^2 for (a) $\phi = \lambda/b$ and (b) $\phi = \lambda/2b$. In (a), the two superimposed intensity patterns are lined up, maxima on maxima, minima on minima. In (b), maxima and minima of the two patterns superimpose, and no fringe pattern appears.

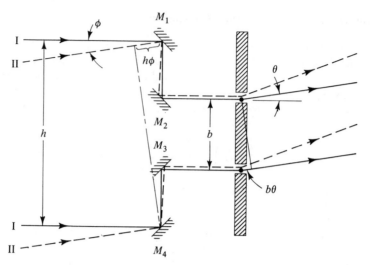

Fig. 2.39
Michelson's stellar interferometer. The changes with respect to Figure 2.37 are the introduction of mirrors M_1 to M_4, where h is the distance from M_1 to M_4. ϕ is the angle of incidence, and θ is the angle to the interference pattern. The angles ϕ and θ are exaggerated.

are superimposed in such a way that a fringe pattern results; in the case of $\phi = \lambda/2b$, no fringe pattern results.

For demonstration purposes, we used ϕ, the angle we wish to measure. But ϕ is a constant. To demonstrate an observable change in the fringe pattern, Michelson introduced a new parameter h and four more mirrors (Figure 2.39). The introduction of mirrors M_1 to M_4 results in dependence on h of the path difference of the waves from star II participating in the double-slit interference pattern. For the small-angle approximation, the intensities for patterns I and II are now described by

$$u_I^2 = A^2 \cos^2 \pi \frac{b\theta}{\lambda} \tag{2.83}$$

and

$$u_{II}^2 = A^2 \cos^2 \pi \frac{b\theta - h\phi}{\lambda} \tag{2.84}$$

By changing the distance from M_1 to M_4, that is, by changing h, we can go from fringe pattern to no fringe pattern. The peak-to-peak distances of patterns I and II are equal. To go from a fringe pattern at a certain h to an absence of fringe pattern at another h', we need only move the maximum to the next minimum of pattern II. In other words, for the same m,

$$\text{max for II: } \pi \frac{b\theta - h\phi}{\lambda} = \frac{\pi}{2}(2m) = \pi m \tag{2.85}$$

$$\text{next min for II: } \pi \frac{b\theta - h'\phi}{\lambda} = \frac{\pi}{2}(2m + 1) = \pi\left(m + \frac{1}{2}\right) \tag{2.86}$$

2. Interference of Light

The difference $|\Delta h| = |h' - h|$ is

$$|\Delta h| = \frac{\lambda}{2\phi} \tag{2.87}$$

By measuring h, we obtain ϕ.

The availability of the Michelson stellar interferometer (see Figure 2.39) combined with our knowledge of coherence conditions and fringe patterns due to faraway extended sources (see Figure 2.35) enables us to determine the angular diameter of single stars. This was done in 1920 at the Mount Wilson Observatory for Betelgeuse, located in the Orion constellation (see also problem 12).

In this chapter, we have discussed fringe patterns. We saw that the superposition of waves having a fixed phase relation results in amplitude patterns having maxima and minima. The fixed phase relation was then related to coherence. The corresponding observable intensity pattern extends in its magnitude between maxima and zero.

We also discussed the observation of fringes and the nonobservation of fringes by superposition of two intensity patterns. The observation of fringe systems was related to coherence. This should not be confused with the fixed phase relation necessary to obtain an amplitude interference pattern.

We discussed the superposition of light having a certain wavelength. White light contains a range of different wavelengths, and to discuss interference patterns obtained with white light, we must consider each wavelength separately and the resulting intensity pattern. Then we can take the superposition of all the different intensity patterns for all different wavelengths. This results in the observation of color spectra, as seen on soap bubbles (see problem 11).

We have frequently referred to Huygens' principle for the generation of spherical waves by a plane wave arriving at a small opening. In the next chapter, we will discuss Huygens' principle in detail.

Problems

1. *Time average.* Calculate the time average in equation 2.12. The second factor in equation 2.12 is

$$a^2 = \cos^2\left[2\pi\left(\frac{x_1}{\lambda} - \frac{t}{T}\right) - 2\pi\frac{\delta/2}{\lambda}\right]$$

Make an appropriate choice for x_1 and show that we obtain

$$a_v^2 = \frac{1}{T}\int_0^T \cos^2(\ldots)\,dt = \frac{1}{2}$$

Ans. $a_v^2 = \frac{1}{2}$

2. *Wedge-shaped film: fringes of equal thickness.* Consider the fringes produced by a wedge-shaped film. This configuration can be produced by two microscope slides and a thin wire (diameter b). If we assume that the wedge angle α is small, then the bending of light rays can be neglected. The interference between light rays reflected from surfaces 2 and 3 results in a fringe pattern (see Section 3E).

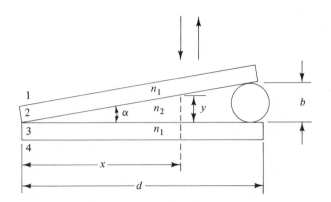

a. Use the derivation in the text and find the difference $\Delta x_m = x_{m+1} - x_m$ between successive maxima and the change $\Delta y_m = y_{m+1} - y_m$ in the film thickness from one maximum to the next.

b. Calculate the diameter b of the wire in micrometers if $\Delta x_m = 2$ mm, $\lambda = 0.5890$ μm, and $d = 8$ cm.

Ans. a. $\Delta x_m = x_{m+1} - x_m = \lambda/(2 \tan \alpha)$, $\Delta y_m = \lambda/2$; b. $b = 12$ μm

3. *Newton's rings.* The figure shows how fringes of equal thickness may be produced (Newton's rings). A spherical surface of index n_1 rests on a flat surface of the same index. A gap of variable thickness and index n_2 produces the interference fringes.

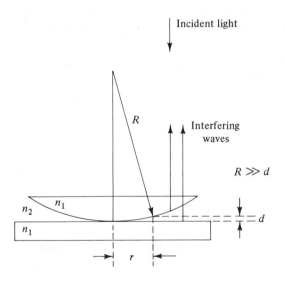

a. For constructive interference, give the dependence of d on m and λ.

b. Calculate r_m, the radius of the mth fringe as a function of m, λ, R, and n_2, and use the approximation $R \gg d$.

c. For $\lambda = 0.5$ μm, $n_2 = 1$, and $R = 5$ m, calculate the radius of the first two rings.

Ans. a. $d_m = 1/2\{\lambda/n_2[m - (1/2)]\}$; b. $r_m = \sqrt{\lambda R(m - 1/2)/n_2}$; c. $r_{m=1} = 1.12$ mm,

$r_{m=2} = 1.94$ mm

2. Interference of Light

4. *Complex numbers.* We saw how the superposition of two waves in the cosine form may be accomplished with complex numbers to obtain the square of the amplitude. Here we summarize the amplitude factor in question.

$$y_1 = A \cos \alpha_1$$
$$y_2 = A \cos(\alpha_1 - \Delta)$$

For the sum $y_1 + y_2$, we get

$$y = y_1 + y_2 = 2 A \cos \frac{\Delta}{2} \cos\left(\alpha_1 - \frac{\Delta}{2}\right)$$

Using complex numbers

$$z_1 = A e^{i\alpha_1}$$
$$z_2 = A e^{i(\alpha_1 - \Delta)}$$

we have

$$z = z_1 + z_2 = 2 A \cos \frac{\Delta}{2} e^{i(\alpha_1 - \Delta/2)}$$

Now let us consider waves in the sine form.

$$y_1 = A \sin \alpha_1, \quad y_2 = A \sin(\alpha_1 - \Delta)$$

a. Show that by adding y_1 and y_2 and using

$$\sin \gamma + \sin \beta = 2 \sin \frac{\gamma + \beta}{2} \cos \frac{\gamma - \beta}{2}$$

the same amplitude factor is obtained. Underline it.

b. Use complex numbers. By complementing y_1 and y_2 to become complex numbers

$$z_1 = A e^{i\alpha_1} \quad \text{and} \quad z_2 = A e^{i(\alpha_1 - \Delta)}$$

we obtain the result given above,

$$z = z_1 + z_2 = 2 A \cos \frac{\Delta}{2} e^{(\alpha_1 - \Delta/2)}$$

Show that the result of (a) is the imaginary part of z in (b).

5. *General case: different amplitudes, different phases, sine notation.* We want to calculate the superposition of the following two functions having different amplitudes and different phases:

$$u_1 = u_{10} \sin(\omega t - kx_1 + \delta_1)$$

and

$$u_2 = u_{20} \sin(\omega t - kx_2 + \delta_2)$$

If we attempt to evaluate $u^2 = (u_1 + u_2)^2$ directly, we encounter some complications. To avoid this, let $\alpha_i = -kx_i + \delta_i$, $i = 1, 2$, and calculate

$$u = u_1 + u_2 = u_{10} \sin(\omega t + \alpha_1) + u_{20} \sin(\omega t + \alpha_2) = u_0 \sin(\omega t + \alpha)$$

where

$$u_0 \cos \alpha = u_{10} \cos \alpha_1 + u_{20} \cos \alpha_2$$

and

$$u_0 \sin \alpha = u_{10} \sin \alpha_1 + u_{20} \sin \alpha_2$$

Since

$$(u_0 \cos \alpha)^2 + (u_0 \sin \alpha)^2 = u_0^2$$

we get

$$u_0^2 = u_{10}^2 + u_{20}^2 + 2u_{10}u_{20} \cos(\alpha_1 - \alpha_2)$$

and therefore

$$u = [u_{10}^2 + u_{20}^2 + 2u_{10}u_{20} \cos(\alpha_1 - \alpha_2)]^{1/2} \sin(\omega t + \alpha)$$

a. Fill in all the missing steps in the calculations leading to the final equation.
b. Express $\alpha_1 - \alpha_2$ in terms of k, x_i and δ_i, where $i = 1, 2$.
c. The conditions for interference maxima and minima are determined by the amplitude factor in the final equation for u, that is, the square root term. Find these conditions of u and obtain a general expression involving k, x_i, and δ_i for these two cases.
d. For the special case of $u_{10} = u_{20}$ (i.e., equal amplitude waves), show that the expression for u leads to the same result for the intensity u^2 as was obtained in equation 2.36 using complex notation.

6. *N-source interference pattern.* Consider the case $N = 5$. Make a sketch of the intensity as a function of Y/X. Plot the center and the first two main maxima along with all secondary maxima in between.
Part. ans. Major peaks at $Y/X = (\lambda/b)m$, zeros at $Y/X = m\lambda/5b$ (except at major peaks)

7. *Plane parallel plate. Three media.* A plane parallel plate of thickness D and refraction index n_2 is surrounded by two media with indices n_1 and n_3. Consider only two waves reflected from the interfaces, as indicated in the following figure:

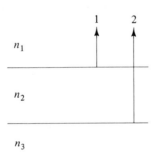

Calculate the path difference and phase difference of waves (1) and (2). For example, for an air gap surrounded by glass, where $n_1 = 1.5$, $n_2 = 1$, and $n_3 = 1.5$, one has

$$\delta = 2D + \frac{\lambda}{2} \quad \text{and} \quad \Delta = \frac{2\pi}{\lambda}\left(2D + \frac{\lambda}{2}\right) = 2\pi\left(\frac{2D}{\lambda}\right) + \pi$$

a. For a glass plate in air: $n_1 = 1$, $n_2 = 1.5$, and $n_3 = 1$
b. For three media with increasing refractive indices, so that $n_2^2 = n_1 \cdot n_3$, with $n_1 = 1$, $n_2 = 1.22$, and $n_3 = 1.5$ (antireflection coating if D is $\lambda/4$).
c. For the case where $n_1 = 1$, $n_2 = 1.5$, and $n_3 = 1.22$.
d. For the case where $n_1 = 1.5$, $n_2 = 1$, and $n_3 = 1.22$.
e. For the case where $n_1 = 1.5$, $n_2 = 1.22$, and $n_3 = 1$.
Ans. a, c, d. $\delta = 2D + \lambda/2$, $\Delta = 2\pi(2D/\lambda) + \pi$; b, e. $\delta = 2D$, $\Delta = 2\pi(2D/\lambda)$

8. *Michelson interferometer circular fringes.* The interference of the beams in a Michelson interferometer was discussed for two beams incident and reflected under 45° to the normal of the beam splitter. We will call these two beams the axial beams. For the intensity, we obtained (equation 2.21),

$$y^2 = 4A^2 \cos^2\left(2\pi \frac{\delta/2}{\lambda}\right)$$

and the conditions for maxima were

$$\delta = 2(x_2 - x_1) = m\lambda$$

and for minima

$$\delta = 2(x_2 - x_1) = \left(m + \frac{1}{2}\right)\lambda$$

Now we want to discuss the interference of beams with slightly larger or smaller angles than 45° leading to a ring pattern.

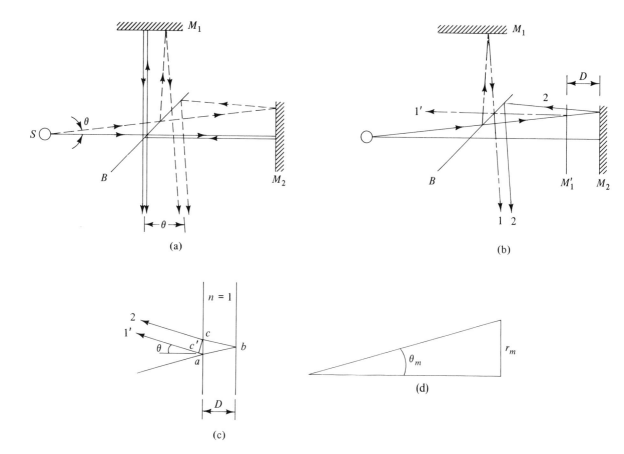

(a)

(b)

(c)

(d)

Part (a) of the accompanying figure shows two beams leaving the source at different angles. For the axial beams, the superposition produces the usual result. For the beams leaving the source with angle θ, the resulting superposition of the beams has the angle θ with the axial beams (in direction of the observation screen). To obtain the conditions for maxima and minima for the superposition of the nonaxial beam, we look at part (b). Only the superposition of two nonaxial beams are shown. The plane M_1' indicates the distance beam 1 travels (with respect to M_2) compared to beam 2. If we "fold" the path of beam 1 around the beam splitter, we see that M_1 and M_1' are at the same position. Therefore, beams 1 and 1' are equivalent. The difference in optical path is shown in figure (c) and is the same as obtained for the plane parallel plate (see equation 2.58),

$$\delta = 2D \cos \theta$$

where $n = 1$.

a. Find the conditions for maxima and minima. Beams emitted with the same angle θ from the source form a cone. The superposition that results also forms a cone in the direction of the observation screen. Since the optical path difference depends on θ, we observe a ring pattern at the screen. At a plane parallel plate, the corresponding fringes are called fringes of equal inclination. The same term is also used here.

b. Calculate all allowed values of θ of maxima for the following cases:
   (i) $D/\lambda = 1$
   (ii) $D/\lambda = 2$
   (iii) $D/\lambda = 4$

c. Make a sketch of the ring pattern. The radius is proportional to $\sin \theta$.
   Note: The fringe with the highest m is closest to the axis. This is in contrast to the interference pattern of N point sources.

9. *Plane parallel plate: multiple-beam interference.* For the transmitted intensity (equation 2.69),

$$I_t = \frac{1}{1 + g^2 \sin^2 \Delta/2}$$

where g is defined as $g = 2r/(1 - r^2)$.

This formula is also valid for the case of two highly reflecting plates; that is, r may have values close to 1. For large values of r, the device is called a Fabry-Perot etalon. The quantity Δ is defined as (equation 2.71)

$$\Delta = \frac{2\pi}{\lambda} 2nD \cos \theta_2$$

We would like to investigate I_t as a function of θ, holding λ and r constant. This will

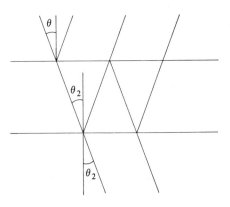

result in fringes of constant inclination. Consider light incident on a Fabry-Perot etalon at an angle θ with respect to the normal (see the figure). The two surfaces of the Fabry-Perot etalon have constant reflectivity r and are separated by a distance D. The index of refraction between the two surfaces is assumed to be a little larger than 1 (to remain a denser medium); we take $\theta \approx \theta_2$.

a. Determine the angle θ at which constructive interference is observed on a distant screen, that is, when $I_t = 1$. Find all possible angles (of fringes) for the following cases:

   (i) $D/\lambda = 1$ (ii) $D/\lambda = 2$ (iii) $D/\lambda = 4$

b. Plot m versus θ for case (iii). Note that the smallest number of m corresponds to the largest angle.

Ans. (i) $\cos\theta_2 = m/2$, $\theta_2 = 90°, 60°, 0°$

10. *Plane parallel plate: width of half height.* For the plane parallel plate, transmitted intensity is given as

$$I_t = \frac{1}{1 + g^2 \sin^2 \Delta/2}$$

where

$$g^2 = \frac{4r^2}{(1 - r^2)^2}$$

and r is the reflection coefficient of the amplitude at one surface. For Δ we have

$$\frac{\Delta}{2} = \frac{2\pi}{\lambda} nD \cos\theta_2$$

where n is the refractive index, D is the thickness of the plate, λ is the wavelength, and θ_2 is the angle with respect to the normal in the plate.

a. Make a sketch of I_t versus λ for $\lambda = 2$ mm to $\lambda = \frac{1}{4}$ mm. Set $n = 1$, and $\theta_2 = 0$, and use $D = 1$ mm. Compare with Figure 2.24, where we plotted I_t versus $\Delta/2 \propto D/\lambda$. Note that in Figure 2.24, the width at half height of the lines depends on r.

b. Show that we have for the width $\Delta\lambda$ at half height of the lines

$$\frac{\lambda}{\Delta\lambda} = \frac{m\pi g}{2} = \frac{m\pi r}{1 - r^2}$$

(The calculation is similar to that shown in Chapter 8 for the resolution of a Fabry-Perot spectrometer)

Ans. a. $\lambda = 2, 1, \frac{2}{3}, \frac{1}{2}, \frac{2}{5}, \frac{1}{3}, \frac{2}{7}, \frac{1}{4}$

11. *Interference of white light on a thin film (soap bubble).* Consider the formula of the transmitted light for the plane parallel plate with constant g,

$$I_t = \frac{1}{1 + g^2 \sin^2 \Delta/2}$$

The minima and maxima of I_t are obtained from various values of

$$\frac{\Delta}{2} = \frac{2\pi}{\lambda} nD \cos\theta_2$$

where θ_2 is the angle of the refracted light in the plate. If we take $n = 1.33$ (i.e., the refractive index of water in the visible spectrum), we have

$$\frac{\Delta}{2} = \pi 5.32 \frac{D}{\lambda} \cos\theta_2$$

Consider a wedge-shaped water film between two microscope slides of index $n = 1.5$. If light of various wavelengths is incident on a wedge-shaped film, we will see different colors for different thicknesses of the wedge, depending on the interference condition for maximum. Assume that $\theta_2 = 0$, and find the minimum thickness ($\neq 0$) of the portion of the wedge for the following:

a. $\lambda = 0.4\ \mu m$ (blue)
b. $\lambda = 0.5\ \mu m$ (green)
c. $\lambda = 0.7\ \mu m$ (red)

Ans. a. 0.075 μm; b. 0.094 μm; c. 0.132 μm

12. *A double star and the Michelson stellar interferometer.* We want to use the stellar interferometer to estimate the angle ϕ between the two components of a double star. Assume that $\lambda = 0.5\ \mu m$ and $h = 100$ in. Calculate ϕ.

Ans. $\phi = 0.02$ s.

Diffraction | 3

1. INTRODUCTION

In Chapter 2, we discussed the superposition of light waves and the resulting interference patterns. We also considered the different origins of light waves and the experimental setups to observe fringe patterns. In Young's experiment, we saw the generation of secondary waves by a plane wave at an aperture due to **Huygens' principle**. The study of these secondary waves is the topic called *diffraction*.

In this chapter, we will discuss additional consequences of Huygens' principle. Historically, Huygens' principle was formulated to explain various interference-diffraction experiments. Later Gustav Kirchhoff and Augustin Jean Fresnel reformulated Huygens' principle using electromagnetic wave theory. Interference-diffraction phenomena can also be described by solving Maxwell's equations for the boundary value problem of the aperture. However, Maxwell's equations are mathematically very difficult; this is why Huygens' principle has survived for the description of interference-diffraction experiments.

In the last chapter, we stated Huygens' principle as follows: A plane wave arriving at an aperture with small openings generates spherical waves. At a large distance, these spherical waves again appear as plane waves. We also gave some numerical examples for "small" and "large." In this chapter, we will also discuss what happens if we eliminate the restrictions of "small" and "large."

Huygens' principle tells us that an incident wave generates new waves in the plane of the aperture. The newly generated waves have a fixed phase relation with the incident wave and therefore also with one another. They propagate only in the forward direction. The origins of the newly generated waves are continuous over the area of the aperture, and the phase difference of adjacent waves is infinitesimally small.

Let us consider the diffraction at a slit. Figure 3.1 shows the dependence of the observed diffraction pattern on the width of the slit. A large slit will almost give us the geometrical shadow (Figure 3.1a); a smaller slit will bend the light and a fringe pattern will occur (Figure 3.1b); and for a very small slit, the fringe

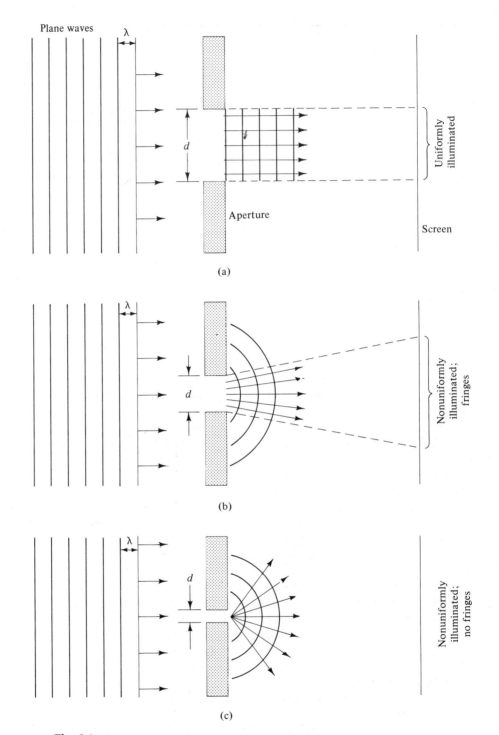

Fig. 3.1
Conditions for diffraction on a single slit. (a) $d \gg \lambda$, and there is no appreciable diffraction. (b) $d \sim \lambda$, and diffraction is observed (fringes). (c) $d \ll \lambda$, nonuniformly illuminated, but no fringes.

3. Diffraction

pattern will disappear but the intensity will fall off with larger angles (Figure 3.1c). If the source is far enough away, we have essentially plane waves incident on the aperture; if we are observing at a screen that is also far away from the slit, the waves arriving at the screen are again plane waves, and we have what is called **Fraunhofer diffraction**. Later we will discuss the case in which the incident and the diffracted waves are not plane waves, a phenomenon called **Fresnel diffraction**.

A frequently observed phenomenon of diffraction is the pattern we observe at night when we look at a street lamp through a window screen. We observe a cross pattern of different colors extending along the symmetry of the screen's wires. It is significant that the light is bent out of its straight path at large angles and that this phenomenon is not due to refraction.

We will also see how white light generates color pattern, as we mentioned in Chapter 2. The waves with different wavelengths are considered separately, and the intensity patterns are then added to produce the observed intensity color pattern.

2. FRAUNHOFER DIFFRACTION

A. Single Slit Aperture

According to Huygens' principle, every point on the incident wave front arriving at the aperture is a source of secondary waves. It is the superposition of these **Huygens wavelets** in the forward direction that results in the diffraction pattern.

Let us assume that the aperture is a long, narrow, rectangular opening and that cylindrical waves are generated. For the mathematical description, we assume that the incident wave is a plane wave and that we can use harmonic waves generated at each point within the aperture and traveling in all directions (Figure 3.2). If we divide the aperture into M sections, each representing one

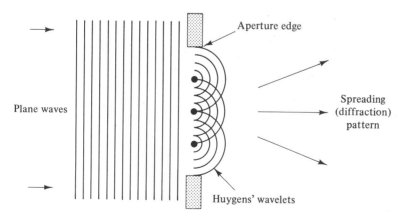

Fig. 3.2
General optical diagram for the diffraction of light at a single rectangular slit, assuming Huygens' principle. Shown are three Huygens wavelets.

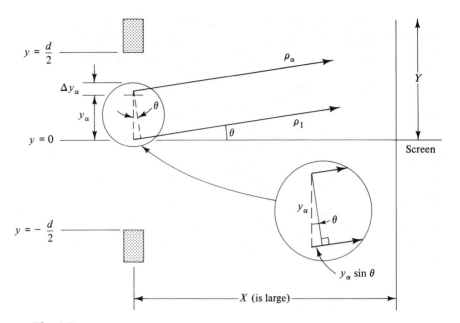

Fig. 3.3
Geometry for the observation of a single-slit diffraction pattern with parameters defined. Incident waves are from the left, and Fraunhofer diffraction is observed.

source, we have a similar problem as discussed in Chapter 2, Section 4A. We observe the superposition of all the newly generated waves at a faraway screen. However, we must note that our choice of M is arbitrary and that each wave is a fictitious Huygens wavelet. We may choose any value for Mi; later we will make M very large so that the summation process of the superposition can be replaced by integration.

The geometry and coordinates for the superposition of the waves are shown in Figure 3.3; each of the M sections is called Δy_α. The αth interval is located at $+y_\alpha$ with respect to the central interval at $y = 0$. A generated wave travels a distance ρ_α from source to screen. The optical path difference between the αth wave and the reference wave ($\alpha = 1$) is $y_\alpha \sin\theta$. Thus, for the αth wave, we have

$$\rho_\alpha = \rho_1 - y_\alpha \sin\theta \tag{3.1}$$

and for the phase factor

$$\cos 2\pi \left(\frac{\rho_\alpha}{\lambda} - \frac{t}{T} \right) \tag{3.2}$$

or

$$\exp\left[i2\pi \left(\frac{\rho_\alpha}{\lambda} - \frac{t}{T} \right) \right] = e^{i(k\rho_\alpha - \omega t)} = e^{i(k\rho_1 - \omega t)} e^{-iky_\alpha \sin\theta} \tag{3.3}$$

where t is the time and λ and T are the wavelength and period of the oscillation, respectively. In equation 3.2, the wave is written in real terms; in equation 3.3,

3. Diffraction

as a complex number. The first term in equation 3.3 emphasizes the order of magnitude of ρ_α; in the other terms, the wave vector magnitude $k = 2\pi/\lambda$ and the angular frequency $\omega = 2\pi/T$ are used. The Fraunhofer condition is implied by the use of $\rho_\alpha = \rho_1 - y_\alpha \sin \theta$. As in the case of interference, the waves are considered to travel in almost parallel directions from the aperture or to travel in parallel directions and meet in the focal plane of a lens (see Section 3, Chapter 2).

Because the incident wave is assumed to be a plane wave, the intensity C^2 reaching the aperture is uniformly distributed across the slit. Then the amplitude of the αth secondary source, of length Δy_α, is the fraction $\Delta y_\alpha/d$ of C, and the newly generated wave is

$$u_\alpha = \left(C \frac{\Delta y_\alpha}{d} \right) e^{i(k\rho_1 - \omega t)} e^{-iky_\alpha \sin \theta} \tag{3.4}$$

Summing up all contributions from the M sources across the slit gives us for the superposition of the amplitude

$$u = \frac{C}{d} e^{i(k\rho_1 - \omega t)} \sum_{\alpha=1}^{M} \Delta y_\alpha e^{-iky_\alpha \sin \theta} \tag{3.5}$$

We now let M become very large; that is, the slit is now composed of an infinite number of infinitesimally narrow sources. The sum in equation 3.5 passes to an integral over dy, and we have

$$u = \frac{C}{d} e^{i(k\rho_1 - \omega t)} \int_{-d/2}^{+d/2} e^{-iky \sin \theta} dy \tag{3.6}$$

The integral in equation 3.6 can be evaluated directly. Using $\beta = k \sin \theta$, we have

$$\int_{-d/2}^{+d/2} e^{-i\beta y} dy = 2i \frac{e^{i\beta(d/2)} - e^{-i\beta(d/2)}}{(i\beta)2i} = \frac{1}{i\beta} 2i \sin \left(\beta \frac{d}{2} \right) = d \frac{\sin[k(d/2)\sin \theta]}{k(d/2)\sin \theta} \tag{3.7}$$

The amplitude becomes

$$u = C e^{i(k\rho_1 - \omega t)} \frac{\sin[k(d/2)\sin \theta]}{[k(d/2)\sin \theta]} \tag{3.8}$$

and for the intensity we get

$$uu^* = C^2 \frac{\sin^2[k(d/2)\sin \theta]}{[k(d/2)\sin \theta]^2} \tag{3.9}$$

We can now set $C^2 = 1$, which serves only to normalize the central maximum to unit intensity; that is, $|u|^2 = 1$ at $Y = 0$ $(\theta = 0)$. Using the small-angle approximation, where

$$\sin \theta \simeq \theta \simeq \tan \theta = \frac{Y}{X} \tag{3.10}$$

and substituting, we have

$$u^2 = \frac{\sin^2[k(d/2)(Y/X)]}{[k(d/2)(Y/X)]^2} = \frac{\sin^2[\pi(d/\lambda)(Y/X)]}{[\pi(d/\lambda)(Y/X)]^2} \tag{3.11}$$

2. Fraunhofer Diffraction

131

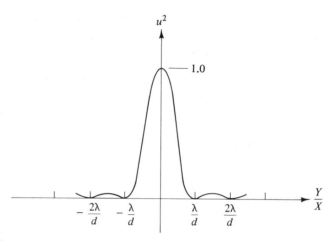

Fig. 3.4
Intensity of a single-slit diffraction pattern for the small-angle approximation.

In Figure 3.4, we plot u^2 versus Y/X. The intensity minima occur at $Y/X = \pm\lambda/d, \pm 2\lambda/d$, and so on. The secondary maxima are located between these values but are not exactly equidistant. To locate the intensity of the secondary maxima, we differentiate equation 3.11, and with $b = \pi(d/\lambda)(Y/X)$, we have

$$\frac{du}{db} = \frac{d}{db}\left(\frac{\sin b}{b}\right) = \frac{b\cos b - \sin b}{b^2} \tag{3.12}$$

Setting $du/db = 0$ gives us $b = \tan b$, a transcendental equation that we can solve graphically by locating the intersections of the curves b and $\tan b$ (see problem 2). From Figure 3.4, it is clear that the solution must give roots "close" to

$$\frac{Y}{X} = \left(m + \frac{1}{2}\right)\frac{\lambda}{d}$$

so

$$\pi\frac{d}{\lambda}\left(\frac{Y}{X}\right) = \frac{1}{2}\pi, \frac{3}{2}\pi, \frac{5}{2}\pi, \ldots \tag{3.13}$$

as expected. Using a calculator, we can easily find the first few exact roots of $b = \tan b$. Using the result of equation 3.13 as a guide, the first three roots are $b = 1.43\pi, 2.459\pi$, and 3.471π. We would use these values in equation 3.9 to get the exact positions for the secondary maxima.

The central maximum at $Y/X = 0$ gives an indeterminate form $|u|^2 = 0/0$. Using l'Hôpital's rule, we get

$$\lim_{b\to 0}\frac{d\sin b/db}{db/db} = \lim_{b\to 0}\frac{\cos b}{1} = 1 \tag{3.14}$$

for the amplitude expression. This shows that for the central maximum, we have $|u|^2 = 1$. Figure 3.5 shows the diffraction pattern formed by a single slit.

Fig. 3.5
Photo of a diffraction pattern formed by a single slit. (From Cagnet, Françon, and Thrierr, *Atlas of Optical Phenomena.*)

B. The Fourier Transformation of Slit Functions

Let us reconsider equation 3.6 and rewrite the integral for the small angle approximation as

$$u(Y) = \frac{C}{d} \lambda e^{i(k\rho_1 - \omega t)} \int_{-d/2\lambda}^{+d/2\lambda} e^{-i2\pi(y/\lambda)(Y/X)} d\left(\frac{y}{\lambda}\right) \tag{3.15}$$

Substituting $\bar{y} = y/\lambda$, $\bar{Y} = Y/X$, $a = d/(2\lambda)$, and

$$\bar{C} = \frac{C}{d} \lambda e^{i(k\rho_1 - \omega t)}$$

in equation 3.15 gives

$$u(\bar{Y}) = \bar{C} \int_{-a}^{+a} e^{-i2\pi\bar{y}\bar{Y}} d\bar{y} \tag{3.16}$$

If we introduce a function $P(\bar{y})$ with the property that

$$P(\bar{y}) = \begin{cases} 1 & \text{where } -a \leq \bar{y} < a \\ 0 & \text{otherwise} \end{cases} \tag{3.17}$$

then we can rewrite equation 3.16 with the limits ranging from $-\infty$ to $+\infty$:

$$u(\bar{Y}) = \bar{C} \int_{-\infty}^{+\infty} P(\bar{y}) e^{-i2\pi\bar{y}\bar{Y}} d\bar{y} \tag{3.18}$$

The integral on the right-hand side of equation 3.18 is a **Fourier integral** and $u(\bar{Y})$ is the **Fourier transform** of $P(\bar{y})$. Evaluation of equation 3.18 yields a form similar to that in equation 3.8, only now expressed in terms of the variable $\bar{Y}$:

$$u(\bar{Y}) = \bar{C}' \frac{\sin 2\pi a \bar{Y}}{2\pi a \bar{Y}} \tag{3.19}$$

Here, $\bar{C}'$ is a constant containing all other quantities not explicitly displayed. The

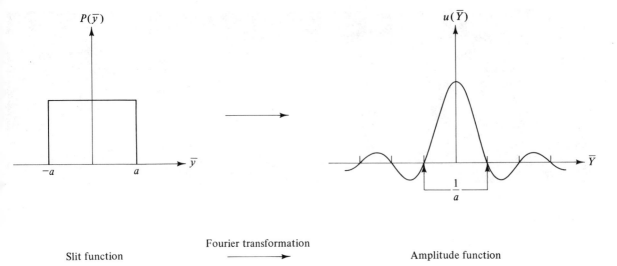

Fourier transformation

Slit function $\longrightarrow$ Amplitude function

Fig. 3.6
Fourier transformation of the "slit" function.

usefulness of this approach is that it schematically represents the transformation of the "slit" function $P(\bar{y})$ into the amplitude function $u(\bar{Y})$.

Under certain mathematical assumptions, we can reverse the Fourier transformation and then obtain the slit function as the Fourier transformation of the amplitude function. The optical process of diffraction is mathematically represented by a Fourier transformation. If we apply the diffraction process twice, we again get the slit function. How this is accomplished will be discussed in Chapter 10. There is a reciprocal relationship between the width of the functions $P(\bar{y})$ and $u(\bar{Y})$. From Figure 3.6 we see that for small a, $P(\bar{y})$ is narrow but $u(\bar{Y})$ is wide, and for large a, $P(\bar{y})$ is wide but $u(\bar{Y})$ is narrow.

C. Rectangular Aperture

The rectangular aperture in Figure 3.7 has dimensions a and d. The integral for the amplitude on the screen is formulated in a way similar to the preceding discussion. That is, the amplitude is given by using the small angle approximation

$$u = \bar{C} \int_{y=-d/2}^{d/2} \int_{z=-a/2}^{a/2} e^{-i2\pi[(y/\lambda)(Y/X)+(z/\lambda)(Z/X)]} \, dy \, dz$$

or

$$u = \bar{C} \int_{-d/2}^{d/2} e^{-2\pi i(y/\lambda)(Y/X)} \, dy \int_{-a/2}^{a/2} e^{-2\pi i(z/\lambda)(Z/X)} \, dz \qquad (3.20)$$

Note that the factor $e^{i(k\rho_1-\omega t)}$ has been omitted from the formulas, since it is always the same for all secondary waves, does not contain variables affecting the integral, and vanishes upon calculation of uu^*.

The result expressed in equation 3.20 is not difficult to interpret physically;

3. Diffraction

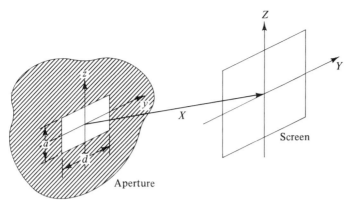

Fig. 3.7
Geometry for diffraction at a rectangular aperture, with parameters defined in the diagram.

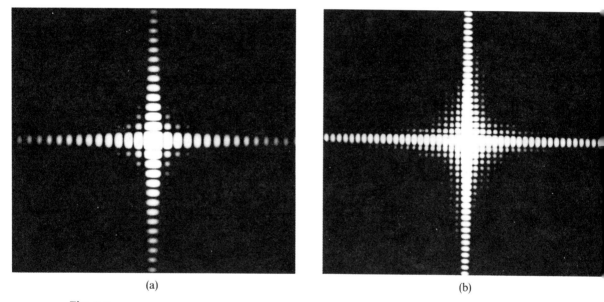

(a)   (b)

Fig. 3.8
(a) Fraunhofer pattern formed by a square aperture. (b) The same pattern further exposed to bring out some of the faint details of the diffraction pattern. (From Hecht and Zajac, *Optics*, p. 348.)

we have a product of two integrals, each of which describes the diffraction associated with one of the two perpendicular directions. For the intensity, we have

$$uu^* = u_0^2 \frac{\sin^2[k(d/2)(Y/X)]}{[k(d/2)(Y/X)]^2} \cdot \frac{\sin^2[k(a/2)(Z/X)]}{[k(a/2)(Z/X)]^2} \qquad (3.21)$$

where u_0^2 is the intensity at $Y = Z = 0$.

In the preceding result, we have used the small-angle approximation. If we

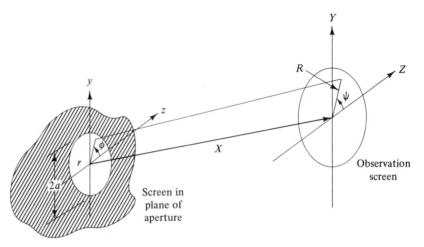

Fig. 3.9
Geometry for diffraction at a circular aperture, with parameters defined in the diagram.

want to formulate the result for any angle, we must substitute $Y/X = \sin\theta$ and $Z/X = \sin\phi$, where θ and ϕ are defined by $\tan\theta = Y/X$ and $\tan\phi = Z/X$. The result can then be written as

$$uu^* = u_0^2 \frac{\sin^2(k(d/2)\sin\theta)}{(k(d/2)\sin\theta)^2} \frac{\sin^2(k(a/2)\sin\phi)}{(k(a/2)\sin\phi)^2} \qquad (3.22)$$

In Figure 3.8, we see the pattern for the special case where $d = a$, that is, a square aperture. Frequently we observe many orders of diffraction appearing under a small angle. Such patterns are well described by equation 3.21. On the other hand, equation 3.22 includes the experimental situation of large angle diffraction, which can be realized, for example, at a diffraction grating.

D. Circular Apertures

Lenses have a round circumference, and light is refracted by a lens according to principles of geometrical optics discussed in Chapter 1. The fact that the lens has a finite opening results in diffraction at the rim of the lens. We will now briefly discuss the diffraction pattern of a circular aperture obtained in Fraunhofer observation.

Figure 3.9 shows an aperture of radius a. The coordinates y and z are used for the aperture, while the coordinates Y and Z are used at the observation screen a distance X away. We assume that Y and Z are small compared to X and we write the diffracted amplitude as

$$u = u_0 \int_{\substack{\text{over} \\ \text{opening}}} e^{-i2\pi[(y/\lambda)(Y/X)+(z/\lambda)(Z/X)]} \, dy \, dz \qquad (3.23)$$

This integral is well known from the rectangular aperture. However, it is now to be calculated over the round aperture of radius a. This is difficult to do, and

3. Diffraction

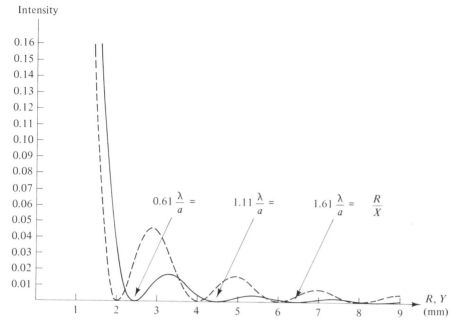

Fig. 3.10
Graph of the intensity versus R in mm for a circular aperture of radius $a = .5$ mm,
$\lambda = 500$ nm, and $X = 4$ m in Fraunhofer observation. Broken lines indicate the
intensity of a slit of $d = 2a = 1$ mm, for both $\lambda = 500$ nm and $X = 4$ m. The patterns
are normalized to 1 at the center (not shown). The first three zeros are indicated.

Bessel functions are used for the analytical expression. We will do this in problem
7. The result of the calculation is shown in Figure 3.10, where we plot the intensity
as a function of R/X, just as we did as a function of Y/X for the slit in Figure
3.4. Figure 3.11 shows a photograph of the diffraction pattern of a circular
aperture, called an **Airy disk**.

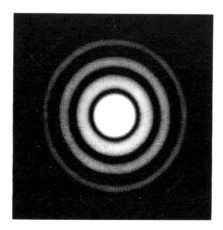

Fig. 3.11
Photograph of the diffraction pattern at a round aperture (Airy-disc). (From Cagnet,
Françon, and Thrierr, *Atlas of Optical Phenomena*.)

The first minimum of the diffraction pattern plays an important role in the resolution of optical instruments, as we will see in Chapter 9. By comparing the diffraction of the slit with that of the round aperture, we see that the slit of width equal to the diameter of the round aperture produces a narrower diffraction pattern (see Figure 3.10):

E. Double Slits

Having discussed the diffraction pattern due to a single slit, we now turn to the study of the pattern produced by two equal slits. This is Young's experiment in more detail than was treated in Chapter 2. The coordinate system is shown in Figure 3.12. We must integrate over both slits, similar to the way discussed for one slit, and obtain the sum of two integrals, each taken over one slit (see equation 3.6).

$$u = \frac{C}{d} e^{i(k\rho_1 - \omega t)} \int_{-d/2}^{d/2} e^{-iky \sin\theta} \, dy$$

$$+ \frac{C}{d} e^{i(k\rho_1 - \omega t)} \int_{-a-d/2}^{-a+d/2} e^{-iky \sin\theta} \, dy \tag{3.24}$$

We introduce the abbreviation $\gamma = (2\pi/\lambda)\sin\theta$; and for the small-angle approximation, $\gamma = (2\pi/\lambda)(Y/X)$, and we obtain

$$u = \frac{C}{d} e^{i(k\rho_1 - \omega t)} \left(\int_{-d/2}^{+d/2} e^{-i\gamma y} \, dy + \int_{-a-d/2}^{-a+d/2} e^{-i\gamma y} \, dy \right)$$

if we let $\eta = y + a$ for the second slit, then

$$u = \frac{C}{d} e^{i(k\rho_1 - \omega t)} \left(\int_{-d/2}^{+d/2} e^{-i\gamma y} \, dy + \int_{-d/2}^{+d/2} e^{-i\gamma(\eta - a)} \, d\eta \right)$$

The limits of the integrals are the same; so we can condense by factoring the integrands and use y instead of η in the last integral for the integration variable:

$$u = \frac{C}{d} e^{i(k\rho_1 - \omega t)} \left(\int_{-d/2}^{+d/2} e^{-i\gamma y} \, dy (1 + e^{i\gamma a}) \right)$$

Factoring out $e^{i\gamma(a/2)}$ and then integrating,

$$u = C e^{i(k\rho_1 - \omega t + \gamma a/2)} [e^{-i\gamma(a/2)} + e^{i\gamma(a/2)}] \left(\frac{\sin\gamma(d/2)}{\gamma(d/2)} \right)$$

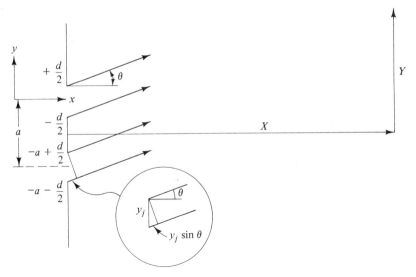

Fig. 3.12
Geometry for diffraction at two slits, with the parameters defined in the diagram.

Writing the cosine function for the expression enclosed within square brackets,

$$u = C2e^{i[k\rho_1 - \omega t + \gamma(a/2)]}\left(\cos\gamma\frac{a}{2}\right)\frac{\sin\gamma(d/2)}{\gamma(d/2)} \tag{3.25}$$

And for the intensity, we obtain

$$uu^* = \frac{\sin^2\gamma(d/2)}{[\gamma(d/2)]^2}\left(4C^2\cos^2\gamma\frac{a}{2}\right) \tag{3.26}$$

The first factor represents the intensity distribution due to a single slit. We call it the **diffraction factor**. The second factor is the intensity distribution of two superimposed waves having a phase difference of

$$\delta = \left(\gamma\frac{a}{2}\right) = \frac{2\pi}{\lambda}\cdot\frac{a\sin\theta}{2} \tag{3.27}$$

We call it the **interference factor**. For this factor, we have

$$4C^2\cos^2\left(\pi\frac{a\sin\theta}{\lambda}\right) = \begin{cases}4C^2 & \text{for } a\sin\theta = m\lambda \\ 0 & \text{for } a\sin\theta = (m + \frac{1}{2})\lambda\end{cases} \tag{3.28}$$

This is the same result we obtained in Section 3, Chapter 2, for the interference of waves from two sources traveling in direction θ. If we set $C^2 = A^2$ and $a = b$, we are back to equation 2.15.

Any pair of waves originating in different slits having the distance a will contribute to the interference factor of the final intensity. If we made the slits "small," we would approach the case of simple interference of two waves. For this case, the diffraction factor approaches 1, since d is close to zero (Figure 3.13a).

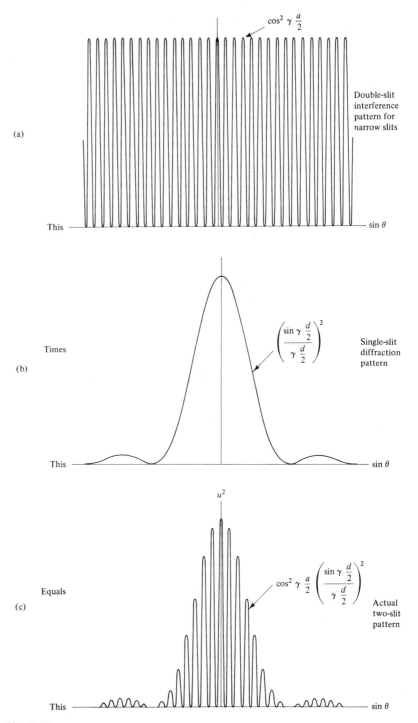

Fig. 3.13
Double-slit pattern as developed through the product of (a) double-slit interference pattern for two point sources at distance a; (b) single slit diffraction pattern of slit width d; (c) superposition for $a > d$.

Fig. 3.14
(a) Diffraction pattern of a double slit for the case where the separation a of the two slits is much larger than the width d of the slits. (b) The central part of the diffraction pattern shows an interference pattern just as observed in Young's experiment. (From Cagnet, Françon, and Thrierr, *Atlas of Optical Phenomena.*)

On the other hand, we can also have a approach zero and obtain 1 for the interference factor. We have a confluence of the two slits and are back to the single-slit pattern (Figure 3.13b).

The superposition of the two patterns, represented by multiplication of the two factors, is shown in Figure 3.13c for the case where a is comparable but larger than d. For a much larger than d, we get the photograph shown in Figure 3.14.

In Chapter 2, we used the word *interference* for the superposition of two or more distinct waves. We have just seen, in the double-slit experiment, that the second factor represents the interference of two waves originating at a distance a apart. The word *diffraction* describes what is represented by the first factor: the superposition of an infinite number of fictitious secondary waves with infinitely small phase differences between adjacent waves. While the diffraction factor was obtained by integration, the interference factor presented just a summation. If we associate interference with superposition, we need interference for any fringe pattern. Diffraction, on the other hand, describes the superposition of the fictitious Huygens wavelets produced in accordance with the shape of the aperture.

F. Multislit Diffraction Patterns

We now consider diffraction of waves due to an array of N regularly spaced finite apertures. As shown in Figure 3.15, the slit width is d and the repeat distance is a. As in the two-slit case, we use $\gamma = (2\pi/\lambda)\sin\theta$ or, for small angles, $(2\pi/\lambda)(Y/X)$. The amplitude of the resulting pattern on the screen is obtained by summing the contributions from secondary waves made by all N openings. That is, we must integrate terms like

$$u_j = \bar{C}\int_{j\text{th slit}} e^{-i2\pi(y_j/\lambda)\sin\theta}\,dy_j = \bar{C}\int_{j\text{th slit}} e^{-i\gamma y_j}\,dy_j \qquad (3.29)$$

for each slit to get the total amplitude. Using the labeling scheme shown in Figure

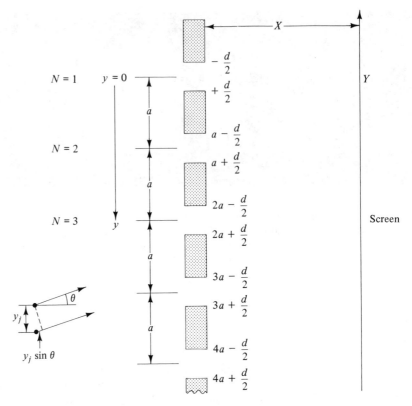

Fig. 3.15
Geometry for diffraction at a multislit arrangement, with the parameters defined in the diagram.

3.15, we write

$$u = \sum_{j=1}^{N} \bar{C} \int e^{-i\gamma y_j} \, dy_j \tag{3.30}$$

which gives

$$u = \bar{C} \int_{-d/2}^{+d/2} e^{-i\gamma y} \, dy + \bar{C} \int_{a-d/2}^{a+d/2} e^{-i\gamma y} \, dy + \cdots + \bar{C} \int_{(N-1)a-d/2}^{(N-1)a+d/2} e^{-i\gamma y} \, dy \tag{3.31}$$

The first integral yields $(2\bar{C}/\gamma) \sin(\gamma(d/2))$. In the second integral, we change variables such that $y' = y - a$. Then the second integral becomes

$$\bar{C} e^{-i\gamma a} \int_{-d/2}^{d/2} e^{-i\gamma y'} \, dy' = \frac{2\bar{C}}{\gamma} e^{-i\gamma a} \sin \gamma \frac{d}{2} \tag{3.32}$$

Performing the integrations for the third and other integrals similarly yields

$$u = 2 \frac{\bar{C}}{\gamma} \sin \gamma \frac{d}{2} (1 + e^{-i\gamma a} + e^{-2i\gamma a} + \cdots + e^{-i(N-1)\gamma a}) \tag{3.33}$$

The series in parentheses was encountered previously in the discussion of the

3. Diffràction

interference pattern produced by N sources (equation 2.43):

$$\sum_{m=0}^{N-1} e^{im\gamma a} = \frac{1 - e^{-iN\gamma a}}{1 - e^{-i\gamma a}}$$ (3.34)

With this result, the amplitude becomes

$$u = \bar{C}d\frac{\sin\gamma(d/2)}{\gamma(d/2)}\left(\frac{1 - e^{-iN\gamma a}}{1 - e^{-i\gamma a}}\right)$$ (3.35)

and the intensity is

$$uu^* = \bar{C}^2 d^2\left(\frac{\sin\gamma(d/2)}{\gamma(d/2)}\right)^2\left(\frac{1 - e^{iN\gamma a}}{1 - e^{i\gamma a}}\right)\left(\frac{1 - e^{-iN\gamma a}}{1 - e^{-i\gamma a}}\right)$$

or using twice the formula

$$(e^{ix} - 1)(e^{-ix} - 1) = 4\sin^2\frac{x}{2}$$

we get

$$
\begin{aligned}
uu^* &= u_0^2\frac{\sin^2\gamma(d/2)}{[\gamma(d/2)]^2} \cdot \frac{\sin^2 N\gamma(a/2)}{\sin^2\gamma(a/2)} \\
&= u_0^2\frac{\sin^2[\pi(d/\lambda)\sin\theta]}{[\pi(d/\lambda)\sin\theta]^2} \cdot \frac{\sin^2[N\pi(a/\lambda)\sin\theta]}{\sin^2[\pi(a/\lambda)\sin\theta]}
\end{aligned}
$$ (3.36)

where u_0^2 is the intensity for $y = 0$. The first factor in equation 3.36 arises from the treatment of diffraction due to a single slit of finite width. The second factor describes the superposition of N waves with fixed phase differences (see Chapter 2, Section 4). In obtaining these results, we summed the contributions from the N separate slits, but integrated the contributions of all secondary Huygens waves from each aperture, as discussed for the double slit. If we set $N = 2$, we again obtain equation 3.26 for the double slit. We also see how the interference of N openings could be experimentally realized, as discussed in Chapter 2, Section 4. Through Huygens' principle, the phase relation of the N sources is guaranteed. If we make the openings "small," the diffraction factor is almost 1 and the interference factor is dominant.

As the number of lines in the diffracting aperture increases, additional interesting effects evolve. For example, consider the case for $N = 4$. If we choose $a = 4d$, equation 3.36 becomes

$$uu^* = u_0^2\frac{\sin^2\gamma(d/2)}{[\gamma(d/2)]^2} \cdot \frac{\sin^2 4\gamma(a/2)}{\sin^2\gamma(a/2)}$$ (3.37)

Now

$$\sin^2 4\gamma\frac{a}{2} = 0 \quad \text{for} \quad 4\gamma\frac{a}{2} = 0, \pi, 2\pi, \ldots$$

and

$$\sin^2\gamma\frac{a}{2} = 0 \quad \text{for} \quad \gamma\frac{a}{2} = 0, \pi, 2\pi, \ldots$$

and it can be shown that we get 1 if uu^* is 0/0. A plot of the interference factor

$$\left(\frac{\sin 4\gamma(a/2)}{\sin \gamma(a/2)}\right)^2$$

versus $\gamma a/2 = \beta$ is shown in Figure 3.16a for the small-angle approximation $\sin \theta \approx Y/X$. The principal maxima are at $\gamma a/2 = m\pi$, where $m = 0, 1, 2, \ldots$, and secondary maxima are approximately midway between the three zeros separating the principal maxima. For larger values of N, we would observe that these peaks "sharpen," that is, the principal maxima become narrower and the number of secondary peaks between principal maxima increases. In Figure 3.16b we plot the single-slit diffraction factor

$$\left(\frac{\sin \gamma(d/2)}{\gamma(d/2)}\right)^2$$

versus $\gamma d/2 = \alpha$ and in the small-angle approximation $Y/X = m(\lambda/\alpha)$. This factor has zeros at $\gamma(d/2) = \pi, 2\pi$, and so on. The product of both factors in equation 3.37 gives the curve shown in Figure 3.16c. At $Y/X = 4\lambda/a$, corresponding to $\gamma d/2 = \pi$ or $\gamma a/2 = 4\pi$, one principal maximum has been eliminated. This effect also occurs for $\gamma(d/2) = 2\pi, 3\pi$, and so on, and is called the **missing order**.

The diffraction patterns produced by two, three, four, and five slits are shown in Figure 3.17. The $N - 2$ side maxima and $N - 1$ side minima are evident. The intensity decreases to both sides as a result of the diffraction factor. The ratio of d/a was chosen in such a way that no missing orders can be seen.

G. Amplitude Diffraction Gratings

A device like the one shown schematically in Figure 3.15 but with a large value of N is called an **amplitude diffraction grating**. A typical grating for use in the visible part of the electromagnetic spectrum (4×10^{-5} cm $< \lambda < 7 \times 10^{-5}$ cm) has approximately 12,000 lines per inch. This corresponds to a grating spacing of

$$a = \frac{1}{1.2 \times 10^4 \text{ lines/in.}} \cdot 2.54 \frac{\text{cm}}{\text{in.}} \simeq 2 \times 10^{-4} \text{ cm}$$

By comparison, a grating to be used at infrared wavelengths near 400 μm has 2 lines per millimeter, or 50 lines per inch, or a spacing of approximately 5×10^{-2} cm. Gratings for other regions of the electromagnetic spectrum have spacings that similarly correspond to the wavelength of the radiation to be diffracted. If monochromatic light is normally incident on the surface of a grating and the diffracted light is observed on a screen far from the grating plane (i.e., $Y/X = \tan \theta \simeq \sin \theta$), then the positions of the diffraction intensity maxima for a given grating constant a can be obtained from our general-result equation 3.36. That is,

$$\left(\frac{\sin N\gamma(a/2)}{\sin \gamma(a/2)}\right)^2 = N^2 = \text{maximum}$$

and is found using l'Hôpital's rule when $\gamma a/2 = \pi m$, where $m = 0, 1, 2$, and so on.

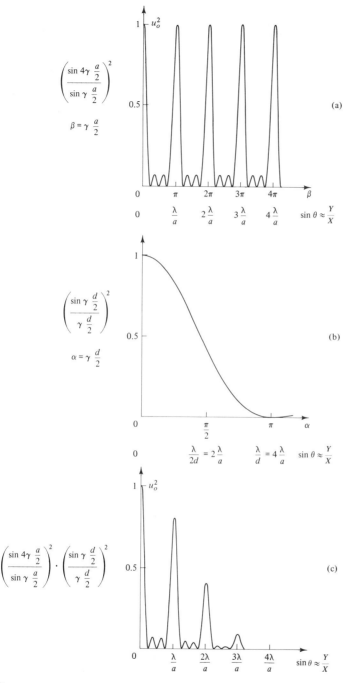

Fig. 3.16
Diffraction pattern for $N = 4$ when $a = 4d$. (a) The multiple-slit interference pattern. (b) The single-slit diffraction term. (c) The resultant intensity pattern. At $4\lambda/a$ an "order" is missing.

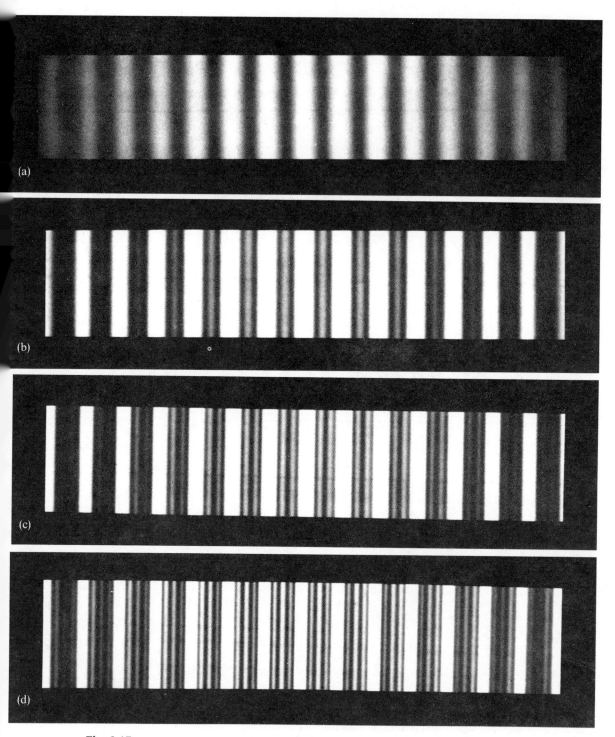

Fig. 3.17
Diffraction patterns produced by (a) two slits, (b) three slits, (c) four slits, and (d) five slits.
(From Cagnet, Françon, and Thrierr, *Atlas of Optical Phenomena*.)

3. Diffraction

Since

$$\gamma \frac{a}{2} = \left(\frac{2\pi}{\lambda} \sin \theta\right) \frac{a}{2}$$

we have $2\pi a \sin \theta / 2\lambda = m\pi$, or

$$m\lambda = a \sin \theta \qquad (3.38)$$

Equation 3.38 is called the **grating equation**. The integer m labels the *order* of the principal maxima. For polychromatic light, equation 3.38 gives the position of each component wavelength. Diffraction gratings are useful radiation dispersion devices capable of good spectral resolution. This topic will be discussed again in Chapter 8.

H. Phase Gratings

The type of grating discussed in the preceding section is called an amplitude grating. The incident light is blocked off periodically at certain portions of the grating. The phase difference for diffraction results from the phase difference of the Huygens waves generated at the periodic areas of the illuminated grating.

Phase gratings are transparent periodic structures of refractive index n. The phase difference for diffraction is again obtained from the Huygens waves generated at the surface of the grating. The surface is now fully illuminated but not flat. Phase gratings may have a periodic structure of steps, a zigzag profile, a sinusoidal structure, or a similar periodic profile.

To calculate the diffracted intensity, we proceed as we did for the plane amplitude grating (see equation 3.36). We consider the steps on one side of a plane parallel plate (see Figure 3.18). Such a grating is called an **echelette grating**.

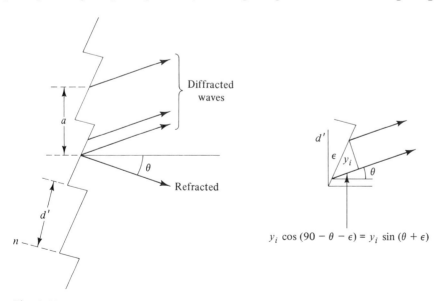

$$y_i \cos(90 - \theta - \epsilon) = y_i \sin(\theta + \epsilon)$$

Fig. 3.18
Coordinates for diffraction at an echelette grating, with the parameters defined in the diagram.

We choose the angle of incidence so that the refracted light beam leaves perpendicular to the surface of the facet. Diffracted waves from one facet and from the periodic arrangement of facets are shown as arrows in Figure 3.18.

For the interference factor (i.e., the second factor in equation 3.36), we have to count the path difference leading to the grating equation. As a consequence of the choice of the coordinate system in Figure 3.18, we obtain the same result as for the plane grating: $a \sin \theta$. So the second factor of equation 3.36 remains the same. For the diffraction factor (i.e., the first factor in equation 3.36), we have a different result, since the area to be integrated over is now tilted by the angle ε.

The phase difference for two Huygens waves on one facet is

$$y_i \sin(\theta + \varepsilon) \tag{3.39}$$

(see Figure 3.18). Integration from $-d'/2$ to $+d'/2$ gives us the following diffraction factor:

$$d' \frac{\sin[\pi(d'/\lambda)\sin(\theta + \varepsilon)]}{\pi(d'/\lambda)\sin(\theta + \varepsilon)} \tag{3.40}$$

For the diffracted intensity, we have

$$uu^* = u_0^2 \frac{\sin^2[\pi(d'/\lambda)\sin(\theta + \varepsilon)]}{[\pi(d'/\lambda)\sin(\theta + \varepsilon)]^2} \cdot \frac{\sin^2[N\pi(a/\lambda)\sin\theta]}{\sin^2\pi(a/\lambda)\sin\theta} \tag{3.41}$$

Let us assume that the angles ε and θ are small and that we have approximately $d' = a$. The first factor (diffraction) in equation 3.41 has zeros for $(a/\lambda)(\theta + \varepsilon) = m$, where m is an integer, except for $m = 0$. For $m = 0$, we have $\theta = -\varepsilon$ and the first factor is 1. This is indicated in Figure 3.19. The second factor (interference) has maxima at $(a/\lambda)\theta = m'$, where m' is an integer. For a given ε, we choose λ in such a way that $(a/\lambda)(-\varepsilon) = -1$, or $\lambda/a = \varepsilon$, so that the second factor always has maxima for θ equal to any multiples of ε. The difference between the maxima of the second factor is equal to the difference between the minima of the first factor (excluding $m = 0$). The difference for the second factor is

$$\frac{a}{\lambda}(\theta_{m'+1} - \theta_{m'}) = 1 \tag{3.42}$$

and the difference for the first factor is ($d' = a$)

$$\frac{a}{\lambda}(\theta_{m+1} + \varepsilon - \theta_m - \varepsilon) = 1 \tag{3.43}$$

In Figure 3.19, we indicate the values of the first factor for $m = 1$ and the second factor for $m' = 0$. The first factor is zero, the second factor is N^2, and the product is zero. The same is true for all other corresponding combinations of m and m', except for $m = 0$ and $m' = -1$ when the product is N^2. This is also indicated in Figure 3.19.

What has been accomplished is that all of the diffracted intensity will appear in just one order, the order $m' = -1$ in the coordinate systems we have chosen. We have a directional redistribution of the light into one angle. In infrared spectroscopy, where light sources are weak, we must get as much diffracted light into the detector as possible. Since suitable transmissive materials are not readily

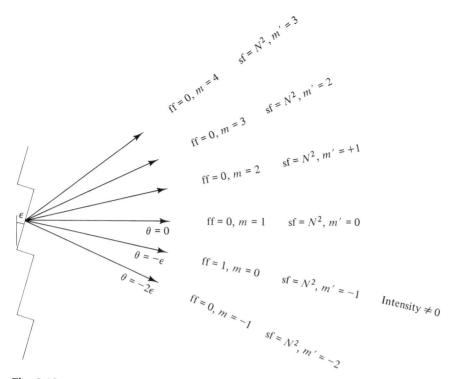

Fig. 3.19
Orders of diffraction at an echelette grating. The first factor in equation 3.41 is indicated as ff, and the second factor as sf.

available, we use **reflection echelette gratings**. We also speak of **blazed gratings** at a certain wavelength and certain angle ε. For example, a grating of groove spacings of 50 μm, having a blaze angle ε of 25°, is blazed at wavelength 42 μm.

A transmission phase grating with a sinusoidal profile can be used for scanners in nonimpact printers, but is also used for bar-code reading. In this application, it is desirable to have a laser beam diffracted in such a way that the angles of incidence and diffraction are 45° (Figure 3.20). By choosing the appropriate periodicity and profile of the sinusoidal grating and the appropriate wavelength of the laser light, we can achieve 90 percent intensity in the diffracted beam.

I. The Question of Periodicity

Having previously treated a regular array of slits, of which the diffraction grating was an example, we ask the question, How important is the regularity of the slit spacing to the achievement of a diffraction pattern? Our result, equation 3.36, can be expressed schematically as

$$u^2 = u_0^2 (\text{single-slit diffraction factor})^2 \cdot (N\text{-source interference factor})^2$$

In the case of the N-source interference problem treated in Chapter 2, Section 4,

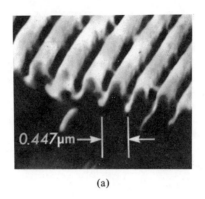

0.447μm →

(a)

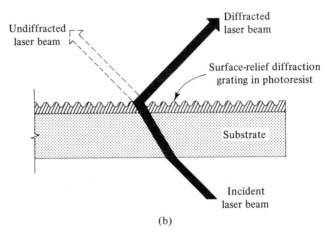

Undiffracted laser beam

Diffracted laser beam

Surface-relief diffraction grating in photoresist

Substrate

Incident laser beam

(b)

Fig. 3.20
(a) Photomicrograph of a surface-relief grating. (b) Enlarged cross section schematic of a transmission phase grating. (Courtesy of Holotek, Ltd., Rochester, N.Y.)

we have already seen that removal of the periodicity requirement reduces the interference factor to a constant; that is, no interference fringes are produced. If we allow the N slits in our diffraction grating to have a random distribution, we will obtain

$$uu^* = u_0^2(\text{single-slit diffraction factor})^2(\text{const})N \qquad (3.44)$$

that is, essentially just N times the single-slit result. For example, the $N = 4$ slit pattern shown in Figure 3.16c would degenerate into the single-slit pattern of Figure 3.5 upon randomizing the distances between the four slits.

If we randomize the array of the round apertures shown in Figure 3.21a (see Figure 3.21c), the interference factor becomes a constant and we observe, in Figure 3.21d, only the superposition of the intensities of the diffraction pattern due to round holes. Thus, the pattern in Figure 3.21d resembles that made from N times the intensity of a round aperture (Figure 3.11).

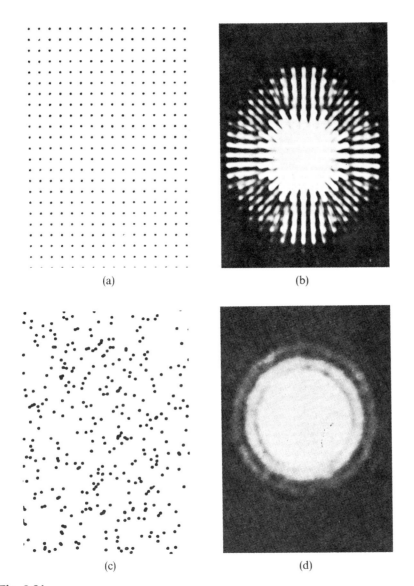

Fig. 3.21
Effect of periodicity on diffraction patterns produced by a two-dimensional array of circular apertures. (a) Regular array of circular apertures. (b) Pattern produced by (a). (c) Random array of circular apertures. (d) Pattern produced by (c). (From Hecht and Zajac, *Optics*, pp. 362–363.)

In Figure 3.22, we see the transition from a diffraction pattern that is a regular array to a pattern that is a random array, but now demonstrated with rectangular apertures. Here, the two-dimensional array is visualized as an extension of a linear diffraction grating, now, however, with rectangular holes. The diffraction pattern resembles two crossed N-slit patterns (see Figure 3.22b). If we randomize this array (Figure 3.22c), it is reduced to the diffraction pattern of a

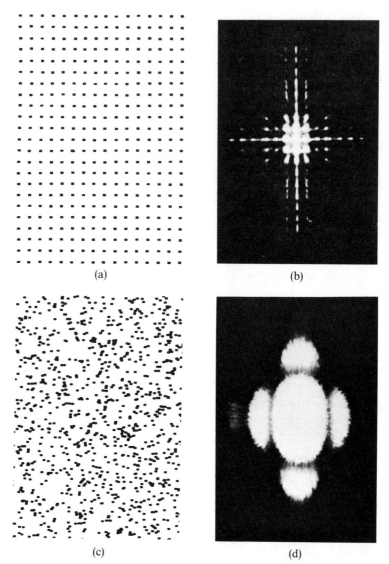

Fig. 3.22
Effect of periodicity on diffraction patterns produced by a two-dimensional array of rectangular apertures. (a) Regular array of rectangular apertures. (b) Pattern produced by (a). (c) Random array of rectangular apertures. (d) Pattern produced by (c). (From Hecht and Zajac, *Optics*, pp. 362–363.)

rectangular aperture (see Figure 3.8) produced by the superposition of the intensities of the N random rectangles.

All these observations can be explained when we consider the phase relation between interfering waves. In the case of the regular array (e.g., a grating), the phases of waves are added up for waves originating from one opening as well as for the waves from different openings. The final amplitude is obtained by superposition, and the intensity is calculated. In the randomized case, we determine

3. Diffraction

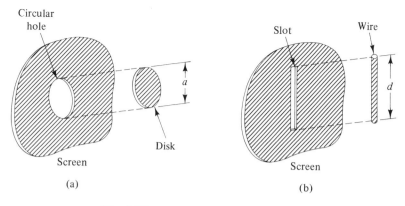

Circular hole

Disk

Screen

(a)

Slot   Wire

Screen

(b)

Fig. 3.23
Two examples of complementary screens.

the amplitude of the waves coming from one opening. The phase contribution from all other openings cancel because of the random arrangement; thus, the intensity is obtained from the amplitude of one opening. The final intensity is obtained by adding the same intensity for all openings.

J. Babinet's Principle

Complementary screens are defined as two screens such that one is transparent where the other is opaque. An example of complementary screens is an opaque screen with an aperture of radius a and an opaque disk with radius a, both with collinear axes (Figure 3.23a). If the two complementary screens are placed together in the same plane, no aperture will exist.

It is interesting to calculate the amplitude of the diffraction pattern generated by two complementary screens. In Figure 3.24, we see a laser beam illuminating a screen A and the diffraction pattern that is observed. The same setup is used for the complementary screen B. We can calculate the amplitude of the waves reaching the observation screen at point P from

$$u(P) = C \int_{opening} e^{-i\gamma y}\, dy \tag{3.45}$$

where the integration must be done over the open illuminated area. The integral is essentially the same as was given in equation 3.6.

Using screen A, we have

$$u_A(P) = \int_{\substack{open\ area \\ of\ A}} e^{-i\gamma y}\, dy \tag{3.46}$$

and for screen B, we get

$$u_B(P) = \int_{\substack{open\ area \\ of\ B}} e^{-i\gamma y}\, dy \tag{3.47}$$

If we add the integrals, we get

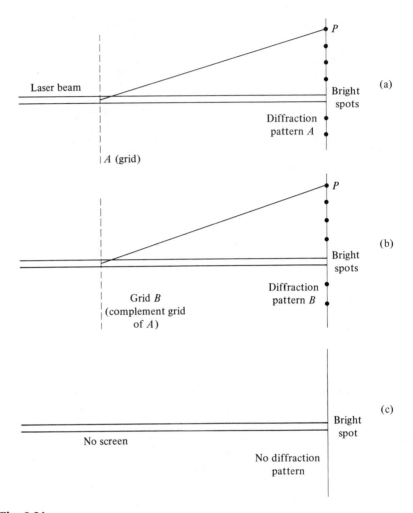

Fig. 3.24
(a) Diffraction at screen A. (b) Diffraction at complementary screen B. (c) No screens, no diffraction.

$$u_A(P) + u_B(P) = \int_{\substack{\text{open area} \\ \text{of } A \text{ and } B}} e^{-i\gamma y}\, dy \qquad (3.48)$$

We must integrate over the fully illuminated cross section; that is, no screen is in the path of the beam. We observe a bright spot on the observation screen as a result of direct illumination, but no diffraction pattern appears. The amplitude $u_A(P) + u_B(P)$ is zero for any point P on the observation screen outside the bright spot. We can express this as

$$u_A(P) + u_B(P) = 0 \quad \text{or} \quad u_A(P) = -u_B(P) \qquad (3.49)$$

Equation 3.49 describes **Babinet's principle**, which states that the *amplitude of the diffraction pattern* of two complementary screens has the same magnitude but

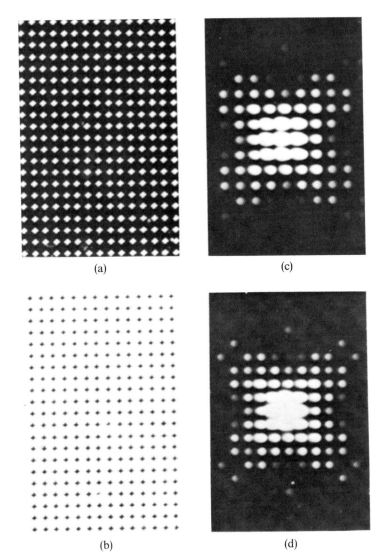

Fig. 3.25
(a, b) Regular arrays of apertures and complementary obstacles in the form of rounded plus signs. (c, d) Diffraction pattern. (From Hecht and Zajac, *Optics*, p. 389.)

opposite phase. We have derived this result for all points P outside of the central bright spot. Applications of Babinet's principle will be discussed in Chapter 11.

From equation 3.49, the intensities are

$$|u_A|^2 = |u_B|^2$$

Figure 3.25 shows the diffraction patterns for two complementary screens. The similarities in the patterns are evident.

In the experiment shown in Figure 3.24, the aperture screens were larger than the diameter of the illuminating beam. Figure 3.26 shows the reverse case,

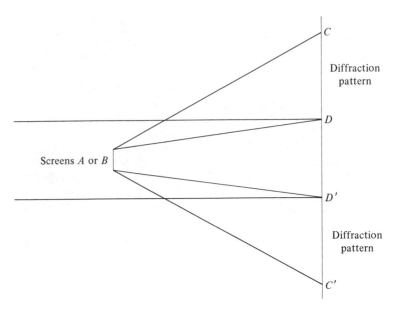

Fig. 3.26
Insertion of complementary screen *A* or *B* results in the same diffraction pattern in the regions *CD* and *D'C'*.

where the diameter of the illuminating beam is larger than the size of the complementary screens. In the "not directly" illuminated areas *CD* and *D'C'*, we again observe the same diffraction pattern. We exclude the directly illuminated area *DD'* from this application of Babinet's principle.

3. FRESNEL DIFFRACTION

A. The Kirchhoff-Fresnel Integral

So far, we have discussed the diffraction of waves under Fraunhofer conditions, that is, when the source and observation screen are very far away from the aperture. We will now drop these restrictions and consider the case where the source and observer are much closer to the aperture. An immediate consequence is that incoming light can no longer be represented by plane waves. We now have wavefronts that are spherical (or possibly cylindrical, for an extended source). A spherical wave emitted from a point source can be represented by

$$u(R) = A \frac{e^{i(kR-\omega t)}}{R} \tag{3.50}$$

where R is the radial distance measured from the source located at $R = 0$. The amplitude A/R decreases as R increases.

We assume that $u(R)$ is traveling in the direction of R, away from the source. Moreover, we omit the time-dependent factor for the reasons given earlier.

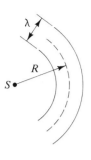

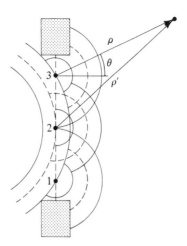

——— Maxima intensities

– – – Minima intensities

Fig. 3.27
Spherical wave diffracted at an aperture. The source is at S, and the wavefront is spherical with radius R. In the plane of the aperture, the wavefront of the incident wave is not coincident with the plane of the aperture.

When a spherical wavefront strikes a plane aperture, the wavefront and the plane of the aperture are not coincident (Figure 3.27). The curvature of the wavefront causes different sections of it to intersect the aperture plane at different times. According to Huygens' principle, secondary spherical waves are created in the plane of the aperture.

In Figure 3.27, we see that the secondary spherical waves 1 and 3 are created at the maxima of the incoming wave, while wave 2 is created at the minimum. We describe this situation by letting the secondary waves be represented by $(1/\rho)e^{ik\rho}$, where ρ is the distance to the observation point. Since the primary wave $(A/R)e^{ikR}$ excites the secondary waves $(1/\rho)e^{ik\rho}$, we can treat the primary wave as the driving wave. We can see a similarity to the case of an externally driven damped harmonic oscillator, and we express the phase difference between the two waves as $e^{i\alpha}$. We can now write the amplitude for each generated wave as

$$A\frac{e^{ikR}}{R}e^{i\alpha}\frac{e^{ik\rho}}{\rho} \tag{3.51}$$

Huygens' principle also tells us that secondary waves only propagate in the forward direction. We can account for this on a temporary *ad hoc* basis by summing the contributions of only those waves traveling toward the screen. From experiments, we observe that the intensity of the forward-diffracted waves is a maximum in the direction normal to the aperture. Destructive interference of the secondary Huygens wavelets causes reduced intensity in directions parallel to the aperture. To account for this directionality effect, we include a $\cos\theta$

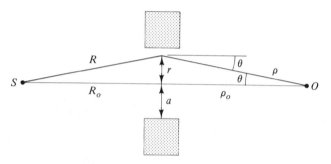

Fig. 3.28
Geometry for Fresnel diffraction of a circular aperture of radius a, with the source and screen equally distant from the plane of the opening.

factor as a directional modulator of the amplitude. The amplitude of the waves diffracted by the aperture can now be represented by

$$u = \int_{\text{aperture}} \frac{A}{R} e^{ikR} e^{i\alpha} \frac{1}{\rho} e^{ik\rho} \cos\theta \, d\sigma \tag{3.52}$$

where $d\sigma$ is the surface element of the area of the aperture. Equation 3.52, known as the **Kirchhoff-Fresnel integral**, provides us with the prescription for summing up the secondary Huygens wavelets from an arbitrarily shaped aperture with due consideration for phase and directional effects. It has been developed here on a heuristic basis using plausible arguments. A rigorous derivation of equation 3.52 using the wave equation and appropriate boundary conditions is presented in the Solutions Manual. An outline is given at the end of this chapter.

B. The Circular Opening

We will apply equation 3.52 to two important cases, the circular opening and the circular obstacle, or stop. For both calculations,* we will employ a simplified coordinate system and assume that the source and screen are at equal distances from the plane of the aperture. Also, the source point S and observation point O in Figure 3.28 are symmetrically positioned as a result of the circular geometry. Our assumptions result in the following relations:

$$R_0 = \rho_0, \quad R = \rho, \quad \cos\theta = \frac{\rho_0}{\rho}, \quad \text{and} \quad d\sigma = 2\pi r \, dr$$

For the circular aperture, equation 3.52 becomes

$$u = A \int_{\text{aperture}} e^{i\alpha} \frac{e^{ik(R+\rho)}}{R\rho} \left(\frac{\rho_0}{\rho}\right)(2\pi r \, dr) \tag{3.53}$$

Since only the region $0 \leq r \leq a$ contributes to secondary Huygens wavelets, the

---

* Following A. Sommerfeld, *Vorlesungen über theoretische Physik*, Band IV. Dieterich'sche Verlagsbuchhandlung, Wiesbaden, 1950.

limits on the integral are immediately known. Also, from $r^2 + \rho_0^2 = \rho^2$, we get $r\,dr = \rho\,d\rho$, and choosing ρ as the variable of integration, equation 3.53 becomes

$$u = Ae^{i\alpha}2\pi\rho_0 \int_{\rho_0}^{\sqrt{a^2+\rho_0^2}} \frac{e^{2ik\rho}}{\rho^2}\,d\rho \tag{3.54}$$

where the phase difference has been assumed constant. Integration by parts with $u = 1/\rho^2$, $dv = e^{i2k\rho}\,d\rho$, and $v = e^{2ik\rho}/2ik$ gives us

$$\int_{\rho_0}^{\sqrt{a^2+\rho_0^2}} \frac{e^{2ik\rho}}{\rho^2}\,d\rho = \left(\frac{1}{\rho^2}\right)\left(\frac{e^{2ik\rho}}{2ik}\right)\bigg|_{\rho_0}^{\sqrt{a^2+\rho_0^2}} + \frac{1}{ik}\int_{\rho_0}^{\sqrt{a^2+\rho_0^2}} \frac{e^{2ik\rho}}{\rho^3}\,d\rho \tag{3.55}$$

The integral on the right-hand side of equation 3.55 is similar to the one on the left-hand side, that is, the original integral, except that the power of ρ in the denominator is increased by 1. A second integration by parts will yield a ρ^{-4} term in the $-\int v\,du$ term. Repeated integrations will decrease the power of this term so that the mth partial integration results in a $\rho^{-(m+2)}$ term in $-\int v\,du$. Since ρ is "large," we can neglect the ρ^{-3} term relative to the ρ^{-2} term and consequently all higher-order terms as well. Therefore, only the constant term on the right-hand side of equation 3.55 needs to be retained, and we get

$$u = \frac{2\pi Ae^{i\alpha}\rho_0}{2ik}\left(\frac{e^{2ik\sqrt{a^2+\rho_0^2}}}{a^2+\rho_0^2} - \frac{e^{2ik\rho_0}}{\rho_0^2}\right) \tag{3.56}$$

Since $\rho_0 \gg a$, we can neglect the a^2 term in the denominator of the first term, but not in the more sensitive exponential term. There we can use an expansion of the square root

$$\sqrt{a^2 + \rho_0^2} \simeq \rho_0\left(1 + \frac{a^2}{2\rho_0^2}\right)$$

With these approximations, equation 3.56 becomes

$$u = \frac{2\pi Ae^{i\alpha}\rho_0}{2ik}\frac{e^{2ik\rho_0}}{\rho_0^2}\left[e^{ik(a^2/\rho_0)} - 1\right] \tag{3.57}$$

To obtain the intensity, we use the identity

$$(e^{ix} - 1)(e^{-ix} - 1) = 4\sin^2\frac{x}{2}$$

where $x = ka^2/\rho_0$.

Using $u_0^2 = (A/\rho_0)^2$, we finally get

$$uu^* = u_0^2\lambda^2\sin^2\frac{ka^2}{2\rho_0} \tag{3.58}$$

for the intensity. This result shows that the positions of the minima and maxima on the symmetry axis of the system depend on the ratio a^2/ρ_0. Figure 3.29 shows the development of these minima and maxima for increasing values of a. It is important to recognize that our result, equation 3.58, is only applicable to points

3. Fresnel Diffraction

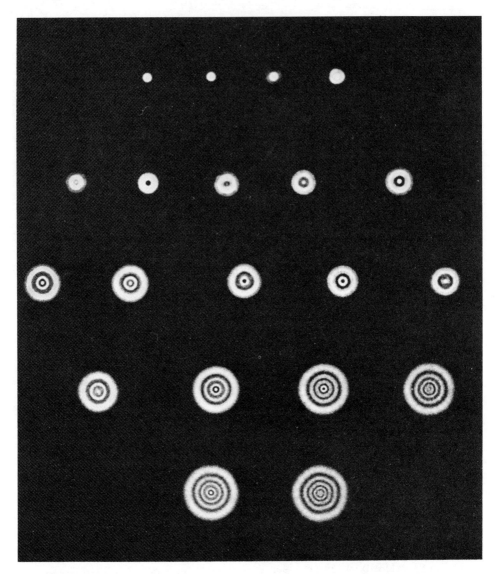

Fig. 3.29
Diffraction pattern for a circular aperture. The radius a is increasing from figure to figure. Black and white centers are observed according to changes in a (see equation 3.58). (From Hecht and Zajac, *Optics*, p. 373.)

on the symmetry axis. An expression for values of $|u^2|$ at arbitrary (nonaxial) points can be developed, but only with considerable increase in complexity.

C. The Circular Stop

The coordinate system, assumptions, and starting point for the case of the circular stop are the same as for the circular aperture. The only difference is

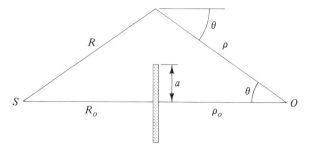

Fig. 3.30
Geometry for Fresnel diffraction produced by a circular stop of radius a, with the source S and observation screen O at an equal distance from the stop.

the region of space considered for the integration: In the present case, $r > a$. We are treating the complement of the circular aperture (Figure 3.28).

We must now evaluate

$$u = A e^{i\alpha} 2\pi \rho_0 \int_{\sqrt{a^2+\rho_0^2}}^{\infty} \frac{e^{2ik\rho}}{\rho^2} d\rho \tag{3.59}$$

In analogy with equation 3.54, integration by parts yields

$$\int_{\sqrt{a^2+\rho_0^2}}^{\infty} \frac{e^{2ik\rho}}{\rho^2} d\rho = \left(\frac{1}{\rho^2}\right)\left(\frac{e^{2ik\rho}}{2ik}\right)\bigg|_{\sqrt{a^2+\rho_0^2}}^{\infty} + \frac{1}{ik}\int_{\sqrt{a^2+\rho_0^2}}^{\infty} \frac{e^{2ik\rho}}{\rho^3} d\rho \tag{3.60}$$

Again neglecting the last integral, we get

$$u = \frac{2\pi A e^{i\alpha} \rho_0}{2ik}\left(-\frac{e^{2ik\sqrt{a^2+\rho_0^2}}}{a^2+\rho_0^2}\right) \tag{3.61}$$

Introducing u_0^2 as

$$u_0^2 = \frac{A^2}{\rho_0^2+a^2} \simeq \frac{A^2}{\rho_0^2}$$

we obtain for the intensity

$$uu^* = u_0^2 \frac{\lambda^2}{4} \tag{3.62}$$

Equation 3.62 tells us that at $\rho_0 = 0$, at the symmetry axis, we have always a maximum, independent of the size of the stop. This is shown in Figure 3.31.

Experimentally, we can observe the **Poisson spot** (or **Arago spot**) shown in Figure 3.32.* An interesting bit of history is associated with this spot. In 1818, when Augustin Fresnel submitted his wave theory of diffraction, Simeon Poisson attempted to contradict Fresnel's explanation. Poisson deduced from the wave

---

*For experimental details, see W. A. Hilton and R. Sandquist, *American Journal of Physics* **36**, 4, ix (1968).

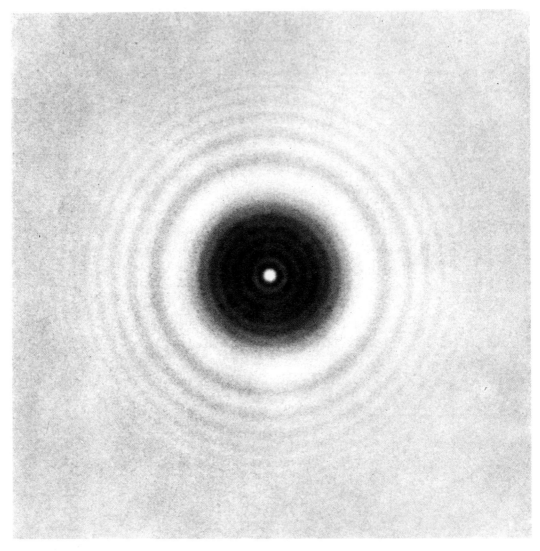

Fig. 3.31
Photograph of the diffraction pattern produced by a round stop. In the middle appears the Poisson spot. (From Cagnet, Françon, and Thrierr, *Atlas of Optical Phenomena*.)

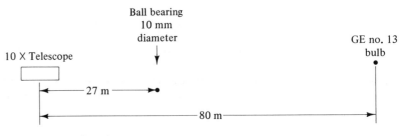

Fig. 3.32
Parameters for the observation of the Poisson spot.

3. Diffraction

theory that a maximum should be observed on the axis behind any circular stop. He believed that this conclusion would disprove the wave theory. D. F. J. Arago performed the experiment and found the spot; instead of disproving the wave theory, Poisson's prediction reinforced it.

We can combine the results of the previous two discussions concerning the circular aperture and circular stop and write the intensity from equations 3.58 and 3.62 as

$$uu^* = u_0^2 \lambda^2 \begin{cases} \sin^2 \dfrac{ka^2}{2\rho_0} & \text{for circular aperture} \\[2ex] \dfrac{1}{4} & \text{for circular stop} \end{cases} \tag{3.63}$$

As we will see in Section 5C, Babinet's principle is also valid for Fresnel diffraction. For the circular aperture and circular stop, the analytical expressions just derived cannot be used for "checking it out," since we employed some approximations in the derivation.

4. FRESNEL VERSUS FRAUNHOFER DIFFRACTIONS

A. Approximations (Discussed for a Slit)

We will now compare the assumptions leading to the two types of diffraction phenomena. To do this, we consider an "asymmetric" geometry in which the source is very far from the aperture plane but the observation screen is at a much closer distance; in other words, we are using a Fresnel diffraction observation. Our starting point is equation 3.52. The factor Ae^{ikR}/R can be taken outside the integral, the phase factor $e^{i\alpha}$ can be assumed to be a constant, and we assume that $\cos\theta = \rho_0/\rho$. Equation 3.52 is then reduced to

$$u = C_1 \int_{\text{aperture}} \frac{e^{ik\rho}}{\rho^2} d\sigma \tag{3.64}$$

where C_1 contains all constant factors.

In Figure 3.33, we consider three arbitrary, equally spaced secondary waves that are excited by the incident plane waves. With respect to the central ray 2, the optical paths of rays 1 and 3 are $\rho_1 = \rho_2 + \varepsilon$ and $\rho_3 = \rho_2 - \varepsilon$, respectively. The integrands of equation 3.64 will then be of the form

$$\frac{e^{ik(\rho_2+\varepsilon)}}{(\rho_2+\varepsilon)^2} \quad \text{and} \quad \frac{e^{ik(\rho_2-\varepsilon)}}{(\rho_2-\varepsilon)^2} \tag{3.65}$$

respectively. Now ε is a quantity of the order of λ (approximately 10^{-6} m), whereas ρ_2 is approximately $10^6\lambda$, or of the order of 1 meter. This means that we can neglect the ε in the denominator of each such term, but *not* in the exponential, where the phase angle $2\pi(\rho_2 \pm \varepsilon)/\lambda$ is measured in multiples of λ.

If we now consider the aperture to be subdivided into an arbitrary number of rays, instead of just three, then the factor ρ^{-2} in the integrand will be, at worst, a slowly varying function of Y. We can therefore consider ρ^{-2} approximately

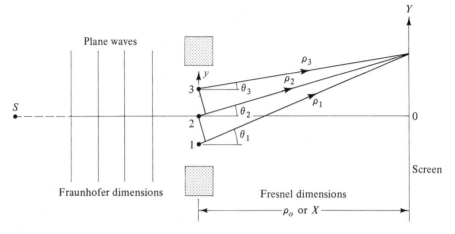

Plane waves

ρ_3

ρ_2

ρ_1

3 θ_3

S

θ_2

2

1 θ_1

0

Y

Screen

Fraunhofer dimensions

Fresnel dimensions

ρ_o or X

Fig. 3.33
Optical diagram for diffraction at a slit with a mixture of far-field (Fraunhofer) and
near-field (Fresnel) observation. Three centers of Huygens' wavelets are shown.

constant and rewrite equation 3.64 as

$$u = C_2 \int_{\text{aperture}} e^{ik\rho}\, d\sigma$$

where C_2 contains ρ^{-2} as well as the other factors. Expressing ρ in terms of
y, X, and Y (Figure 3.34) and taking $d\sigma = dy$ for the one dimensional opening,

$$u = C_2 \int_{\text{aperture}} e^{i(2\pi/\lambda)\sqrt{X^2+(Y-y)^2}}\, dy \tag{3.66}$$

What distinguishes Fresnel from Fraunhofer conditions rests in the exponential in equation 3.66. We first develop the square root term as

$$\sqrt{X^2 + (Y-y)^2} = X\sqrt{1 + \left(\frac{Y-y}{X}\right)^2} \simeq X\left[1 + \frac{1}{2}\left(\frac{Y-y}{X}\right)^2\right]$$
$$\simeq X + \frac{1}{2X}(Y^2 - 2yY + y^2) \tag{3.67}$$

All we have assumed so far is that $|(Y-y)/X| \ll 1$. (That corresponds to our
small-angle approximation.)

B. Conditions for Fraunhofer and Fresnel Observations; the Fresnel Integrals

Fraunhofer condition. If the observation screen is far from the aperture,
then $Y \gg y$ and we can neglect the y^2 term in equation 3.67, giving us

$$u = C_2 \int_{\text{aperture}} \exp\left[i\frac{2\pi}{\lambda}\left(X + \frac{Y^2 - 2Yy}{2X}\right)\right] dy$$

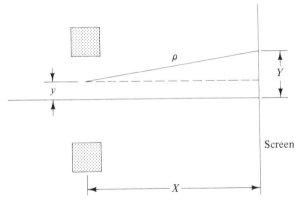

Fig. 3.34
Parameters for discussion of Fraunhofer and Fresnel diffractions.

By factoring the constant terms out of the integral and including them in C_3, we have

$$u = C_3 \int_{\text{aperture}} e^{-i(2\pi/\lambda)(Y/X)y}\,dy \qquad (3.68)$$

This result is equivalent to equations 3.6 and 3.15 and presents Fraunhofer diffraction.

Fresnel conditions. If the screen is not at a great distance from the aperture, we cannot neglect the y^2 term in equation 3.67, but we still have $|(Y-y)/X| \ll 1$. Using equation 3.67, we must evaluate

$$u = \bar{C}_2 \int_{\text{aperture}} \exp\left[i\frac{2\pi}{\lambda}\left(X + \frac{(Y-y)^2}{2X}\right)\right]dy$$

or

$$u = \bar{C}_3 \int_{\text{aperture}} \exp\left[i\left(\frac{2\pi}{\lambda}\right)\left(\frac{(Y-y)^2}{2X}\right)\right]dy \qquad (3.69)$$

where the factor $e^{i(2\pi/\lambda)X}$ is included in $\bar{C}_3$. If we change variables by substituting $\eta^2 = 2(Y-y)^2/\lambda X$ and restrict the discussion to a slit of width d, then $dy = (\text{const})\,d\eta$ and equation 3.69, with all constants included in C_4, becomes

$$u = C_4 \int_{\eta_1}^{\eta_2} e^{i(\pi/2)\eta^2}\,d\eta \qquad (3.70)$$

where

$$\eta = (Y-y)\sqrt{\frac{2}{\lambda X}}, \quad \eta_1 = \left(Y + \frac{d}{2}\right)\sqrt{\frac{2}{\lambda X}}$$

and

$$\eta_2 = \left(Y - \frac{d}{2}\right)\sqrt{\frac{2}{\lambda X}} \qquad (3.71)$$

4. Fresnel Versus Fraunhofer Diffractions   **165**

Using the **Euler identity**, $e^{ix} = \cos x + i \sin x$, we can rewrite this with $C_4 = u_0$

$$u = u_0 \left[\int_{\eta_1}^{\eta_2} \cos\left(\frac{\pi}{2}\eta^2\right) d\eta + i \int_{\eta_1}^{\eta_2} \sin\left(\frac{\pi}{2}\eta^2\right) d\eta \right] \qquad (3.72)$$

Table 3.1
Fresnel integrals

η	$C(\eta)$	$S(\eta)$	η	$C(\eta)$	$S(\eta)$
0.00	0.0000	0.0000	4.50	0.5261	0.4342
0.10	0.1000	0.0005	4.60	0.5673	0.5162
0.20	0.1999	0.0042	4.70	0.4914	0.5672
0.30	0.2994	0.0141	4.80	0.4338	0.4968
0.40	0.3975	0.0334	4.90	0.5002	0.4350
0.50	0.4923	0.0647	5.00	0.5637	0.4992
0.60	0.5811	0.1105	5.05	0.5450	0.5442
0.70	0.6597	0.1721	5.10	0.4998	0.5624
0.80	0.7230	0.2493	5.15	0.4553	0.5427
0.90	0.7648	0.3398	5.20	0.4389	0.4969
1.00	0.7799	0.4383	5.25	0.4610	0.4536
1.10	0.7638	0.5365	5.30	0.5078	0.4405
1.20	0.7154	0.6234	5.35	0.5490	0.4662
1.30	0.6386	0.6863	5.40	0.5573	0.5140
1.40	0.5431	0.7135	5.45	0.5269	0.5519
1.50	0.4453	0.6975	5.50	0.4784	0.5537
1.60	0.3655	0.6389	5.55	0.4456	0.5181
1.70	0.3238	0.5492	5.60	0.4517	0.4700
1.80	0.3336	0.4508	5.65	0.4926	0.4441
1.90	0.3944	0.3734	5.70	0.5385	0.4595
2.00	0.4882	0.3434	5.75	0.5551	0.5049
2.10	0.5815	0.3743	5.80	0.5298	0.5461
2.20	0.6363	0.4557	5.85	0.4819	0.5513
2.30	0.6266	0.5531	5.90	0.4486	0.5163
2.40	0.5550	0.6197	5.95	0.4566	0.4688
2.50	0.4574	0.6192	6.00	0.4995	0.4470
2.60	0.3890	0.5500	6.05	0.5424	0.4689
2.70	0.3925	0.4529	6.10	0.5495	0.5165
2.80	0.4675	0.3915	6.15	0.5146	0.5496
2.90	0.5624	0.4101	6.20	0.4676	0.5398
3.00	0.6058	0.4963	6.25	0.4493	0.4954
3.10	0.5616	0.5818	6.30	0.4760	0.4555
3.20	0.4664	0.5933	6.35	0.5240	0.4560
3.30	0.4058	0.5192	6.40	0.5496	0.4965
3.40	0.4385	0.4296	5.45	0.5292	0.5398
3.50	0.5326	0.4152	6.50	0.4816	0.5454
3.60	0.5880	0.4923	6.55	0.4520	0.5078
3.70	0.5420	0.5750	6.60	0.4690	0.4631
3.80	0.4481	0.5656	6.65	0.5161	0.4549
3.90	0.4223	0.4752	6.70	0.5467	0.4915
4.00	0.4984	0.4204	6.75	0.5302	0.5362
4.10	0.5738	0.4758	6.80	0.4831	0.5436
4.20	0.5418	0.5633	6.85	0.4539	0.5060
4.30	0.4494	0.5540	6.90	0.4732	0.4624
4.40	0.4383	0.4622	6.95	0.5207	0.4591

3. Diffraction

The Fresnel integrals. The integrals in equation 3.72 can be evaluated using the **Fresnel integrals**, which are defined by

$$C(\eta) = \int_0^\eta \cos\left(\frac{\pi\omega^2}{2}\right) d\omega \quad \text{and} \quad S(\eta) = \int_0^\eta \sin\left(\frac{\pi\omega^2}{2}\right) d\omega \qquad (3.73)$$

where $C(-\eta) = -C(\eta)$ and $S(-\eta) = -S(\eta)$. These functions have been studied extensively and their values tabulated (Table 3.1). To obtain the amplitude of the diffraction pattern, we write

$$u = u_0[C(\eta) + iS(\eta)]\Big|_{\eta_1}^{\eta_2} \qquad (3.74)$$

For uu^*, we get

$$uu^* = u_0^2\{[C(\eta_2) - C(\eta_1)] + i[S(\eta_2) - S(\eta_1)]\} \cdot$$
$$\{[C(\eta_2) - C(\eta_1)] - i[S(\eta_2) - S(\eta_1)]\} \qquad (3.75)$$
$$= u_0^2\{[C(\eta_2) - C(\eta_1)]^2 + [S(\eta_2) - S(\eta_1)]^2\}$$

We can obtain the solution from Table 3.1. Figure 3.35 shows the calculated diffraction pattern for a slit of width d, where the observation screen is at distance X and $u_0 = 1$. A comparison of the Fraunhofer and Fresnel diffractions produced by a slit will be considered in problem 9. (Note that the first minimum for Fresnel diffraction is not zero, while the others are approximately zero.)

5. CORNU'S SPIRAL

A. Cornu's Spiral and the Intensity of the Diffraction Pattern

We can also solve diffraction problems by using the **Cornu spiral**. The Cornu spiral is obtained by plotting $S(\eta)$ against $C(\eta)$ and using η as the parameter (Figure 3.36). The intensity of the diffraction pattern can be obtained by giving an interpretation of the expression

$$u^2 = u_0^2\{[C(\eta_2) - C(\eta_1)]^2 + [S(\eta_2) - S(\eta_1)]^2\} \qquad (3.76)$$

that is, the square of the distance between the points with parameters η_2 and η_1 (Figure 3.37). We can plot the intensity of the diffraction pattern by measuring the distance between the parameters η_1 and η_2 and then squaring it (see problem 10). The parameters η_1 and η_2 depend on the geometry of the aperture (through y), the wavelength, and the coordinates of the observation point on the screen.

B. The Fresnel Diffraction at an Edge

Let us calculate the diffraction pattern at the aperture of an edge by using Cornu's spiral. We apply the diffraction integral for a slit (equation 3.70) to a slit with increasing extension (Figure 3.38). From equation 3.70, we have

$$u = C_4 \int_{\eta_1}^{\eta_2} e^{i(\pi/2)\eta^2} d\eta \qquad (3.77)$$

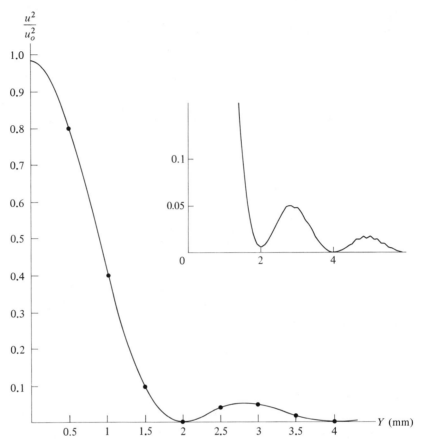

Fig. 3.35
Fresnel diffraction pattern produced by a slit, using equation 3.75 and Table 3.1. The width d is 1 mm, the distance to the observation screen is $X = 4$ m, and $\lambda = 500$ nm (5×10^{-7} m). The insert shows a computer calculation plotted on an enlarged intensity scale. The intensity is not zero at $Y = 2$.

where

$$\eta_1 = Y\sqrt{\frac{2}{\lambda X}} \quad \text{and} \quad \eta_2 = (Y - y_2)\sqrt{\frac{2}{\lambda X}}$$

Because of the large opening of the slit, we have approximately $y_2 \approx \infty$ and $\eta_2 \simeq -\infty$. Using the separation of real and imaginary parts (see equation 3.72) and the definition of Fresnel's integrals, we obtain

$$u = C_4 [C(\eta) + iS(\eta)] \Big|_{\eta_1}^{\eta_2 \simeq -\infty} \tag{3.78}$$

which is similar to equation 3.74. Obtaining the intensity from equation 3.75, we get

$$uu^* = u_0^2\{[C(\eta_2 \simeq -\infty) - C(\eta_1)]^2 + [S(\eta_2 \simeq -\infty) - S(\eta_1)]^2\} \tag{3.79}$$

where u_0^2 is a constant.

168   **3. Diffraction**

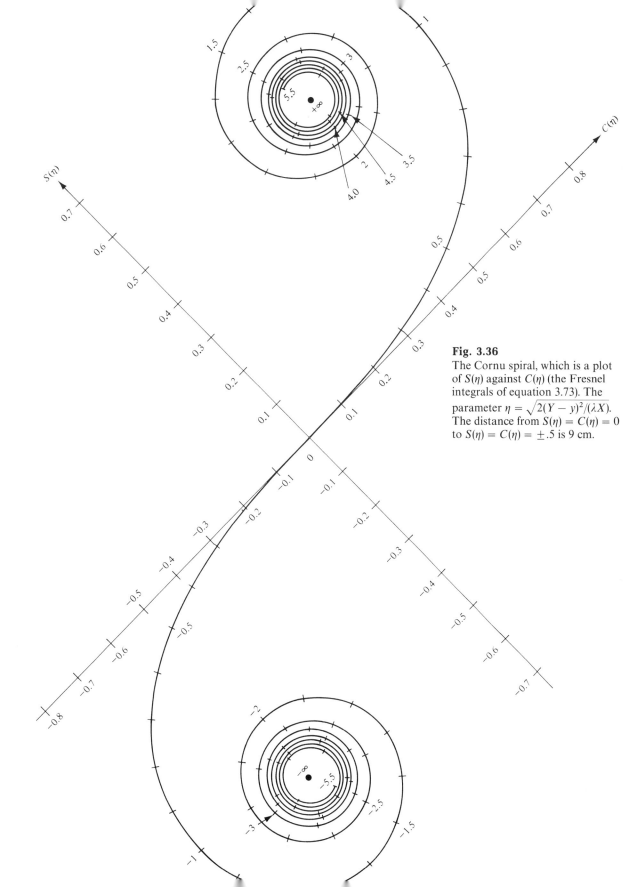

Fig. 3.36
The Cornu spiral, which is a plot of $S(\eta)$ against $C(\eta)$ (the Fresnel integrals of equation 3.73). The parameter $\eta = \sqrt{2(Y - y)^2/(\lambda X)}$. The distance from $S(\eta) = C(\eta) = 0$ to $S(\eta) = C(\eta) = \pm.5$ is 9 cm.

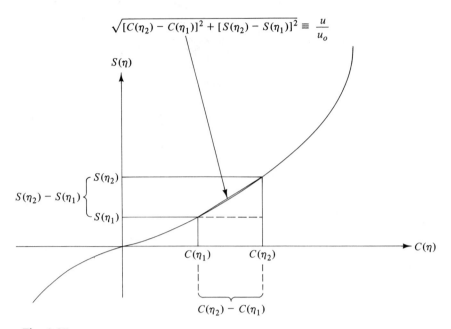

$$\sqrt{[C(\eta_2) - C(\eta_1)]^2 + [S(\eta_2) - S(\eta_1)]^2} \equiv \frac{u}{u_o}$$

Fig. 3.37
Partial plot of the Cornu spiral with indication of the parameters η_1 and η_2. Note that equation 3.76 squares the differences of the C's and S's, as shown on the graph. The distance from $S(\eta) = C(\eta) = 0$ to $S(\eta) = C(\eta) = \pm 5$ is 9 cm.

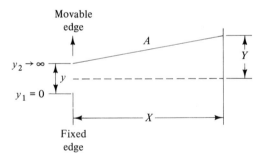

Fig. 3.38
Diagram of a slit with a movable edge, for calculation of the Fresnel diffraction at an edge. The edge is produced by having $y_1 = 0$ and by allowing y_2 approach a large number.

We have assumed before that $|(Y - y)/X| \ll 1$ and therefore y_2 must be chosen accordingly. For large values of $-\eta$, both $S(-\eta)$ and $C(-\eta)$ are close to -0.5, while for $\eta = -\infty$, we have exactly $S(-\eta) = C(-\eta) = -0.5$. Therefore, by extending the integration limit y_2 to infinity, only slightly different values are used in equation 3.79.

We obtain the diffraction pattern by identifying the $\eta_2 = -\infty$ point in the negative plane and the y_1 position in the positive plane formed by $S(\eta)$ and $C(\eta)$

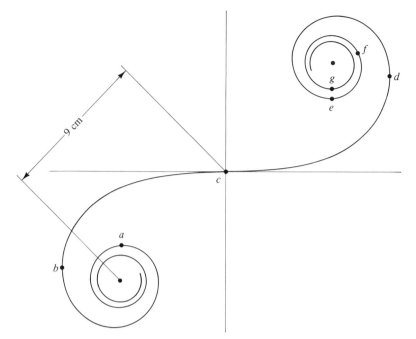

Fig. 3.39
Schematic diagram of a Cornu spiral used for the calculation of the diffraction at an edge. The distance from $\eta = 0$ to $\eta = -\infty$ is 9 cm on Figure 3.36 and may serve for normalization. The points a, b, c, and so on, correspond to values of η_1. These are points in the diffraction pattern of Figure 3.40.

(see Figure 3.36). The η_2 point is fixed, and since $y_1 = 0$, η_1 depends only on Y, the coordinate on the screen. To use Cornu's spiral, we must get the distance from $\eta_2 = -\infty$ to η_1 as a function of Y and square it. This is schematically shown in Figure 3.39.

For example, assume that the Cornu spiral of Figure 3.36 has a distance of 9 cm for $\eta = 0$ to $\eta = -\infty$. The intensity corresponding to this pair of η-values is

$$u_1^2 = u_0^2\{[C(0) - C(-\infty)]^2 + [S(0) - S(-\infty)]^2\}$$
$$= u_0^2(9 \text{ cm})^2 = u_0^2 81 \text{ cm}^2 \tag{3.80}$$

This value may be used for normalization of u^2 for all pairs of $\eta_2 = -\infty$ and η_1 by division:

$$I = \frac{u^2}{u_1^2} = \frac{(\text{distance from } \eta_1 \text{ to } \eta_2 = -\infty \text{ on Cornu spiral})^2}{81 \text{ cm}^2} \tag{3.81}$$

Using $\eta_1 = Y\sqrt{2/\lambda X}$ with $X = 4$ m, and $\lambda = 5 \times 10^{-7}$ m results in $\eta_1 = Y \times 10^3$ m^{-1} or $\eta_1 = Y_{[mm]}$. Using the Cornu spiral of Figure 3.36 and plotting the normalized intensity I for $\eta_1 = -2, -1, 0, +1, \ldots, +4$, we obtain Figure 3.40. Diffraction fringes from an edge are shown in Figure 3.41.

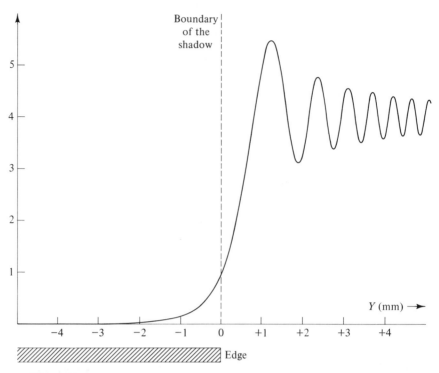

Fig. 3.40
Diffraction pattern at an edge. The intensity is unnormalized and plotted as a function of Y_1 in mm corresponding to η_1 ($X = 4000$ mm; $L = 0.0005$ mm).

Fig. 3.41
The diffraction pattern observed from an edge. (From Cagnet, Françon, and Thrierr, *Atlas of Optical Phenomena*.)

3. Diffraction

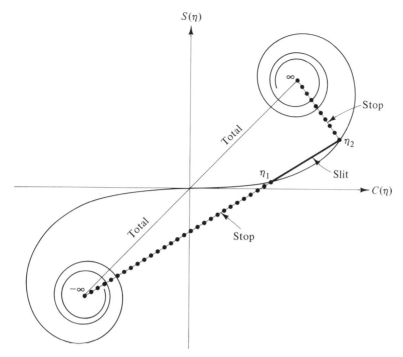

Fig. 3.42
Babinet's principle and Cornu's spiral. The intensity is proportional to the square
of the distance between two points on the spiral. We have the same intensity for
"total" as we have for "stop" plus "slit."

C. Babinet's Principle and the Cornu Spiral

We will consider two complementary apertures, a slit and a long flat stop, and
apply equation 3.76. For the slit, we have

$$u_{\text{slit}} = C \int_{\eta_1}^{\eta_2} e^{i(\pi/2)\eta^2} \, d\eta \tag{3.82}$$

and for the stop, we have

$$u_{\text{stop}} = C \int_{-\infty}^{\eta_1} e^{i(\pi/2)\eta^2} \, d\eta + C \int_{\eta_2}^{\infty} e^{i(\pi/2)\eta^2} \, d\eta \tag{3.83}$$

If no aperture were present, we would have

$$u_{\text{total}} = u_{\text{slit}} + u_{\text{stop}} = C \int_{-\infty}^{+\infty} e^{i(\pi/2)\eta^2} \, d\eta \tag{3.84}$$

where C is a constant.

From equation 3.84, we again have Babinet's principle similar to that discussed in the Fraunhofer case (equation 3.48). The Cornu spiral shows the
connection of the diffraction patterns in an interesting way (Figure 3.42). However, there are no points free of illumination for both apertures. The two diffrac-

tion patterns for slit and stop do not have the same appearance. Also, we should note that when using the Cornu spiral in this way we are dealing with intensities only, and no phase information is involved.

We have discussed the mathematical formulation of Huygens' principle. Secondary waves assumed to be generated in the opening of an aperture and integration over the aperture results in the intensity at the observation screen. We distinguish between two types of diffraction patterns. In the Fraunhofer case, we assume that the source and the observation screen are far away from the aperture, and a relation to the Fourier transformation can be obtained. In the Fresnel case, we do not assume such large distances, and the resulting diffraction pattern is more complicated to obtain through the mathematical process.

We also discussed Babinet's principle, which states that complementary apertures will give the same intensity diffraction pattern (outside the directly illuminated area). However, the amplitude diffraction pattern has a 180° phase difference. This fact will be used in a later chapter when we discuss phase-sensitive imaging.

Maxwell's theory can also describe the diffraction process. However, extremely complicated mathematical methods must be employed for a precise description. Still, an approximate description can be obtained from the wave equation and appropriate boundary conditions. And since only one component of the electrical field is used, this approach is called the scalar theory.

We can derive the Kirchhoff-Fresnel integral used extensively in this chapter. This derivation is outlined in the next section, and a complete calculation is presented in the Solutions Manual. So far we have not discussed how valid the diffraction theory is if we compare the opening with the wavelength. All the theory we discussed is equivalent to the Kirchhoff-Fresnel scalar theory and holds only if $\lambda/a \leq 0.3$.

6. SCALAR DIFFRACTION THEORY AND THE KIRCHHOFF-FRESNEL INTEGRAL

The waves u and v used in the following approach are solutions of the three-dimensional wave equations

$$\nabla^2 u + k^2 u = 0 \quad \text{and} \quad \nabla^2 v + k^2 v = 0$$

Since u and v are scalar functions, the theory developed is called the **scalar theory of diffraction**. (The electromagnetic plane wave of Maxwell's theory has two components; see Chapter 4.) The function u describes the waves incident from the source on the aperture; the function v describes the Huygens wavelets. The result of the integration of these waves over a defined volume in space containing the aperture (Figure 3.43) is the Kirchhoff-Fresnel integral

$$u_p = \frac{ik}{4\pi} \int_S \frac{e^{ikR}}{R} [\cos(\mathbf{R}, \mathbf{n}) - \cos(\boldsymbol{\rho}, \mathbf{n})] \frac{e^{ik\rho}}{\rho} d\rho \qquad (3.85)$$

The calculation is presented in detail as a sample calculation in the Solutions Manual. The amplitude u_p is the diffracted amplitude at the observation point

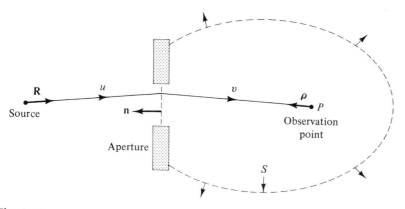

Fig. 3.43
Source point, observation point, and integration surface for integration of equation 3.85. Note the direction of **n**.

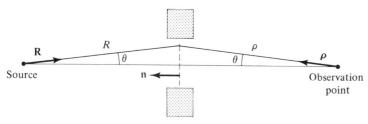

Fig. 3.44
Coordinate system for calculating the diffraction on a circular opening (see also Figure 3.28).

P. The integration is done over the closed surface S containing the opening of the aperture (see Figure 3.43). The unit vectors **R**, **ρ**, and **n** and the distances R and ρ are also indicated.

We will now apply the Kirchhoff-Fresnel integral to the round aperture and show that the integral of equation 3.53 results. From Figure 3.44, we see that

$$\cos(\mathbf{R}, \mathbf{n}) = \cos(\pi - \theta) = -\cos\theta \qquad (3.86)$$

and

$$\cos(\mathbf{\rho}, \mathbf{n}) = \cos\theta \qquad (3.87)$$

Then, using $\rho = R$ as before, equation 3.85 becomes

$$u_p = \frac{i}{2\lambda} \int \frac{e^{ik(2\rho)}}{\rho^2} (-2\cos\theta) \, d\sigma \qquad (3.88)$$

or

$$u_p = \frac{1}{i\lambda} \int \frac{e^{2ik\rho}}{\rho^2} \cos\theta \, d\sigma$$

Comparing these results with equations 3.52 and 3.53, we see that the $\cos \theta$ factor arises naturally from equation 3.85 and that the phase factor is $e^{i\alpha} = 1/i$. The imaginary factor $1/i$ means that the diffracted waves undergo a phase change of 90° with respect to the incident wave (see the forced oscillator in Chapter 7).

The factor $\cos(\mathbf{R}, \mathbf{n}) - \cos(\boldsymbol{\rho}, \mathbf{n})$ is called the **obliquity factor**, and in addition to supplying the basis for the $\cos \theta$ factor introduced in equation 3.52, it shows us why the light waves are not diffracted into the backward direction. That is, for "back diffraction," we have $\cos(\mathbf{R}, \mathbf{n}) = \cos(\boldsymbol{\rho}, \mathbf{n})$ and the obliquity factor vanishes.

In summary, the Kirchhoff-Fresnel formula accounts for all essential factors that were intuitively introduced into the diffraction formula, equation 3.52. It explains the absence of backward diffracted waves, allows for calculation of the amplitude at nonaxial observation points, and predicts a phase change of $\pi/2$. These characteristics are not derivable from Huygens' principle.

Problems

1. *Slit aperture.* For the small angle approximation, the intensity was shown to be given by

$$u^2 = u_0^2 \left(\frac{\sin\left[(2\pi/\lambda)(d/2)(Y/X)\right]}{(2\pi/\lambda)(d/2)(Y/X)} \right)^2$$

If we set $u_0^2 = 1$ at $Y = 0$, then the central maximum has unit intensity. Use the following numerical values: $d = 1$ mm, $\lambda = 500$ nm (visible region), and $X = 4$ m. Plot u^2 versus Y. Choose a scale such that $|u|^2 = 0.01$ corresponds to 1 cm and $Y = 1$ mm corresponds to 2 cm. Then, for graph paper with a grid of 10 lines per centimeter, the central maxima will be off scale, that is, off the upper end of the graph paper. (We are interested in the location of the side maxima.)
Part. ans. Minima at $Y = 2, 4, 6, 8, \ldots$ mm

2. *Secondary maxima of the diffraction pattern of a slit.* We found the relation $b = \tan b$ [with $b = \pi(d/\lambda)(Y/X)$] for the location of the secondary maxima of a slit. Now using $\lambda = 500$ nm, $d = 1$ mm, $X = 4$ m, and Y in millimeters, graphically determine the values of the secondary maxima. Compare with plot of problem 1.

3. *Double slit: $d \ll a$.* Consider the double-slit experiment and assume that $d \ll a$.

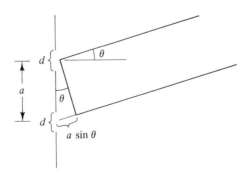

For the intensity of the diffraction pattern, we obtained

$$u^2 = 4 \frac{\sin^2\left[(2\pi/\lambda)\sin\theta(d/2)\right]}{\left[(2\pi/\lambda)\sin\theta(d/2)\right]^2} \cdot \cos^2\left[(2\pi/\lambda)\sin\theta(a/2)\right]$$

The first factor describes the diffraction on the single slit; the second factor describes the interference effect. Derive the conditions for maxima (approximate) and minima that are equal to what we derived by elementary derivation.
Part. ans.

Peaks of single slit (approximately):

$$2d\sin\theta = (2m+1)\lambda$$

Peaks of interference term:

$$a\sin\theta = m\lambda$$

4. *Double slit: a = 2d.* Consider the double-slit experiment and assume that $a = 2d$.

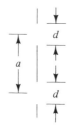

The intensity is given as

$$u^2 = 4 \frac{\sin^2\left[(2\pi/\lambda)\sin\theta(d/2)\right]}{\left[(2\pi/\lambda)\sin\theta(d/2)\right]^2} \cdot \cos^2\left[(2\pi/\lambda)\sin\theta(a/2)\right]$$

a. Make a sketch of the diffraction factor, labeling the minima.
b. Make a sketch of the interference factor, labeling the minima and maxima.
c. Make a sketch of u^2 and indicate the "missing order."

5. *Diffraction on a double slit for two different wavelengths.* Consider the pattern of a double slit for two different wavelengths (not in the apparatus at the same time). For the intensity, we have

$$u^2 = u_0^2 \frac{\sin^2\gamma(d/2)}{\left[\gamma(d/2)\right]^2} \cos^2\gamma(a/2)$$

Set $u_0^2 = 1$, $a = 5d$, $d = 2.5\ \mu m$, and $\gamma = (2\pi/\lambda)\sin\theta$. Plot the pattern as a function of θ in degrees.
a. For blue light, use $\lambda = 0.4\ \mu m$ and find the maxima and minima for the two factors.

$$\frac{\sin^2\gamma(d/2)}{\left[\gamma(d/2)\right]^2}$$

0 Max:
1 Min:
1 Max:
2 Min:
2 Max:

$$\cos^2 \gamma \frac{a}{2} \qquad \begin{array}{l} \text{0 Max:} \\ \text{1 Min:} \\ \text{1 Max:} \\ \text{2 Min:} \\ \text{2 Max:} \end{array}$$

   b. Repeat the calculation for red light, where $\lambda = 0.7 \ \mu m$.

   c. Draw the two patterns on the same figure, one with a blue pencil, one with a red pencil.

6. *Diffraction on a plane grating: oblique incident light.* Consider the diffraction on a plane grating with periodicity constant a and slit opening d. Assume that the incident light makes the angle ϕ with respect to the normal of the grating. The diffraction angle θ is as shown in the figure.

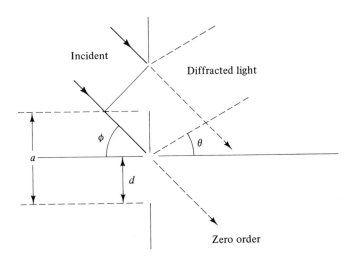

   a. Derive the intensity formula for the diffracted light. The result must be the same as in equation 3.36 for $\phi = 0$.

   b. Show that for the choice of $\sin \theta + \sin \phi = m(\lambda/a)$, the interference factor is equal to N^2.

   c. Calculate the ratio of the intensities for $m = 1$ with respect to $m = 0$ for the case where $a = 8d$.

7. *Circular aperture.* An important practical example is the diffraction pattern due to a circular aperture. For example, a circular lens interrupting a wave produces diffraction, and the resolving power of the lens is limited by diffraction considerations. This application provides us with a chance to use Bessel functions. Using the following coordinate system and, in analogy with the rectangular aperture, consider the integral

$$u = u_0 \int_{\substack{\text{over} \\ \text{opening}}} \exp\left\{ i 2\pi \left[\left(\frac{y}{\lambda}\right)\left(\frac{Y}{X}\right) + \left(\frac{z}{\lambda}\right)\left(\frac{Z}{X}\right) \right] \right\} dy \, dz$$

Since rectangular coordinates are not appropriate for circular geometry, the polar coordinates shown in the figure will be used instead.

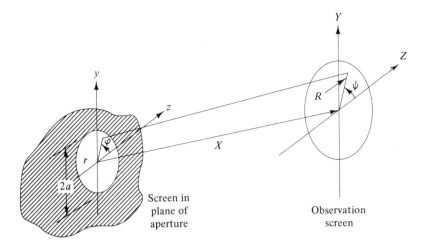

Screen in plane of aperture

Observation screen

These variables satisfy the following transformation equations:

$$y = r \sin \phi \qquad Y = R \sin \psi$$
$$z = r \cos \phi \qquad Z = R \cos \psi$$

where $0 \le r \le a$, $0 \le \phi \le 2\pi$, and $0 \le \psi \le 2\pi$.

a. Write $[(y/\lambda)(Y/X) + (z/\lambda)(Z/X)]$ and the area element $dy\,dz$ in coordinates r, R, ϕ, and ψ and obtain the following integral:

$$u = u_0 \int_{-\pi}^{+\pi} \int_0^a e^{i2\pi(r/\lambda)(R/X)\cos(\psi-\phi)} r\,dr\,d\phi$$

b. Rewrite the integral by using the definition of the Bessel function of order zero

$$J_0(q) = \frac{1}{2\pi} \int_{-\pi}^{+\pi} e^{iq\cos(\phi-\psi)}\,d\phi$$

and arrive at

$$u = u_0 \left(\frac{\lambda X}{2\pi R}\right)^2 2\pi \int_{q=0}^{q=2\pi(aR/\lambda X)} J_0(q)q\,dq$$

Provided is a graph of J_0 and J_1 as a function of $q(1 \to 18)$.

c. To carry out the integration, use an identity for Bessel functions relating the zeroth order $J_0(q)$ and the first order $J_1(q)$:

$$\int_0^{q'} J_0(q)q\,dq = q'J_1(q')$$

($'$ is not a derivative). Arrive at the following:

$$u(R) = u_0 2\pi a^2 \frac{J_1[2\pi(a/\lambda)(R/X)]}{2\pi(a/\lambda)(R/X)}$$

d. Find the value of $u^2(R)$ for $R = 0$. To do this, consider the series expansion of $J_1(q)$

$$J_1(q) = \frac{q}{2}\left[1 - \frac{1}{2}\left(\frac{q}{2}\right)^2 + \frac{1}{12}\left(\frac{q}{2}\right)^4 + \cdots\right]$$

If you set $R = 0$, you will get 0/0. Application of L'Hôpital's rule will give you the desired result:

$$\lim_{R \to 0} |u|^2 = 1$$

e. Find the zeros of $J_1(q)$ from the graph (or tables you have access to), and give the position of the zero of the function in R/X.

	R/X
1 zero $q =$	
2 zero $q =$	
3 zero $q =$	

The first minimum is located at $R/X = 0.61\lambda/a$ or, if we introduce the diameter d of the aperture, at $R/X = 1.22\lambda/d$. This relationship is used in the discussion of the resolution of optical instruments. The secondary maxima, as was found in the single-slit case, are almost, but not exactly, located between the minima points. The central maximum, containing more than 80 percent of the total diffracted energy, is called the Airy disk and corresponds to the region $|R/X| < 3.83\lambda/2\pi a$. This central maximum is surrounded by concentric rings, which reduce in intensity as R/X increases.

Intuitively, you might expect the circular aperture diffraction pattern to resemble that of the slit pattern when the latter is rotated about its central maximum. Indeed, this is observed. Quantitatively, however, the intensities of the secondary maxima of the circular aperture pattern are less than their single-slit counterparts. Also, the positions of the minima are further from the central maxima for the circular aperture than for the slit.

f. Make a plot of the diffraction patterns of a slit, using (see problem 1)

$$u^2 = u_0^2 \left(\frac{\sin(2\pi/\lambda)(d/2)(Y/X)}{2\pi/\lambda(d/2)(Y/X)} \right)^2$$

For $u_2^0 = 1$, $d = 1$ mm, $\lambda = 500$ nm, and $X = 4$ m. Plot u^2 versus Y, and choose a scale such that $|u|^2 = 0.01$ corresponds to 1 cm on the graph paper and $Y = 1$ mm corresponds to 2 cm on the paper. (For typical graph paper, the central maximum will be off scale, but we are not interested in that position.)

g. On the same graph paper, plot the circular aperture, using

$$u^2 = 4 \left(\frac{J_1(2\pi)(a/\lambda)(R/X)}{(2\pi)(a/\lambda)(R/X)} \right)^2$$

Take $2a = d = 1$ mm, $\lambda = 500$ nm, and $X = 4$ m. Read off your graph the first two minima for both cases and the first secondary maxima (between them) and intensities, and fill in the Table.

	First minima	Intensity	Next maxima	Intensity	Second minima	Intensity
Slit						
Round aperture						

3. Diffraction

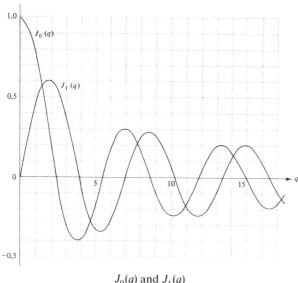

$J_0(q)$ and $J_1(q)$

8. *Wave equation.* Show that $u(r) = e^{ikr}/r$ is a solution of the wave equation $\nabla^2 u + k^2 u = 0$.

9. *Fresnel diffraction, slit, table.* Use Table 3.1 and plot the Fresnel diffraction pattern of a slit. Use $d = 1$ mm, $\lambda = 500$ nm, $X = 4$ m, and Y in millimeters from $Y = 0$ to 5 mm in steps of 0.5 mm.

10. *Fresnel diffraction, slit, Cornu spiral.* Plot the Fresnel diffraction pattern of a single slit by using the Cornu spiral as provided in the Solutions Manual. Use the following values: $d = 1$ mm, $\lambda = 500$ nm, $X = 4$ m, and Y from 0 to 4.5 mm in increments of 0.25 mm. Use

$$u = C \int_{\eta_1}^{\eta_2} e^{i(\pi/2)\eta^2} \, d\eta$$

and equations 3.71. The distance on the Cornu spiral between the parameters η_1 and η_2 is

$$\frac{u}{u_0} = \sqrt{[C(\eta_2) - C(\eta_1)]^2 + [S(\eta_2) - S(\eta_1)]^2} = \sqrt{\frac{I}{I_0}}$$

To normalize the intensity to unity at $Y = 0$, find the values of η_1 and η_2 corresponding to $Y = 0$. Fill in the table for η_1 and η_2. Note that the diffraction intensity is never 0.

For the plot, use graph paper in centimeter divisions and scale u^2 and Y as indicated below (see also Section 5B):

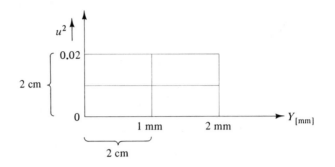

Y mm	η_1	η_2	Distance	Normalized Intensity
0	0.5	−0.5	18.0	1.000
0.5	1.0	0	16.6	0.853

11. *Fresnel diffraction, half-plane.* Plot the diffraction pattern obtained in the illuminated half-plane. Use the values $\lambda = 500$ nm and $X = 4$ m, and locate the half-plane such that the projection of the edge onto the screen is at $Y = 0$. Normalize the pattern to unity for $Y \to \infty$ (in the illuminated area; see Section 5B). Choose values of Y from $+2$ mm to -5 mm in increments of 1 mm. Compare with the Cornu spiral to check if all minima and maxima are found.

12. *Sample calculation.* The Kirchhoff-Fresnel integral. (See the Solutions Manual.)

Maxwell's Theory $\quad$ 4

1. INTRODUCTION

In the preceding three chapters, we were able to explain a considerable number of optical phenomena with just a few laws. In Chapter 1, we used only the laws of reflection and refraction. In Chapter 2, we used the superposition of scalar harmonic waves. And in Chapter 3, we used Huygens' principle and the superposition of Huygens secondary waves. There are, however, other phenomena that cannot be described with what we have learned so far.

For example, consider the case of light waves reflected successively by two glass plates, arranged as shown in Figure 4.1. The two angles α and β at the two plates are equal and have a specific value. No light is observed after the second reflection. The refractive index of the glass plates determines the values of these angles, which are called **Brewster's angles**. We can not explain this phenomenon with what we have discussed so far about reflectivity of light on materials. We need a more elaborate theory to describe this phenomenon.

In this chapter, we will derive Fresnel's formulas describing quantitatively the reflection and transmission of light at an interface between two dielectrics. Fresnel derived these formulas from a mechanical wave theory long before Maxwell's theory was known. The same formulas also follow from **Maxwell's electromagnetic theory**.

For our purposes here, we will use Maxwell's theory, which is the accepted theory for the description of optical phenomena. We can derive from Maxwell's theory what we have used so far for the description of optical phenomena: the laws of reflection and refraction and the superposition principle of waves. The Kirchhoff-Fresnel integral is a mathematical formulation of Huygens' principle and is obtained as an approximation. But Maxwell's theory allows us to treat the diffraction problems exactly, though at the price of considerable mathematical involvement. We will see that Maxwell's theory is the umbrella theory for all optical phenomena, except for certain quantum phenomena. We will use it for the derivation of Fresnel's formulas by applying certain properties of the electromagnetic theory as it is presented in introductory physics textbooks.

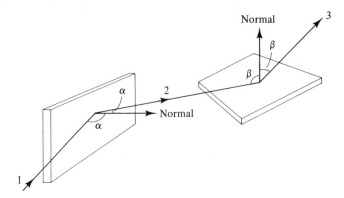

Fig. 4.1
Reflection of light by two glass plates arranged in such a way that α and β are Brewster angles. No light is reflected. The plane of incidence and reflection of waves 1 and 2 is perpendicular to the one formed by waves 2 and 3.

2. ELECTROMAGNETIC WAVES

A. The Wave Equation and Waves in Free Space

In Chapter 2, we used harmonic waves depending on time and on one space coordinate for the description of various interference phenomena. We mentioned the fact that the amplitude of the wave may be associated with one component of the electric field vector **E**. In this respect, we dealt only with scalar waves. We now associate the amplitude with the Y component of **E** and have

$$E_Y = E_{Y_0} e^{i(kX - \omega t)} \tag{4.1}$$

as shown in Figure 4.2 (X direction of propagation). The E_Y component is the solution of a differential equation in X, t, which may be derived from Maxwell's equations. To obtain the differential equation, we take the second derivative of E_Y with respect to t and X, giving

$$\frac{\partial^2 E_Y}{\partial t^2} = E_{Y_0}(-i\omega)^2 e^{i(kX - \omega t)} = -\omega^2 E_Y$$

$$\frac{\partial^2 E_Y}{\partial X^2} = E_{Y_0}(ik)^2 e^{i(kX - \omega t)} = -k^2 E_Y$$

We combine the two equations to get

$$\frac{\partial^2 E_Y}{\partial t^2} = -\omega^2 E_Y = -\omega^2 \left(\frac{1}{-k^2} \right) \left(\frac{\partial^2 E_Y}{\partial X^2} \right)$$

or

$$\frac{\partial^2 E_Y}{\partial t^2} = \frac{\omega^2}{k^2} \frac{\partial^2 E_Y}{\partial X^2}$$

This is a special case of the wave equation that will be derived from Maxwell's equations in Chapter 5. It describes the wavelike propagation of the E_Y com-

4. Maxwell's Theory

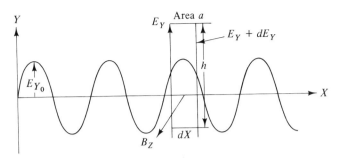

Fig. 4.2
E_Y and B_Z are the components of the electric field and the magnetic induction vector, respectively, for the application of Faraday's law.

ponent in the X direction. In satisfying the wave equation with the plane wave solution (equation 4.1), we have the relation

$$\frac{\omega}{k} = \frac{2\pi v}{2\pi/\lambda} = \lambda v = c \tag{4.3}$$

where c is the velocity of light in vacuum.

We can similarly associate the amplitude with the Z component of $\mathbf{E}$ and have

$$E_Z = E_{Z_0} e^{i(kX - \omega t)}$$

The wave equation then becomes

$$\frac{\partial^2 E_Z}{\partial t^2} = c^2 \frac{\partial^2 E_Z}{\partial X^2} \tag{4.4}$$

What has been done for the E_Y and E_Z components may not be repeated for the E_X component (in free space). From Maxwell's theory (through div $\mathbf{E} = 0$), it follows that we have only transverse waves and $E_X = 0$ for waves propagating in the X direction. For such a wave, the field vector $\mathbf{E}$ may have any direction in the Y-Z plane. If only the E_Y or E_Z component is present, we speak of Y or Z **polarization** (Figure 4.3).

What we will now discuss should not be confused with the notion of simultaneously having both E_Y and E_Z components in a wave propagating in the X direction. We will, for the moment, consider only the E_Y component. We know from **Faraday's law of electromagnetic induction** that there is a connection between the electrical field vector $\mathbf{E}$ and the magnetic induction vector $\mathbf{B}$, which we may write for the E_Y and B_Z components (see Figure 4.2) as

$$\oint \mathbf{E}_Y \cdot \mathbf{d}\ell = -\frac{\partial}{\partial t}(B_Z a) \tag{4.5}$$

where a is the area of the surface with boundary ℓ around which the line integral is extended. $\mathbf{E}_Y \cdot \mathbf{d}\ell$ results from the dot product $(\mathbf{E} \cdot \mathbf{d}\ell)$, where only the Y component of $\mathbf{E}$ has been considered. If we take the area indicated in Figure 4.2, we have

2. Electromagnetic Waves   **185**

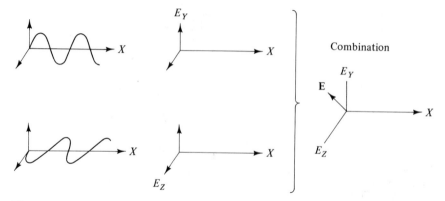

Fig. 4.3
Mutually perpendicular polarization of E_Y and E_Z components of the electromagnetic field vector.

$$-E_Y h + \text{top} + (E_Y + dE_Y)h - \text{bottom} = -\frac{\partial}{\partial t} B_Z h \partial X \qquad (4.6)$$

where the "top" and "bottom" parts of the line integral are zero because E_Y has no component along the line (direction) of integration. Thus, we have

$$\frac{\partial E_Y}{\partial X} = -\frac{\partial B_Z}{\partial t} \qquad (4.7)$$

We introduce the plane wave solution (equation 4.1) and have

$$ik E_Y = -\frac{\partial B_Z}{\partial t} \qquad (4.8)$$

or

$$B_Z = -ik \int E_{Y_0} e^{i(kX - \omega t)} \partial t$$

which yields

$$B_Z = \frac{ik}{i\omega} E_Y = \frac{1}{c} E_Y \qquad (4.9)$$

We see that the magnitude of the E_Y component is connected to the magnitude of the B_Z component. There exists for B_Z a wave equation as well:

$$\frac{\partial^2 B_Z}{\partial X^2} = \frac{1}{c^2}\left(\frac{\partial^2 B_Z}{\partial t^2}\right) \qquad (4.10)$$

The electromagnetic wave can be depicted as shown in Figure 4.4. For the E_Z component we have a similar result, as shown in Figure 4.5.

We see that the vector (cross) product of **E** and **B** always points in the direction of propagation.

4. Maxwell's Theory

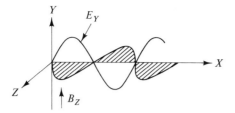

Fig. 4.4
The E_Y and B_Z waves propagating in the X direction. The cross product $(E_Y) \times (B_Z)$ points into the direction of propagation.

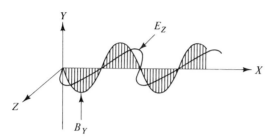

Fig. 4.5
The E_Z and B_Y waves propagating in the X direction. The cross product $(E_Z) \times (B_Y)$ points into the direction of propagation.

The vector

$$\mathbf{S} = \frac{1}{\mu_0}(\mathbf{E} \times \mathbf{B})$$

is called the **Poynting vector**, and it presents the power per area in watts per square meter (Figure 4.6).

We can rewrite **S** for our one-component example as

$$\mathbf{S} = \frac{1}{\mu_0} E_Y B_Z = \frac{1}{\mu_0 c} E_Y^2 \tag{4.11}$$

This is the flow per unit area. If we wish to express it with the maximum value of E_{Y_m}, we introduce the factor $\frac{1}{2}$ because of time averaging:

$$\overline{\mathbf{S}} = \frac{1}{2\mu_0 c} E_{Y_m}^2 \tag{4.12}$$

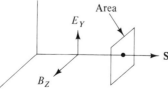

Fig. 4.6
The Poynting vector $\mathbf{S} = 1/\mu_0(\mathbf{E} \times \mathbf{B})$ points into the direction of propagation.

The averaging process is similar to the one discussed in problem 3 of Chapter 2. Our assumption in Chapter 2 that the intensity is proportional to the square of the amplitude is well justified.

B. Waves in a Dielectric Medium

From the wave equation for a vacuum,

$$\frac{\partial^2 E_Y}{\partial X^2} = \frac{1}{c^2}\left(\frac{\partial^2 E_Y}{\partial t^2}\right) \tag{4.13}$$

we get the equation for a dielectric medium,

$$\frac{\partial^2 E_Y}{\partial X^2} = \frac{1}{v^2}\left(\frac{\partial^2 E_Y}{\partial t^2}\right) \tag{4.14}$$

The wave now moves with speed v, and we know that the ratio $c/v = n$ is the refractive index. A similar result holds for the E_Z component.

Also, for the Poynting vector (one component) in the dielectric medium, we have

$$\bar{S} = \frac{1}{2\mu_0 v} E_{Y_m}^2 \tag{4.15}$$

We are now interested in describing the dielectric medium by making the connection between the refractive index n and the dielectric constant used in electromagnetic theory for the characterization of the dielectric medium. Recall the relation between the displacement vector **D** and the electric field **E**:

$$\mathbf{D} = \varepsilon\mathbf{E} = K\varepsilon_0\mathbf{E} = (1 + \chi)\varepsilon_0\mathbf{E}$$

where ε is permittivity of the medium; ε_0 is the permittivity of free space; $K = \varepsilon/\varepsilon_0$, which is the dielectric constant; and χ is the electric susceptibility.

The wave equation in free space can be written (for the Y component only, with propagation in the X direction)

$$\frac{\partial^2 E_Y}{\partial X^2} = \varepsilon_0\mu_0\frac{\partial^2 E_Y}{\partial t^2}; \tag{4.16}$$

For the dielectric medium, it is

$$\frac{\partial^2 E_Y}{\partial X^2} = \varepsilon\mu\frac{\partial^2 E_Y}{\partial t^2}.$$

Here

$$c^2 = \frac{1}{\varepsilon_0\mu_0} \quad \text{and} \quad v^2 = \frac{1}{\varepsilon\mu} \tag{4.17}$$

where μ is the permeability of the medium and μ_0 is the permeability of free space. Since in most dielectrics we have $\mu \simeq \mu_0$ (to about 1 part in 10^5), the desired

4. **Maxwell's Theory**

connection is obtained as

$$n = \frac{c}{v} = \sqrt{\frac{\varepsilon}{\varepsilon_0}} = \sqrt{K} \qquad (4.18)$$

Using these relations, we can rewrite the Poynting vector (one component only) with

$$v = \frac{1}{\sqrt{\varepsilon \mu_0}} \qquad (4.19)$$

as

$$\bar{S} = \frac{1}{2\mu_0 v} E_{Y_m}^2 = \frac{1}{2} \varepsilon v E_{Y_m}^2 \qquad (4.20)$$

and for vacuum as

$$\bar{S} = \tfrac{1}{2} \varepsilon_0 c E_{Y_m}^2 \qquad (4.21)$$

For the connection of B_Z and E_Y in the dielectric medium, we replace c by v in equation 4.9 and have

$$B_Z = \frac{1}{v} E_Y \quad \text{or} \quad B_Z = \frac{n}{c} E_Y \qquad (4.22)$$

3. FRESNEL'S EQUATIONS

A. Boundary Conditions

We will now discuss the transmission and reflection of a transverse electromagnetic wave at an interface of two dielectric media. This is related to the refraction problem discussed in Chapter 1. Using the electromagnetic theory, we will obtain not only the laws of refraction and reflection but also a quantitative result with respect to how much light is reflected and refracted at certain angles. To do this, we must take into account simultaneously the incident, transmitted, and reflected waves. What values can we assume for these three waves in the plane of the interface where they meet? This question is of interest, since we have waves that are dependent on time and space coordinates. At a certain value of the space coordinate and a certain value of the time coordinate, the amplitude can have any value between A and $-A$ if A is the maximum amplitude. Maxwell's theory tells us that the tangential components of the electric field vector must be the same on both sides of the interface. The same is required for the tangential components of the magnetic field vector. We express this by saying that

$$\mathbf{E}\text{(tang) and } \mathbf{B}\text{(tang) must be continuous} \qquad (4.23)$$

where (tang) indicates that the component of the vector parallel to the plane of the interface must be taken.

We consider the electromagnetic wave

$$E_s = A_s e^{i(k_1 x - \omega_1 t)} \qquad (4.24)$$

incident at an air-dielectric interface as shown in Figure 4.7. We use x, x', x'' as

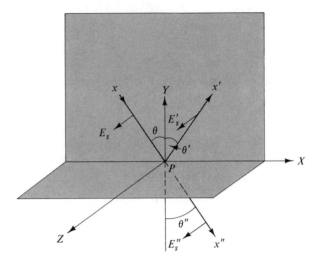

Fig. 4.7
Incident, reflected, and refracted waves at a dielectric interface. The **E** vectors are perpendicular to the plane of incidence (X-Y plane).

the coordinates along the incident, reflected, and transmitted waves, respectively. The electrical vector is oriented parallel to the Z direction, that is, perpendicular to the plane of incidence. According to common notation, we call it E_s. The reflected wave is expressed as

$$E_s' = A_s' e^{i(k_1 x' - \omega_1 t)} \tag{4.25}$$

and the transmitted wave is expressed as

$$E_s'' = A_s'' e^{i(k_2 x'' - \omega_2 t)} \tag{4.26}$$

We have distinguished between the k and ω values inside and outside of the boundary. Application of the boundary condition "E(tang) must be continuous at the interface" results in

$$A_s e^{i(k_1 x - \omega_1 t)} + A_s' e^{i(k_1 x' - \omega_1 t)} = A_s'' e^{i(k_2 x'' - \omega_2 t)} \tag{4.27}$$

On the left side of equation 4.27 is the superposition of the fields outside; this must be equal to the field inside the dielectric (on the right side of the equation).

Note that the boundary condition

$$\text{field (outside)} = \text{field (inside)} \tag{4.28}$$

is drastically different from energy conservation, where we have

$$S_{\text{incident}} = S_{\text{reflected}} + S_{\text{transmitted}} \tag{4.29}$$

Equation 4.27 should hold at all times; that is, we must have

$$\omega_1 = \omega_2$$

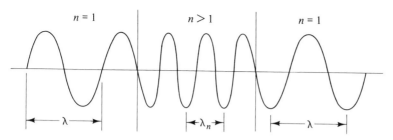

Fig. 4.8
Change in wavelength as the wave enters and leaves a dielectric medium.

This means that the frequency of the wave inside and outside is the same. But since the speed is different inside and outside, the wavelength must change (Figure 4.8). Since $\lambda v = c$ in a vacuum, we have $\lambda_n v = v$ in the dielectric. For λ_n, with $v/\lambda_n = c/\lambda$, we have

$$\lambda_n = \frac{v}{c} \lambda = \frac{\lambda}{n} \tag{4.30}$$

We can now write equation 4.27 as

$$A_s e^{ik_1 x} + A'_s e^{ik_1 x'} = A''_s e^{ik_2 x''} \tag{4.31}$$

with $k_1 = 2\pi/\lambda_1$ and $k_2 = 2\pi/\lambda_2$. To fulfill the boundary conditions, we first choose the space coordinates x' and x'' so that they are zero at P. We can also choose the origin of x in such a way that the factor $e^{ik_1 x} = 1$ at point P. Then, at point P, equation 4.31 gives us

$$A_s + A'_s = A''_s \tag{4.32}$$

B. Energy Conservation and Fresnel's Formulas for the s and p Cases

We have already distinguished between boundary conditions and energy conservation. In our case, energy conservation is equivalent to power conservation. Since the Poynting vector **S** has the dimension power/area, we must consider **S** · area. The area on the interface common to incident, reflected, and transmitted power is Q (Figure 4.9). The projections of Q that are normal to the direction of power flow for the incident, reflected, and transmitted power are $q = Q \cos \theta$, $q' = Q \cos \theta'$, and $q'' = Q \cos \theta''$, respectively.

Now consider the wave traveling from a medium characterized by ε_1 and v_1 into a medium characterized by ε_2 and v_2. For the conservation of power, equation 4.20 gives us

$$\varepsilon_1 v_1 A_s^2 Q \cos \theta = \varepsilon_1 v_1 A_s'^2 Q \cos \theta' + \varepsilon_2 v_2 A_s''^2 Q \cos \theta'' \tag{4.33}$$

Using

$$\theta = \theta' \quad \text{and defining} \quad \alpha = \frac{\varepsilon_2 v_2}{\varepsilon_1 v_1} \frac{\cos \theta''}{\cos \theta} \tag{4.34}$$

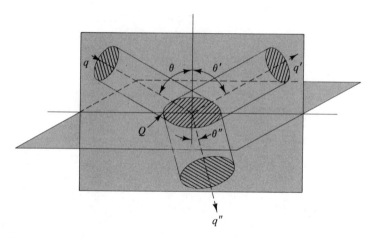

Fig. 4.9
Areas q, q', and q'' of the power flow of incident, reflected, and refracted light. Q is the common area on the interface.

we can write

$$A_s^2 = A_s'^2 + \alpha A_s''^2 \qquad (4.35)$$

We know from our earlier discussion that

$$\frac{v_2}{v_1} = \frac{n_1}{n_2}$$

and that

$$\frac{\varepsilon_2}{\varepsilon_1} = \frac{n_2^2}{n_1^2}$$

Therefore, we have

$$\alpha = \frac{n_2^2}{n_1^2}\left(\frac{n_1}{n_2}\right)\left(\frac{\cos \theta''}{\cos \theta}\right) = \frac{n_2}{n_1}\left(\frac{\cos \theta''}{\cos \theta}\right) \qquad (4.36)$$

Fresnel's formulas for the s case. For the calculation of the amplitude reflection and transmission coefficients,

$$r_s = \frac{A_s'}{A_s} \quad \text{and} \quad t_s = \frac{A_s''}{A_s}$$

respectively, we have from equation 4.32

$$1 + \frac{A_s'}{A_s} = \frac{A_s''}{A_s}$$

and from equation 4.35

$$\left(1 - \frac{A_s'}{A_s}\right)\left(1 + \frac{A_s'}{A_s}\right) = 1 - \frac{A_s'^2}{A_s^2} = \alpha \frac{A_s''^2}{A_s^2}$$

4. Maxwell's Theory

then

$$1 + r_s = t_s$$
$$1 - r_s = \alpha t_s \tag{4.37}$$

or

$$r_s - t_s = -1$$
$$r_s + \alpha t_s = 1 \tag{4.38}$$

We now solve this system of linear equations by using determinants:

$$r_s = \frac{\begin{vmatrix} -1 & -1 \\ 1 & \alpha \end{vmatrix}}{\begin{vmatrix} 1 & -1 \\ 1 & \alpha \end{vmatrix}} = \frac{1-\alpha}{1+\alpha} \qquad t_s = \frac{\begin{vmatrix} 1 & -1 \\ 1 & 1 \end{vmatrix}}{\begin{vmatrix} 1 & -1 \\ 1 & \alpha \end{vmatrix}} = \frac{2}{1+\alpha} \tag{4.39}$$

Introducing

$$\alpha = \frac{n_2}{n_1}\left(\frac{\cos \theta''}{\cos \theta}\right)$$

and rearranging, we obtain

$$r_s = \frac{n_1 \cos \theta - n_2 \cos \theta''}{n_1 \cos \theta + n_2 \cos \theta''} \qquad t_s = \frac{n_1 2 \cos \theta}{n_1 \cos \theta + n_2 \cos \theta''} \tag{4.40}$$

In equation 4.35, we gave the formula for the conservation of energy. We can now read off the reflected intensity as $R_s = r_s^2$ and the transmitted intensity as $T_s = \alpha t_s^2$. We are again representing intensity by an expression that is proportional to the square of the amplitude of the electric field. If we needed to express the intensity in watts per square meter, we would multiply by $\frac{1}{2}\varepsilon v$.

Fresnel's formulas for the p case. Before discussing the result for E_s (where the wave is perpendicular polarized to the plane of incidence), we will treat the case of E_p, where the electric vector is parallel to the plane of incidence. Here we have a problem of orientation of the electrical vectors for the incident, reflected, and transmitted waves. The problem comes from the fact that the phase factor of the reflected light picks up a factor π on reflection. The presence or absence of this factor in the definition will affect the sign of the amplitude of the reflected wave. Moreover, whether we start at one point or another, we will still get into difficulties. This is why different authors use different definitions. Ours is close to the one used by M. Born and E. Wolf, but avoids the difficulty they mention in their book.*

In the previous case, we held that E_s, E_s', and E_s'' are all parallel for any angle θ including normal incidence. In a similar way, we define for the case where the E_p vectors are parallel to the plane of incidence that E_p, E_p', and E_p'' are pointing in similar directions for smaller angles θ, θ', and θ''; for normal incidence they are parallel (Figure 4.10).

The three waves to be considered are

---

*M. Born and E. Wolf, *Principles of Optics*, 2nd ed. (Elmsford, N.Y.: Pergamon Press, 1964), p. 41 (footnote).

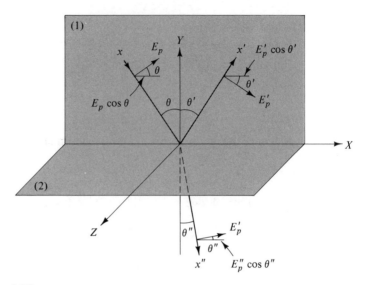

Fig. 4.10
Incident, reflected, and refracted waves at a dielectric interface. The **E** vectors are
parallel to the plane of incidence ($Y = X$ plane).

$$E_p = A_p e^{ik_1 x}, \quad E_p' = A_p' e^{ik_1 x'}, \quad \text{and} \quad E_p'' = A_p'' e^{ik_2 x''} \tag{4.41}$$

The time factor has been omitted. (We have $\omega_1 = \omega_2$ as before.)

To apply the boundary condition, we must compare the components of E_p,
E_p', and E_p'' in the plane of incidence. We again choose x, x', and x'' in the
way discussed before. The components are $A_p \cos \theta$, $A_p' \cos \theta'$, and $A_p'' \cos \theta''$,
measured in the X direction (see Figure 4.10). Then we have

$$A_p \cos \theta + A_p' \cos \theta' = A_p'' \cos \theta'' \tag{4.42}$$

The equation for energy conservation is similar to the one developed for the A_s
component:

$$A_p^2 = A_p'^2 + \alpha A_p''^2 \tag{4.43}$$

We again introduce the amplitude reflection and transmission coefficients

$$r_p = \frac{A_p'}{A_p} \quad \text{and} \quad t_p = \frac{A_p''}{A_p} \tag{4.44}$$

and use the abbreviation

$$\beta = \frac{\cos \theta''}{\cos \theta} \tag{4.45}$$

We then have from equation 4.42, using $\theta = \theta'$,

$$1 + r_p = \beta t_p \tag{4.46}$$

From equation 4.43,

4. **Maxwell's Theory**

$$1 - r_p^2 = \alpha t_p^2 \quad \text{or} \quad (1 - r_p)(1 + r_p) = \alpha t_p^2 \tag{4.47}$$

Using equation 4.46, we get

$$1 - r_p = \frac{\alpha}{\beta} t_p \tag{4.48}$$

We then write equations 4.46 and 4.48 as

$$-r_p + \beta t_p = 1$$
$$r_p + \frac{\alpha}{\beta} t_p = 1 \tag{4.49}$$

Solving this system of linear equations we have

$$r_p = \frac{\begin{vmatrix} 1 & \beta \\ 1 & \dfrac{\alpha}{\beta} \end{vmatrix}}{\begin{vmatrix} -1 & \beta \\ 1 & \dfrac{\alpha}{\beta} \end{vmatrix}} = \frac{\dfrac{\alpha}{\beta} - \beta}{-\dfrac{\alpha}{\beta} - \beta} = -\frac{\dfrac{n_2}{n_1} - \dfrac{\cos\theta''}{\cos\theta}}{\dfrac{n_2}{n_1} + \dfrac{\cos\theta''}{\cos\theta}} \tag{4.50}$$

$$t_p = \frac{\begin{vmatrix} -1 & 1 \\ 1 & 1 \end{vmatrix}}{\begin{vmatrix} -1 & \beta \\ 1 & \dfrac{\alpha}{\beta} \end{vmatrix}} = \frac{-2}{-\dfrac{\alpha}{\beta} - \beta} = \frac{2}{\dfrac{n_2}{n_1} + \dfrac{\cos\theta''}{\cos\theta}} \tag{4.51}$$

or finally

$$r_p = -\frac{n_2 \cos\theta - n_1 \cos\theta''}{n_2 \cos\theta + n_1 \cos\theta''} \quad \text{and} \quad t_p = \frac{n_1 2\cos\theta}{n_2 \cos\theta + n_1 \cos\theta''} \tag{4.52}$$

The reflected intensity is $R_p = r_p^2$, and the transmitted intensity is $T_p = \alpha t_p^2$. For normal incidence, that is, $\theta = \theta'' = 0$, we have

$$r_p = \frac{n_1 - n_2}{n_1 + n_2} \quad \text{and} \quad t_p = \frac{2n_1}{n_1 + n_2}$$

The same results hold for r_s and t_s (equation 4.40): at normal incidence, the two cases are indistinguishable. The intensity of unpolarized light is equal to the sum of the intensities of the two polarized components.

C. Fresnel's Formulas as Functions of the Angle of Incidence

Equations 4.40 and 4.52 relate the fraction of incident light reflected or transmitted as a function of (i) the angle of incidence θ, (ii) the angle of refraction θ'', and (iii) the indices of refraction. We can eliminate the angle of refraction by using the law of refraction $n_1 \sin\theta = n_2 \sin\theta''$. Then for Fresnel's formulas we have

$$r_s = r_{E_\perp} = \frac{\cos\theta - (n_2/n_1)\sqrt{1 - [(n_1/n_2)\sin\theta]^2}}{\cos\theta + (n_2/n_1)\sqrt{1 - [(n_1/n_2)\sin\theta]^2}} \qquad (4.53)$$

$$t_s = t_{E_\perp} = \frac{2\cos\theta}{\cos\theta + (n_2/n_1)\sqrt{1 - [(n_1/n_2)\sin\theta]^2}} \qquad (4.54)$$

$$r_p = r_{E_\parallel} = \frac{-(n_2/n_1)\cos\theta + \sqrt{1 - [(n_1/n_2)\sin\theta]^2}}{(n_2/n_1)\cos\theta + \sqrt{1 - [(n_1/n_2)\sin\theta]^2}} \qquad (4.55)$$

$$t_p = t_{E_\parallel} = \frac{2\cos\theta}{(n_2/n_1)\cos\theta + \sqrt{1 - [(n_1/n_2)\sin\theta]^2}} \qquad (4.56)$$

where we have used the subscript $E_\perp$ for s and $E_\parallel$ for p to emphasize the orientation of the **E** vector with respect to the plane of incidence.

Next we will plot reflection and transmission coefficients (equations 4.53–4.56) as functions of the angle of incidence. We will consider the two cases $n_2 > n_1$ and $n_1 > n_2$ separately. To help distinguish between the two cases, we will use the notation $N_2 > n_1$ and $N_1 > n_2$, respectively. That is, N will always represent the larger of the two indices.

D. The Case of $N_2 > n_1$

We will first consider the case where the incident light travels toward a denser medium (e.g., air to glass). In Figure 4.11, we plot the reflection and transmission coefficients depending on the angle of incidence $\theta(0°$ to $90°)$. We also show the squares of the respective quantities, which are proportional to the intensities. However, we should remember that $r^2 + t^2 \neq 1$.

The two extreme cases of $\theta = 0°$ and $\theta = 90°$ are compiled in Table 4.1. Note that while $r_{E_\perp}$ is always negative, $r_{E_\parallel}$ becomes positive at some angle θ_B. At this angle, called Brewster's angle, only light having a component of its electric field vector perpendicular to the plane of incidence is reflected; that is $(r_{E_\perp})^2 \neq 0$, but $(r_{E_\parallel})^2 = 0$. Thus, if light incident upon the dielectric boundary at an angle θ_B contains both $E_\perp$ and $E_\parallel$ components (i.e., unpolarized light), the reflected beam will contain only waves for which the electric field is oscillating perpendicular to the plane of incidence (i.e., polarized light).

The reflection process may be repeated at a second plate. Let us consider the $E_{r_\perp}^{(1)}$ component reflected from the first plate. We can orient the second plate in such a way that the polarization direction of $E_{r_\perp}^{(1)}$ is parallel to the plane of incidence. At the second plate, we have $E_{r_\perp}^{(1)} \rightarrow E_{\text{inc}_\parallel}^{(2)}$ for the incident wave. If the reflection angle is Brewster's angle, then we have no light reflected off the second plate. This is the experiment presented at the beginning of this chapter and illustrated in Figure 4.1.

To find Brewster's angle θ_B, we set $r_{E_\parallel} = 0$ in equation 4.55 (no reflected light in the parallel direction), and we have

$$\frac{-N_2}{n_1}\cos\theta_B + \sqrt{1 - \left(\frac{n_1}{N_2}\sin\theta_B\right)^2} = 0 \qquad (4.57)$$

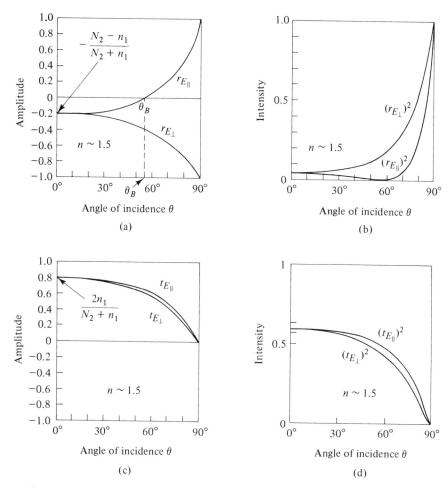

Fig. 4.11
Dielectric interface where $N_2 > n_1$: (a) amplitude reflection coefficient; (b) intensity reflection coefficient; (c) amplitude transmission coefficient; (d) intensity transmission coefficient.

Table 4.1
The reflection and transmission coefficients for $\theta = 0°$ and $\theta = 90°$

$\theta = 0°$	$\theta = 90°$
$r_{E_\perp} = \dfrac{1 - (N_2/n_1)}{1 + (N_2/n_1)} = -\dfrac{N_2 - n_1}{N_2 + n_1}$	$r_{E_\perp} = \dfrac{-(N_2/n_1)\sqrt{1 - (n_1/N_2)^2}}{(N_2/n_1)\sqrt{1 - (n_1/N_2)^2}} = -1$
$r_{E_\parallel} = \dfrac{1 - (N_2/n_1)}{1 + (N_2/n_1)} = -\dfrac{N_2 - n_1}{N_2 + n_1}$	$r_{E_\parallel} = +1$
$t_{E_\perp} = \dfrac{2}{1 + (N_2/n_1)} = \dfrac{2n_1}{N_2 + n_1}$	$t_{E_\perp} = 0$
$t_{E_\parallel} = \dfrac{2}{1 + (N_2/n_1)} = \dfrac{2n_1}{N_2 + n_1}$	$t_{E_\parallel} = 0$

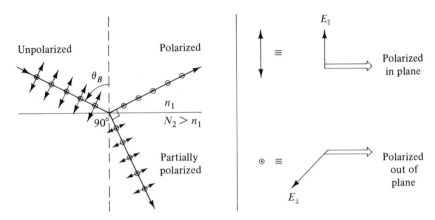

Fig. 4.12
Illustrating polarization by reflection at Brewster's angle.

Letting $a = N_2/n_1$, we can write equation 4.57 as

$$a^2 \cos^2 \theta_B = 1 - \frac{\sin^2 \theta_B}{a^2} = \cos^2 \theta_B + \sin^2 \theta_B - \frac{\sin^2 \theta_B}{a^2} \qquad (4.58)$$

or

$$(a^2 - 1)\cos^2 \theta_B = (1 - a^{-2})\sin^2 \theta_B = \left(\frac{a^2 - 1}{a^2}\right)\sin^2 \theta_B \qquad (4.59)$$

Since $N_2 > n_1$, we have $a > 1$ and can cancel the $(a^2 - 1)$ terms and get

$$\cos^2 \theta_B = \frac{\sin^2 \theta_B}{a^2}$$

or

$$\tan \theta_B = \frac{N_2}{n_1} \qquad (4.60)$$

We immediately see that Brewster's angle is larger than $45°$, since we assumed $N_2 > n_1$. For example, for an air-glass interface ($N_2 = 1.5, n_1 = 1$), we have $\theta_B = \tan^{-1}(1.5) = 56.3°$.

In Figure 4.12, we illustrate the polarization properties of the reflected and transmitted light at Brewster's angle for unpolarized incident light. For angles other than Brewster's angle, the reflected and transmitted light is partially polarized. We can see this in Figure 4.11. At all angles, the reflection and transmission coefficients are different for the parallel and perpendicular cases.

Brewster's angle used for the determination of the refractive index. There are many methods to determine the refractive index. One is the application of measurements made at Brewster's angle. This can be done by measuring the amount of reflected p-polarized light as a function of the angle of incidence. At Brewster's angle we may not find exactly zero intensity because of stray light. Taking the angle of the minimum as Brewster's angle will yield the refractive index n if we use $\tan \theta_B = n$.

4. **Maxwell's Theory**

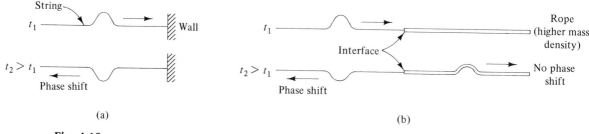

Fig. 4.13
Phase shift upon reflection of a mechanical wave on a string. (a) The corresponding case of a mirror (reflection only). (b) The corresponding case of a dielectric medium (transmission and reflection).

Finally, if we combine the law of refraction with equation 4.60, we obtain

$$\frac{N_2}{n_1} = \tan \theta_B = \frac{\sin \theta_B}{\cos \theta_B} = \frac{(\sin \theta'')(N_2/n_1)}{\cos \theta_B}$$

or

$$\sin \theta'' = \cos \theta_B$$

from which follows

$$\theta'' = 90° - \theta_B$$

or

$$\theta'' + \theta_B = 90° \tag{4.61}$$

That is, the reflected and transmitted beams are perpendicular to each other at Brewster's angle.

Phase shift upon reflection. Negative values for $r_{E_\parallel}$ and $r_{E_\perp}$ mean that upon reflection, the amplitude undergoes a $180°$ phase shift relative to the incident wave. A phase change of $180°$ can be accounted for by a factor $e^{i\pi} = -1$. This phase shift is analogous to that found when a mechanical wave on a string is incident toward a denser medium (e.g., a string attached to a rope). The analogy is illustrated in Figure 4.13 with a single pulse instead of a continuous wave train. Such a phase change was introduced in our discussion on interference by the statement that reflected light picks up a $180°$ (or $\lambda/2$) phase shift if reflected on the optically denser medium. This statement has been used for normal as well as for grazing incidence. From Table 4.1, we find that for $\theta = 0$, $r_{E_\perp}$ and $r_{E_\parallel}$ both have negative values for reflection on a denser medium. But for $\theta = 90°$, $r_{E_\perp}$ remains negative while $r_{E_\parallel}$ is positive. This seems to be in disagreement with the statement that a phase change occurs on the denser medium. The phase shift for the parallel component is shown in Figure 4.14.

We can resolve the discrepancy by taking a closer look at our choice of coordinate system (Figure 4.15). The coefficient $r_{E_\perp}$ is negative for $\theta \to 0°$ as well as for $\theta \to 90°$, and we obtain a phase shift of $180°$ at reflection on the denser medium. The coefficient $r_{E_\parallel}$ is negative for $\theta \to 0°$ but positive for $\theta \to 90°$.

3. Fresnel's Equations 199

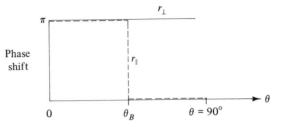

Fig. 4.14
Phase change upon reflection of the $r_\perp$ and $r_\parallel$ components of Figure 4.11.

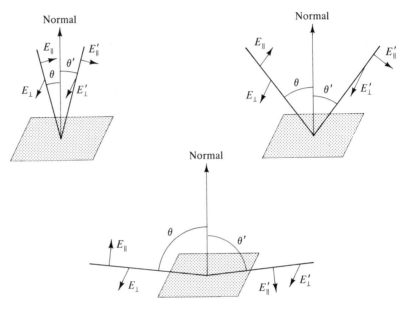

Fig. 4.15
Direction of $E_\parallel$, $E_\parallel'$ and $E_\perp$, $E_\perp'$ for various angles of incidence θ. The angle of reflection θ' is between $0°$ and $90°$.

However, the directions of the incident and reflected vectors for $E_\parallel$ are assumed in our choice of the coordinate system to be parallel for $\theta \to 0°$ and antiparallel for $\theta \to 90°$. Therefore, we again obtain the result that on reflection at the denser medium $r_{E_\parallel}$ shows a phase shift of $180°$. In this coordinate system, the jump from positive to negative values comes at Brewster's angle when $r_{E_\parallel} = 0$. (The intensities are, of course, independent of these definitions and are handled more easily for all different approaches.)

E. The Case of $N_1 > n_2$

We now consider the case where light is incident upon the boundary of a medium having a lower refractive index than the medium in which it was traveling. We might expect the discussion to be much the same as in the case where $N_2 > n_1$.

4. Maxwell's Theory

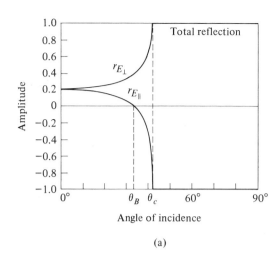 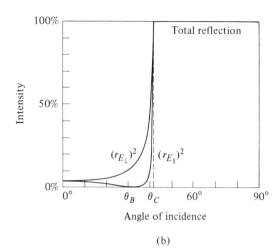

Fig. 4.16
Plot of $r_{E_\perp}$ and $(r_{E_\perp})^2$ and of $r_{E_\parallel}$ and $(r_{E_\parallel})^2$ depending on θ between $0°$ and $90°$.

While this is somewhat true, there are important and unexpected results to be deduced.

From equations 4.53 and 4.55, we get, for normal incidence ($\theta = 0°$),

$$r_{E_\perp} = r_{E_\parallel} = \frac{N_1 - n_2}{N_1 - n_2} \tag{4.62}$$

which tells us that no phase shift occurs upon reflection from a less dense medium. For larger angles θ, we again have Brewster's angle given by $\tan \theta_B = n_2/N_1$. This time, θ_B is less than $45°$. If we then increase the angle of incidence further, we see that at some point $\sqrt{1 - [(N_1/n_2)\sin\theta]^2}$ will vanish and then become imaginary. (This occurs for $\theta \leq 90°$, where $\sin\theta \leq 1$). The value of θ for which the square root vanishes is called the **critical angle** θ_c and is given by

$$\sin \theta_c = \frac{n_2}{N_1} \tag{4.63}$$

In our glass-air interface example ($N_1 = 1.5, n_2 = 1$), we have $\theta_c = 41.8°$. For $0° \leq \theta \leq \theta_c$, the behavior of the wave at the boundary is similar to the case of $N_2 > n_1$. For $\theta_c \leq \theta \leq 90°$, however, both $r_{E_\perp}$ and $r_{E_\parallel}$ have absolute values of unity. This can be easily seen, since both quantities are now complex and we can write for either quantity $r = (a - ib)/(a + ib)$. Then

$$|r| = \sqrt{rr^*} = \sqrt{\frac{a - ib}{a + ib} \cdot \frac{a + ib}{a - ib}} = 1 \tag{4.64}$$

where a and b are real numbers. Thus, we have

$$|r_{E_\perp}| = |r_{E_\parallel}| = 1 \quad \text{for} \quad \theta_c \leq \theta \leq 90° \tag{4.65}$$

Figure 4.16 shows graphs of $r_{E_\perp}$, $r_{E_\parallel}$, $(r_{E_\perp})^2$ and $(r_{E_\parallel})^2$. At angles $\theta_c \leq \theta \leq 90°$ for both directions of polarization, we see that all the incident light is reflected

3. Fresnel's Equations 201

Fig. 4.17
Demonstration of the critical angle and total internal reflection when a laser beam is incident at the boundary of a less dense medium. (From *PSSC Physics*, 2nd edition, 1965; D. C. Heath and Company with Education Development Center, Inc., Newton, Mass.)

back into the denser medium. That is, the boundary acts as a mirror. Figure 4.17 shows a laser beam, coming from a denser medium, incident at various angles onto an interface. The onset of total internal reflection is well demonstrated.

The phase changes for $r_{E_\perp}$ and $r_{E_\parallel}$ are shown in Figure 4.18. In the interval $0° \leq \theta \leq \theta_B$, the phase shift $\delta_\parallel$ for $r_{E_\parallel}$ is zero. At θ_B, it jumps abruptly to π and stays there until $\theta = \theta_c$. Then it decreases to become zero at $\theta = 90°$. The phase shift $\delta_\perp$ for $r_{E_\perp}$ is zero up to θ_c, at which point it decreases, reaching $-\pi$ at $\theta = 90°$. To see this behavior qualitatively, we can discuss two particular cases: $\theta = \theta_c$ and $\theta = 90°$. Since $(N_1/n_2) \sin \theta > 1$, the radical in equations 4.53 and 4.55 is written as

$$i \sqrt{\left(\frac{N_1}{n_2} \sin \theta\right)^2 - 1} = ia$$

Also, letting $b = \cos \theta$ and $c = n_2/N_1$, we have

4. Maxwell's Theory

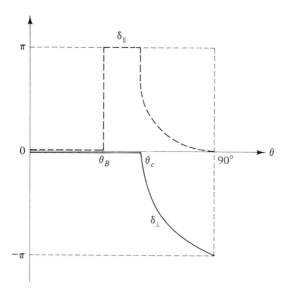

Fig. 4.18

Phase changes $\delta_\parallel$ and $\delta_\perp$ for $r_{E_\parallel}$ and $r_{E_\perp}$, respectively, for the case where $N_1 > n_2$. See sample calculation, problem 11.

$$r_{E_\perp} = \frac{b - ica}{b + ica} \quad \text{and} \quad r_{E_\parallel} = \frac{-cb + ia}{cb + ia} \tag{4.66}$$

At the critical angle $\theta = \theta_c$, $a = 0$ and we get $r_{E_\perp} = +1$, and $r_{E_\parallel} = -1$ (a phase shift of π). At $\theta = 90°$, $a = i\sqrt{(1/c)^2 - 1}$ and $b = 0$, so that $r_{E_\perp} = -1$ and $r_{E_\parallel} = 1$. Between these two particular values of θ, we expect the curves to join smoothly, as shown in Figure 4.18.

F. Applications of Brewster's Angle

Brewster Windows. Consider a plane parallel plate and unpolarized light incident at Brewster's angle (Figure 4.19). The plate is in air and has the refractive index n. The incident light is refracted and is incident on the second interface with an angle θ''. After refraction, it is parallel to the original incident light. At the first interface, the $E_\parallel$ component is not reflected but is transmitted. If θ'' is Brewster's angle at the second interface, then the $E_\parallel$ component is again not reflected but transmitted. To show that θ'' is Brewster's angle for reflection on the less dense medium, we start with $\tan \theta_B = n$ and use $\theta'' + \theta_B = 90°$ from equation 4.61. We then have

$$\tan \theta'' = \tan(90° - \theta_B) = \cot \theta_B = \frac{1}{\tan \theta_B} = \frac{1}{n}$$

This is the condition for θ'' to be Brewster's angle in the case of reflection on the less dense medium.

Brewster windows are often used in laser technology. Flat plates are used to seal the tubes of gas lasers, with the plates mounted in such a way that light is

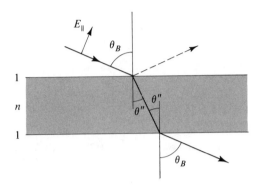

Fig. 4.19
Plane parallel plate with incident and emerging light at Brewster's angle.

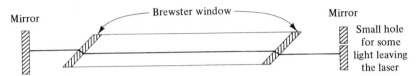

Fig. 4.20
Schematic of a laser tube with Brewster windows and mirrors outside the tube.

incident at Brewster's angle (Figure 4.20). Mirrors mounted outside the tube reflect the light forth and back through the tube. It is essential for the laser to function so that light waves can move with minimum attenuation between the mirrors. Because light is generated in the tube, a small hole in one mirror may serve to let some light out.

Since the windows are tilted to the axis to make the light incident at Brewster's angle, the $E_{\parallel}$ component can traverse forth and back through the tube with minimal attenuation. The $E_{\perp}$ component is only partially transmitted at each traversal, and after many traversals, it is strongly attenuated. As a result, the light leaving the laser is polarized in the parallel direction. The orientation can be obtained from evaluation of the Brewster window.

The pile-of-plates polarizer. We can use a pile of plates (Figure 4.21). To obtain polarized light in transmission, the light is incident at Brewster's angle at each plate. The $E_{\parallel}$ component is transmitted, and the $E_{\perp}$ component is partially transmitted. This happens at every plate, and the $E_{\parallel}$ component increases while the $E_{\perp}$ component decreases. After many plates have been traversed, the light is practically polarized in the parallel direction.

Fig. 4.21
Pile of plates polarizer.

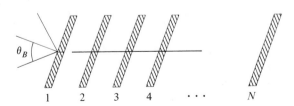

　　　4. Maxwell's Theory

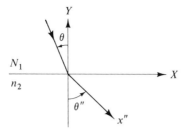

Fig. 4.22
Coordinates for the light penetrating into the less dense medium.

G. Wave Penetration Under Total Reflection Conditions

Now that we have investigated the relationship between reflected and transmitted intensities for the case when light is incident from a more dense to a less dense medium, we may ask: How absolute is the total reflection? That is, is there any penetration of the wave into the second medium? According to equation 4.65,

$$|r_{E_\perp}| = |r_{E_\parallel}| = 1 \quad \text{for} \quad \theta_c \le \theta \le 90°$$

To address our question, we consider the solution of the wave equation in the medium n_2, that is,

$$E'' = A'' e^{i(k_2 x'' - \omega t)} \tag{4.67}$$

If we express the coordinate x'' in terms of X, Y and θ'' as

$$x'' = X \sin \theta'' - Y \cos \theta''$$

(see, Chapter 1, Section 6B, and Figure 4.22) then equation 4.67 becomes

$$E'' = A'' e^{i[k_2(X \sin \theta'' - Y \cos \theta'') - \omega t]} \tag{4.68}$$

The law of refraction may be expressed as

$$\frac{2\pi}{\lambda} n_1 \sin \theta_1 = \frac{2\pi}{\lambda} n_2 \sin \theta_2$$

or

$$k_1 \sin \theta = k_2 \sin \theta''$$

Using this relation and $\cos \theta'' = \sqrt{1 - \sin^2 \theta''}$, we eliminate θ'' from equation 4.68 and have

$$E'' = A'' \exp \left\{ i \left[k_1 X \sin \theta - k_2 Y \sqrt{1 - \left(\frac{k_1}{k_2} \sin \theta \right)^2} - \omega t \right] \right\} \tag{4.69}$$

Since $k_1 > k_2$ because of $(N_1 > n_2)$, we write the radical as

$$i \sqrt{\left(\frac{N_1}{n_2} \sin \theta \right)^2 - 1}$$

and for equation 4.69 we have

$$E'' = A'' \cdot \{\exp[i(k_1 X \sin\theta - \omega t)]\} \exp\left[Yk_2 \sqrt{\left(\frac{N_1}{n_2}\sin\theta\right)^2 - 1} \right] \quad (4.70)$$

Note that Y is negative.

The first factor on the right-hand side is just the expression for a harmonic wave traveling in the X direction. The second factor contains a term that exponentially decreases in magnitude for negative values of Y (in agreement with the geometry of Figure 4.22).

The amplitude of the wave traveling in the X direction is very quickly attenuated in the Y direction. The depth of the penetration ℓ is defined as the value of $-Y$ for which the amplitude falls to $1/e$ of its value at the surface $Y = 0$; that is,

$$|\ell| = \frac{1}{k_2\sqrt{[(N_1/n_2)\sin\theta]^2 - 1}} \quad (4.71)$$

For an air-glass interface where $\lambda = 5000$ Å $= 5 \times 10^{-7}$m, $N_1 = 1.5$, $n_2 = 1.0$, and $\theta = 45°$, we have $|\ell| \sim 2 \times 10^{-7}$m $\sim \lambda/2$. That is, the wave penetrates into the medium a distance of the order of the wavelength. (For numerical values, see Figure 4.23.) This type of wave is called an **evanescent wave**; it is not absorbed by a lossless medium, but turns around and reemerges from the surface. In his treatise on optics, A. Sommerfeld parallels this behavior to that of an army arriving at the edge of a wooded area. The main part of the army changes direction at the interface, but a small patrol reconnoiters for a short distance into the wooded area and then rejoins the main army.

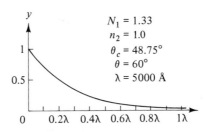

Fig. 4.23
Calculated penetration into the less dense medium. $N_1 = 1.33$, $n_2 = 1.0$, $\theta_c = 48.75°$, $\theta = 60$ and $\lambda = 5000$ Å.

4. THIN FILMS AND OPTICAL COATING

A. Introduction to the Treatment of Reflection and Transmission at Multilayer Dielectrics

In Section 3, we derived Fresnel's formulas by applying the boundary condition to the interface of two media. We found an expression for the reflected and transmitted fraction of the incident light. If we want to treat the plane parallel plate in the same way, we must apply the boundary condition to both interfaces

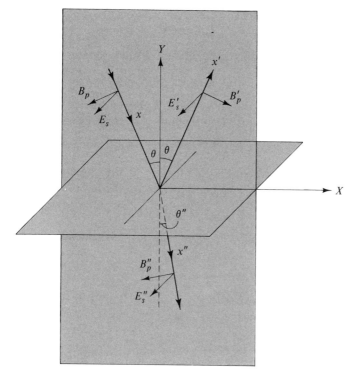

Fig. 4.24
Incident, reflected, and refracted light for the components of B_p and E_s.

and solve the resulting equations. In Chapter 2; on the other hand, we summed all the reflected and transmitted rays and took care of the phase difference between them. We also arrived at a formula for the reflected and transmitted light. Of course, both methods will give us the same result. The method used here is called the **boundary value method**; the method used in Chapter 2 is called the **summation method**.

We now wish to study the reflection and transmission on several boundaries, each between media of different refractive indices. The summation method is clearly impractical here, and we will develop a matrix method that will allow us to study the transmission and reflection through such a set of stratified media. One of the most important applications of such a set is the production of highly reflecting dielectric mirrors for lasers. These mirrors are made from thin films with high and low refractive indices.

Consider an electromagnetic wave whose direction makes an angle θ with the normal to such an interface. The **E** vector for the incident, reflected, and refracted wave is assumed perpendicular to the plane of incidence (Figure 4.24). The orientation of the **B** vectors is then obtained by considering the Poynting vector, which must point in the direction of propagation.

Application of the boundary condition for **E** and **B** at $Y = 0$ yields

$$E_s + E'_s = E''_s \tag{4.72}$$

and (with $\theta = \theta'$)

$$-B_p \cos \theta + B'_p \cos \theta = -B''_p \cos \theta'' \qquad (4.73)$$

In equation 4.72, we introduce harmonic wave solutions and have

$$E_s + E'_s = A_1 e^{i(k_1 x - \omega t)} + A'_1 e^{i(k_1 x' - \omega t)} = E''_s = A_2 e^{i(k_2 x'' - \omega t)} \qquad (4.74)$$

Expressing coordinates x, x', and x'' in the X, Y coordinate system (again using the rotation matrix of Chapter 1) and omitting the time factor, we get

$$A_1 e^{i[k_1(X \sin \theta - Y \cos \theta)]} + A'_1 e^{i[k_1(X \sin \theta' + Y \cos \theta')]} = A_2 e^{i[k_2(X \sin \theta'' - Y \cos \theta'')]} \qquad (4.75)$$

The boundary condition requires that for $Y = 0$, the phase factor for all three terms must be equal. As a consequence, we derive the law of refraction

$$k_1 \sin \theta = k_2 \sin \theta''$$

For our present discussion, we need only normal incidence; that is, $\theta = \theta' = \theta'' = 0°$. From equation 4.72 we have

$$A_1 + A'_1 = A_2 \qquad (4.76)$$

From equation 4.73, we have with $\theta = \theta' = \theta'' = 0°$ and $B_p = (n/c)E_s$ (see equation 4.22), and

$$-n_1 A_1 + n_1 A'_1 = -n_2 A_2 \qquad (4.77)$$

The left side of equation 4.76 is the electrical field in medium 1, E_1, and the left side of equation 4.77 is the induction field, B_1, at the boundary where the light is incident. The right side of the two equations represent the field on the "other side" of the boundary (see Figure 4.24).

We can write equations 4.72 and 4.73 or equations 4.76 and 4.77 in vector notation as follows:

$$\begin{pmatrix} E_1 \\ B_1 \end{pmatrix} = \begin{pmatrix} A_1 + A'_1 \\ -n_1 A_1 + n_1 A'_1 \end{pmatrix} = \begin{pmatrix} A_2 \\ -n_2 A_2 \end{pmatrix} = \begin{pmatrix} E_2 \\ B_2 \end{pmatrix} \qquad (4.78)$$

At $Y = 0$ At $Y = 0$
(left boundary)   (right boundary)

B. Matrix Formulation for One Interface

Assume that the medium on the right side of the first boundary contains not only the refracted wave, but also a wave incident toward the boundary. Inclusion of such a wave will make our approach to both sides of the boundary more symmetric.

For

$$\begin{pmatrix} E_1 \\ B_1 \end{pmatrix} = \begin{pmatrix} E_2 \\ B_2 \end{pmatrix} \qquad (4.79)$$

$Y = 0$ $Y = 0$
(left boundary,   (right boundary,
medium 1)   medium 2)

we now have the equations

$$A_1 + A_1' = A_2 + A_2' = E_2$$
$$-n_1 A_1 + n_1 A_1' = -n_2 A_2 + n_2 A_2' = B_2$$

(4.80)

We can write this in matrix notation as

$$\begin{pmatrix} E_1 \\ B_1 \end{pmatrix}_{\text{left}} = \begin{pmatrix} 1 & 1 \\ -n_1 & n_1 \end{pmatrix}\begin{pmatrix} A_1 \\ A_1' \end{pmatrix} = \begin{pmatrix} 1 & 1 \\ -n_2 & n_2 \end{pmatrix}\begin{pmatrix} A_2 \\ A_2' \end{pmatrix} = \begin{pmatrix} E_2 \\ B_2 \end{pmatrix}_{\text{right}}$$

(4.81)

This result can also be expressed as a matrix T acting on

$$\begin{pmatrix} A_1 \\ A_1' \end{pmatrix}$$

to yield

$$\begin{pmatrix} A_2 \\ A_2' \end{pmatrix}$$

Thus,

$$\begin{pmatrix} A_2 \\ A_2' \end{pmatrix} = \underbrace{\begin{pmatrix} 1 & 1 \\ -n_2 & n_2 \end{pmatrix}^{-1}\begin{pmatrix} 1 & 1 \\ -n_1 & n_1 \end{pmatrix}}_{T}\begin{pmatrix} A_1 \\ A_1' \end{pmatrix}$$

(4.82)

Note that the -1 power on the first matrix indicates that it is the inverse matrix.

The coefficients of the incident and reflected waves in medium 1 are transformed into the coefficients of the incident and reflected waves in medium 2 by the matrix T. The inverse matrix of

$$\begin{pmatrix} 1 & 1 \\ -n_2 & n_2 \end{pmatrix}$$

will be calculated in problem 13. The result is

$$\begin{pmatrix} 1 & 1 \\ -n_2 & n_2 \end{pmatrix}^{-1} = \begin{pmatrix} \dfrac{1}{2} & -\dfrac{1}{2n_2} \\ \dfrac{1}{2} & \dfrac{1}{2n_2} \end{pmatrix}$$

(4.83)

The matrix T is therefore

$$\frac{1}{2}\begin{pmatrix} 1 & -\dfrac{1}{n_2} \\ 1 & \dfrac{1}{n_2} \end{pmatrix}\begin{pmatrix} 1 & 1 \\ -n_1 & n_1 \end{pmatrix} = \frac{1}{2}\begin{pmatrix} 1+\dfrac{n_1}{n_2} & 1-\dfrac{n_1}{n_2} \\ 1-\dfrac{n_1}{n_2} & 1+\dfrac{n_1}{n_2} \end{pmatrix}$$

(4.84)

and from equation 4.82 we have

$$\begin{pmatrix} A_2 \\ A_2' \end{pmatrix} = \frac{1}{2} \begin{pmatrix} 1 + \dfrac{n_1}{n_2} & 1 - \dfrac{n_1}{n_2} \\ 1 - \dfrac{n_1}{n_2} & 1 + \dfrac{n_1}{n_2} \end{pmatrix} \begin{pmatrix} A_1 \\ A_1' \end{pmatrix} \tag{4.85}$$

We may check this procedure by applying this matrix to obtain the well-known coefficients A_1'/A_1 and A_2'/A_1, that is, the reflection and transmission coefficients for normal incidence on one interface as obtained from Fresnel's formulas. We do not consider a backward-running wave in medium 2, that is, $A_2' = 0$. We have

$$\begin{pmatrix} A_2 \\ 0 \end{pmatrix} = \frac{1}{2} \begin{pmatrix} 1 + \dfrac{n_1}{n_2} & 1 - \dfrac{n_1}{n_2} \\ 1 - \dfrac{n_1}{n_2} & 1 + \dfrac{n_1}{n_2} \end{pmatrix} \begin{pmatrix} A_1 \\ A_1' \end{pmatrix} \tag{4.86}$$

Writing the two equations out, we have

$$A_2 = \frac{1}{2}\left(1 + \frac{n_1}{n_2}\right)A_1 + \frac{1}{2}\left(1 - \frac{n_1}{n_2}\right)A_1'$$
$$0 = \frac{1}{2}\left(1 - \frac{n_1}{n_2}\right)A_1 + \frac{1}{2}\left(1 + \frac{n_1}{n_2}\right)A_1' \tag{4.87}$$

We obtain

$$\frac{A_1'}{A_1} = -\frac{\left(1 - \dfrac{n_1}{n_2}\right)}{\left(1 + \dfrac{n_1}{n_2}\right)} = -\frac{n_2 - n_1}{n_2 + n_1} \tag{4.88}$$

(i.e., r_E of Table 4.1) and

$$\frac{A_2}{A_1} = \frac{2n_1}{n_1 + n_2} \tag{4.89}$$

(i.e., t_E of Table 4.1). These results are exactly what we expected.

C. Matrix Formulation for Two Interfaces

Now let us assume that there is a third medium with the index of refraction n_3. This medium is located at a distance d from the n_1/n_2 interface, as shown in Figure 4.25. We assume that in this third medium, there are forward- and backward-traveling waves.

Fig. 4.25
Indication of fields at boundaries $Y = 0$ and $Y = d$.

$$\begin{matrix} n_1 & & n_2 & & & n_2 & & n_3 \\ \begin{pmatrix} E_1 \\ B_1 \end{pmatrix}_0 & & \begin{pmatrix} E_2 \\ B_2 \end{pmatrix}_0 & & & \begin{pmatrix} E_2 \\ B_2 \end{pmatrix}_d & & \begin{pmatrix} E_3 \\ B_3 \end{pmatrix}_d \\ & Y = 0 & & & & Y = d & \end{matrix}$$

4. Maxwell's Theory

For the boundary conditions at $Y = d$, and using

$$k_2 = \frac{2\pi}{\lambda / n_2}$$

we have

$$E_2 = A_2 e^{ik_2 d} + A_2' e^{-ik_2 d} = A_3 e^{ik_3 d} + A_3' e^{-ik_3 d} = E_3 \qquad (4.90)$$

and

$$B_2 = -n_2 A_2 e^{ik_2 d} + n_2 A_2' e^{-ik_2 d} = -n_3 A_3 e^{ik_3 d} + n_3 A_3' e^{-ik_3 d} = B_3 \qquad (4.91)$$

The negative sign in the exponentials $e^{-ik_3 d}$ expresses our assumption of a backward-traveling wave in medium n_3. Using matrix notation, we write

$$\begin{pmatrix} E_2 \\ B_2 \end{pmatrix}_d = \begin{pmatrix} e^{ik_2 d} & e^{-ik_2 d} \\ -n_2 e^{ik_2 d} & n_2 e^{-ik_2 d} \end{pmatrix} \begin{pmatrix} A_2 \\ A_2' \end{pmatrix} = \begin{pmatrix} e^{ik_3 d} & e^{-ik_3 d} \\ -n_3 e^{ik_3 d} & n_3 e^{-ik_3 d} \end{pmatrix} \begin{pmatrix} A_3 \\ A_3' \end{pmatrix} = \begin{pmatrix} E_3 \\ B_3 \end{pmatrix}_d$$
$$(4.92)$$

Substituting equation 4.82 for

$$\begin{pmatrix} A_2 \\ A_2' \end{pmatrix}$$

we can get the connection of the fields at $Y = 0$ and $Y = d$, as illustrated in Figure 4.25.

By first representing the field at $Y = d$, we have

$$\begin{pmatrix} E_3 \\ B_3 \end{pmatrix}_d = \begin{pmatrix} e^{ik_3 d} & e^{-ik_3 d} \\ -n_3 e^{ik_3 d} & n_3 e^{-ik_3 d} \end{pmatrix} \begin{pmatrix} A_3 \\ A_3' \end{pmatrix}$$

$$\overbrace{\begin{pmatrix} A_2 \\ A_2' \end{pmatrix}}$$

$$(4.93)$$

$$= \underbrace{\begin{pmatrix} e^{ik_2 d} & e^{-ik_2 d} \\ -n_2 e^{ik_2 d} & n_2 e^{-ik_2 d} \end{pmatrix} \overbrace{\begin{pmatrix} \dfrac{1}{2} & -\dfrac{1}{2n_2} \\ \dfrac{1}{2} & \dfrac{1}{2n_2} \end{pmatrix}}}_{M_2} \underbrace{\begin{pmatrix} 1 & 1 \\ -n_1 & n_1 \end{pmatrix} \begin{pmatrix} A_1 \\ A_1' \end{pmatrix}}_{\begin{pmatrix} E_1 \\ B_1 \end{pmatrix}_0}$$

Equation 4.83 was used to calculate the inverse matrix in the expression for

$$\begin{pmatrix} A_2 \\ A_2' \end{pmatrix}$$

and now the matrices have been regrouped as

$$M_2 \quad \text{and} \quad \begin{pmatrix} E_1 \\ B_1 \end{pmatrix}_0$$

The latter is the field at $Y = 0$. Using the matrix M_2, we can write

$$\begin{pmatrix} E_3 \\ B_3 \end{pmatrix}_d = M_2 \begin{pmatrix} E_1 \\ B_1 \end{pmatrix}_0 \qquad (4.94)$$

where

$$M_2 = \begin{pmatrix} \cos k_2 d & -\dfrac{i}{n_2} \sin k_2 d \\ -n_2 i \sin k_2 d & \cos k_2 d \end{pmatrix} \qquad (4.95)$$

Equation 4.94 tells us that we can obtain the field at the boundary d in medium 3 from the field in medium 1 at the boundary 0 by multiplication with the matrix M_2.

If we have many different media, we can write

$$\begin{pmatrix} E_f \\ B_f \end{pmatrix}_{\text{final}} = M_{f-1} \cdots M_3 \cdot M_2 \begin{pmatrix} E_1 \\ B_1 \end{pmatrix}_0 \qquad (4.96)$$

To illustrate these results, consider the antireflection coating that is applied to camera lenses or binoculars. The principal color we see on the surface of the lens results from a thin film of thickness $\lambda/4$, where λ is the wavelength of visible light in the medium of the coating. If the refractive index is chosen correctly and we use only one wavelength in the visible spectrum, no light is reflected. To determine this refractive index, we assume that on one side of the film the refractive index is n_1 (1.0 for air) and on the other side n_3 (1.5 for glass). We also assume that there is no backward-traveling wave in medium 3, that is, $A'_3 = 0$, and that the film thickness is $\lambda/4$; that is, at the boundary (between mediums 2 and 3) we have

$$e^{ik_3 d} = e^{ik_2 d} = i$$

or $\cos k_2 d = 0$ and $\sin k_2 d = 1$.

From equations 4.93–4.95, we have

$$\begin{pmatrix} A_3 e^{ik_3 d} \\ -n_3 A_3 e^{ik_3 d} \end{pmatrix} = \begin{pmatrix} 0 & -\dfrac{i}{n_2} \\ -n_2 i & 0 \end{pmatrix} \begin{pmatrix} A_1 + A'_1 \\ -n_1 A_1 + n_2 A'_1 \end{pmatrix} \qquad (4.97)$$

From equation 4.97, we get

$$iA_3 = A_1 \frac{n_1}{n_2} i - A'_1 \frac{n_2}{n_2} i \qquad (4.98)$$

$$-n_3 i A_3 = -A_1 n_2 i - A'_1 n_2 i$$

or

$$\frac{A_3}{A_1} = \frac{n_1}{n_2} - \frac{A'_1}{A_1}$$

$$\frac{n_3 A_3}{A_1} = n_2 + n_2 \frac{A'_1}{A_1} \qquad (4.99)$$

　　　　4. Maxwell's Theory

For no reflection to occur, $A_1'/A_1 = 0$. Then it follows that

$$\frac{A_3}{A_1} = \frac{n_1}{n_2} = \frac{n_2}{n_3} = \frac{A_3}{A_1} \rightarrow n_1 n_3 = n_2^2 \qquad (4.100)$$

Thus, the refractive index of the film should be $n_2 = \sqrt{n_1 n_3}$. For our example, $n_1 = 1.0$, $n_3 = 1.5$, and we have $n_2 = 1.22$. A suitable material with this refractive index is difficult to find. In practice, cryolite ($n = 1.35$) or magnesium flouride ($n \simeq 1.38$) is employed.

For high-reflecting laser mirrors, a large number of thin films of alternating high and low refractive indices are used. Examples of this are discussed in problems 14 to 16.

5. MAXWELL'S EQUATIONS

In this chapter, we have derived all our results from the laws of the electromagnetic theory. Maxwell formulated the electromagnetic theory in a set of differential equations. The results we have obtained may be deduced from the following simplified set of equations:

$$\nabla \times \mathbf{E} = -\frac{\partial \mathbf{B}}{\partial t}$$

$$\nabla \times \mathbf{H} = \frac{\partial \mathbf{D}}{\partial t} \qquad (4.101)$$

$$\nabla \cdot \mathbf{E} = 0$$

$$\nabla \cdot \mathbf{H} = 0$$

where $\nabla = \mathbf{i}(\partial/\partial x) + \mathbf{j}(\partial/\partial y) + \mathbf{k}(\partial/\partial z)$ and $\mathbf{i}$, $\mathbf{j}$, and $\mathbf{k}$ are the unit vectors in x, y, z coordinate systems. The field quantities are as follows:

$\mathbf{E}$ is the electrical field vector, $\mathbf{B}$ is the magnetic induction vector, $\mathbf{D}$ is the displacement vector, and $\mathbf{H}$ is the magnetic field vector. The relations between these field vectors are

$$\mathbf{D} = \varepsilon \mathbf{E}$$

and

$$\mathbf{B} = \mu \mathbf{H}$$

where ε is the permittivity and μ the permeability of the medium.

From this set of Maxwell's equations, we can obtain the wave equation in three dimensions. The solution of the wave equation and the use of the boundary conditions applied to the problems we have discussed will yield all the results we have presented. We will not deduce the wave equation from Maxwell's equation here. This will be done for the more complicated case of a uniaxial crystal in Chapter 5. The wave equation obtained in Chapter 5 may easily be specialized for a homogeneous medium and it then applies to all cases in Chapter 4.

If we follow the line of development starting from Maxwell's equations, we will obtain the laws of refraction and reflection, which are the foundation of geometrical optics. (We obtained the law of refraction in Section 3.)

The representation of light as waves and the superposition principle for waves also follow from Maxwell's equations. From the derivation of Fresnel's equations we obtained the phase shift upon reflection at an optically denser medium, and we saw that the contraction of the wavelength in the medium by the factor $1/n$ is also obtained. This is all we used in Chapter 2.

All the results of diffraction can also be obtained from Maxwell's theory. One way to do this is by taking the solutions as power series, applying the boundary conditions, and determining the coefficients. The final solution will then appear as a power series. This process is mathematically complicated but is currently achieved with the aid of computers. The approximate diffraction theory, which is called scalar theory and which leads to the Kirchhoff-Fresnel integral, was discussed in Chapter 3. Clearly, the fundamental laws of geometrical optics, interference, and diffraction all follow from Maxwell's theory.

Problems

1. *Reflection at an optically denser medium.*
   a. Plot $r_{E_\perp}$, $r_{E_\parallel}$, $t_{E_\perp}$ and $t_{E_\parallel}$ for $N_2 = 1.5$ and $n_1 = 1$, where the angle of incidence θ is equal to 20°, 40°, 60°, and 80°.
   b. Find the angle for which $r_{E_\parallel}$ is zero (read off your graph). Compare this with $\tan \theta_1 = N_2/n_1$.

2. *Dielectric plate and Brewster's angle.* Unpolarized light is incident on a dielectric plate ($N_2 = 1.45$) at Brewster's angle.
   a. Calculate $r_\parallel^2$, $r_\perp^2$, $\alpha t_\parallel^2$ and $\alpha t_\perp^2$, where

$$\alpha = \frac{N_2 \cos \theta''}{n_1 \cos \theta}$$

   b. Show that $r_\perp^2 + \alpha t_\perp^2 = 1$ and $r_\parallel^2 + \alpha t_\parallel^2 = 1$.

   c. Give the percentage of the transmitted light of the incident unpolarized light.

   d. For a check, calculate the percentage for the reflected light.
   Ans. a. $r_\parallel^2 = 0$, $r_\perp^2 = .127$, $\alpha t_\parallel^2 = 1.00$, $\alpha t_\perp^2 = .874$; c. 93.7

3. *Transmission close to Brewster's angle for glass.* In section 3F, we gave Brewster's angle for glass ($n = 1.5$) as 56.3°.
   a. Calculate $r_\parallel$ for $\theta = 56°$, and calculate $r_\parallel^2$.
   b. Calculate the transmitted light $I_\parallel = 1 - r_\parallel^2$ and compare with 1, which is the value for the exact Brewster's angle. (Use five digits after decimal point.)
   c. Calculate $I_\parallel^{10}$, $I_\parallel^{100}$, $I_\parallel^{10,000}$, where $I_\parallel$ is the transmitted intensity. (I^n is equivalent to n times transversal across the interface.)
   Ans. a. $r_\parallel = -0.00323$, $r_\parallel^2 = 0.0001$; b. $I_\parallel = 1 - r_\parallel^2 = 0.99999$; c. $I_\parallel^{10} = 0.99990$, $I_\parallel^{100} = 0.999$, $I_\parallel^{10,000} = 0.90484$

4. *Amplitude reflection coefficient.* In Chapter 2, we saw that the amplitude reflection coefficient r for reflection at the optically denser medium and the amplitude reflection coefficient r' for reflection on the optically less dense medium are related by $r = -r'$. This was obtained from Stoke's consideration. Show that it also follows from Fresnel's formulas.

5. *Reflection on the optically less dense medium.*
   a. Plot $r_{E_\perp}$, $r_{E_\parallel}$, $t_{E_\perp}$, and $t_{E_\parallel}$ for $n_2 = 1.0$ and $N_1 = 1.5$, where $\theta_2 = 20°$, $40°$, $60°$, and $80°$.
   b. Find the angle for which $r_{E_\parallel}$ is zero (read off your graph). Compare this with $\tan \theta_2 = n_2/N_1$.
   c. What happens in the range where you get imaginary numbers?

6. *Brewster's angle at different media.* When light is incident at Brewster's angle on a plane parallel plate of glass, such as used for a laser window, it hits the glass-air interface also at Brewster's angle.
   a. Show that the law of refraction is obeyed by calculation with formulas.
   b. Show that the law of refraction is obeyed by numerically calculating Brewster's angle and inserting it into the law of refraction.

7. *Degree of polarization.*
   a. The degree of polarization q may be defined as

   $$q = \frac{I_p - I_s}{I_p + I_s}$$

   where I_p and I_s are the transmitted intensities for the p and s polarization, respectively. If we define

   $$a = \frac{n_1 \cos \theta'' + n_2 \cos \theta}{n_1 \cos \theta + n_2 \cos \theta''}$$

   show that q can be written as

   $$q = \frac{1 - a^2}{1 + a^2}$$

   b. Calculate q for the angles of incidence $20°$, $40°$, $60°$, and $80°$, $(n_1 = 1, n_2 = 1.5)$ and make a plot.

8. *Partly polarized light, parallel component.* Consider light passing through a plate of glass $(n = 1.5)$ under angles of incidence of $20°$, $40°$, $60°$, and $80°$.
   a. Calculate the angle of refraction θ''.
   b. The angles θ'' are the angles of incidence when the ray hits the second (inner) surface of the glass plate. Using Fresnel's formula, calculate the power $1 - r_p^2 = \alpha t_p^2$ transmitted through the *first* interface for the component parallel to the plane of incidence.
   c. Calculate the transmitted intensity through the second interface for the parallel component.
   d. Combine (b) and (c) and calculate the total intensity P_p transmitted for the parallel component.

9. *Partly polarized light, perpendicular component.* Calculate, in the same way as in problem 8, the transmitted intensity for the perpendicular component.
   a. Use the angles of refraction as calculated in problem 8a.
   b. Calculate $(1 - r_s^2)_1$, transmitted through the first interface.
   c. Calculate $(1 - r_s^2)_2$, transmitted through the second interface.
   d. Calculate $(1 - r_s^2)_1 \cdot (1 - r_s^2)_2$ and plot it on the same graph with the data of $(1 - r_p^2)_1 \cdot (1 - r_p^2)_2$ from problem 8d.

10. *Critical angle for glass.* We want to couple a light beam into a plane parallel plate using a prism, as shown.

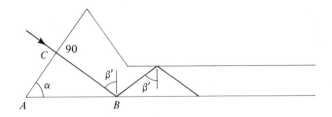

The light should be at normal incidence on the prism.
a. What is the relation between α and β''?
b. How should we choose α so that the waves get internally reflected? The refractive index is $n = 1.5$ (glass).
Ans. a. $\alpha = \beta'$; b. $\theta_C = 41.8$, $41.8 \leq \alpha \leq 90°$

11. *Phase calculation for total reflection (sample calculation).* Calculate the phase angle of the reflected waves $r_{E_\parallel}$ and $r_{E_\perp}$ for reflection at the less dense medium. Take $N_1 = 1.5$, $n_2 = 1$, and find the phase angles $\delta_\perp$ and $\delta_\parallel$ for $\theta_c < 0 < 90°$; make a plot.

12. *Attenuation at total reflection in the less dense medium.* The function

$$y = A e^{-|Y|(2\pi/\lambda)\sqrt{(n \sin \theta)^2 - 1}}$$

is the penetration depth at angle $\theta > \theta_c$ for a wave incident on a glass-air interface. Plot y as a function of $|Y|$. To do this, take $y = A = 1$ for $|Y| = 0$, $\lambda = 0.5 \cdot 10^{-4}$ cm, and $n = 1.5$ for glass. Plot y for $\theta = \theta_c + 2°$ for $|Y| = 0.1\lambda$, 0.3λ, 0.5λ, 0.7λ, and 0.9λ.

13. *Inverse matrix.* Find the inverse matrix of the following 2×2 matrix.

$$\begin{pmatrix} 1 & 1 \\ -n_2 & n_2 \end{pmatrix}$$

14. *Fabry-Perot filter: I.* Consider a device called a Fabry-Perot filter, which is composed of many layers of films, each with the same thickness $\lambda/4$ but with different indices of refraction. The films have either a high index n_H or a low index n_L. The corresponding matrices are

$$M_H = \begin{pmatrix} 0 & -\dfrac{i}{n_H} \\ -in_H & 0 \end{pmatrix} \qquad M_L = \begin{pmatrix} 0 & -\dfrac{i}{n_L} \\ -in_L & 0 \end{pmatrix}$$

Assume there is an alternating sequence of layers, symmetrically arranged with respect to the center and having an even number of layers on each side:

$$|n_H|n_L|n_H|n_L| \ldots |n_H|n_L|n_L|n_H| \ldots |n_L|n_H|n_L|n_H|$$
$$\uparrow$$
$$\text{Center}$$

The matrix representation of this device is

$$M_H M_L M_H M_L \ldots M_H M_L M_L M_H \ldots M_L M_H M_L M_H$$

Show that the product of all the matrices is the unit matrix

$$\begin{pmatrix} 1 & 0 \\ 0 & 1 \end{pmatrix}$$

(*Hint:* Start with the calculation of $M_L M_L$ in the middle.) The result tells us that for

4. Maxwell's Theory

the particular wavelength $\lambda = 4d$, the film is completely transparent. Why do we need so many layers? (To eliminate the other wavelengths.)

15. *Fabry-Perot filter: II.* Consider a filter made of alternating layers (a Fabry-Perot filter). Assume that there are N of each kind of layers. The sequence of the refractive indices is then

$$n = 1 \qquad \underbrace{|n_H|n_L|n_H|\ldots|n_L|}_{N \text{ times}} \qquad n = 1$$

Show that for the matrix product we get

$$\begin{pmatrix} \left(-\dfrac{n_L}{n_H}\right)^N & 0 \\ 0 & \left(-\dfrac{n_H}{n_L}\right)^N \end{pmatrix}$$

That is, we have

$$\begin{pmatrix} A_f \\ -A_f \end{pmatrix} = \begin{pmatrix} \left(-\dfrac{n_L}{n_H}\right)^N & 0 \\ 0 & \left(-\dfrac{n_H}{n_L}\right)^N \end{pmatrix} \begin{pmatrix} A_1 + A_1' \\ -A_1 + A_1' \end{pmatrix}$$

16. *Fabry-Perot filter: III.*
   a. Show that we obtain for the reflection and transmission coefficients

$$\frac{A_f}{A_1} = \frac{2}{[-(n_H/n_L)]^N + [-(n_L/n_H)]^N} \quad \text{and} \quad \frac{A_1'}{A_1} = \frac{[-(n_L/n_H)]^N - [-(n_H/n_L)]^N}{[-(n_L/n_H)]^N + [-(n_H/n_L)]^N}$$

   b. Show that the reflected intensity is equal to

$$R = \left[\frac{(n_H/n_L)^{2N} - 1}{(n_H/n_L)^{2N} + 1}\right]^2$$

   c. Show that if $n_H = 2.5$, $n_L = 1.5$, and $N = 20$, then the reflected intensity is $R = 0.99999999$.

Electromagnetic Waves in Crystals | 5

1. INTRODUCTION

In Chapter 4, we investigated the propagation of electromagnetic waves in free space and in dielectrics, as well as the behavior of the waves at dielectric interfaces. There we assumed that the media were homogeneous (uniform), isotropic (same properties in all directions), and lossless (no absorption mechanism). The next step in our study of light and matter is to consider an anisotropic medium.

Most crystals exhibit different electrical properties in different directions, a result of the atomic structure of the crystal. The electrical field of a wave traveling through a material polarizes the electrical charges of the atoms of that material in the direction of the wave's oscillation. The electrical polarization of the material is then in the direction of the polarization of the wave. The electrical polarization is described by the permittivity ε of the medium and therefore also by the index of refraction n. The permittivity ε and the index of refraction n depend on the atomic structure. For crystals, they may be different for different directions. Then they are called **anisotropic**.

In this chapter, we will continue to treat the medium as lossless so that attenuation considerations may be neglected.

As a model for the anisotropic properties of matter, we consider an electron bound to an atom in the crystal lattice. Schematically, we represent the attractive forces acting on the electron by springs, as shown in Figure 5.1. For a lossless medium, the electron interacts with the electric field of the wave and is driven at the frequency of the applied field. Since the incoming wave usually oscillates in an arbitrary direction, the electron will oscillate with displacement vector components along all three mutually perpendicular directions. The electrostatic restoring forces acting on the oscillating electron are represented by the spring force constants f_x, f_y, and f_z. If the values for the f's are all different, the restoring forces, hence the optical constants, will be different for the X, Y, and Z directions. This is the case for an anisotropic crystal. If all the f values are equal, then the crystal is isotropic and the optical properties are independent of directions. As

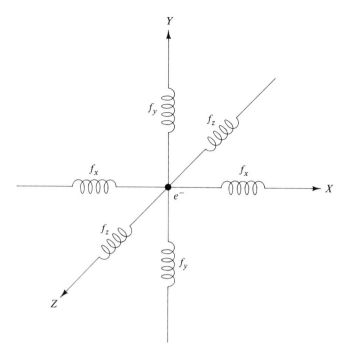

Fig. 5.1
Representation of an electron bound by different forces in three directions.

a consequence of anisotropy, the permittivity and refractive index are functions of direction.

The velocity of the wave is related to ε and n:

$$\frac{1}{v^2} = \varepsilon\mu = \frac{n^2}{c^2}$$

(see equation 4.17). Therefore, we expect the wave in the crystal to have different velocities in different directions. We would now like to see how ε and n are related to the magnitude and direction of the velocity v as a function of the different directions of the forces in the crystal. This is done in the next section for the simple case of a **uniaxial crystal**, where there are only two different values for both ε and n.

2. UNIAXIAL CRYSTALS

A. Propagation Direction and Direction of the k Vector

Consider a crystal with a molecular structure that has planar symmetry. That is, the molecular layers may be considered to be like pancakes and stacked up parallel to the X-Y plane (Figure 5.2).

The Z direction is a symmetry axis. Rotation about the Z axis will transform the positions of equivalent atoms in the X-Y plane into one another. The

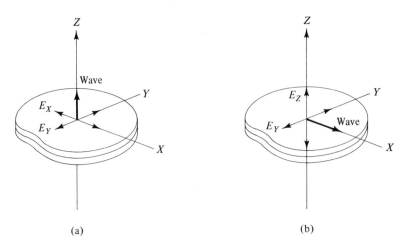

Fig. 5.2
Propagation with respect to the optic axis (Z), shown by black arrows. (a) Wave propagating parallel to the Z axis. Possible directions of the oscillating electric field are along the X and Y axes. (b) Wave propagating perpendicular to the Z axis, for example, in the direction of the X axis. Possible directions of the oscillating electric field are along the Y and Z axis.

Z axis is called the **optic axis**. Polarization by the electrical vector in the X-Y plane is the same in all directions. Polarization along the Z direction, however, is different. An electromagnetic wave moving along the Z axis has two components of polarization in the X-Y plane, one may be in the X direction and the other in the Y direction. Since permittivity ε_1 and refractive index n_1 are the same for all directions in the X-Y plane, the polarization of the two components is the same. Given the relations $n_1 = \sqrt{\varepsilon_1/\varepsilon_0}$, $\lambda_1 = \lambda_0/n_1$, and $k_1 = k_Z = 2\pi/\lambda_1$, where λ_0 is the wavelength in vacuum, we have for the electric field components E_X and E_Y oscillating in the X and Y direction

$$E_X = A_X e^{i(k_Z Z - \omega t)} \tag{5.1}$$

$$E_Y = A_Y e^{i(k_Z Z - \omega t)} \tag{5.2}$$

where A_X and A_Y are the amplitudes of the waves.

The situation is different for waves moving in the X or Y direction. For the X direction, we have a wave polarized either in the Y direction or in the Z direction. The wave with E_Y oscillates in the direction of electrical polarization of the crystal with permittivity ε_1 and index of refraction n_1. With the relations $n_1 = \sqrt{\varepsilon_1/\varepsilon_0}$, $\lambda_1 = \lambda_0/n_1$, and $k_1 = k_X = 2\pi/\lambda_1$, we have

$$E_Y = A_Y e^{i(k_X X - \omega t)} \tag{5.3}$$

But the wave with E_Z oscillates in the direction of electrical polarization of the crystal with $\varepsilon_2 \neq \varepsilon_1$. We have $n_2 = \sqrt{\varepsilon_2/\varepsilon_0}$, $\lambda_2 = \lambda_0/n_2$, and $k_2 = \tilde{k}_X = 2\pi/\lambda_2$. Therefore,

$$E_Z = A_Z e^{i(\tilde{k}_X X - \omega t)} \tag{5.4}$$

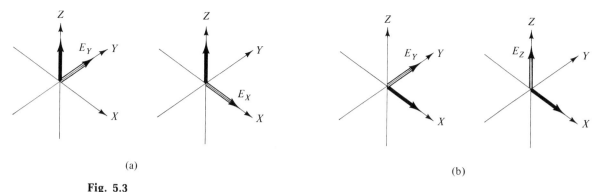

(a)   (b)

Fig. 5.3
(a) Propagation parallel to the optic axis, and vibration perpendicular to the optic axis.
(b) Propagation perpendicular to the optic axis, and vibration perpendicular and parallel to the optic axis.

where A_Z is the maximum amplitude of the wave. For the wave moving in the Y direction, we have a similar result.

B. The Two k Surfaces

We see from equations 5.3 and 5.4 that we have two waves propagating in the X direction with different k vectors, depending on the electrical environment in which the electrical field oscillates. These two waves with k_X and $\tilde{k}_X$ have different wavelengths and different velocities while propagating in the same direction (Figure 5.3). A similar situation exists, of course, for waves propagating in the Y direction.

The direction of oscillation determines the ε or n value used in the k value of the exponential. This determines the wavelength and velocity of propagation, as illustrated in Figure 5.4.

For the Z direction, we have

$$k_Z \to \frac{2\pi}{\lambda_0}\left(\frac{c}{v_1}\right) = n_1 \frac{2\pi}{\lambda_0} \quad \text{for } E_X \text{ and } E_Y \tag{5.5}$$

For the X direction, we have

$$k_X \to \frac{2\pi}{\lambda_0}\left(\frac{c}{v_1}\right) = n_1 \frac{2\pi}{\lambda_0} \quad \text{for } E_Y$$

$$\tilde{k}_X \to \frac{2\pi}{\lambda_0}\left(\frac{c}{v_2}\right) = n_2 \frac{2\pi}{\lambda_0} \quad \text{for } E_Z \tag{5.6}$$

Note that k_X has the same value as k_Z, $\tilde{k}_X$ is different from k_Z.

We must distinguish carefully between the propagation direction of a wave and the vibration direction of the **E** fields. Let us plot k_X and k_Z in X, Y, and Z space, depending on the propagation direction of the different waves. We assume $n_2 > n_1$.

First we consider propagation in the Z direction, where k_Z is the same for both directions of vibration (Figure 5.5). Then we consider propagation in the

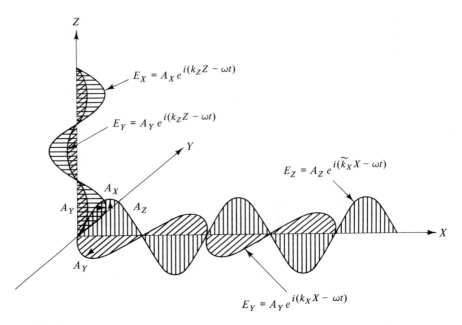

Fig. 5.4
Waves traveling along the Z direction (optic axis). The waves traveling along the X direction have a different wavelength (not shown).

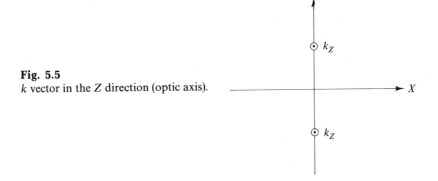

Fig. 5.5
k vector in the Z direction (optic axis).

X direction, where we have k_X for Y vibration and $\tilde{k}_X$ for Z vibration and where $\tilde{k}_X > k_X$ (Figure 5.6). What is the value of k for a wave propagating in an arbitrary direction in the X-Z plane? For the limiting case of Y vibration, we have $k_X = k_Z$ independent of the direction of propagation in the X-Z plane. The corresponding k values are located on a circle. For the limiting case of vibration in the X-Y plane, the k value of the wave is between the values of $\tilde{k}_X$ and k_Z and is located on an ellipse (Figure 5.7). A mathematical formulation is given in Section 4 of this chapter.

For a uniaxial crystal, we have rotation symmetry around the Z axis; therefore the situation is similar for the Y direction. The circle and ellipse in the X, Z space of Figure 5.7 becomes a sphere and an ellipsoid in X, Y, Z space. For

5. Electromagnetic Waves in Crystals

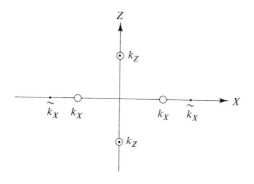

• For Z vibration   ○ For Y vibration

Fig. 5.6
The k vectors in the Z and X directions.

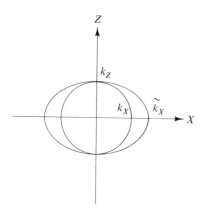

Fig. 5.7
The locus of the k vector is in the X-Z plane.
There are two different k_X vectors,
depending on the direction of vibration.

all waves propagating in directions having k values on the circle, we have ordinary behavior, as if the wave were in an isotropic medium. For all directions of propagation having k values on the ellipse, we have extraordinary behavior except at the points where the circle and ellipse intersect. Ordinary waves are associated with vibrations of the amplitude in the X-Y plane. Extraordinary waves have their amplitudes vibrating *not* in the X-Y plane.

We may have either $\tilde{k}_X > k_X$ (positive crystal) or $\tilde{k}_X < k_X$ (negative crystal). This is shown in Figure 5.8.

A wave propagating along the optic axis will have the velocity

$$v_1 = \sqrt{\frac{1}{\varepsilon_1 \mu}}$$

The index of refraction corresponding to ε_1 is called the **ordinary index of refraction** $n_o = c/v_1$.

A wave propagating perpendicular to the optic axis may have two different velocities. If it vibrates in the X-Y plane, it has ordinary velocity and the ordinary

2. Uniaxial Crystals

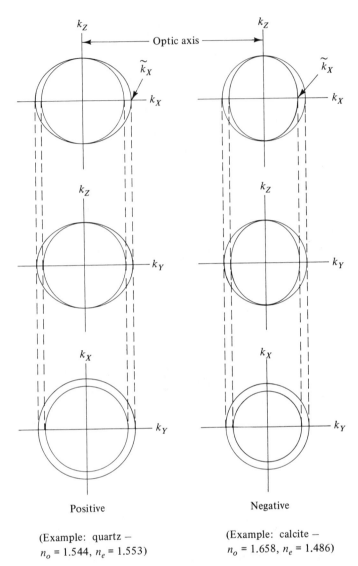

Positive

(Example: quartz —
$n_o = 1.544$, $n_e = 1.553$)

Negative

(Example: calcite —
$n_o = 1.658$, $n_e = 1.486$)

Fig. 5.8
Wave vector surfaces for positive and negative uniaxial crystals.

index of refraction applies. If it vibrates in the Z direction, then the velocity is

$$v_2 = \sqrt{\frac{1}{\varepsilon_2 \mu}}$$

and the corresponding index of refraction is called the **extraordinary index of refraction** $n_e = c/v_2$. For vibration directions that do not make a right angle with the Z axis, we expect the velocity of the wave to have a value between v_1 and v_2. For the velocities of a wave in a uniaxial crystal, there are additional complications. We must distinguish between the phase velocity and the ray velocity. The

5. Electromagnetic Waves in Crystals

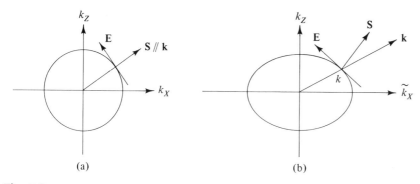

Fig. 5.9
(a) Positions of **E**, **k**, and **S** for the case where $k_X = k_Z$. (b) Positions of **E**, **k**, and **S** for the case where $\tilde{k}_X > k_Z$. The magnetic induction vector **B** points out of the paper.

phase velocity of a wave

$$E = E_0 e^{i(kX - \omega t)}$$

is the velocity of a given phase (the value within the parentheses). For example, if the phase is zero, then $kX = \omega t$ and we have

$$\frac{X}{t} = \frac{\omega}{k} = \frac{2\pi v}{2\pi/\lambda_n} = v\lambda_n = v_X \tag{5.7}$$

The *ray velocity* is the velocity in the direction in which the energy is transported, that is, the direction of the Poynting vector **S**. In Section 4C, we will see that there are propagation directions in the crystal for which the phase velocity and the ray velocity differ.

We have no difficulty with the case where the k value is between k_X and k_Z, and both k_X and k_Z have the same absolute value. The phase velocity and the ray velocity are in the same direction (Figure 5.9a). When k_Z and $\tilde{k}_X$ are different in absolute value, then the locus of the intermediate k values lies on an ellipse. The **k** vector's direction is in a straight line from the origin. The **S** vector must be in the direction of the cross product $\mathbf{E} \times \mathbf{B}$. We have **E** tangent to the ellipse and **B** perpendicular to the plane of the paper (Figure 5.9b). As a result, **S** is not parallel to **k**. This will be shown in mathematical terms in Section 4C.

C. Refraction at Three Different Cuts of a Uniaxial Crystal

Now we will look at three different cases of refraction at the interface of air and a uniaxial crystal. The cases will differ by the orientation of the optic axis with respect to the plane of incidence and the plane of the interface. We will assume for all three cases that $n_e > n_o$, that is, a positive crystal. The optic axis may be either in the plane of the interface or perpendicular to it. The vibration of the wave is either parallel or perpendicular to the plane of incidence.

Optic axis in the plane of the interface and perpendicular to the plane of incidence. Consider the case where the Y axis is the normal to the crystal interface and is in the plane of incidence. The plane of incidence is the

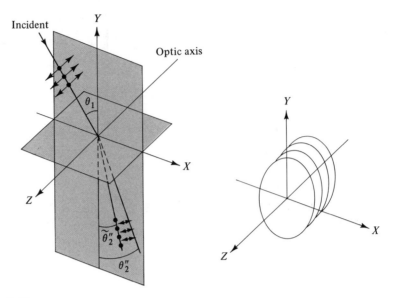

Fig. 5.10
Refraction at an interface of the crystal containing the optic axis Z. The plane of incidence is perpendicular to the optic axis. The small arrows indicate vibration in the plane of incidence (and perpendicular to the optic axis), and the small, dark circles indicate vibration perpendicular to the plane of incidence (and parallel to the optic axis). Using the pancake model, the plane of the pancakes is the plane of incidence. The vibration direction indicated by the small arrows is in the plane of the pancakes.

X-Y plane. The optic axis Z is in the plane of the interface and perpendicular to the plane of incidence (Figure 5.10).

For light vibrating perpendicular to the Z axis (in the plane of incidence), Snell's law gives

$$n_1 \sin \theta_1 = n_2 \sin \theta_2'' \tag{5.8}$$

where $n_1 = 1$ and n_2 is n_o. This is the case of ordinary refraction. Multiplication of equation 5.8 by $k = 2\pi/\lambda_0$ (where λ_0 is the wavelength in vacuum) results in

$$k_1 \sin \theta_1 = k_2 \sin \theta_2'' \tag{5.9}$$

where k_1 contains n_1 and k_2 contains n_o.

For light vibrating parallel to the optic axis (perpendicular to the plane of incidence), we have

$$k_1 \sin \theta_1 = \tilde{k}_2 \sin \tilde{\theta}_2'' \tag{5.10}$$

where $\tilde{k}_2$ contains n_2, which is n_e. Here the vibration is exposed to a different environment, resulting in **extraordinary refraction.**

Since the left sides of equations 5.9 and 5.10 are equal, we may write

$$k_2 \sin \theta_2'' = \tilde{k}_2 \sin \tilde{\theta}_2'' \tag{5.11}$$

5. Electromagnetic Waves in Crystals

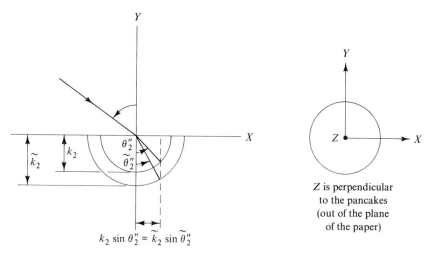

Fig. 5.11
The k surfaces, which correspond to n surfaces, are indicated in the plane of incidence. A parallel line to Y through the point where the ordinary refracted ray meets the circle of the ordinary refractive index also cuts through the point on the circle where the extraordinary refractive index meets the extraordinary refracted ray. The plane of the pancakes is the plane of incidence.

We know that

$$k_2 = \frac{2\pi}{\lambda_0} n_o \quad \text{and} \quad \tilde{k}_2 = \frac{2\pi}{\lambda_0} n_e$$

and, since $n_e > n_0$ ($\tilde{k}_2 > k_2$), we have $\theta_2'' > \tilde{\theta}_2''$. Equation 5.11 can be interpreted as follows: In planes perpendicular to the Z axis, the k_X and k_Y values, and the $\tilde{k}_X$ and k_Y values, are on circles of different radii (see Figure 5.8). The magnitude of the wave vectors k_2 and $\tilde{k}_2$ are independent of direction; they are constants for all angles of incidence and refraction. The same is true for n_e and n_o. In Figure 5.11, we plot the circles corresponding to k_2 and $\tilde{k}_2$ (proportional to n_o and n_e) using the same scale.

We see from equation 5.11 that the projections of k_2 and $\tilde{k}_2$ on the X axis, that is, $k_2 \sin \theta_2''$ and $\tilde{k}_2 \sin \tilde{\theta}_2''$, are equal. Using the ordinary law of refraction and knowing n_o and n_e, we may graphically determine the angle $\tilde{\theta}_2''$.

Optic axis in the plane of the interface and in the plane of incidence.
The optic axis is again in the plane of the interface, but now it is also contained in the plane of incidence (Figure 5.12).

For light vibrating perpendicular to the optic axis (in the Y direction), we have ordinary behavior, where

$$k_1 \sin \theta_1 = k_2 \sin \theta_2'' \tag{5.12}$$

For light vibrating in the plane of incidence, the vibration is neither perpendicular nor parallel to the optic axis:

$$k_1 \sin \theta_1 = \tilde{k}_3 \sin \tilde{\theta}_3'' \tag{5.13}$$

Since this is not ordinary behavior, it is called extraordinary behavior.

2. Uniaxial Crystals

227

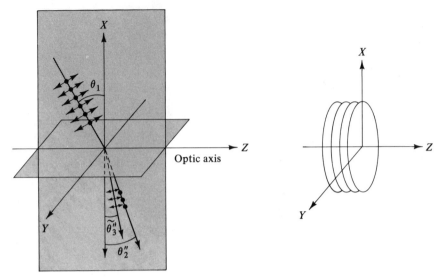

Fig. 5.12
Refraction at a plane containing the optic axis. The plane of incidence also contains
the optic axis. The small arrows indicate vibration in the plane of incidence, and the
small, dark circles indicate vibration perpendicular to the plane of incidence (and
perpendicular to the optic axis). In the pancake model, the plane of the pancakes is
perpendicular to the interface and perpendicular to the plane of incidence.

Again, we superimpose the circle of k_X and k_Z and the ellipse of $\tilde{k}_X$ and k_Z
to get the result shown in Figure 5.13. Note that while k_2 (the circle) is independent of direction, $\tilde{k}_3$ (the ellipse) is not. Equations 5.12 and 5.13 tell us that the
projections along the Z direction of the wave vectors in the crystal are equal, as
before. However, the usual law of refraction is not valid for the extraordinary
case, since $\tilde{k}_3$ is not constant. We also see that $\tilde{k}_3$ is usually different from $\tilde{k}_2$
because a different section of the ellipse is involved.

**Optic axis perpendicular to the plane of interface and parallel to
the plane of incidence.** Since this third case is similar to the case just considered, we will merely summarize the results for all orientations. In Figure 5.14,
the optic axis is denoted by a dot if it is normal to the plane of the page and by
a line if it is in the plane.

3. APPLICATIONS

A. The Nicol, Glan-Foucault, Rochon, and Wollaston Prisms

The most important uniaxial crystals are quartz ($n_o = 1.544$, $n_e = 1.553$) and
calcite ($n_o = 1.658$, $n_e = 1.486$). The refractive indices correspond to $\lambda = 589$ nm.
Materials that have the properties of a uniaxial crystal are called **birefringent**
materials. Their properties include the fact that in specific directions light propagates at two different velocities with different k values.

5. Electromagnetic Waves in Crystals

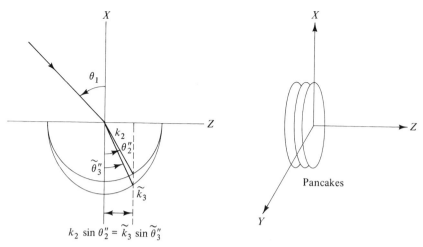

Fig. 5.13
The k surfaces, corresponding to n surfaces, are indicated in the plane of incidence.
Since the vibration in the plane of incidence is neither parallel nor perpendicular to the
optic axis, the corresponding k surface is an ellipse. A parallel line to X through the
point where the ordinary refracted ray meets the circle of n_o also cuts through the point
on the ellipse of n_e where the extraordinary ray cuts through. The plane of the pancakes
is perpendicular to the interface and the plane of incidence.

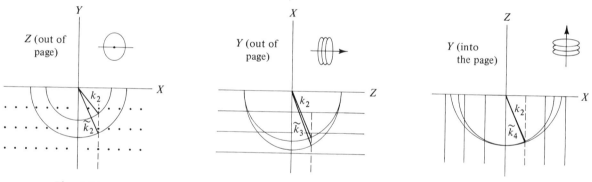

Fig. 5.14
Schematic summary of refraction on the three different surfaces of a uniaxial crystal and indication
of k surfaces and k vectors. If the optic axis is in the plane of the paper, it is represented by a
line; if perpendicular, by a dot.

Some plastics become birefringent when stretched. The molecular chains are
oriented in one direction, and the electric polarization is different for oscillations
parallel or perpendicular to the direction of the molecular chains.

We will limit our discussion to the applications of calcite crystals ($CaCO_3$).
In such crystals, the molecules form layers perpendicular to the optic axis
(pancake model), but the natural cleavage planes do not contain the optic axis.

The prisms formed by calcite crystals show the different behavior of the
ordinary and extraordinary waves for the polarizing action observed in uniaxial
crystals. Consider light inside a calcite crystal and traveling toward the crystal-air

3. Applications

229

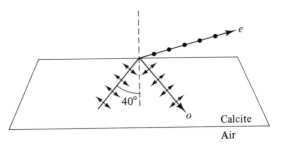

Fig. 5.15
Ordinary ray (*o*) refracted and extraordinary ray (*e*) totally internally reflected at a crystal-air surface. The angle of incidence is 57°.

surface. If the angle of incidence exceeds the critical angle, we expect to observe total internal reflection. But since the light is composed of both an ordinary wave (n_o) and an extraordinary wave (n_e), a certain angle of incidence may give us the situation in which the extraordinary wave is refracted at the surface while the ordinary wave is totally internally reflected (Figure 5.15 and problem 3).

The Nicol prism. The **Nicol prism** is used to separate the ordinary and extraordinary rays of refraction. The calcite crystal is cut into two triangular cross sections and then cemented together, as illustrated in Figure 5.16. The incident light is decomposed into light vibrating perpendicular to the optic axis (ordinary ray) and light vibrating in the plane of the optic axis. The refraction angles are different for these two polarization directions. The angles of the triangular sections have been chosen so that the perpendicularly polarized component is internally reflected at the prism-cement layer (the critical angle for this process is 69°). The other component is not reflected, and polarized light emerges from the prism.

The Glan-Foucault prism. As in the Nicol prism, two triangular cross sections are cemented together to form the **Glan-Foucault prism**. The optic axis

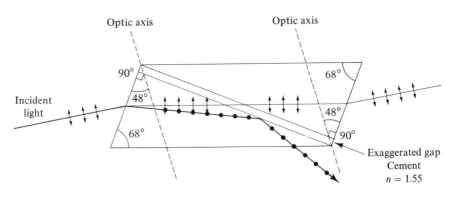

Fig. 5.16
The Nicol prism of calcite. The incident light is refracted at two different angles. The light polarized perpendicular to the optic axis (*o* ray) is internally reflected. The optic axis is rrot in the plane of the paper.

5. Electromagnetic Waves in Crystals

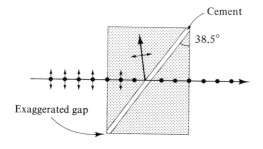

Fig. 5.17
Internal reflection of the *o* ray with the Glan-Foucault prism (calcite). The optic axis is perpendicular to the plane of the paper, as well as the plane of incidence.

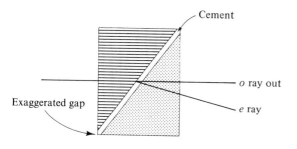

Fig. 5.18
The Rochon prism. The optic axis is indicated by lines in one section and by dots (perpendicular to the plane of the paper) in the second section.

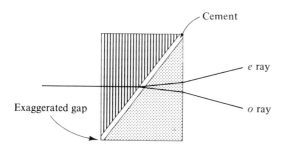

Fig. 5.19
The Wollaston prism (calcite). The optic axis is indicated by lines and dots (perpendicular to the plane of the paper) in each section of the prism.

is, in both sections, perpendicular to the plane of incidence. The angles are chosen so that the *o* ray is again internally reflected (Figure 5.17).

Note that in the first section, the two rays travel along the same line, since the incident light is normal to the interface. (This does not happen in the Nicol prism.) But each ray vibrates in a different electronic polarization environment; therefore, the critical angles at the cement layer are different. The result is that one ray is reflected, while the other is transmitted.

The Rochon and Wollaston prisms. The **Rochon prism** also uses two triangular cross sections, but the optic axis is oriented perpendicular to the plane of incidence in one section and parallel in the other (Figure 5.18). The incident light travels with ordinary behavior in the first section. In the second section, the ordinary ray continues to travel without deviation, but the extraordinary ray is deviated.

A similar arrangement occurs in the **Wollaston prism**. Again, the incident light travels in a straight line; but in this case, both emerging rays are deviated (Figure 5.19).

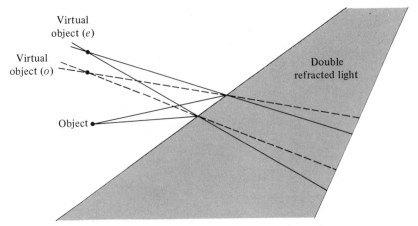

Fig. 5.20
Exaggerated schematic of double refraction. The light rays of the two polarization components coming from the same object are refracted differently. Tracing back to the "appearance of the object," two virtual object points are seen, one for the o rays and one for the e rays. (From Hecht and Zajac, *Optics*, p. 233.)

B. Double Refraction, Biaxial Crystals, and the Use of White Light

Double refraction. The phenomenon of **double refraction** can be observed if a calcite crystal is placed on a page of a book. The printing in the book can be read twice along parallel lines. Double refraction is also seen when the crystal is placed in many different positions. This phenomenon is more difficult than the simple cases we discussed in Section 2C, since the optic axis of a calcite prism goes diagonally through the crystal and is not parallel to the natural cleavage planes. It can be explained, however, by the different refractions of the o and e rays; see Figure 5.20 and the color table at the end of this book. In Figure 5.21, we see the propagation of light through a calcite crystal. The two beams in the crystal are mutually perpendicularly polarized.

Biaxial crystals. In Chapter 4, we discussed electromagnetic waves in dielectrics. These dielectrics were assumed to be homogeneous; that is, the electric permittivity ε was the same in all directions. In uniaxial crystals, there are two different permittivities. Crystals with three different permittivities are called **biaxial crystals**. The mathematical treatment is very involved.*

Mica is a good example of a biaxial crystal with its two axes almost parallel to the cleavage plane. It is used in thicknesses of about 50 μm as a half-wave plate (see Chapter 6, Section 3C).

The use of white light. In this chapter, we have assumed that the light being used had a particular wavelength. We saw that the k vector in the crystal (inversely proportional to the wavelength in vacuum divided by the appropriate

---

*See, for example, M. Born and E. Wolf, *Principles of Optics*, 2nd ed. (Elmsford, N.Y.: Pergamon Press, 1964), p. 678.

5. Electromagnetic Waves in Crystals

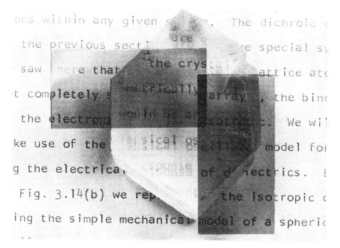

Fig. 5.21
Double refraction by a calcite crystal. Two polarizers (mutually perpendicular) show the polarization of the *o* and *e* rays. The originator of this photo, E. Hecht, recommends: "Take a long look, there is a lot in this one." (From Hecht and Zajac, *Optics*, p. 233.)

refractive index) plays a dominant role in the description of the phenomena discussed. **White light** is composed of a continuous range of wavelength from about 4000 Å to 7000 Å. Each wavelength must be considered separately with respect to its polarization and refraction properties. The intensities of all waves of the white light are added at the observation point. The beautiful color patterns we see when crystals are studied are a result of analysis with white light and polarizers.

C. The Kerr Effect

If certain liquids are brought into a static electric field, we observe different refractive indices for light vibrating parallel and perpendicular to the direction of the electric field (Figure 5.22). This is called the **Kerr effect**. The light wave

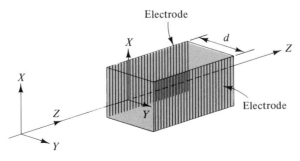

Fig. 5.22
Coordinate system and direction of the electric field for observation of the Kerr effect.

Table 5.1
The Kerr constant for some compounds

Compound	Formula	$K \left[\dfrac{cm}{V^2} \right]$
Benzene	C_6H_6	0.7×10^{-12}
Carbon disulfide	CS_2	3.5×10^{-12}
Chloroform	$CHCl_3$	-3.8×10^{-12}
Water	H_2O	5.5×10^{-12}
Chlorobenzene	C_6H_5Cl	11×10^{-12}
Nitrotoluene	$C_7H_7NO_2$	136×10^{-12}
Nitrobenzene	$C_6H_5NO_2$	224×10^{-12}

travels in the Z direction and vibrates parallel to the X and Y directions. The external electric field E is parallel to the Y direction. Incident light vibrating parallel to the Y direction also vibrates parallel to the E field. Incident light vibrating parallel to the X direction vibrates perpendicular to the E field. The E field polarizes the medium, and two different refractive indices—called $n_\parallel$ and $n_\perp$—result for light vibrating parallel and perpendicular to the E field. From experiments, we know that the difference between $n_\parallel$ and $n_\perp$ is

$$n_\parallel - n_\perp = K\lambda_0 E^2$$

or

$$n_\parallel - n_\perp = K\lambda_0 \frac{V^2}{d^2} \tag{5.14}$$

where λ_0 is the wavelength in vacuum, d is the separation of the electrodes in centimeters, V is the potential in volts, and K is the **Kerr constant** in centimeters per square volt (Table 5.1).

In the presence of an electric field, the liquid acts as a uniaxial or birefringent medium. Of particular interest is the possible continuous change, with an increasing E field, from an isotropic medium to a birefringent medium. The response of the medium to a change in the E field is very fast. An alternating E field with a frequency of 10^{10} Hz may be used to change the medium from isotropic to birefringent and back. This property is employed for light modulation and will be discussed in Chapter 15.

When the E field is applied, the molecules in the liquid are polarized and line up in the direction of the field. Their polarization properties parallel and perpendicular to the E field are different, which explains the different refractive indices. The difference between the refractive indices is not proportional to E but to E^2. In some crystals we can observe a similar effect, called the **Pockels effect**, which is dependent on the symmetry of the crystal being proportional to either E or E^2.

The linear Pockels effect appears in crystals of the symmetry class $\bar{4}2m$, examples of which will be given shortly. Molecules making up the crystals of this class have a certain asymmetry. The quadratic Pockels effect appears in crystals of a more symmetric class (centrosymmetric). In liquids, where the molecules are in no fixed position, the polarization of the E field is directional independent, and the polarization depends on E^2.

5. **Electromagnetic Waves in Crystals**

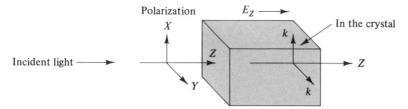

Fig. 5.23
Linear Pockels effect, no field. Incident light vibrates in X and Y directions and propagates in Z direction. The electric field and the optic axis of the crystal are also in Z direction. [Coordinate system according to A. Yariv, *Quantum Electronics* (New York: Wiley, 1967), p. 311.]

D. The Linear Pockels Effect

Crystals of the symmetry class $\overline{4}2m$ show a linear dependence of $n_{\parallel} - n_{\perp}$ on the electric field:

$$n_{\parallel} - n_{\perp} = (\text{const}) \frac{E}{\lambda_0}$$

or

$$n_{\parallel} - n_{\perp} = (\text{const}) \frac{V}{\lambda L} \tag{5.15}$$

where L is the length of the cell (see Figure 5.24). Examples of crystals of this symmetry class for which the linear Pockels effect has been observed are KDP (KH_2PO_4), RbDP(RbH_2PO_4), and CsDA(CsH_2AsO_4).

In this effect (in contrast to the Kerr effect), the electric field is in the direction of propagation, as is the orientation of the optic axis of the uniaxial crystal (Figure 5.23).

In the absence of an electric field (see Figures 5.6 and 5.7), light traveling along the optic axis (Z direction) has the same k value for vibrations along the X and Y directions. In the presence of an electric field, the polarization of the medium is changing in such a way that light vibrating along the X' and Y' axes "feels" different polarizations; this results in two different indices of refraction (Figure 5.24). The axes X' and Y' are rotated by 45° with respect to X and Y.

The two different refractive indices corresponding to vibrations along the X' and Y' axes can be expressed in terms of the ordinary refractive index n_o (i.e., n for vibrations in the X-Y plane without an electric field), the applied electric field E, and a constant r_{63}. For X',

$$n_o - \frac{n_o^3}{2} r_{63} E_Z = n_o - \hat{n} \tag{5.16}$$

and for Y',

$$n_o + \frac{n_o^3}{2} r_{63} E_Z = n_o + \hat{n} \tag{5.17}$$

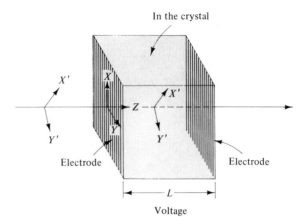

Fig. 5.24
The applied field changes the polarization of the molecules in the crystal. Light polarized along X' propagates with k', and light polarized along Y' propagates with k''. The refractive indices are $n_o - \hat{n}$ and $n_o + \hat{n}$, respectively (for $\hat{n}$, see equations 5.16 and 5.17).

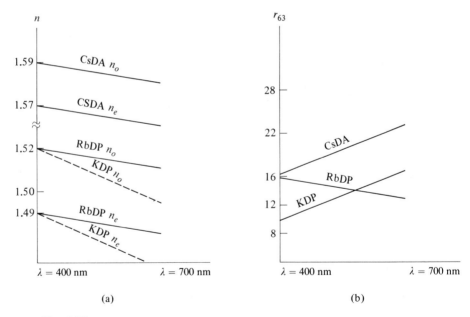

Fig. 5.25
(a) Dependence of the refractive indices n_o and n_e on wavelength, for three crystals of the $\bar{4}2m$ symmetry class. (b) Dependence of the coefficient r_{63} on wavelength, for the crystals given in (a). [Data from R. S. Adhav, *J. Opt. Soc. Am.* **59**, 414 (1969); J. Dennis and R. H. Kingston, *Appl. Opt.* **2**, 1334 (1963).]

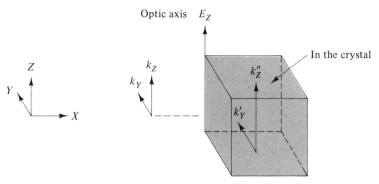

Optic axis E_Z

In the crystal

Fig. 5.26
Quadratic Pockels effect. Both k'_Y and k''_Z depend on E^2. [Coordinate system according to A. Yariv, *Quantum Electronics* (New York: Wiley, 1967), p. 315.]

For the three crystals mentioned earlier, Figure 5.25 shows the dependence of n and r_{63} on the wavelength in the visible region. Also, n_e is given for comparison. The dependence of n_o, n_e, and r_{63} on the wavelength is called **dispersion** and will be discussed in detail in Chapter 7.

E. The Quadratic Pockels Effect

As already mentioned, birefringent crystals with centrosymmetric symmetry will not show the linear dependence of the change in refractive index on the field E; rather their change in refractive index will depend on E^2. For these birefringent crystals, both the ordinary and extraordinary indices of refraction are changed by the electric field applied in the Z direction (optic axis) (Figure 5.26). k'_Y and k''_Z both depend on E^2_Z, and we have for the difference $n' - n''$

$$\frac{\lambda}{2\pi}k'_Y - \frac{\lambda}{2\pi}k''_Z = n' - n'' = \frac{n_o^3}{2}(g_{11} - g_{12})(\varepsilon - \varepsilon_0)^2 E_Z^2 \tag{5.18}$$

where $g_{11} - g_{12}$ is a constant. This constant is related to the effect of E_Z on the polarization of the molecules in the different crystal directions.

Consider, for example, the crystal KTN ($KTa_{0.65}Nb_{0.35}O_3$), which is used in the modulation of light (Chapter 6, Section 3H). For a field in which $E_z = 2 \times 10^5$ V/m, $n_o = 2.29$, and $g_{11} - g_{12} = 0.174$ m^4/C^2 (C $\equiv$ coulomb),

$$n' - n'' = \frac{(2.29)^3}{2}(0.174)(8.85)^2 \times 10^{-24} \times 10^8 \times 4 \times 10^{10} = 0.34 \times 10^{-3}$$

where we used

$$(\varepsilon - \varepsilon_0)^2 = \varepsilon_0^2\left(\frac{\varepsilon}{\varepsilon_0} - 1\right)^2$$

with $\varepsilon/\varepsilon_0 \approx 10^4$ and $\varepsilon_0 = 8.85 \times 10^{-12}$ C^{2}N^{-1}m^{-2} (N $\equiv$ newton). (See Yariv.)[*]

---

[*]A. Yariv, *Quantum Electronics* (New York: Wiley, 1967), p. 315.

4. MATHEMATICAL TREATMENT OF UNIAXIAL CRYSTALS

A. Plane Waves in a Uniaxial Crystal

In Chapter 4, Section 2B, we saw that for an isotropic dielectric, the relative permittivity (dielectric constant) $K = \varepsilon/\varepsilon_0 = 1 + \chi$ was a constant; thus $\mathbf{D}$ was proportional to E. For a uniaxial crystal, we have the case where ε is not a constant. For the different directions, we may write $\mathbf{D} = \tilde{\varepsilon}\mathbf{E}$, or

$$\begin{pmatrix} D_X \\ D_Y \\ D_Z \end{pmatrix} = \begin{pmatrix} \varepsilon_{XX} & \varepsilon_{XY} & \varepsilon_{XZ} \\ \varepsilon_{YX} & \varepsilon_{YY} & \varepsilon_{YZ} \\ \varepsilon_{ZX} & \varepsilon_{ZY} & \varepsilon_{ZZ} \end{pmatrix} \begin{pmatrix} E_X \\ E_Y \\ E_Z \end{pmatrix} \tag{5.19}$$

Here, $\tilde{\varepsilon}_{ij}$ is called the **dielectric tensor**. This relation should not cause undue anxiety, for it merely tells us that, for example, D_X is not produced by E_X only, but is the result of E_X, E_Y, and E_Z. For most applications, however, $\tilde{\varepsilon}_{ij}$ will have a number of zero elements. For example, if all the nondiagonal elements are zero,

$$\tilde{\varepsilon} = \begin{pmatrix} \varepsilon_{XX} & 0 & 0 \\ 0 & \varepsilon_{YY} & 0 \\ 0 & 0 & \varepsilon_{ZZ} \end{pmatrix} \tag{5.20}$$

We have a nonisotropic medium in which D_X depends only on E_X, etc. If in addition, $\varepsilon_{XX} = \varepsilon_{YY} = \varepsilon_{ZZ}$, the medium is isotropic. It can be shown that in general, the dielectric tensor is symmetric, so that $\varepsilon_{ij} = \varepsilon_{ji}$.

B. The Wave Equation in Anisotropic Dielectric Media

The basis for formulating our discussion of wave properties here is Maxwell's equations and the resulting wave equation. For our model, consider a crystal with two different permittivities, ε_1 in the X and Y directions and ε_2 in the Z direction. Accordingly, the refractive index and velocity of propagation will be different in the Z direction than in the X and Y directions. We will study the propagation of plane waves in the X and Z directions only.

Maxwell's equations for the special case of a nonconducting anisotropic medium are

$$\nabla \times \mathbf{E} = -\frac{\partial \mathbf{B}}{\partial t} \tag{5.21}$$

$$\nabla \times \mathbf{B} = \mu_0 \frac{\partial \mathbf{D}}{\partial t} = \mu_0 \varepsilon_0 \frac{\partial \mathbf{E}}{\partial t} + \mu_0 \frac{\partial \mathbf{P}}{\partial t} \tag{5.22}$$

$$\nabla \cdot \mathbf{D} = 0 \tag{5.23}$$

$$\nabla \cdot \mathbf{B} = 0 \tag{5.24}$$

where $\mathbf{B}$ is the magnetic induction vector and $\mathbf{P}$ is the electric polarization vector

5. Electromagnetic Waves in Crystals

$\mathbf{P} = \mathbf{D} - \varepsilon_0 \mathbf{E}$. Writing out equation 5.23 gives

$$\frac{\partial D_X}{\partial X} + \frac{\partial D_Y}{\partial Y} + \frac{\partial D_Z}{\partial Z} = \varepsilon_1 \frac{\partial E_X}{\partial X} + \varepsilon_1 \frac{\partial E_Y}{\partial Y} + \varepsilon_2 \frac{\partial E_Z}{\partial Z} = 0 \qquad (5.25)$$

(In this special case of a medium, where $\varepsilon_1 = \varepsilon_2$, equation 5.23 reduces to $\nabla \cdot \mathbf{E} = 0$, as we know.) Also, from equation 5.22, we get

$$(\nabla \times \mathbf{B})_X = \mu_0 \varepsilon_1 \frac{\partial E_X}{\partial t}$$

$$(\nabla \times \mathbf{B})_Y = \mu_0 \varepsilon_1 \frac{\partial E_Y}{\partial t} \qquad (5.26)$$

$$(\nabla \times \mathbf{B})_Z = \mu_0 \varepsilon_2 \frac{\partial E_Z}{\partial t}$$

as the components of $\nabla \times \mathbf{B}$ instead of, for example,

$$(\nabla \times \mathbf{B})_X = \mu_0 \varepsilon_0 \frac{\partial E_X}{\partial t}$$

for the vacuum case.

We obtain the wave equation in cartesian coordinates by taking the curl of equation 5.21 and using the relation

$$\nabla \times (\nabla \times \mathbf{A}) = \nabla(\nabla \cdot \mathbf{A}) - \nabla^2 \mathbf{A}$$

and equation 5.22, to obtain

$$\nabla(\nabla \cdot \mathbf{E}) - \nabla^2 \mathbf{E} = -\mu_0 \frac{\partial^2 \mathbf{D}}{\partial t^2} \qquad (5.27)$$

We introduce plane-wave trial solutions into equation 5.27, specifically

$$E_Y = A_Y e^{i(k_X X - \omega t)} \quad \text{and} \quad E_Z = A_Z e^{i(k_X X - \omega t)} \qquad (5.28)$$

for the wave traveling in the X direction. For the wave traveling in the Z direction,

$$E_X = A_X e^{i(k_Z Z - \omega t)} \quad \text{and} \quad E_Y = A_Y e^{i(k_Z Z - \omega t)} \qquad (5.29)$$

Note that if we differentiate with respect to X, we obtain, for example,

$$\frac{\partial E_Y}{\partial X} = \frac{\partial}{\partial X} A_Y e^{i(k_X X - \omega t)} = ik_X E_Y$$

or

$$\frac{\partial}{\partial X} \rightarrow ik_X$$

That is, differentiating these plane wave functions with respect to X is equivalent to multiplication by ik_X.

In general, if the wave components E_X, E_Y, and E_Z are functions of X, Y, and

Z and if $\mathbf{k} = \hat{i}k_X + \hat{j}k_Y + \hat{k}k_Z$, then the curl of $\mathbf{E}$ becomes

$$\nabla \times \mathbf{E} = \begin{vmatrix} \hat{i} & \hat{j} & \hat{k} \\ \dfrac{\partial}{\partial X} & \dfrac{\partial}{\partial Y} & \dfrac{\partial}{\partial Z} \\ E_X & E_Y & E_Z \end{vmatrix} = \begin{vmatrix} \hat{i} & \hat{j} & \hat{k} \\ ik_X & ik_Y & ik_Z \\ E_X & E_Y & E_Z \end{vmatrix} = i(\mathbf{k} \times \mathbf{E}) \quad (5.30)$$

Thus, taking the curl of a harmonic vector field is equivalent to taking the cross product of $i\mathbf{k}$ with that field vector. The procedure may be repeated to give:

$$\nabla \times (\nabla \times \mathbf{E}) = \nabla \times (i\mathbf{k} \times \mathbf{E}) = i\mathbf{k} \times (i\mathbf{k} \times \mathbf{E}) \quad (5.31)$$

Using the vector identity

$$\mathbf{A} \times (\mathbf{B} \times \mathbf{C}) = \mathbf{B}(\mathbf{A} \cdot \mathbf{C}) - \mathbf{C}(\mathbf{A} \cdot \mathbf{B}) \quad (5.32)$$

equation 5.31 becomes

$$\nabla \times (\nabla \times \mathbf{E}) = -\mathbf{k}(\mathbf{k} \cdot \mathbf{E}) + \mathbf{E}(\mathbf{k} \cdot \mathbf{k}) \quad (5.33)$$

We may now employ equation 5.33 for the left side of equation 5.27:

$$-\mathbf{k}(\mathbf{k} \cdot \mathbf{E}) + \mathbf{E}(\mathbf{k} \cdot \mathbf{k}) = -\mu_0 \frac{\partial^2 \mathbf{D}}{\partial t^2} \quad (5.34)$$

For the right side of equaton 5.34 we use

$$\mu_0 \frac{\partial^2 D_X}{\partial t^2} = \mu_0 \varepsilon_1 \frac{\partial^2 E_X}{\partial t^2}$$

$$\mu_0 \frac{\partial^2 D_Y}{\partial t^2} = \mu_0 \varepsilon_1 \frac{\partial^2 E_Y}{\partial t^2} \quad (5.35)$$

$$\mu_0 \frac{\partial^2 D_Z}{\partial t^2} = \mu_0 \varepsilon_2 \frac{\partial^2 E_Z}{\partial t^2}$$

Using equations 5.28 and 5.29, the relations $\mu_0 \varepsilon_i = n_i^2/c^2$ $(i = 1, 2)$, and the fact that $k_Y = 0$ for our case where the waves are traveling in the X and Z directions, equation 5.34 gives three component equations:

$$-k_X(k_Z E_Z) + E_X k_Z^2 - n_1^2 \left(\frac{\omega}{c}\right)^2 E_X = 0$$

$$E_Y(k_X^2 + k_Z^2) - n_1^2 \left(\frac{\omega}{c}\right)^2 E_Y = 0 \quad (5.36)$$

$$-k_Z(k_X E_X) + E_Z k_X^2 - n_2^2 \left(\frac{\omega}{c}\right)^2 E_Z = 0$$

We rewrite these equations as

$$\left[k_Z^2 - n_1^2 \left(\frac{\omega}{c}\right)^2\right] E_X \qquad -k_X k_Z E_Z \qquad\qquad +0E_Y = 0$$

$$-k_Z k_X E_X + \left[k_X^2 - n_2^2 \left(\frac{\omega}{c}\right)^2\right] E_Z \qquad\qquad +0E_Y = 0 \quad (5.37)$$

$$+0E_X \qquad\qquad +0E_Z + \left[k_X^2 + k_Z^2 - n_1^2 \left(\frac{\omega}{c}\right)^2\right] E_Y = 0$$

5. Electromagnetic Waves in Crystals

This system of linear homogeneous equations has a nontrivial solution if the determinant of the coefficients vanishes, that is, if

$$
\begin{vmatrix}
k_z^2 - n_1^2\left(\dfrac{\omega}{c}\right)^2 & -k_x k_z & 0 \\[2em]
-k_z k_x & k_x^2 - n_2^2\left(\dfrac{\omega}{c}\right)^2 & 0 \\[2em]
0 & 0 & k_x^2 + k_z^2 - n_1^2\left(\dfrac{\omega}{c}\right)^2
\end{vmatrix} = 0
\qquad (5.38)
$$

Equation 5.38 factors into

$$
\left[k_z^2 - n_1^2\left(\frac{\omega}{c}\right)^2\right]\left[k_x^2 - n_2^2\left(\frac{\omega}{c}\right)^2\right] = k_x^2 k_z^2
$$

$$
k_x^2 k_z^2 - k_z^2 n_2^2\left(\frac{\omega}{c}\right)^2 - k_x^2 n_1^2\left(\frac{\omega}{c}\right)^2 + n_1^2 n_2^2 \frac{\omega^4}{c^4} = k_x^2 k_z^2
$$

$$(5.39)$$

and

$$
k_x^2 + k_z^2 - n_1^2 \frac{\omega^2}{c^2} = 0 \qquad (5.40)
$$

We can rewrite these as

$$
\frac{k_x^2}{(n_2\omega/c)^2} + \frac{k_z^2}{(n_1\omega/c)^2} = 1 \qquad (5.41)
$$

and

$$
\frac{k_x^2}{(n_1\omega/c)^2} + \frac{k_z^2}{(n_1\omega/c)^2} = 1 \qquad (5.42)
$$

In Figure 5.8, we see the plot of k_x and $\tilde{k}_x$ versus k_z for $n_e > n_o$ ($n_2 > n_1$ or $\varepsilon_2 > \varepsilon_1$). We see that equation 5.42 describes a circle, while equation 5.41 describes an ellipse. The result is a single value of k_z on the vertical axis but two values of k_x on the horizontal axis. Two values of k_x imply two values for the velocity of the wave.

At the beginning of this chapter, we discussed the refractive indices and k surfaces of uniaxial crystals. We now see that the case of different refractive indices for different propagation directions in a uniaxial crystal may be analyzed by the application of Maxwell's equations to the propagation of electromagnetic waves in such a crystal.

C. Energy Flow in a Uniaxial Crystal

We have mentioned that an extraordinary ray does not usually propagate along the direction of its k vector. To see this, we must find the direction of energy flow as represented by the Poynting vector. To proceed, we start with equation 5.34,

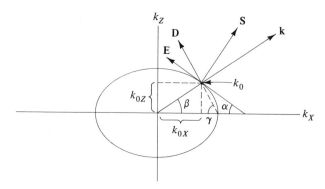

Fig. 5.27
Ellipse of the k vector in the X-Z plane. At k_0, **E**, **D**, **S**, and **k** are indicated; **B** points out of the plane of the paper.

which, for harmonic waves, becomes (ω^2 from differentiation)

$$-\mathbf{k}(\mathbf{k} \cdot \mathbf{E}) + \mathbf{E}(\mathbf{k} \cdot \mathbf{k}) = \mu_0 \omega^2 \mathbf{D} \qquad (5.43)$$

If we take the dot product of equation 5.43 with **k**, we get $\mathbf{k} \cdot \mathbf{D} = 0$; that is, **k** and **D** are perpendicular. However, if we take the dot product of equation 5.43 with **E**, we get

$$-(\mathbf{k} \cdot \mathbf{E})^2 + \mathbf{E}^2 \mathbf{k}^2 = \mu_0 \omega^2 \mathbf{D} \cdot \mathbf{E} \qquad (5.44)$$

which tells us that $\mathbf{k} \cdot \mathbf{E} \neq 0$ and **k** and **E** are not perpendicular. Further, since our model has $D_X = \varepsilon_1 E_X$ and $D_Z = \varepsilon_2 E_Z$, **E** and **D** are in the same X-Z plane, but not parallel. These quantities are shown in Figure 5.27. Not shown in this diagram is **B**, which points out of the page. The Poynting vector $\mathbf{S} = \mathbf{E} \times \mathbf{B}/\mu_0$ also lies in the X-Y plane, but perpendicular to **E**. Therefore, unless **D** and **E** are parallel (as for an isotropic crystal and certain directions of the uniaxial crystal), **S** and **k** will not be parallel and the ray will not travel along the **k** direction.

We would like to know the orientation of **S** with respect to the elliptical surface. For an isotropic crystal, the wave vector surface is always a sphere. Using this result as a guide, we might guess that for our case, **S** would be normal to the surface of the ellipse at the point where **k** intersects the surface (k_0 in Figure 5.27). If this is so, then **E** should be tangent to the ellipse at point k_0. To demonstrate this, we differentiate the equation describing the ellipse (equation 5.41),

$$\frac{k_X^2}{\varepsilon_2 \mu_0 \omega^2} + \frac{k_Z^2}{\varepsilon_1 \mu_0 \omega^2} = 1 \qquad (5.45)$$

and get

$$\frac{dk_Z}{dk_X} = -\frac{\varepsilon_1 k_X}{\varepsilon_2 k_Z} \qquad (5.46)$$

A tangent to the surface at point k_0 intersects the k_X axis at an angle α such that

$$\tan \alpha = \frac{dk_Z}{dk_X} = -\frac{\varepsilon_1 k_{0X}}{\varepsilon_2 k_{0Z}} \qquad (5.47)$$

5. Electromagnetic Waves in Crystals

In Figure 5.27, $\gamma + \beta = \pi/2$ (since $\mathbf{k} \cdot \mathbf{D} = 0$),

$$\tan \beta = \frac{k_{0z}}{k_{0x}} = \tan\left(\frac{\pi}{2} - \gamma\right) = \cot \gamma = \frac{1}{\tan \gamma}$$

or

$$\tan \gamma = \frac{k_{0x}}{k_{0z}} \qquad (5.48)$$

But we can also write

$$\tan \gamma = -\frac{D_z}{D_x} = -\frac{\varepsilon_2 E_z}{\varepsilon_1 E_x} \qquad (5.49)$$

Combining equations 5.48 and 5.49 gives us

$$\frac{k_{0x}}{k_{0z}} = -\frac{\varepsilon_2 E_z}{\varepsilon_1 E_x} \qquad (5.50)$$

Now let α' be the angle between the line of action of $\mathbf{E}$ and the k_x axis. Then

$$\tan \alpha' = -\frac{E_z}{E_x} \qquad (5.51)$$

From equation 5.50,

$$\frac{E_z}{E_x} = -\frac{\varepsilon_1 k_{0x}}{\varepsilon_2 k_{0z}}$$

and evaluating equation 5.47 at point k_0, we see that

$$\tan \alpha' = +\frac{\varepsilon_1 k_{0x}}{\varepsilon_2 k_{0z}} = -\tan \alpha = \tan(180 - \alpha)$$

Thus, we see that $\mathbf{E}$ is tangent to the ellipse at point k_0, and consequently, $\mathbf{S}$ is normal to that tangent or normal to the ellipse at that point. Therefore, for uniaxial crystals, $\mathbf{S}$ and $\mathbf{k}$ are, in general, not parallel. The energy carried by the wave travels in a direction normal to the $k_x - k_z$ ellipse, and that direction is determined by the orientation of $\mathbf{k}$ (but is not in the direction of $\mathbf{k}$).

Finally, we repeat that our particular model chosen in this discussion was a simple and restricted one. For example, we could now consider the case where the incoming light makes an arbitrary angle with the optic axis and the interface. Another possible extension would be to treat a crystal with three different dielectric constants instead of just two. This was briefly discussed earlier.

D. Maxwell's Equations and Isotropic Dielectric Media

The mathematical description of an electromagnetic wave propagating in a uniaxial crystal may also be applied to homogeneous media. All that need be done is to set $\varepsilon_1 = \varepsilon_2 = \varepsilon$, that is, $\mathbf{D} = \varepsilon\mathbf{E}$; the displacement vector $\mathbf{D}$ is now proportional to $\mathbf{E}$, and no tensor for ε is needed. This lets us obtain the wave equation from Maxwell's equations. All the properties of the electromagnetic waves in vacuum and in dielectrics that we have derived less rigorously in Chapter 4 may then be obtained. We see again why Maxwell's equations are viewed as the "umbrella" theory of optics. This is demonstrated on T-shirts and buttons by

And God said

$$\nabla \times \mathbf{E} = -\frac{\partial \mathbf{B}}{\partial t}$$

$$\nabla \times \mathbf{B} = \mu_0 \frac{\partial \mathbf{D}}{\partial t}$$

$$\nabla \cdot \mathbf{D} = 0$$

$$\nabla \cdot \mathbf{B} = 0$$

and there was light.

Problems

1. *Wavelength in crystal.* Unpolarized light of wavelength $\lambda_0 = 0.5 \times 10^{-4}$ cm is incident normally on a crystal quartz plate with $n_o = 1.544$ and $n_e = 1.553$. The optic axis is in the plane of the interface. Calculate the wavelength of the ordinary and extraordinary waves in the quartz plate.
   Ans. $\lambda_{n_o} = 0.3238 \times 10^{-4}$ cm
   $\lambda_{n_e} = 0.3220 \times 10^{-4}$ cm

2. *Ordinary and extraordinary rays in calcite.* Consider a calcite prism with $n_o = 1.658$ and $n_e = 1.486$. The prism has the shape shown in the figure, with apex angle A. The light is incident parallel from the left. The optic axis is perpendicular to the plane of the paper, indicated by dots.

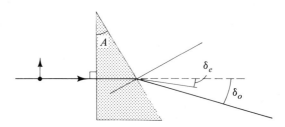

   a. Calculate the angle of deviation δ_o for the ordinary ray. Do this for $A = 10°$, $20°$, and $30°$.
   b. Calculate the angle of deviation δ_e for the extraordinary ray, again for $A = 10°$, $20°$, and $30°$.
   c. Find $\Delta = \delta_o - \delta_e$ for the three angles.

3. *Critical angle, calcite, and quartz.* Consider a plate of calcite with the optic axis perpendicular to the plane of the paper (indicated by dots in the accompanying figure). An unpolarized beam of light is entering the part of the plate that is shaped

like a prism. The light enters the prism at normal incidence. (This relates α and β'.) The refractive indices are $n_o = 1.658$ and $n_e = 1.486$.

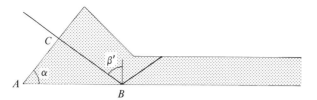

a. Calculate the critical angle β' for the ordinary and extraordinary rays.
b. If α is close to $40°$ (that is if the light is incident close to $40°$ at B), which ray will escape from the plate?
c. For quartz, $n_o = 1.544$ and $n_e = 1.553$. Calculate the critical angles.
d. Give a value of the angle α such that one ray stays in the plate and the other escapes. Which stays in the plate?

Ans. a. $\theta_{co} = 37.09°$, $\theta_{ce} = 42.29°$; b. Extraordinary ray; c. $\theta_{co} = 40.37°$, $\theta_{ce} = 40.08°$; d. $\alpha = 40.22°$, extraordinary ray stays in plate

4. *Double refraction: I.* For the case where the optic axis is perpendicular to the plane of incidence and in the plane of the interface, we have n_o and n_e independent of θ. The angle of refraction of the extraordinary ray can be calculated from the relation

$$\sin \theta_1 = n_o \sin \theta_2'' = n_e \sin \tilde{\theta}_2''$$

Calculate θ_2'' and $\tilde{\theta}_2''$ for $\theta_1 = 20°$, $40°$, and $60°$ for both calcite ($n_o = 1.658$; $n_e = 1.486$) and quartz ($n_o = 1.544$; $n_e = 1.553$). Compare the values for calcite and quartz.

5. *Double refraction: II.* For the case where the optic axis is in the plane of the interface and in the plane of incidence, determine graphically the angle of refraction $\tilde{\theta}_2''$ for calcite and quartz for $\theta_1 = 20°$, $40°$, and $60°$. See problem 4 for the values of n_o and n_e, and use similar formulas.

6. *Wollaston prism.* Consider a Wollaston prism with an angle of $20°$ (see accompanying figure). Calculate the angles α_1 and α_2 between the normal of the prism faces and the emerging ordinary and extraordinary waves. The prism is made of calcite ($n_o = 1.658$ and $n_e = 1.486$).

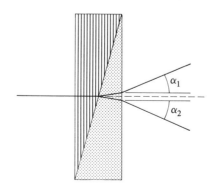

Ans. $\alpha_1 = 3.61°$, $\alpha_2 = 3.57°$

7. *Kerr effect.* For carbon disulfide, the Kerr constant $K = 3.5 \times 10^{-12}$ cm/V^2; for chloroform, $K = -3.8 \times 10^{-12}$ cm/V^2. As a result, one of these compounds has $n_\parallel > n_\perp$. The velocity of the waves in the medium is $v = c/n$.
   a. For which compound do we have a higher velocity for the wave vibrating parallel to the E field compared to the wave vibrating perpendicular?
   b. Calculate $\Delta n_K = n_\parallel - n_\perp$ for nitrobenzene. Assume that $V = 3 \times 10^4$ V, $d = 1$ cm, $\lambda = 0.5 \times 10^{-4}$ cm, and $K = 224 \times 10^{-12}$ cm/V^2.

8. *Linear Pockels effect.* The difference between the refractive indices for vibrations along the X' an Y' axes is $\Delta n_p = -n_o^3 r_{63} E_Z$.
   a. Estimate n_o and r_{63} for CsDA and KDP from Figure 5.25.
   b. For $\lambda_0 = 0.5 \times 10^{-4}$ cm and $E_Z = 3 \times 10^3$ V/cm, calculate Δn_p for CsDA and KDP.

9. *Wave equation.* Show that we obtain the wave equation 5.27,

$$\nabla(\nabla \cdot \mathbf{E}) - \nabla^2 \mathbf{E} = -\mu_0 \frac{\partial^2 \mathbf{D}}{\partial t^2}$$

by taking the curl of equation 5.21,

$$\nabla \times \mathbf{E} = -\frac{\partial \mathbf{B}}{\partial t}$$

and using the relation

$$\nabla \times (\nabla \times \mathbf{A}) = \nabla(\nabla \cdot \mathbf{A}) - \nabla^2 \mathbf{A}$$

10. *Equivalence of $\partial/\partial X$ and multiplication by ik_X.* Show that for plane waves we have the equivalence of the operations $\nabla \times \mathbf{E}$ with multiplication by $i(\mathbf{k} \cdot \mathbf{E})$ (equation 5.30). Write the expressions out in differential form and then do the substitution.

11. *Equation for $\mathbf{E}$ and $\mathbf{k}$ in the crystal.*
   a. Use

$$-\mathbf{k}(\mathbf{k} \cdot \mathbf{E}) + \mathbf{E}(\mathbf{k} \cdot \mathbf{k}) = -\mu_0 \frac{\partial^2 \mathbf{D}}{\partial t^2}$$

and

$$\mu_0 \frac{\partial^2 D_X}{\partial t^2} = \mu_0 \varepsilon_1 \frac{\partial^2 E_X}{\partial t^2}$$

$$\mu_0 \frac{\partial^2 D_Y}{\partial t^2} = \mu_0 \varepsilon_1 \frac{\partial^2 E_Y}{\partial t^2}$$

$$\mu_0 \frac{\partial^2 D_Z}{\partial t^2} = \mu_0 \varepsilon_2 \frac{\partial^2 E_Z}{\partial t^2}$$

and

$$\mu_0 \varepsilon_i = \frac{n_i^2}{c^2} \quad (i = 1, 2)$$

and assume that $k_Y = 0$. Arrive at

$$-k_X(k_Z E_Z) + E_X k_Z^2 - n_1^2 \left(\frac{\omega}{c}\right)^2 E_X = 0$$

$$E_Y(k_X^2 + k_Z^2) - n_1^2 \left(\frac{\omega}{c}\right)^2 E_Y = 0$$

$$-k_Z k_X E_X + E_Z k_X^2 - n_2^2 \left(\frac{\omega}{c}\right)^2 E_Z = 0$$

b. Write the equations obtained in (a) as

$$a_{11} E_X + a_{12} E_Z + 0 E_Y = 0$$
$$a_{21} E_X + a_{22} E_Z + 0 E_Y = 0$$
$$0 E_X + 0 E_Z + a_{33} E_Y = 0$$

The determinant of this system of equations must vanish. Show that we get equation 5.38 for the determinant.

Polarized Light | 6

1. INTRODUCTION

In Chapter 4, we saw that light is described by transverse electromagnetic waves. As a consequence, we have directions of polarization (vibration) perpendicular to the direction of propagation. Two directions of polarization can be chosen perpendicular to the direction of propagation and perpendicular to each other. Light polarized in these directions have different transmission and reflection properties at dielectric interfaces.

In Chapter 5, we studied how light is refracted at an interface of a uniaxial crystal and how it traverses through it. We found that in such a crystal, the light may travel along certain directions with two different velocities, one for each direction of polarization.

In Chapter 2, we studied the superposition of light vibrating in only one direction. Phase differences were introduced, and we dealt with interference phenomena.

Now, in this chapter, we will study the superposition (vector addition) of light mutually perpendicularly polarized and having a phase difference between its components. As a simple and nontrivial example, we will consider again light traversing a uniaxial crystal. We will see that linearly polarized, circularly polarized, and elliptically polarized light may be produced. For simplicity, we will only consider light with one wavelength. White light (consisting of many different wavelengths) could also be analyzed by studying each wavelength separately, and then superimposing the results for all wavelengths present leads to more complex situations.

2. LINEARLY POLARIZED LIGHT

A. Superposition of Linearly Polarized Light

Let us begin by considering two mutually perpendicular polarized components of light.

$$\mathbf{E}_Y = \mathbf{j}A_{Y_0}e^{i(kX-\omega t)} \tag{6.1}$$

$$\mathbf{E}_Z = \mathbf{k}A_{Z_0}e^{i(kX-\omega t)} \tag{6.2}$$

where $\mathbf{j}$ and $\mathbf{k}$ are unit vectors along the Y and Z directions, respectively. The two waves propagate in the X direction, $\mathbf{E}_Y$ vibrates in the Y direction, and $\mathbf{E}_Z$ vibrates in the Z direction. They have the same k values. In Figure 6.1, the waves are presented as cosine waves. Equations 6.1 and 6.2 can be combined by addition or subtraction to give us

$$\mathbf{E}_+ = (\mathbf{j}A_{Y_0} + \mathbf{k}A_{Z_0})e^{i(kX-\omega t)} \tag{6.3}$$

$$\mathbf{E}_- = (\mathbf{j}A_{Y_0} - \mathbf{k}A_{Z_0})e^{i(kX-\omega t)} \tag{6.4}$$

Let us look along the X direction toward the Y, Z plane. Note that waves $\mathbf{E}_+$ and $\mathbf{E}_-$ vibrate along the lines making the angles θ and θ', respectively, with the Y axis (Figure 6.2). These angles are given by

$$\tan\theta = \frac{A_{Z_0}}{A_{Y_0}} \quad \text{and} \quad \tan\theta' = \frac{-A_{Z_0}}{A_{Y_0}} \tag{6.5}$$

If we set $A_{Y_0} = A_{Z_0} = A$, we obtain $\theta = 45°$ and $\theta' = -45°$.

B. Linear Polarizers

There are several methods of producing polarized light. In Chapter 4, we saw that light incident upon the interface between two dielectrics at Brewster's angle θ_B will be plane-polarized upon reflection. In Chapter 5, we saw that uniaxial crystals can be used to separate components of linearly polarized light. Still another method is to use a wire grid. Such a grid may be produced by winding

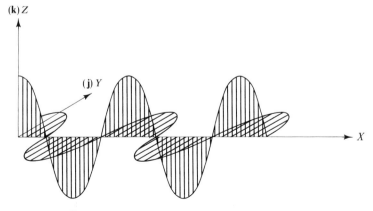

Fig. 6.1
$\mathbf{E}_Y$ and $\mathbf{E}_Z$ waves propagating in the X direction.

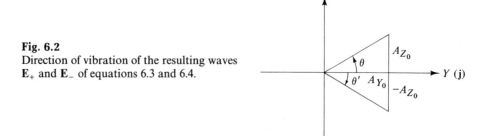

Fig. 6.2
Direction of vibration of the resulting waves
E_+ and E_- of equations 6.3 and 6.4.

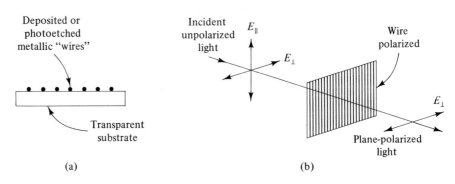

(a)   (b)

Fig. 6.3
(a) Wire grid polarizer and (b) transmitted or reflected components.

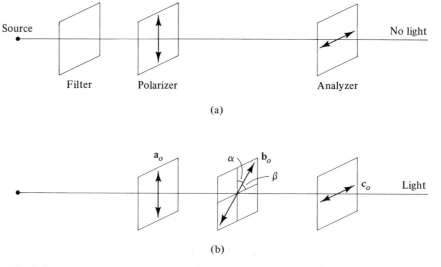

Fig. 6.4
Cross polarizers. (a) Light passing the filter is linearly polarized by the first polarizer and cannot pass the second (crossed) polarizer. (b) An additional polarizer polarizes the light in a direction different from the first polarizer. A component of the light can now pass the second polarizer.

wire around two rods positioned at a certain distance. After all the wire has been wound, they are soldered onto the rods. Then one layer is cut off and each second wire removed. Such wire **polarizers** are only useful for the long-wavelength infrared. For shorter wavelengths, metal is vacuum deposited on a transparent substrate, as shown in Figure 6.3a, or photoetching methods are used. The metallic grid reflects or absorbs electric field components parallel ($\parallel$) to the "wires," but transmits the perpendicular ($\perp$) components (Figure 6.3b).

Another method is to produce selective absorption internally, such as in commercially available polarizing sheets. This material, a microscopic analog to the wire grid polarizer, is a polymer (polyvinyl alcohol) whose chains are aligned mechanically by stretching and then impregnated with an iodine dye. The iodine attaches itself to the polymer chain sites and contributes conduction electrons that are mobile only along the chain axis. These electrons can then absorb energy if the electric field is parallel to the chain axis. Thus, this "one-dimensional conductor" mimics the deposited metal in the wire grid polarizer.

C. Crossed Polarizers

We consider light incident on a linear polarizer. A filter selects one wavelength whose amplitude when passing the polarizer is A. We add another polarizer, called the **analyzer**, and orient it in such a way that the directions of polarizer and analyzer are mutually perpendicular (Figure 6.4a). No light can pass these **crossed polarizers**. We now insert an additional polarizer with its polarization direction at an angle α with the polarizer and β with the analyzer (Figure 6.4b). Light can now pass through the entire setup. The amount of light can be determined simply by introducing the unit vectors $\mathbf{a}_0$, $\mathbf{b}_0$, and $\mathbf{c}_0$ along the polarization directions of the polarizer, additional polarizer, and analyzer, respectively. The amplitude passing the polarizer is now called $A\mathbf{a}_0$. Forming the dot product with $\mathbf{b}_0$ represents the projection of $A\mathbf{a}_0$ onto $\mathbf{b}_0$, that is for the amplitude*

$$A(\mathbf{a}_0 \cdot \mathbf{b}_0) = A \cos \alpha \tag{6.6}$$

This passing component is now projected onto the $\mathbf{c}_0$ direction, and we have

$$A(\mathbf{a}_0 \cdot \mathbf{b}_0)(\mathbf{b}_0 \cdot \mathbf{c}_0) = A \cos \alpha \cos \beta \tag{6.7}$$

Since $\beta = 90° - \alpha$, we have

$$A \cos \alpha \sin \alpha = A \tfrac{1}{2} \sin 2\alpha \tag{6.8}$$

From this we see that the maximum amount of light emerging is obtained for $\alpha = 45°$, as we would expect by considering the symmetry of the setup as well.

D. Phase Difference Between Polarization Components

Let us return to equations 6.1 and 6.2. The two waves are vibrating in the Y and Z directions and propagating in the X direction, and they have the same amplitude A and wave vector component k. But now we assume that there is a constant

---

*For the intensity one has $I = I_0 \cos^2 \alpha$; this is called Malus' Law.

phase difference of δ:

$$\mathbf{E}_Y = \mathbf{j}A_{Y_0}e^{i(kX-\omega t)} \tag{6.9}$$

$$\mathbf{E}_Z = \mathbf{k}A_{Z_0}e^{i(kX-\omega t+\delta)} \tag{6.10}$$

We saw in Chapter 2 that a phase difference of δ may be attributed to the space fraction or time fraction of the phase factor. At this point, we will consider δ as part of the space fraction.

For $\delta = 0$ and $\delta = \pi$ (where $e^{i\pi} = -1$), and assuming the same amplitude A for both waves, we have $A_{Y_0} = A$, $A_{Z_0} = A$ and $A_{Y_0} = A$, $A_{Z_0} = -A$, respectively. We are back to linearly polarized light (equations 6.1 and 6.2). In general, δ will have the effect of shifting the two waves against each other along the X axis (Figure 6.5). To draw a sketch shown in Figure 6.5, we assume that a particular time instant is chosen such that the time dependence is a maximum, for example, $e^{-i\omega t} = 1$. The resulting wave is expressed, using equations 6.9 and 6.10, as

$$\mathbf{E} = \mathbf{E}_Y + \mathbf{E}_Z = A(\mathbf{j} + \mathbf{k}e^{i\delta})e^{i(kX-\omega t)} \tag{6.11}$$

The intensity is then

$$\begin{aligned} E^2 &= A^2(\mathbf{j} + \mathbf{k}e^{i\delta})(\mathbf{j} + \mathbf{k}e^{-i\delta})e^{i(kX-\omega t)}e^{-i(kX-\omega t)} \\ &= A^2(\mathbf{j}^2 + \mathbf{k}^2) = 2A^2 \end{aligned} \tag{6.12}$$

This result is drastically different from what we obtained in Chapter 2. Here the superposition of the two waves results in a constant value, independent of the phase difference.

3. POLARIZED LIGHT IN UNIAXIAL CRYSTALS

A. Phase Difference Originating from Different k Values

Again let us consider equations 6.1 and 6.2 with $\delta = 0$. We assume that $A_{Y_0} = A_{Z_0} = A$ but that the two wave vectors are different ($k_1 < k_2$):

$$\mathbf{E}_Y = \mathbf{j}Ae^{i(k_1X-\omega t)} \tag{6.13}$$

$$\mathbf{E}_Z = \mathbf{k}Ae^{i(k_2X-\omega t)} \tag{6.14}$$

Because $k_1 \neq k_2$, we expect the two waves to have different velocities of propagation. This is the same situation as two waves propagating along a direction perpendicular to the optic axis in a uniaxial crystal (Chapter 5).

For equation 6.14, we can write

$$\begin{aligned} \mathbf{E}_Z &= \mathbf{k}Ae^{i(k_1X-\omega t+k_2X-k_1X)} \\ &= \mathbf{k}Ae^{i(k_1X-\omega t+\phi_X)} \end{aligned} \tag{6.15}$$

where $\phi_X = (k_2 - k_1)X$. Now, for further consideration, we have

$$\mathbf{E}_Y = \mathbf{j}Ae^{i(k_1X-\omega t)} \tag{6.16}$$

$$\mathbf{E}_Z = \mathbf{k}Ae^{i(k_1X-\omega t+\phi_X)} \tag{6.17}$$

Equations 6.16 and 6.17 bear a close resemblance to equations 6.9 and 6.10.

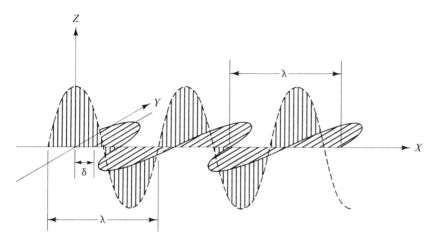

Fig. 6.5
Waves vibrating in E_Y and E_Z directions with a phase difference of δ. The waves are drawn at a time instant where $e^{-i\omega t} = 1$.

But here the phase factor depends on X and is introduced through different k_1 and k_2 values. Also, the two waves depicted by equations 6.16 and 6.17 or 6.13 and 6.14 have different wavelengths, unlike those of equations 6.9 and 6.10, shown in Figure 6.5.

B. The Half-Wave Plate

Let us consider the case of waves traveling in a uniaxial crystal perpendicular to the optic axis. When the polarized light enters the crystal, the two components have a phase difference $\phi_X = 0$. After traveling a certain distance X, they have $\phi_X = (k_2 - k_1)X$. For a specific $X = L_h$, $\phi_X = \pi$ and $e^{i\pi} = -1$. When equations 6.16 and 6.17 are added, the resulting waves are

$$\mathbf{E}_0 = A(\mathbf{j} + \mathbf{k})e^{i(k_1 X - \omega t)} \quad \phi_X = 0$$

and

$$\mathbf{E}_\pi = A(\mathbf{j} - \mathbf{k})e^{i(k_1 X - \omega t)} \quad \phi_X = \pi \tag{6.18}$$

Formally, these equations would be the same as equations 6.3 and 6.4 if we had used $A_{Y_0} = A_{Z_0} = A$.

Entering the crystal plate is linearly polarized light composed of the components E_Y and E_Z with a resultant at 45°. Emerging from the plate of thickness L the light has components $-E_Z$ and E_Y and a resultant at $-45°$ (Figure 6.6a and b).

The distance L_h necessary to produce a phase change of π is

$$L_h = \frac{\pi}{k_2 - k_1} \tag{6.19}$$

as can also be seen from $\phi_X = (k_2 - k_1)X \to \pi$.

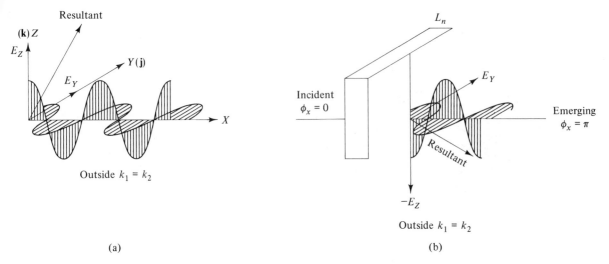

Fig. 6.6
Two components of linearly polarized light passing through a half-wave plate: (a) incident;
(b) emerging.

Making a sketch of how the Z component advances while the waves move along the X axis in the crystal is somewhat complicated and will not be presented here. But Table 6.1 lists the various quantities involved.

We know that both components vibrate with the same frequency, but the Z polarization component has a smaller wavelength than the Y component. After a certain number of vibrations, the Y component has advanced by $\lambda/2$, so that the net phase difference corresponds to $\phi = \pi$. This is the reason that such a plate is called a **half-wave plate**.

We see that a uniaxial crystal of thickness

$$L_h = \frac{\pi}{k_2 - k_1}$$

produces a phase shift $\phi = L_h(k_2 - k_1) = \pi$ between the two components. For the visible part of the spectrum ($\lambda = 10^{-7}$ m), such a plate would be quite thin. However, a plate of thickness

$$L_h' = \frac{\pi + 2\pi m}{k_2 - k_1}$$

Table 6.1
Relations of k, n, λ, and v for Y and Z polarization

j, Y		**k, Z**
k_1	$<$	k_2
n_1	$<$	n_2
λ_1	$>$	λ_2
v_1	$>$	v_2

6. Polarized Light

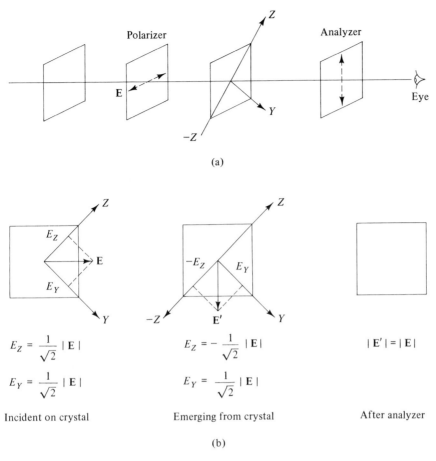

(a)

$E_Z = \dfrac{1}{\sqrt{2}} \, |\,E\,|$

$E_Y = \dfrac{1}{\sqrt{2}} \, |\,E\,|$

Incident on crystal

$E_Z = -\dfrac{1}{\sqrt{2}} \, |\,E\,|$

$E_Y = \dfrac{1}{\sqrt{2}} \, |\,E\,|$

Emerging from crystal

$|\,E'\,| = |\,E\,|$

After analyzer

(b)

Fig. 6.7
Light passing a half-wave plate. The light passing the polarizer may be considered
to have the two components E_Z and E_Y. When the light leaves the half-wave plate,
the component E_Z is changed into $-E_Z$. The total amount of light passing the
polarizer is the same as that passing the analyzer. (a) Polarization with respect to
polarizer and analyzer. (b) View in direction of the light.

(where m is an integer) would serve the same purpose and is more feasible to
produce.

To see if a crystal is a half-wave plate, we position it between crossed
polarizers (Figure 6.7a). The optic axis of the half-wave plate is positioned at 45°
to the polarizer. As a result, wave components polarized parallel and perpendicular to the optic axis are incident on the half-wave plate (Figure 6.7b). Leaving
the half-wave plate, the E_Z component is turned by 180°, and the resulting wave
is now parallel to the analyzer and may pass through it. The absolute value of
the amplitude passing the analyzer is the same as that passing the polarizer.
Clearly, we would not observe any light passing the analyzer if we had oriented
the half-wave plate with its Z axis either parallel or perpendicular to the direction
of the polarizer.

The light vibrating along two perpendicular axes of a uniaxial crystal has

3. Polarized Light in Uniaxial Crystals

an advancing component and a retarding component. The direction of polarization that is advanced is called the **fast axis**; the direction that is retarded is called the **slow axis**. For a positive uniaxial crystal such as quartz, with n_e larger than n_o, the extraordinary ray is polarized along the slow axis. For negative uniaxial crystals such as calcite, we have n_e smaller than n_o, and the extraordinary ray is polarized along the fast axis.

C. Applications of the Half-Wave Plate

One application of the half-wave plate is concerned with laser light. We know that lasers may use Brewster's angles to produce linearly polarized output. To produce light perpendicularly polarized to the laser output, a half-wave plate may be used. This procedure is important when the lasers themselves cannot be rotated.

A half-wave plate may be crudely constructed in the laboratory by layering pieces of transparent cellophane tape on a microscope slide. By trial and error, we can find the correct number of layers so that the "net" phase difference for the device is approximately π, as required. Other plastics may also be used. If stretched, many of them become uniaxial crystals, as discussed in Chapter 5.

A half-wave plate can also be obtained by using any natural cleavage pieces of mica, which is a biaxial crystal. The two axes are almost in the cleavage planes and make, for different types of mica, angles of $0°$–$50°$ with each other. The refractive indices along these two axes are very similar (1.599 and 1.594 for $\lambda = 589.29$ nm) and about the same for different types of mica. The thickness is found by trial and error. In the arrangement of crossed polarizers, we can determine the correct device by rotation of the analyzer. If the analyzer is perpendicular to the polarizer, maximum light should pass; if parallel to the polarizer, no light should pass.

Crystal quartz may also be utilized for a half-wave plate. Quartz has a large wavelength range in the visible and near-infrared, as well as for the far-infrared. The refractive indices for $\lambda = 5790.66$ Å are $n_o = 1.54467$ and $n_e = 1.55379$. For $\lambda = 52.4$ μm, we have $n_o = 2.1941$ and $n_e = 2.2502$.

D. The Quarter-Wave Plate

We will now consider a phase change of only $\phi'_X = 0$ to $\phi'_X = \pi/2$ for light traveling in a uniaxial crystal. Since this is half the phase change encountered for the half-wave plate, such a plate is called a **quarter-wave plate**.

We begin by returning to equations 6.16 and 6.17. Consider the following equations:

$$\mathbf{E}_Y = \mathbf{j}Ae^{i(k_1 X - \omega t)} \tag{6.20}$$

$$\mathbf{E}_Z = \mathbf{k}Ae^{i(k_1 X - \omega t + \phi'_X)} \tag{6.21}$$

For these waves, $\phi'_X = 0$ for the light incident on the crystal. After the waves travel the distance $X = L'$, we have

$$\phi'_X = (k_2 - k_1)L' = \frac{\pi}{2}$$

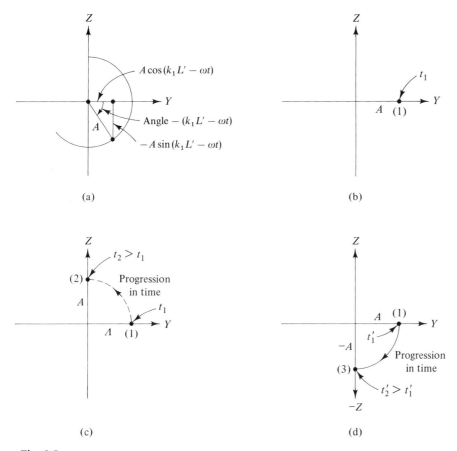

(a)

(b)

(c)

(d)

Fig. 6.8
(a) Components in the Y and Z directions of left polarized light (counterclockwise rotation). The X direction points out of the plane of the page. (b) Time instant where $E_Y = A$, $E_Z = 0$. (c) Time instant later where $E_Y = 0$, $E_Z = A$. (d) Right polarized light. Two positions (1) and (3) of the resultant are indicated (clockwise rotation).

Entering the crystal plate at $\phi'_X = 0$, the waves have a resultant vibrating along the direction of $\mathbf{j} + \mathbf{k}$, that is, at a $45°$ angle to the Y axis.

For the emerging light at the plane $X = L'$, and taking only the real parts of equations 6.20 and 6.21, we get

$$\mathbf{E}_Y = \mathbf{j}A \cos(k_1 L' - \omega t) \tag{6.22}$$

$$\mathbf{E}_Z = \mathbf{k}A \cos\left(k_1 L' - \omega t + \frac{\pi}{2}\right) = -\mathbf{k}A \sin(k_1 L' - \omega t) \tag{6.23}$$

where $\cos(\alpha + \pi/2) = -\sin\alpha$.

The E_Y component vibrates as a cosine function, and the E_Z component as a sine function, both in the same plane $X = L'$ (Figure 6.8a).

At the instant $t_1 = k_1 L'/\omega$, we have $E_Y = A$ and $E_Z = 0$, the resultant is parallel to the Y direction (Figure 6.8b). At some time later,

$$t_2 = \frac{k_1 L'}{\omega} + \frac{1}{\omega}\left(\frac{\pi}{2}\right)$$

we have $E_Y = 0$ and $E_Z = A$, as can be seen from

$$-A \sin\left\{k_1 L' - \omega\left[\frac{k_1 L'}{\omega} + \frac{1}{\omega}\left(\frac{\pi}{2}\right)\right]\right\} = -A \sin\left(-\frac{\pi}{2}\right) = +A$$

During this time, the resultant has moved in a counterclockwise rotation in the plane $X = L'$ (looking toward the Y-Z plane and the source of light) from (1) to (2), as shown in Figure 6.8c. The path of the resultant in the $X = L'$ plane is obtained by eliminating the time dependence of equations 6.22 and 6.23:

$$E_Y^2 + E_Z^2 = A^2 \cos^2(k_1 L' - \omega t) + A^2 \sin^2(k_1 L' - \omega t) = A^2 \qquad (6.24)$$

This is the equation of a circle.

When the resultant rotates in a counterclockwise direction, we speak of **left circularly polarized** light.

Right circularly polarized light is obtained for $\phi_x = 3\pi/2$. From equations 6.20 and 6.21, we have looking toward the Y, Z plane and the source of light

$$\mathbf{E}_Y = \mathbf{j}A \cos(k_1 L'' - \omega t) \qquad (6.25)$$

$$\mathbf{E}_Z = \mathbf{k}A \cos\left(k_1 L'' - \omega t + \frac{3\pi}{2}\right) = \mathbf{k}A \sin(k_1 L'' - \omega t) \qquad (6.26)$$

where

$$\cos(\alpha + 270) = \cos(\alpha + 180 + 90) = -\sin(\alpha + 180) = \sin\alpha$$

For $t_1' = k_1 L''/\omega$, we again have $E_Y = A$ and $E_Z = 0$. Later, at time

$$t_2' = \frac{k_1 L''}{\omega} + \frac{1}{\omega}\left(\frac{\pi}{2}\right)$$

we have $E_Y = 0$ and $E_Z = -A$, as can be shown from

$$E_Z = A \sin\left\{k_1 L'' - \omega\left[\frac{k_1 L''}{\omega} + \frac{1}{\omega}\left(\frac{\pi}{2}\right)\right]\right\} = A \sin\left(-\frac{\pi}{2}\right) = -A$$

During this time, the resultant has moved in a clockwise direction in the plane $X = L''$ from (1) to (3), as illustrated in Figure 6.8d.

The amplitude A in equations 6.22 and 6.23 were assumed to be equal. For different amplitudes in the Y and Z directions, we would obtain an ellipse from equation 6.24 and have left elliptically polarized light. Similar considerations apply to right elliptically polarized light.

From

$$\phi_x' = \frac{\pi}{2} = L'(k_2 - k_1)$$

we have for a quarter-wave plate the thickness

$$L' = \frac{\pi/2}{k_2 - k_1} \qquad (6.27)$$

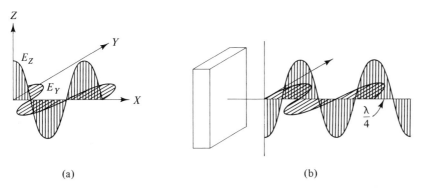

(a)   (b)

Fig. 6.9
Phase relation of the two components of polarized light (a) before entering and
(b) emerging from a quarter-wave plate.

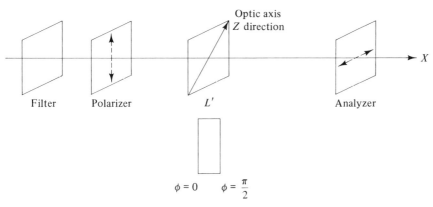

Fig. 6.10
A quarter-wave plate between crossed polarizers.

Similarly, as discussed for the half-wave plate, a plate of thickness

$$L' = \frac{\pi/2 + 2\pi m}{k_2 - k_1}$$

(where m is an integer) would serve the same purpose and is easier to produce.

In Figure 6.9, we show the phase relation of the two components incident and emerging from the quarter-wave plate for a certain time instant t.

The result of testing a quarter-wave plate between crossed polarizers is illustrated in Figure 6.10. The quarter-wave plate is inserted with the optic axis at an angle of $45°$ to the polarizer and analyzer. Since the resulting electric vector is rotated around the X axis in the plane $X = L'$ and emerges as such from the quarter-wave plate, light will pass the analyzer. Rotation of the analyzer does not affect the results in any way. If the quarter-wave plate is rotated so that its axis is parallel to the polarizer, one of the two mutually perpendicular polarized components is eliminated, and the quarter-wave plate will not be able to function.

3. Polarized Light in Uniaxial Crystals   259

Quarter-wave plates may be constructed in the same way as discussed for the half-wave plate, with half the thickness.

E. Elliptically Polarized Light

As already mentioned, we can produce elliptically polarized light by assuming that the amplitudes of the Y and Z components (in equations 6.22 and 6.23) are not equal. But elliptically polarized light can also be produced with equal amplitudes of the Y and Z components. Depending on the value of ϕ_X, that is, the thickness of the uniaxial crystal, we may observe linearly, circularly, or elliptically polarized light.

First we consider ϕ_X between 0 and 2π in increments of $\pi/4$. Already discussed are the results for $\phi_X = 0, \pi/2, \pi, 3\pi/2$, and 2π. The result for 2π is, of course, the same as that for zero because of periodicity. We have compiled these five cases (and several others) in Table 6.2.

To study the case of $\phi_X = \pi/4$, we write equations 6.20 and 6.21 with the general phase term of ϕ and take only the real part:

$$\mathbf{E}_Y = \mathbf{j}A \underbrace{\cos(k_1 L'' - \omega t)}_{\alpha} \tag{6.28}$$

$$\mathbf{E}_Z = \mathbf{k}A \underbrace{\cos(k_1 L'' - \omega t}_{\alpha} + \phi) = \mathbf{k}A(\cos\alpha\cos\phi - \sin\alpha\sin\phi) \tag{6.29}$$

where L'' is the length corresponding to ϕ and $k_1 L'' - \omega t = \alpha$. Now taking $\phi = \pi/4$, we have $\cos\phi = \sin\phi = 1/\sqrt{2}$. We can obtain an equation relating E_Y and E_Z if we eliminate $\cos\alpha$ and $\sin\alpha$. To simplify, we set $A = 1$ and have

$$E_Z = E_Y \frac{1}{\sqrt{2}} - \frac{1}{\sqrt{2}}\sqrt{1 - E_Y^2} \tag{6.30}$$

which can also be written as

$$E_Z^2 - \sqrt{2}E_Z E_Y + E_Y^2 = \tfrac{1}{2} \tag{6.31}$$

This is the equation of an ellipse. We illustrate this by the transformation

$$E_Y = \frac{1}{\sqrt{2}}(E_Y' - E_Z')$$

$$E_Z = \frac{1}{\sqrt{2}}(E_Y' + E_Z') \tag{6.32}$$

From the rotation matrix discussed in Chapter 1, we see that equation 6.31 is a rotation from the Y, Z coordinate system to the new Y', Z' system by $45°$ (Figure 6.11).

Introducing equation 6.32 into equation 6.31 results in

$$\frac{E_Y'^2}{1 + 1/\sqrt{2}} + \frac{E_Z'^2}{1 - 1/\sqrt{2}} = 1 \tag{6.33}$$

Table 6.2
Values of the components E_Y and E_Z for different values of ϕ_X (looking toward the Y-Z plane and the source of light)

ϕ	E_Y	E_Z	Graph
0	$\underbrace{\cos(k_1 L'' - \omega t)}_{\alpha}$	$\underbrace{\cos(k_1 L'' - \omega t)}_{\alpha}$	
$\dfrac{\pi}{4}$	$\cos\alpha$	$\dfrac{1}{\sqrt{2}}(\cos\alpha - \sin\alpha)$	
$\dfrac{\pi}{2}$	$\cos\alpha$	$-\sin\alpha$	
$\dfrac{3\pi}{4}$	$\cos\alpha$	$-\dfrac{1}{\sqrt{2}}(\cos\alpha + \sin\alpha)$	
π	$\cos\alpha$	$-\cos\alpha$	
$\dfrac{5\pi}{4}$	$\cos\alpha$	$\dfrac{1}{\sqrt{2}}(-\cos\alpha + \sin\alpha)$	
$\dfrac{3\pi}{2}$	$\cos\alpha$	$\sin\alpha$	
$\dfrac{7\pi}{4}$	$\cos\alpha$	$\dfrac{1}{\sqrt{2}}(\cos\alpha + \sin\alpha)$	
2π	$\cos\alpha$	$\cos\alpha$	

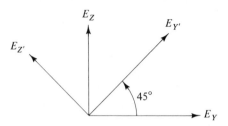

Fig. 6.11
Rotation of $+45°$ to the coordinate system Y', Z' with respect to the Y, Z system.

This ellipse has a larger semiaxis in the Y' direction than in the Z' direction, as indicated in Table 6.2.

If we now look at the case where $\phi_X = 3\pi/4$, we start off with equations 6.28 and 6.29 ($A = 1$):

$$E_Y = \cos \alpha$$
$$E_Z = \cos \alpha \cos \phi - \sin \alpha \sin \phi \tag{6.34}$$

Inserting $3\pi/4$ for ϕ into the equations results in

$$E_Y = \cos \alpha \tag{6.35}$$

$$E_Z = \cos \alpha \left(\frac{-1}{\sqrt{2}}\right) - \sin \alpha \left(\frac{1}{\sqrt{2}}\right) \tag{6.36}$$

Elimination of α gives

$$E_Z = -E_Y \frac{1}{\sqrt{2}} - \frac{1}{\sqrt{2}}\sqrt{1 - E_Y^2} \tag{6.37}$$

and we have

$$E_Z^2 + \sqrt{2}E_Z E_Y + E_Y^2 = \tfrac{1}{2} \tag{6.38}$$

which is quite similar to equation 6.31. Using the same transformation (equation 6.32) will rotate the Y', Z' coordinate system by $45°$, and we get

$$\frac{E_Y'^2}{1 - 1/\sqrt{2}} + \frac{E_Z'^2}{1 + 1/\sqrt{2}} = 1 \tag{6.39}$$

Comparing equation 6.39 with equation 6.33, we see that the two semiaxes have been interchanged. This is also shown in Table 6.2. The cases where $\phi_X = 5\pi/4$ and $\phi_X = 7\pi/4$ are analyzed in a similar manner (see problem 2). All other cases are also compiled in Table 6.2.

With respect to the sense of rotation, we have discussed left circularly polarized light ($\pi/2$) and right circularly polarized light ($3\pi/2$) in detail. Looking at Table 6.2, we expect to find that the ellipses for $\pi/4$ and $3\pi/4$, as well as the left circularly polarized case for $\pi/2$, rotate in the same direction. The sense of rotation does not change as a result of ϕ_X being larger or smaller than $\pi/2$. Only the form of the ellipse changes. This can be proved for these three cases and

6. **Polarized Light**

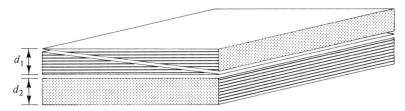

Fig. 6.12
Soleil compensator. Optic axes are indicated by lines and dots. Gaps are exaggerated.

similarly for the other three cases with $\phi_x > \pi$ (opposite sense). The sense of rotation changes, however, at $\phi_x = 0$, π, and 2π, where the ellipse degenerates into a straight line and we can no longer speak of rotation.

F. Applications to Soleil Compensator

As we mentioned earlier, a half-wave plate can be produced by making the thickness at least equal to $\pi/(k_2 - k_1)$ but for practical purposes, we make it equal to $(\pi + 2\pi m)/(k_2 - k_1)$ (where m is an integer). A similar argument can be applied to the quarter-wave plate.

How can we make a plate of variable thicknesses, that is, one that can be used as a quarter-wave plate, a half-wave plate, or a plate able to produce any other phase change? Such a device is shown in Figure 6.12 and called a **Soleil compensator**. Two wedges are placed together so that moving them against each other produces a plate of variable thickness d_1. The optic axis is oriented in the same direction in both wedges, as indicated by lines in Figure 6.12. A second plate of the same material is placed below the wedges, having a fixed thickness d_2, and the optic axis is perpendicular to the optic axes of the wedges. This optic axis is indicated by dots in Figure 6.12.

When light enters the first plate, depending on the thickness employed, a phase difference ϕ_1 is produced:

$$\phi_1 = (k_2 - k_1)d_1 \tag{6.40}$$

Since the optic axis of the second plate is oriented perpendicular to that of the first plate, the two components of the polarized light change their roles, and a phase difference ϕ_2 is produced:

$$\phi_2 = -(k_2 - k_1)d_2 \tag{6.41}$$

The total phase difference (absolute value) can then be obtained as

$$\Delta = \phi_2 - \phi_1 = |k_2 - k_1|(d_2 - d_1) = \left| \frac{2\pi n_2}{\lambda_0} - \frac{2\pi n_1}{\lambda_0} \right| (d_2 - d_1)$$
$$= \frac{2\pi}{\lambda_0} |n_2 - n_1|(d_2 - d_1) \tag{6.42}$$

where λ_0 is the wavelength in vacuum and refractive indices are introduced. For $d_2 = d_1$, $\Delta = 0$.

Fig. 6.13
Babinet compensator (cross section).

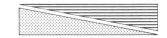

Soleil compensators are made of quartz, magnesium fluoride, cadmium sulfide, or other materials suitable for the spectral region being considered. A simple compensator for a much smaller light beam diameter is shown in Figure 6.13. It is called a **Babinet compensator**.

G. Appearance of Elliptically Polarized Light

To summarize, two mutually perpendicular components of linearly polarized light can be described by

$$\mathbf{E}_Y = \mathbf{j}A_{Y_0}e^{i(kX-\omega t)} \tag{6.43}$$

$$\mathbf{E}_Z = \mathbf{k}A_{Z_0}e^{i(kX-\omega t+\Delta)} \tag{6.44}$$

Upon superposition (or vector addition), we get: for $\Delta = 0$, linearly polarized light; for $A_{Y_0} = A_{Z_0} = A$ and $\Delta \neq 0$, linearly, circularly, or elliptically polarized light, depending on the specific value of Δ; for $A_{Y_0} \neq A_{Z_0}$ and $\Delta \neq 0$, in general, elliptically polarized light, but in specific cases this degenerates into linearly polarized light.

We have seen how the superposition of two waves (equations 6.43 and 6.44) proceeds in a uniaxial crystal. The phase angle Δ depends on the length of propagation X in the crystal; that is, it depends on the length of travel (or the thickness of the crystal plate, which includes the periodicity of $2\pi m$).

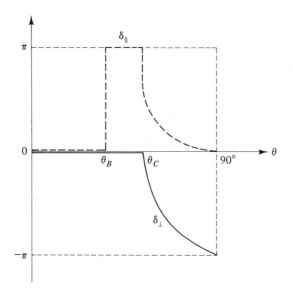

Fig. 6.14
Phase change of $\delta_\parallel$ and $\delta_\perp$ for $\theta > \theta_c$, where θ is the angle of reflection and θ_c is the critical angle.

6. Polarized Light

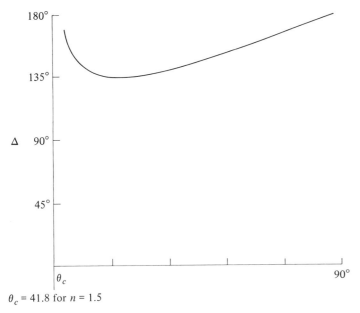

$\theta_c = 41.8$ for $n = 1.5$

Fig. 6.15
Plot of Δ in the range from θ_c to $\theta = 90°$, that is, for total internal reflection.

A slightly different application is the production of all types of polarized light upon internal reflection of light in dielectrics. We assume that the incident light has amplitude $A_{Y_0} = A_{Z_0} = A$. Upon internal reflection, this amplitude value is not changed (see Figure 4.18); but depending on the angle θ of reflection, Δ does change. Consider Figure 6.14, where we repeat part of the graph in Figure 4.18 for the phase change for $\theta > \theta_c$, where θ is the angle of reflection and θ_c is the critical angle. Now the phase angle Δ does not depend on any propagation length, but is given by the difference of the phase change for the parallel and perpendicular components.

The angle δ can be calculated from Fresnel's formulas. For the region of θ where we have total internal reflection, we can write $|r_{E_\perp}|e^{i\delta_\perp}$ and $|r_{E_\parallel}|e^{i\delta_\parallel}$ for the reflected amplitudes, where $|r_{E_\parallel}|$ and $|r_{E_\perp}|$ have the absolute value 1. The angle Δ is given as $\delta_\parallel - \delta_\perp$. The calculation is straightforward, however, and can be simplified if a trick is used (see problem 3). The result is

$$\tan\frac{\Delta}{2} = \frac{\sin^2\theta}{\cos\theta\sqrt{\sin^2\theta - (n_2/n_1)^2}} \tag{6.45}$$

Figure 6.15 shows the plot of Δ for the range θ_c to $\theta = 90°$.

When there are multiple internal reflections, we obtain different types of polarized light after each internal reflection, as shown in Figure 6.16. Such a crystal is called a **Fresnel rhomb**, and problem 4 may be done to calculate the numerical details.

An application similar to the internal reflection in dielectrics, but with a change in the magnitude of the reflected amplitude A, is the reflection on metals. This will be discussed in Chapter 7.

3. Polarized Light in Uniaxial Crystals

265

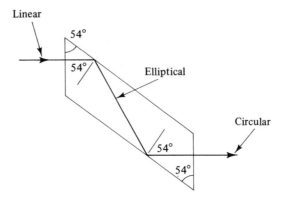

Fig. 6.16
Fresnel rhomb. After two internal reflections, circularly polarized light emerges. Light rays are shown in cross section.

H. Retardation by the Electrooptical Effect: Half-Voltage

The phase difference between the two waves perpendicularly polarized to each other and propagating in the same direction with different **k** vectors is, similar to equation 6.42,

$$\Delta\phi = \frac{2\pi}{\lambda}|\Delta n|\, L \tag{6.46}$$

where L is the length of the medium through which the waves travel. The difference between the refractive indices corresponding to the vibration in the two mutually perpendicular directions is indicated by Δn. In the electrooptical effects, the difference Δn is produced by the application of an electric field E. For the Kerr effect, the direction of the electric field is perpendicular to the direction of propagation; for the linear Pockels effect, it is parallel.

From equation 5.14, the Kerr effect (transverse electrooptical effect), $\Delta n = K\lambda_0\, V^2/d^2$, and we have a phase difference of

$$\Delta\phi_K = \frac{2\pi}{\lambda}\left|K\lambda_0\frac{V^2}{d^2}\right|L \tag{6.47}$$

From equations 5.16 and 5.17, the linear Pockels effect (E in direction of propagation), $\Delta n = n_o^3 r_{63}\, V/d$, and we have a phase difference of

$$\Delta\phi_P = \frac{2\pi}{\lambda}|n_o^3 r_{63}E|\,L = \frac{2\pi}{\lambda}\left|n_o^3 r_{63}\frac{V}{d}\right|L \tag{6.48}$$

In the case of the linear Pockels effect, we may assume that $d = L$. The voltage producing $\Delta\phi = \pi$, that is, for a half-wave plate, is called the **half-voltage**, and we have

$$V_{1/2} = \frac{\lambda}{2n_o^3 r_{63}} \tag{6.49}$$

For KDP (potassium dihydrogen phosphate), with $\lambda = 0.5 \times 10^{-6}$ m, we obtain

　　6. **Polarized Light**

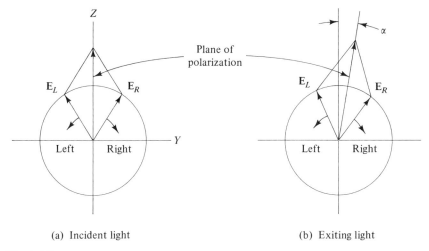

(a) Incident light   (b) Exiting light

Fig. 6.17
Rotation of the plane of polarization in an optically active material. Light is traveling out of the page, toward the viewer.

with $n_o = 1.51$, $r_{63} = 10.6 \times 10^{-12}$, a half-voltage of $V_{1/2} \simeq 700$ V. For ADP (ammonium dihydrogen phosphate), see problem 5.

4. OPTICAL ACTIVITY

The phenomenon of **optical rotation** by materials is described by considering the incident plane parallel polarized light as consisting of equal amounts of left and right circularly polarized light. The compound through which the light travels changes the phase difference between these two components, depending on the length of travel.

Circular **dichroism** is observed when either the left or the right polarized light component is more absorbed than the other.

A. Optical Rotation

A substance that shows optical rotation has the ability to rotate the plane of polarization of light traveling through it. The observed phenomenon is called optical rotation because we measure the angle through which the plane of polarization is rotated as the wave traverses a known path length. For a material to exhibit optical rotation, a helical arrangement of atoms is necessary. Therefore, the molecule has neither a plane nor a center of symmetry.

To describe optical rotation, we consider a plane-polarized light wave incident normally on the surface of the sample. We have previously seen that plane-polarized light can be considered a combination of left and right circularly polarized light of equal amplitudes. Thus, in Figure 6.17a, light incident on the sample is polarized in the vertical plane. (Note that we label the sense of rotation by sighting along the axis toward the Y-Z plane and the source of light. The wave in Figure 6.17 is directed out of the page toward the reader.) After traversing an optically active sample, the light is still plane-polarized, but the plane has been rotated through an angle α (Figure 6.17b). The asymmetry produced by the helical arrangement of atoms causes the left and right circular components to "see" different environments. Consequently, each experiences a different velocity of propagation and index of refraction. The angle of rotation α will depend on the difference between these indices of refraction.

If we assume that the wave is traveling along the X axis toward the sample, the left circularly polarized light has components

$$\mathbf{E}_Y = \mathbf{j}Ae^{i(k_L X - \omega t)} \tag{6.50}$$

and

$$\mathbf{E}_Z = \mathbf{k}Ae^{i(k_L X - \omega t + \pi/2)} \tag{6.51}$$

so that

$$\mathbf{E}_L = Ae^{-i\omega t}\{\mathbf{j}[e^{ik_L X}] + \mathbf{k}[e^{i(k_L X + \pi/2)}]\} \tag{6.52}$$

Similarly, for right circularly polarized light, we have

$$\mathbf{E}_R = Ae^{-i\omega t}\{\mathbf{j}[e^{ik_R X}] + \mathbf{k}[e^{i(k_R X - \pi/2)}]\} \tag{6.53}$$

If we take the real part of the spatially dependent terms in equations 6.52 and 6.53, we can write

$$\mathbf{E} = \mathbf{E}_L + \mathbf{E}_R = Ae^{-i\omega t}\mathbf{j}[\cos(k_L X) + \cos(k_R X)] +$$
$$\mathbf{k}[\cos(k_L X + \pi/2) + \cos(k_R X - \pi/2)]$$

or

$$\mathbf{E} = Ae^{-i\omega t}\mathbf{j}[\cos(k_L X) + \cos(k_R X)] + \mathbf{k}[-\sin(k_L X) + \sin(k_R X)] \tag{6.54}$$

Using appropriate trigonometric identities, we can rewrite this result as

$$\mathbf{E} = Ae^{-i\omega t}[\mathbf{j}\{2\cos[\tfrac{1}{2}(k_R - k_L)X]\} \cdot \{\cos[\tfrac{1}{2}(k_R + k_L)X]\} +$$
$$\mathbf{k}\{2\sin[\tfrac{1}{2}(k_R - k_L)X]\} \cdot \{\cos[\tfrac{1}{2}(k_R + k_L)X]\}]$$

or

$$\mathbf{E} = Ae^{-i\omega t}2\cos[\tfrac{1}{2}(k_R + k_L)X] \cdot \{\mathbf{j}\cos[\tfrac{1}{2}(k_R - k_L)X] + \mathbf{k}\sin[\tfrac{1}{2}(k_R - k_L)X]\} \tag{6.55}$$

The factors preceding the curly brace in equation 6.55 describe a wave traveling in the X direction with a wave vector $(k_R + k_L)/2$ (average). For $X = 0$, equation 6.55 describes a wave that has its amplitude along the Y axis. When X is increased such that $(k_R - k_L)X/2 = \pi/2$, the amplitude lies along the Z axis, and the vector $\mathbf{E}$ has been rotated through an angle of 90°. Therefore, for a substance of length l, the plane of polarization will be rotated through an angle $\alpha = (k_R - k_L)l/2$. The specific rotatory power for this material is given by $\alpha_s = \alpha/l$ and may be expressed

6. Polarized Light

as

$$\alpha_s = \frac{\pi}{\lambda}(n_R - n_L) \tag{6.56}$$

Since both indices are wavelength dependent, α_s will be too.

Optical rotation is observed in gases and liquids, as well as in solids. In crystal quartz,* for example, the silicon dioxide molecules are arranged in either left- or right-handed helical patterns. These two forms are mirror images of each other. In the right-handed form, we get a value of $\alpha_s \simeq 8\pi \times 10^{-5}$ rad/μm at a wavelength of 0.76 μm. This value corresponds to a fractional difference of about 0.004 percent in the refractive indices.

It is perhaps more difficult to see how a liquid or solution can exhibit optical activity. If we consider a solution of helical molecules, such as biological macro-molecules, it is clear that the oscillating electric field of the incident light wave can produce electric polarization in the helix and thus produce refraction effects. Therefore, we should expect the refractive index to be characteristic of a given helix. However, it might be thought that because of the random orientation of the molecules, interactions in solutions cancel one another and no optical rotation is observed. This is not so. As long as a helical arrangement of atoms is present, optical rotation will take place. Further, a right-handed helix will, when viewed from the opposite end, be a right-handed helix, not a left-handed helix. Solutions of right-handed helices will exhibit a different α than left-handed helices. A solution of equal parts of each, however, will be optically inactive. Optical rotation measurements are very important in the study of structures with helical geometry.

5. JONES VECTORS AND JONES MATRICES

We have already seen how linearly polarized, circularly polarized, and elliptically polarized light is produced by a phase difference between two mutually perpendicular components of light. We have always referred the phase difference with respect to one component of the polarized light.

We now come to a useful formalism. Vector notation will be introduced for the discription of polarized light, and matrices will be used to represent the action of linear polarizers, circular polarizers, or elliptical polarizers.

This method is comprehensive and elegant. It leads to a well-presented method of finding the final phase change and, consequently, the final type of polarization of a system of different optical elements.

A. Jones Vectors

The two components of polarized light were presented in equations 6.9 and 6.10 as

$$\mathbf{E}_Y = \mathbf{j}A_{Y_0}e^{i(kX-\omega t)}$$
$$\mathbf{E}_Z = \mathbf{k}A_{Z_0}e^{i(kX-\omega t+\phi)}$$

---

* Fused quartz is isotropic and therefore optically inactive.

As in Chapters 1 and 4, we can introduce a vector notation and write

$$E_Y = \begin{pmatrix} A_{Y_0} \\ 0 \end{pmatrix} e^{i(kX - \omega t)} \quad \text{and} \quad E_Z = \begin{pmatrix} 0 \\ A_{Z_0} \end{pmatrix} e^{i(kX - \omega t + \phi)} \tag{6.57}$$

We assume that $A_{Y_0} = A_{Z_0} = 1$, and write

$$E_Y = \begin{pmatrix} 1 \\ 0 \end{pmatrix} e^{i(kX - \omega t)} \quad \text{and} \quad E_Z = \begin{pmatrix} 0 \\ e^{i\phi} \end{pmatrix} e^{i(kX - \omega t)} \tag{6.58}$$

We saw in the preceding chapters that the phase difference (e.g., δ or ϕ) between the two components is the most important parameter. This phase factor is included in the vector rotation, and in the following equations we may omit the factor $e^{i(kX - \omega t)}$ as redundant. We then have

$$E_Y = \begin{pmatrix} 1 \\ 0 \end{pmatrix} \quad \text{and} \quad E_Z = \begin{pmatrix} 0 \\ e^{i\phi} \end{pmatrix} \tag{6.59}$$

These vectors are called **Jones vectors**.

For the case where $\phi = 0$, we have linearly polarized light:

$$E_Y = \begin{pmatrix} 1 \\ 0 \end{pmatrix} \quad \text{and} \quad E_Z = \begin{pmatrix} 0 \\ 1 \end{pmatrix} \tag{6.60}$$

Adding these two components leads to polarized light along the diagonal at 45° in the Y-Z plane:

$$E_Y + E_Z = \begin{pmatrix} 1 \\ 0 \end{pmatrix} + \begin{pmatrix} 0 \\ 1 \end{pmatrix} = \begin{pmatrix} 1 \\ 1 \end{pmatrix} \tag{6.61}$$

We may see this more clearly by going back and rewriting equation 6.3 with the notation of equation 6.58.

The two vectors

$$\begin{pmatrix} 1 \\ 0 \end{pmatrix} \quad \text{and} \quad \begin{pmatrix} 0 \\ 1 \end{pmatrix}$$

are a simple example of a system of normalized and orthogonal vectors. The condition for normalization of a vector

$$\begin{pmatrix} f \\ g \end{pmatrix}$$

is

$$(f^* \, g^*) \begin{pmatrix} f \\ g \end{pmatrix} = f^*f + g^*g = 1 \tag{6.62}$$

For orthogonality of the two vectors

$$\begin{pmatrix} f \\ g \end{pmatrix} \quad \text{and} \quad \begin{pmatrix} f' \\ g' \end{pmatrix}$$

we have the condition

$$(f^* g^*)\begin{pmatrix} f' \\ g' \end{pmatrix} = f^* f' + g^* g' = 0 \qquad (6.63)$$

The vector

$$(f \quad g)$$

is the transposed vector of

$$\begin{pmatrix} f \\ g \end{pmatrix}$$

The asterisk indicates that the complex conjugate of each element is taken.
   The vectors

$$\begin{pmatrix} 1 \\ 0 \end{pmatrix} \quad \text{and} \quad \begin{pmatrix} 0 \\ 1 \end{pmatrix}$$

fulfill the conditions of equations 6.62 and 6.63, but the vector $\begin{pmatrix} 1 \\ 1 \end{pmatrix}$ is not normalized; since

$$(1 \quad 1)\begin{pmatrix} 1 \\ 1 \end{pmatrix} = 2$$

However, the vector

$$\frac{1}{\sqrt{2}}\begin{pmatrix} 1 \\ 1 \end{pmatrix}$$

is normalized.
   Using the notation of Table 6.2, we have right circularly polarized (rcp) light for $\phi = \pi/2$ and left circularly polarized (lcp) light for $\phi = 3\pi/2$, equivalent to $-\pi/2$. Therefore, we can express circularly polarized light in vector notation as

$$E_{\text{rcp}} = \begin{pmatrix} 1 \\ e^{i(\pi/2)} \end{pmatrix} = \begin{pmatrix} 1 \\ i \end{pmatrix} \qquad (6.64)$$

$$E_{\text{lcp}} = \begin{pmatrix} 1 \\ e^{-i(\pi/2)} \end{pmatrix} = \begin{pmatrix} 1 \\ -i \end{pmatrix} \qquad (6.65)$$

If we add these two vectors, the result is linearly polarized light along the Y direction:

$$\begin{pmatrix} 1 \\ i \end{pmatrix} + \begin{pmatrix} 1 \\ -i \end{pmatrix} = \begin{pmatrix} 1 \\ 0 \end{pmatrix} \qquad (6.66)$$

The two vectors

$$\begin{pmatrix} 1 \\ i \end{pmatrix} \quad \text{and} \quad \begin{pmatrix} 1 \\ -i \end{pmatrix}$$

are orthogonal

$$(1^* \quad i^*)\begin{pmatrix} 1 \\ -i \end{pmatrix} = (1 - i)\begin{pmatrix} 1 \\ -i \end{pmatrix} = 1 - 1 = 0$$

5. Jones Vectors and Jones Matrices   **271**

but not normalized; for example, for $\begin{pmatrix} 1 \\ i \end{pmatrix}$, we have

$$(1 \quad -i)\begin{pmatrix} 1 \\ i \end{pmatrix} = 2$$

But the vectors

$$E_{\text{rcp}} = \frac{1}{\sqrt{2}}\begin{pmatrix} 1 \\ i \end{pmatrix} \quad \text{and} \quad E_{\text{lcp}} = \frac{1}{\sqrt{2}}\begin{pmatrix} 1 \\ -i \end{pmatrix} \tag{6.67}$$

represent right and left circularly polarized light and are both orthogonal and normalized.

B. Jones Matrices

In Chapter 1, we saw how the action of refraction on a curved surface could be represented by a matrix. In a similar way, we can use a matrix to represent the phase change that an optical element imposes on the incident light. The important parameter to be changed is the phase factor ϕ, and since this is usually done on the Z component, let us consider the following matrix,

$$\begin{pmatrix} 1 & 0 \\ 0 & e^{i\phi} \end{pmatrix} \tag{6.68}$$

which is called a **Jones matrix**. If this matrix operates on linearly polarized light and we assume $\phi = \pi/2$ (see Table 6.2), we get right circularly polarized light:

$$\begin{pmatrix} 1 & 0 \\ 0 & e^{i(\pi/2)} \end{pmatrix}\frac{1}{\sqrt{2}}\begin{pmatrix} 1 \\ 1 \end{pmatrix} = \begin{pmatrix} 1 & 0 \\ 0 & i \end{pmatrix}\frac{1}{\sqrt{2}}\begin{pmatrix} 1 \\ 1 \end{pmatrix} = \frac{1}{\sqrt{2}}\begin{pmatrix} 1 \\ i \end{pmatrix} \tag{6.69}$$

Left polarized light is produced when $\phi = 3\pi/2$ (see Table 6.2), which is equivalent to $\phi = -\pi/2$:

$$\begin{pmatrix} 1 & 0 \\ 0 & e^{-i(\pi/2)} \end{pmatrix}\frac{1}{\sqrt{2}}\begin{pmatrix} 1 \\ 1 \end{pmatrix} = \begin{pmatrix} 1 & 0 \\ 0 & -i \end{pmatrix}\frac{1}{\sqrt{2}}\begin{pmatrix} 1 \\ 1 \end{pmatrix} = \frac{1}{\sqrt{2}}\begin{pmatrix} 1 \\ -i \end{pmatrix} \tag{6.70}$$

The matrices

$$\begin{pmatrix} 1 & 0 \\ 0 & i \end{pmatrix} \quad \text{and} \quad \begin{pmatrix} 1 & 0 \\ 0 & -i \end{pmatrix}$$

represent optical elements acting as quarter-wave plates producing right and left circularly polarized light. If these two optical elements were to act one after the other, we would observe no action:

$$\begin{pmatrix} 1 & 0 \\ 0 & i \end{pmatrix}\begin{pmatrix} 1 & 0 \\ 0 & -i \end{pmatrix} = \begin{pmatrix} 1 & 0 \\ 0 & 1 \end{pmatrix} \tag{6.71}$$

The unit matrix does not change any of the vectors representing the incident light.

On the other hand, if each optical element acts twice, we have

$$\begin{pmatrix} 1 & 0 \\ 0 & i \end{pmatrix} \begin{pmatrix} 1 & 0 \\ 0 & i \end{pmatrix} = \begin{pmatrix} 1 & 0 \\ 0 & -1 \end{pmatrix} \tag{6.72}$$

$$\begin{pmatrix} 1 & 0 \\ 0 & -i \end{pmatrix} \begin{pmatrix} 1 & 0 \\ 0 & -i \end{pmatrix} = \begin{pmatrix} 1 & 0 \\ 0 & -1 \end{pmatrix} \tag{6.73}$$

We observe the action of a half-wave plate, regardless of whether we use two right or two left circularly polarizing quarter-wave plates. In more general terms, we can represent the polarizing action of each of the optical elements by one matrix and operate with the product on the incident light. We then obtain the polarization of the emerging light:

$$\begin{pmatrix} 1 & 0 \\ 0 & e^{i\phi_1} \end{pmatrix} \begin{pmatrix} 1 & 0 \\ 0 & e^{i\phi_2} \end{pmatrix} \begin{pmatrix} 1 & 0 \\ 0 & e^{i\phi_3} \end{pmatrix} \cdots \begin{pmatrix} 1 & 0 \\ 0 & e^{i\phi_n} \end{pmatrix} \begin{pmatrix} \text{Polarization of} \\ \text{incident light} \end{pmatrix}$$
$$= \begin{pmatrix} \text{Polarization of} \\ \text{emerging light} \end{pmatrix} \tag{6.74}$$

This matrix method is a good "bookkeeping" method to keep up with the changes of polarization that the light undergoes while passing through an optical system. Examples will be provided as problems.

Problems

1. *Half-wave plate, quarter-wave plate.* The phase difference of an ordinary wave and an extraordinary wave after traveling the distance X is

$$\phi_X = |k_2 - k_1| X$$

   where

$$k_2 = \frac{2\pi}{\lambda_0} n_e \quad \text{and} \quad k_1 = \frac{2\pi}{\lambda_0} n_o$$

   The absolute value bars are introduced to make Δk or Δn positive. We can write ϕ_X as

$$\phi_X = \frac{2\pi}{\lambda_0} |n_e - n_o| X$$

   where X is the coordinate of propagation. We assume that $\lambda = 0.5 \cdot 10^{-3}$ mm and that $n_o = 1.544$ and $n_e = 1.553$ for quartz and $n_o = 1.658$ and $n_e = 1.486$ for calcite.
   a. Calculate the minimum thickness for a half-wave plate of quartz.
   b. Calculate the minimum thickness for a half-wave plate of calcite.
   c. Give the minimum thickness for a quarter-wave plate of quartz and for a quarter-wave plate of calcite.
   d. Suppose that a 1 mm plate of quartz and 1.05 mm plate of calcite are available. How much material must be removed to make half-wave plates of quartz and calcite?
   e. Suppose that the same plates are available as in (d). How much must be taken off to make quarter-wave plates of quartz and calcite?
   Ans. (a) $L = .0278$ mm; (b) $L = .00145$ mm; (d) quartz: 0.0269 mm, calcite: 1.63×10^{-3} mm

2. *General form of the ellipse for elliptically polarized light*
   a. Start with

$$E_Y = A \cos \alpha$$

   and

$$E_Z = A(\cos \alpha \cos \phi - \sin \alpha \sin \phi)$$

   These are equations 6.28 and 6.29, where α is time dependent. Eliminate the time dependence and give the general form of the ellipse. For $\cos \phi = \sin \phi = 1/\sqrt{2}$ and $A = 1$, we must get equation 6.31.
   b. Use the transformation

$$E_Y = (\cos \phi)(E'_Y - E'_Z)$$
$$E_Z = (\cos \phi)(E'_Y + E'_Z)$$

   and get the equation of an ellipse. For $\phi = \pi/4$, where $\cos \phi = \sin \phi = 1/\sqrt{2}$, we must get equation 6.33.
   c. Write the ellipse for the special cases where $\phi = \pi/4$, $3\pi/4$, $5\pi/4$, and $7\pi/4$. Which pairs are the same? Compare with Table 6.2.

3. *Phase difference between internally reflected polarized components.* We want to calculate the phase difference δ between the internally reflected components of total internal reflection. We are interested in the region $\theta_c \leq \theta \leq 90°$. We write Fresnel's formulas for the case of internal total reflection:

$$r_{E_\perp} = \frac{\cos \theta - i\sqrt{\sin^2 \theta - (n_2/n_1)^2}}{\cos \theta + i\sqrt{\sin^2 \theta - (n_2/n_1)^2}}$$

$$r_{E_\parallel} = \frac{-(n_2/n_1)^2 \cos \theta + i\sqrt{\sin^2 \theta - (n_2/n_1)^2}}{(n_2/n_1)^2 \cos \theta + i\sqrt{\sin^2 \theta - (n_2/n_1)^2}}$$

In the region of total internal reflection, we know that $|r_{E_\perp}| = 1$ and $|r_{E_\parallel}| = 1$. The phase difference Δ between the two components can be expressed as

$$\frac{r_{E_\parallel}}{r_{E_\perp}} = \frac{|r_{E_\parallel}|}{|r_{E_\perp}|} e^{i\Delta} = \frac{(-(n_2/n_1)^2 \cos \theta + i\sqrt{\quad})/((n_2/n_1)^2 \cos \theta + i\sqrt{\quad})}{\cos \theta - i\sqrt{\quad}/\cos \theta + i\sqrt{\quad}}$$

   a. Show that we have

$$e^{i\Delta} = \frac{-\sin^2 \theta + i \cos \theta \sqrt{\sin^2 \theta - (n_2/n_1)^2}}{\sin^2 \theta + i \cos \theta \sqrt{\sin^2 - (n_2/n_1)^2}}$$

   To do this, we multiply by

$$\sin^2 \theta - i \cos \theta \sqrt{\quad}$$

   to be able to get the real and imaginary parts:

$$e^{i\Delta} = \frac{-(-\sin^2 \theta + i \cos \theta \sqrt{\quad})^2}{(\sin^2 \theta + i \cos \theta \sqrt{\quad})(\sin^2 \theta - i \cos \theta \sqrt{\quad})}$$

   b. Now call the denominator A, because it is a real number, and show that we have

$$A^2 e^{+i\pi} e^{i\Delta} = (\sin^2 \theta + i \cos \theta \sqrt{\quad})^2$$

   c. Show that we get

$$\tan \frac{\Delta}{2} = \frac{\sin^2 \theta}{\cos \theta \sqrt{\sin^2 \theta - (n_2/n_1)^2}}$$

4. *Elliptically polarized light from total internal reflection: Jones vectors and matrices.* In Chapter 4, we calculated the reflection coefficient for total internal reflection. It turned out to be a complex number. A complex quantity can be written as $Re^{i\phi}$; that is, it can be described by its absolute value and its phase angle. If light polarized parallel and perpendicular to the plane of incidence is totally internally reflected, both components undergo a phase shift. If the phase difference is 90°, for example, we obtain circularly polarized light. We want to investigate what happens to the phase difference if linearly polarized incident light is internally reflected in a plane parallel plate one or more times (see accompanying figure). We are only interested in the part that is internally reflected and will ignore the light lost to the outside. For the reflected light polarized parallel and perpendicular to the plane of incidence,

$$r_{E_\parallel} = -\frac{db - ia}{db + ia} \quad \text{and} \quad r_{E_\perp} = \frac{b - ida}{b + ida}$$

where

$$a = \sqrt{\left(\frac{N_1}{n_2}\right)^2 \sin^2 \theta - 1}, \quad b = \cos\theta, \quad \text{and} \quad d = \frac{n_2}{N_1}$$

and where θ is the angle of incidence, $N_1 = N = 1.5$ (the index of refraction of the glass plate), and $n_2 = 1$ (the index of refraction outside). To find the relative phase difference after reflection, we rewrite one of the two expressions as

$$r_{E_\perp} = \frac{z_\perp^*}{z_\perp}$$

where

$$z_\perp = b + ida = Re^{i\beta}$$
$$z_\perp^* = b - ida = Re^{-i\beta}$$
$$\tan\beta = \frac{da}{b}$$

and we find $r_{E_\perp} = e^{-2i\beta}$.

a. Show that $r_{E_\parallel} = e^{-i(2\alpha - \pi)}$, where $\tan\alpha = a/bd$. For the relative phase difference after one reflection, we have

$$\phi = \arg r_{E_\parallel} - \arg r_{E_\perp} = 2\beta - 2\alpha + \pi$$

If many reflections are involved, or if different devices are involved, each producing a certain phase difference, we can develop a "bookkeeping" method to find out what the final phase difference is.

b. We now go back to our original problem. For one reflection, we obtained

$$\phi = 2\beta - 2\alpha + \pi \quad \text{or} \quad \beta - \alpha = \tfrac{1}{2}(\phi - \pi)$$

We are interested in finding a relation between the angle of incidence θ and the material parameters for the plate. To do this, we need $\tan(\beta - \alpha)$ as

$$\tan(\beta - \alpha) = \frac{\tan \beta - \tan \alpha}{1 + \tan \beta \tan \alpha} = \tan\left(\frac{\phi - \pi}{2}\right) = q$$

Insert the expressions for $\tan \alpha$ and $\tan \beta$ in terms of a, b, and d, and obtain

$$q^2 = \frac{\cos^2 \theta (N^2 \sin^2 \theta - 1)}{N^2 \sin^4 \theta}$$

c. Solve the resulting quadratic equation for $\sin^2 \theta$ and get

$$\sin^2 \theta = \frac{N^2 + 1}{2N^2(q^2 + 1)} \pm \sqrt{\left(\frac{N^2 + 1}{2N^2(q^2 + 1)}\right)^2 - \frac{1}{N^2(q^2 + 1)}}$$

d. We want to obtain, with one reflection, circularly polarized light. For the particular case of $\phi = 90°$ and a glass with $N = 1.5$, find the angle of incidence θ from the equation in part (c); that is, show that $q^2 = 1$ and that the equation for $\sin^2 \theta$ has no real solution.

e. Now we go to two reflections, that is, in Jones matrices,

$$\begin{pmatrix} 1 & 0 \\ 0 & e^{i\phi} \end{pmatrix} \begin{pmatrix} 1 & 0 \\ 0 & e^{i\phi} \end{pmatrix} = \begin{pmatrix} 1 & 0 \\ 0 & e^{2i\phi} \end{pmatrix} = \begin{pmatrix} 1 & 0 \\ 0 & e^{i\delta} \end{pmatrix}$$

where δ is the phase difference of $90°$,

$$\delta = 2\phi = 4\beta - 4\alpha + 2\pi$$

or

$$\beta - \alpha = \frac{\delta}{4}$$

If we write

$$\tan(\beta - \alpha) = \tan\frac{\delta}{4} = q$$

we can use the formulas from before. Show that for glass ($N = 1.5$), the angle of incidence θ has values close to $50°$ and $53°$. A device like this is called a Fresnel rhomb and produces circularly polarized light after two reflections.

5. *Half-voltage.* Calculate the half-voltage for ADP:

$$V_{1/2} = \frac{\lambda}{2n_o^3 r_{63}}$$

Assume that $\lambda = 0.5 \times 10^{-6}$ m, $n_o = 1.52$, and $r_{63} = 8.5 \times 10^{-12}$ m/V.
Ans. 8400 V

6. *Optical rotation in quartz.* Calculate the specific rotatory power α_s for quartz. Use

$$\lambda = 3968 \text{ Å}, n_R = 1.55810, \text{ and } n_L = 1.55821.$$

Ans. 8.7×10^{-4} rad/μm

7. *Optical rotation of an active component in solution: sucrose.* The specific rotation $[\rho]$ (degrees/(g/cm^3)) is defined for a specific wavelength λ as follows. Consider a 10 cm column of solution containing C g/cm^3 of active compound. $[\rho]$ is the angle of rotation of the plane of polarization for 10 cm traversal of the light if 1 g/cm^3 of active compound is contained. The angle α of rotation is therefore given as

$$\alpha = \frac{[\rho] (\text{degrees}/(g/cm^3)) \cdot X_{[cm]} \cdot C_{[g/cm^3]}}{10 \text{ cm}}$$

where X is the distance of travel in centimeters, C is the concentration in g/cm³ and $[\rho]$ is given for a specific wavelength.

Calculate for sucrose using the following information: $[\rho] = 66.5°$ $[1/(\text{g/cm}^3)]$ (for $\lambda = 5893$ Å, sodium D line), $X = 20$ cm, and $C = 0.25$ g/cm³.

Ans. 33.2°

8. *Linear polarizers.* The following Jones matrices represent horizontal, vertical, and 45° polarizers, respectively:

$$\begin{pmatrix} 1 & 0 \\ 0 & 0 \end{pmatrix} \quad \begin{pmatrix} 0 & 0 \\ 0 & 1 \end{pmatrix} \quad \frac{1}{2}\begin{pmatrix} 1 & 1 \\ 1 & 1 \end{pmatrix}$$

Horizontal   Vertical   45°

a. Apply these matrices to vector representation of horizontal, vertical, and 45° polarized light and give an interpretation of the result:

$$\begin{pmatrix} 1 \\ 0 \end{pmatrix} \quad \begin{pmatrix} 0 \\ 1 \end{pmatrix} \quad \begin{pmatrix} 1 \\ 1 \end{pmatrix}$$

Horizontal   Vertical   45°

b. Show that the matrix

$$\frac{1}{2}\begin{pmatrix} 1 & -1 \\ -1 & 1 \end{pmatrix}$$

presents a polarizer of $-45°$. Operate on $+45°$ polarized light,

$$\begin{pmatrix} 1 \\ 1 \end{pmatrix}$$

and on $-45°$ polarized light,

$$\begin{pmatrix} -1 \\ +1 \end{pmatrix}$$

and give an interpretation.

c. Let light first be polarized horizontally, then pass through a $+45°$ polarizer, and then through a $-45°$ polarizer. What is the polarization of the emerging light?

9. *Combination of polarizers: presented in matrices*
   a. What is the action of first a quarter-wave plate and then a half-wave plate?
   b. What is the action of first a half-wave plate and then a quarter-wave plate?
   c. What is the action of a half-wave plate, then a quarter-wave plate, and then another half-wave plate?

Wave Theory in Dispersive Media and Optical Constants | 7

1. INTRODUCTION

In the preceding chapter, we made extensive use of the refractive index n and its relation to the speed of light. In vacuum, the speed is c; in a dielectric medium, the speed is $v = c/n$. Now we want to discuss a model describing the refractive index by the characteristic parameters of the medium. We will obtain descriptions for different materials such as dielectrics and metals.

An electromagnetic wave may traverse the medium and interact with the charges following the oscillations of the wave. The model to describe these oscillations in the material is the damped oscillator driven by the electromagnetic wave. The parameters are the eigenfrequency $\bar{v}_0$ of the oscillator, connected to the force constant f, and the damping constant γ. If the restoring force is small or zero, we have the model describing metals. If the damping constant is small or zero, we have a description of dielectrics. In metals, we have strong attenuation of the wave, that is, strong absorption; in dielectrics, we have almost no absorption in the normal dispersion region.

The connection between the propagating electromagnetic wave in the medium and the parameters describing the medium is accomplished by the refractive index n. For the wave vector, we have

$$k = \frac{2\pi}{\lambda/n}$$

where λ is the wavelength in vacuum.

2. DIELECTRICS

A. Oscillator Models for Dielectrics

We have seen how electromagnetic waves traversing a medium vibrate perpendicular to their direction of propagation. Different electrical environments lead to different permittivities ε and refractive indices n.

As mentioned earlier, if waves are incident on a medium, they are reflected, transmitted, and absorbed. We wish to study a model relating the refractive index n and the attenuation index K to the materials parameters. The electrical polarization and the heating of the material are connected to what happens to the electromagnetic wave, that is, its speed and attenuation.

Our first objective is to relate the macroscopic (i.e., n and K) and microscopic parameters that characterize the medium. The interaction of light with the atomic or molecular species will be described using Maxwell's theory of electromagnetic waves. The **lossy** medium (a medium that absorbs light) is described by a collection of damped oscillators whose interaction with the incident waves accounts for the optical behavior. The macroscopic parameters are then expressed in terms of microscopic oscillator parameters.

For materials that have no free charge carriers (dielectrics), the microscopic response to an applied electric field is polarization of the medium. That is, oppositely charged ions are separated by forces produced by the vibrating field of the traversing electromagnetic wave.

As a simplified model of such a system, we consider the material to consist of N identical one-dimensional oscillators of mass m, charge e, and restoring force constant f. Free oscillators are described by the one-dimensional harmonic oscillator equation

$$m\frac{d^2u}{dt^2} + m\bar{\omega}_0^2 u = 0 \tag{7.1}$$

where u is the displacement of the charge from its equilibrium position and $\bar{\omega}_0^2 = f/m$ is its natural frequency of oscillation. A more realistic model would include a damping term to account for losses. The resulting damped oscillator equation then becomes

$$m\frac{d^2u}{dt^2} + m\gamma\frac{du}{dt} + m\omega_0^2 u = 0 \tag{7.2}$$

where γ is the damping coefficient and ω_0 is the new resonance frequency $(\omega_0 \neq \bar{\omega}_0)$.

The interaction between oscillator and radiation is accounted for by applying a harmonic driving term that oscillates at an angular frequency ω of the light. Therefore, we now consider a collection of forced, damped oscillators, each described by

$$m\frac{d^2u}{dt^2} + m\gamma\frac{du}{dt} + m\omega_0^2 u = eE_0 e^{-i\omega t} \tag{7.3}$$

where $F = eE_0 e^{-i\omega t}$ is the applied force, E_0 is the maximum amplitude of the incident electromagnetic wave, e is the charge of the electron, and ω its frequency. To solve equation 7.3, we substitute a trial solution of the form

$$u(t) = u_0 e^{-i\omega t} \tag{7.4}$$

into equation 7.3 and get

$$-mu_0\omega^2 - im\gamma\omega u_0 + m\omega_0^2 u_0 = eE_0$$

or

$$u_0 = \frac{e(E_0/m)}{\omega_0^2 - \omega^2 - i\gamma\omega}$$

By multiplication with $e^{-i\omega t}$ and by setting $E_0 e^{-i\omega t} = E(t)$, we have

$$u(t) = \frac{e/m}{(\omega_0^2 - \omega^2) - i\gamma\omega} E(t) \tag{7.5}$$

Thus, we see that the displacement $u(t)$ of the driven oscillator depends on the oscillator parameters e, m, ω_0, and γ as well as the incident wave parameters E_0 and ω.

The effect of this displacement is an electrical polarization of the medium. Quantitatively, the polarization P of the medium is defined as the number of **electric dipoles** per unit volume. Since the induced electric dipole moment is just equal to eu, we get

$$P = Neu$$

where N is the number of oscillators per unit volume. Using equation 7.5 for u gives us

$$P = Ne\,u(t) = \frac{Ne^2}{m\varepsilon_0}\left(\frac{1}{(\omega_0^2 - \omega^2) - i\gamma\omega}\right)\varepsilon_0 E(t) = \chi^*\varepsilon_0 E \tag{7.6}$$

where the last equality follows from the relation between polarization P, electric susceptibility χ, and field E (see Chapter 4). We see that as a result of the imaginary damping term in $u(t)$, the susceptibility χ^* is complex. In this chapter, we add an asterisk to indicate that the quantity is complex. Introducing the abbreviation

$$\omega_p = \sqrt{Ne^2/m\varepsilon_0}$$

which is known as the **plasma frequency**, we get

$$\chi^* = \frac{\omega_p^2}{(\omega_0^2 - \omega^2) - i\gamma\omega} \tag{7.7}$$

We will now study the effect of the polarization on the wave traveling through the medium (driving the oscillators). We choose the component E_Y of the electric field traveling in the X direction. From Maxwell's equations, we have for this special case the wave equation in vacuum plus a polarization term:

$$\frac{\partial^2 E_Y}{\partial X^2} = \mu_0\varepsilon_0 \frac{\partial^2 E_Y}{\partial t^2} + \underbrace{\mu_0 \frac{\partial^2 P_Y}{\partial t^2}}_{\text{Polarization term}} \tag{7.8}$$

where ε_0 is the permittivity of free space and μ_0 the permeability of free space. The change of the polarization vector **P** with time represents the effect of the

7. Wave Theory in Dispersive Media and Optical Constants

wave on the medium. Since

$$\mathbf{P} = \chi^* \varepsilon_0 \mathbf{E}$$

we have a wave equation in which

$$v = \frac{1}{\sqrt{\mu_0 \varepsilon_0 (1 + \chi^*)}}$$

represents the speed of the wave in the material:

$$\frac{\partial^2 E_Y}{\partial X^2} = \mu_0 \varepsilon_0 (1 + \chi^*) \frac{\partial^2 E_Y}{\partial t^2} = \frac{1}{v^2} \frac{\partial^2 E_Y}{\partial t^2} \qquad (7.9)$$

The electric susceptibility (and consequently the dielectric constant and permittivity) depends on the frequency of the light. This is called a **dispersion relation**.

Substituting a trial solution

$$E_Y = E_{Y_0} e^{i(kX - \omega t)} \qquad (7.10)$$

into equation 7.9, we obtain

$$(k^*)^2 = \mu_0 \varepsilon_0 (1 + \chi^*) \omega^2 \qquad (7.11)$$

or

$$(k^*)^2 = (1 + \chi^*) \frac{\omega^2}{c^2} \qquad (7.12)$$

An asterisk has been added to k, since it now appears as a complex quantity. Since the wave vector is now complex and $k^* = n^* \omega / c$, we see that for a medium attenuating the traversing wave, a complex index of refraction results:

$$(n^*)^2 = 1 + \chi^* \qquad (7.13)$$

It is customary to express n^* in terms of a real part n and an imaginary part K that we call the **attenuation index**:

$$n^* = n + iK \qquad (7.14)$$

Then

$$(n + iK)^2 = 1 + \frac{\omega_p^2}{(\omega_0^2 - \omega^2) - i\gamma\omega} \qquad (7.15)$$

gives

$$n^2 - K^2 = 1 + \frac{\omega_p^2 (\omega_0^2 - \omega^2)}{(\omega_0^2 - \omega^2)^2 + \gamma^2 \omega^2} \qquad (7.16)$$

and

$$2nK = \frac{\gamma \omega \omega_p^2}{(\omega_0^2 - \omega^2)^2 + \gamma^2 \omega^2} \qquad (7.17)$$

Thus, the presence of a damping term in equation 7.3 has resulted in a complex susceptibility, wave vector, and index of refraction. As a check, for χ^*, if γ goes to zero, we get

$$\chi(\omega) = \frac{\omega_p^2}{\bar{\omega}_0^2 - \omega^2}$$

from equation 7.7; and for n^*, if γ goes to zero, we get

$$n(\omega) = \sqrt{1 + \frac{\omega_p^2}{\bar{\omega}_0^2 - \omega^2}}$$

from equation 7.15. Thus, in the limit of no damping, the parameters are still functions of frequency ω but are all real, that is, lossless. We call $\bar{\omega}_0$, ω_0, and ω_p the model parameters.

We have assumed, in the preceding argument, that the local field E_0 at the site of the oscillators is the same as the applied field. This is strictly true only if the density of the oscillators in the medium is low, as it would be in a gas. (For a dense distribution of oscillators, the effect of the surrounding polarized medium on the oscillator—the so-called Lorentz contribution or local field correction—would have to be accounted for.)

For a low-density medium, the refractive index must be close to 1, the value in vacuum. Therefore, K is small compared to 1, and we may approximate equations 7.16 and 7.17, using $\sqrt{1 + x} \simeq 1 + (x/2)$, as

$$n(\omega) = 1 + \frac{\omega_p^2}{2}\left(\frac{\omega_0^2 - \omega^2}{(\omega_0^2 - \omega^2)^2 + \omega^2\gamma^2}\right) \qquad (7.18)$$

and

$$K(\omega) = \frac{\omega_p^2}{2}\left(\frac{\omega\gamma}{(\omega_0^2 - \omega^2)^2 + \omega^2\gamma^2}\right) \qquad (7.19)$$

B. The Refractive Index Depending on Frequency: Dispersion of a Gas and a Prism

It is instructive to plot $n(\omega)$ and $K(\omega)$ and to consider three specific frequencies: 0, ω_0, and ∞. Consider first the refractive index $n(\omega)$ plotted in Figure 7.1. For $\omega = 0$ and $\gamma = 0$ (corresponding to a lossless medium),

$$n(0) = 1 + \frac{1}{2}\left(\frac{\omega_p}{\omega_0}\right)^2 \quad \text{or} \quad n(0) = 1 + \frac{Ne^2}{2\varepsilon_0 f}$$

Numerically, $n(0)$ should correspond to expected static values for low-density media, that is, $n(0) \simeq 1$. To see this, we choose reasonable values of N and f and evaluate $(\omega_p/\omega_0)^2$; that is,

$$\left(\frac{\omega_p}{\omega_0}\right)^2 = \frac{(10^{30}\ \text{m}^{-3})(1.6 \times 10^{-19}\ \text{coul})^2}{(8.85 \times 10^{-12}\ \text{F/m})(10^8\ \text{N/m})} \simeq 3 \times 10^{-5}$$

Thus, we see that $n(0) \simeq 1$ as expected. For $\omega \to \infty$, $n(\infty) \to 1$, which is the expected high-frequency limit (e.g., in the X-ray region). For $\omega = \omega_0$ (and $\gamma = 0$), $n(\omega)$ is singular. However, this situation is never realized for real systems, since

7. Wave Theory in Dispersive Media and Optical Constants

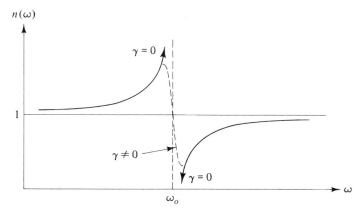

Fig. 7.1
Frequency dependence of refractive index n for a dielectric described by a simplified dispersion model.

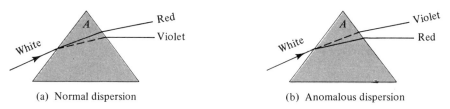

(a) Normal dispersion

(b) Anomalous dispersion

Fig. 7.2
Illustrating (a) normal and (b) anomalous dispersion. A is the apex angle.

some loss mechanism is always present. The effect of the nonzero damping term on $n(\omega)$ is shown as a broken line in Figure 7.1. For this latter case, when $\omega \gtrsim \omega_0$, $n(\omega)$ is greater than 1 and is increasing as ω increases. This is normal dispersion behavior and is illustrated when white light passes through a glass prism, with violet light bending more than red light. Near resonance ($\omega = \omega_0$), however, $n(\omega)$ decreases as ω increases, giving rise to anomalous dispersion. This effect is observed as the opposite of that just described; that is, red light is bent more than blue in traversing the prism (Figure 7.2).

For the case of anomalous dispersion, the speed of light in the medium, $v = c/n$, is larger than c, since n is smaller than 1. This is not a contradiction to the special theory of relativity, which tells us that c is the highest velocity possible. This latter statement refers to an energy-transporting mechanism. The speed of light for the anomalous dispersion region is not of this kind.

Prisms may be used for separating different colors contained in white light. They are used as spectroscopic devices. Depending on the refractive index, the apex angle, and the angle of incidence, the angle of deviation for the different wavelengths may be determined. If the path of the light through the prism is symmetric, the angle of deviation is a minimum (see also Chapter 1, Section 3). The angle of deviation is measured between the directions of the incident and emerging light. The relation between refractive index, apex angle, and angle of minimum deviation is

$$n = \frac{\sin[\frac{1}{2}(A + \delta)]}{\sin A/2} \tag{7.20}$$

This formula will be derived in problem 4.

C. The Absorption Coefficient Depending on Frequency

We will now consider the attenuation index K as a function of ω. From equation 7.19, $K(0) = 0$ and $K(\infty) \to 0$. At resonance, $K(\omega_0)$ has a maximum, the width depending on γ, the damping term. Clearly, for $\gamma = 0$, $K(\omega) = 0$ for all ω. The behavior of $K(\omega)$ is shown in Figure 7.3.

Now let us consider the effect of a complex wave vector **k** on the propagation of a plane electromagnetic wave in a lossy dielectric medium. We insert

$$k^* = \frac{\omega}{c}n^* = \frac{\omega}{c}(n + iK)$$

into the trial solution in one dimension (equation 7.10) and get

$$\begin{aligned} E(X, t) &= E_0 e^{i[(\omega/c)(n+iK)X - \omega t]} \\ &= E_0 e^{i[(\omega/c)nX - \omega t]} e^{-(\omega/c)KX} \end{aligned} \tag{7.21}$$

Since the intensity I of the wave is proportional to the square of the electric field, we get

$$I \propto |E|^2 = |E_0|^2 e^{-2\omega KX/c} \tag{7.22}$$

Letting $I_0 \cong |E_0|^2$ and

$$\bar{\alpha} = 2\alpha = 2\frac{\omega}{c}K \tag{7.23}$$

results in Beer's law,

$$I(X) = I_0 e^{-\bar{\alpha}X} \tag{7.24}$$

Thus, $\bar{\alpha}$ tells us how much a layer X units thick of a medium attenuates a beam of radiation with intensity I_0 incident upon it. In particular, the **absorption**

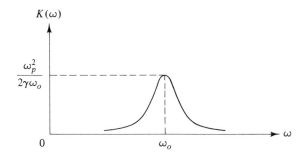

Fig. 7.3
Frequency dependence of the attenuation index K for a dielectric described by a simplified dispersion model.

7. **Wave Theory in Dispersive Media and Optical Constants**

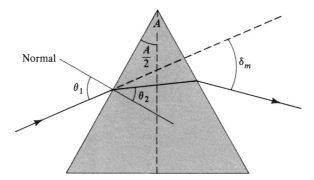

Fig. 7.4
Prism with isosceles triangle cross section and apex angle A. The incident light makes an angle θ_1 with the normal, and the refracted light makes an angle θ_2. The deviation of the emerging wave from the incident wave is given by δ_m.

coefficient $\bar{\alpha}$, measured in units of reciprocal centimeters, is such that in a distance $(\bar{\alpha})^{-1}$, the intensity falls from I_0 to $I = (I_0/e)$. Experimentally, $\bar{\alpha}$ can be measured directly. It is often written as

$$\bar{\alpha} = \frac{4\pi K}{\lambda} \tag{7.25}$$

The units are nepers in cm^{-1}.

D. Determination of Optical Constants

The optical constants n and K can be determined by various experimental methods. The method chosen depends on the magnitude of n and K. In Chapter 4, we saw that the determination of Brewster's angle leads to n. This is the case for a lossless dielectric, that is, where $K \ll 1$.

Here is another method for obtaining n when K is small. We cut prism from the material for which n and K are to be determined, with an isosceles triangle as the cross section (Figure 7.4). We then measure the angle of minimum deviation for a specific wavelength. Using equation 7.20 together with the knowledge of A, we can determine n for the wavelength being considered.

For materials where K is not small compared to 1, we use reflections for the determination of the optical constants. In Chapter 4, we saw that for a lossless medium and angle of incidence $\theta = 0$, the reflection coefficient was

$$r_{E\perp} = r_{E\parallel} = -\frac{N_2 - n_1}{N_2 + n_1} \tag{7.26}$$

Here, N_2 is the refractive index of the medium being considered, and n_1 is the refractive index of the outside medium; essentially, $n_1 = 1$.

To take losses into account, we saw that the real refractive index is replaced by the complex one:

$$N_2 \rightarrow N_2^* = (n + iK)$$

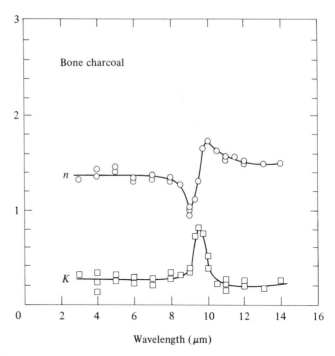

Fig. 7.5
Optical constants of bone charcoal powder. Resonance from vibrations of Ca, P, and O atoms against each other.

and we have

$$r_{E\perp} = r_{E\parallel} = \frac{1 - (n + iK)}{1 + (n + iK)} \tag{7.27}$$

where $r_{E\perp}$ and $r_{E\parallel}$ are now complex numbers of the type $re^{i\delta}$. Determination of r and δ would yield n and K.

It is not easy to measure r and δ quantitatively in reflection at $\theta = 0°$. Therefore, measurements are made at larger angles. In Chapter 4, we found that for such measurements, equations 4.53 and 4.55 hold for the lossless case. For the case with losses, the refractive index is complex. Replacing N_2 by N_2^*, we have complex expressions for $r_{E\perp}$ and $r_{E\parallel}$ and can write

$$|r_{E\perp}|e^{i\delta\perp} = \frac{\cos\theta - N_2^*\sqrt{1 - (\sin\theta/N_2^*)^2}}{\cos\theta + N_2^*\sqrt{1 - (\sin\theta/N_2^*)^2}} = f(\theta, n, K) \tag{7.28}$$

and

$$|r_{E\parallel}|e^{i\delta\parallel} = \frac{-N_2^*\cos\theta + \sqrt{1 - (\sin\theta/N_2^*)^2}}{N_2^*\cos\theta + \sqrt{1 - (\sin\theta/N_2^*)^2}} = g(\theta, n, K) \tag{7.29}$$

The two functions g and f appear as complicated functions of θ, n, and K.

In theory, five quantities can be measured to determine n and K: $|r_{E\perp}|$, $|r_{E\parallel}|$, $\delta_\perp$, $\delta_\parallel$, and θ. There are four equations available, since each of the complex

7. Wave Theory in Dispersive Media and Optical Constants

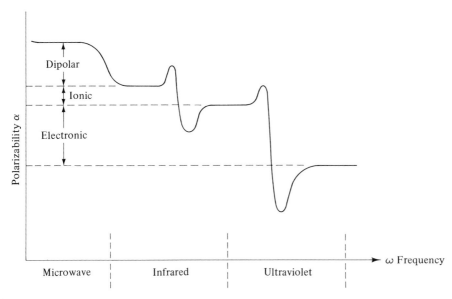

Fig. 7.6
Plot of polarizability of a general dielectric vs. frequency.

equations counts as two. Various combinations are possible among the quantities to be measured and the equations to be considered in order to determine n and K.

One method is the determination of

$$\left| \frac{r_{E_\perp}}{r_{E_\parallel}} \right|^2$$

depending on three different angles. There are tables from which n and K can be determined using these measured values.* Such a measurement is provided in Figure 7.5 for bone charcoal powder pressed into solid pellets. The resonance around 9.5 μm is due to vibrations in calcium phosphate molecules contained in the bone charcoal. Note that n and K, schematically shown in Figures 7.1 and 7.3, respectively, appear as such in Figure 7.5.

E. Extension to Dielectrics with Many Resonance Frequencies

The preceding results were based on a very simple model consisting of a system of oscillators, all having the same frequency. Real materials require more realistic models to predict or describe observed measurements accurately.

In Figure 7.6, we show a plot of the polarizability α over a wide range of frequencies. The resonance just discussed for the bone charcoal is the resonance

---

*A. Vasicek, *Tables of Determination of Optical Constants from the Intensities of Reflected Light* (Progue: Nakladatelstvi Ceskoslovenske Akademie VED, 1964).

shown in the middle, the ionic or infrared resonance. At higher frequencies, we have the electronic resonance, and at lower frequencies, the dipolar resonance.

To represent χ^* for a dielectric, we would try to represent it by three terms instead of one in equation 7.7,

$$\chi^* = \sum_{j=1}^{3} \frac{\omega_p^2 f_j}{(\omega_{0j}^2 - \omega^2) - i\gamma_j \omega} \tag{7.30}$$

With the use of

$$(n^*)^2 = 1 + \chi^* \quad \text{and} \quad n^* = n + iK$$

we obtain (corresponding to equations 7.16 and 7.17)

$$n^2 - K^2 = 1 + \sum_{j=1}^{3} \frac{f_j \omega_p^2 (\omega_{0j}^2 - \omega^2)}{(\omega_{0j}^2 - \omega^2)^2 + \gamma_j^2 \omega^2}$$

$$2nK = \sum_{j=1}^{3} \frac{f_j \gamma_j \omega \omega_p^2}{(\omega_{0j}^2 - \omega^2)^2 + \gamma_j^2 \omega^2} \tag{7.31}$$

where f_j are constants. The method may be extended to more resonance terms for more complicated cases (more degrees of freedoms in the oscillators). Such an analysis has been applied to water, and the obtained values of n and K are shown in Figure 7.7 over a large wavelength region.

The constants n and K are not independent; they are related by the Kramers-Kronig relation.*

3. METALS

A. Drude Model for Metals

We will now look at processes in a medium where the electrons are not bound to certain ionic centers. The incident wave polarizes the material, that is, displaces the electrons. However, now there are no restoring forces for the electrons as there were in the oscillator model. The electrons are almost free, and the incident wave produces microscopic currents, not oscillations. This model is called the **Drude model** for metals.

The motion of the charge in the conducting medium is described by an equation that includes acceleration, damping, and driving terms, but no restoring term. The appropriate modification of equation 7.3 is

$$m\frac{d^2u}{dt^2} + m\gamma\frac{du}{dt} = eE_0 e^{-i\omega t} \tag{7.32}$$

The general solution of equation 7.32 is the sum of the solutions of the homogeneous and inhomogeneous equations. For the homogeneous case,

$$m\frac{d^2u}{dt^2} + m\gamma\frac{du}{dt} = 0 \tag{7.33}$$

---

*A. Yariv, *Quantum Electronics* (New York: Wiley, 1968), Appendix I.

7. Wave Theory in Dispersive Media and Optical Constants

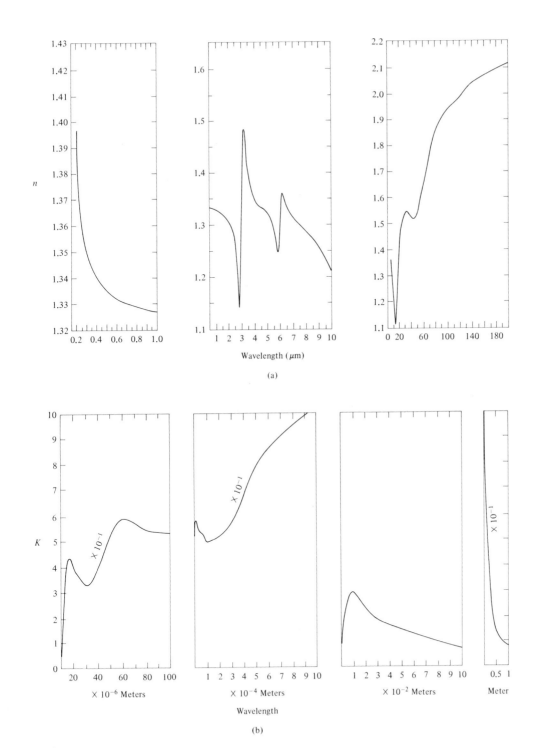

Fig. 7.7
Complex index of refraction of water. (a) Index of refraction of water for the spectral regions 0.2–200 μm. (b) Attenuation coefficient of water for the 20×10^{-6} to 1 m spectral region. [Data from G. M. Hale and M. R. Querry, *Applied Optics* **12**, 555 (1973).]

3. **Metals**   289

Substituting the trial solution

$$u = u_0 e^{-t/\tau} \tag{7.34}$$

where τ is the decay time (relaxation time) results in

$$\gamma = \tau^{-1}$$

Typically, $\tau = 10^{-3}$ sec, and this solution will be neglected in our further discussions.

To treat the inhomogeneous equation, we use the velocity of the charges, $v = du/dt$, and choose a harmonic trial solution

$$v = v_0 e^{-i\omega t} \tag{7.35}$$

Substitution into equation 7.32 gives

$$v_0 = \frac{e/m}{\gamma - i\omega} E_0 \tag{7.36}$$

The current density can be expressed as $J_0 = Nev_0$, and using $\gamma = 1/\tau$, we have

$$J_0 = \frac{(Ne^2/m)\tau}{1 - i\omega\tau} E_0 = \frac{\sigma}{1 - i\omega\tau} E_0 \tag{7.37}$$

where the static conductivity is defined as

$$\sigma = \frac{Ne^2}{m}\tau \tag{7.38}$$

As was discussed for the dielectric case, a term is added to the vacuum wave equation to represent the reaction of the medium on the wave. The effect of microcurrents of the medium on the traversing wave is described by the last term in equation 7.39:

$$\frac{\partial^2 E_Y}{\partial X^2} = \mu_0 \varepsilon_0 \frac{\partial^2 E_Y}{\partial t^2} + \mu_0 \frac{\partial J_Y}{\partial t} \tag{7.39}$$

Introduction of equation 7.37 and the use of the trial solution

$$E_Y = E_{Y_0} e^{i(kX - \omega t)} \quad \text{and} \quad J_Y = J_0 e^{-i\omega t}$$

yields

$$k^2 = \mu_0 \varepsilon_0 \omega^2 + \mu_0 \frac{i\sigma\omega}{1 - i\omega\tau} \tag{7.40}$$

Thus, the wave vector is again complex. Using $k^*/\omega = n^*/c$, we get the complex index of refraction n^*:

$$(n^*)^2 = (\bar{n} + i\bar{K})^2 = c^2 \left[\varepsilon_0 \mu_0 + \frac{\mu_0 \sigma}{\omega} \left(\frac{i - \omega\tau}{1 + \omega^2\tau^2} \right) \right] \tag{7.41}$$

A bar is placed over the optical constants to distinguish them from the n and K obtained for the oscillator model.

7. **Wave Theory in Dispersive Media and Optical Constants**

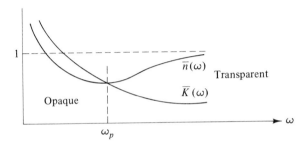

Fig. 7.8
Frequency dependence of the optical constants for a metal.

Writing the real and imaginary parts separately, we have

$$\bar{n}^2 - \bar{K}^2 = 1 - \frac{\sigma}{\varepsilon_0}\left(\frac{\tau}{1 + \omega^2\tau^2}\right) = 1 - \frac{\omega_p^2}{\omega^2 + \gamma^2} \tag{7.42}$$

$$2\bar{n}\bar{K} = \frac{\sigma}{\varepsilon_0\omega}\left(\frac{1}{1 + \omega^2\tau^2}\right) = \frac{\gamma}{\omega}\left(\frac{\omega_p^2}{\omega^2 + \gamma^2}\right) \tag{7.43}$$

where we again use the plasma frequency

$$\omega_p = \left(\frac{Ne^2}{m\varepsilon_0}\right)^{1/2} = \left(\frac{\sigma}{\tau\varepsilon_0}\right)^{1/2} \tag{7.44}$$

Equations 7.42 and 7.43 are drastically different from their dielectric (oscillator model) analog, since no resonance term is present. Consequently, no singularity is present for $\gamma = 0$ except when $\omega = 0$, that is, where metals are opaque (see Figure 7.8). For $\omega \to \infty$, we see from equations 7.42 and 7.43 that $\bar{K} = 0$ is a solution. Metals are transparent when $\bar{n} \to 1$. For example, X rays can penetrate metals.

From Figure 7.8, we see that there is a crossover point for $\bar{n}(\omega)$ and $\bar{K}(\omega)$ at the plasma frequency (see also problem 6).

For large $\bar{K}(\omega < \omega_p)$, metals are opaque and reflect strongly. Here the wave can penetrate only a small distance into the medium before it is completely attenuated. To find this distance, we first consider the following low-frequency approximation: From equation 7.40, we have for small ω the approximation

$$k^* \simeq \sqrt{i\omega\mu_0\sigma} \tag{7.45}$$

We introduce this into $k^* = n^*(\omega/c)$ and use the identity $\sqrt{i} = (1 + i)/\sqrt{2}$ to obtain

$$\frac{\omega}{c}(\bar{n} + i\bar{K}) \simeq \sqrt{\omega\mu_0\sigma}\left(\frac{1 + i}{\sqrt{2}}\right)$$

or

$$\bar{n} \simeq \bar{K} \simeq \sqrt{\frac{\sigma}{2\varepsilon_0\omega}} \tag{7.46}$$

We can use equation 7.22 to obtain the distance into the medium for which the intensity of the wave (assumed to be normal incidence) falls to $1/e$ of its value at

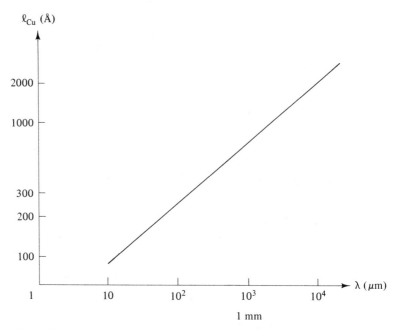

Fig. 7.9
Plot of the skin depth ℓ_{Cu} over the wavelength in the wavelength interval of 1–10,000 μm.

the surface. This parameter is called the **skin depth** and is given by

$$\ell = \frac{1}{\bar{\alpha}} = \frac{c}{2\omega \bar{K}} = \frac{1}{\sqrt{2\mu_0 \omega \sigma}} \qquad (7.47)$$

For a good conductor, such as copper,

$$\sigma \simeq 6 \times 10^7 (\Omega m)^{-1}$$

and

$$\ell_{Cu} = \sqrt{\frac{1}{48\pi\omega}} = \frac{0.081}{\sqrt{\omega}}\,\text{m} \qquad (7.48)$$

In Figure 7.9, we plot ℓ_{Cu} (from equation 7.48) in units of angstroms over the wavelength in micrometers.

If the metal film is even thinner than the skin depth, we can obtain metal films absorbing 50 percent of the incident radiation over a large spectral region. Films made of bismuth that are about 1000 μm thick absorb about 50 percent in the spectral region of 30–3000 μm. Such films are used as absorbers in helium-cooled bolometers.

B. Reflection on Metals

We have seen that metals have high $\bar{K}$ values except for very short wavelengths. Therefore, reflection methods are used to determine $\bar{n}$ and $\bar{K}$ in the visible and infrared spectral regions. The reflectance of metals in the near ultraviolet to the

7. Wave Theory in Dispersive Media and Optical Constants

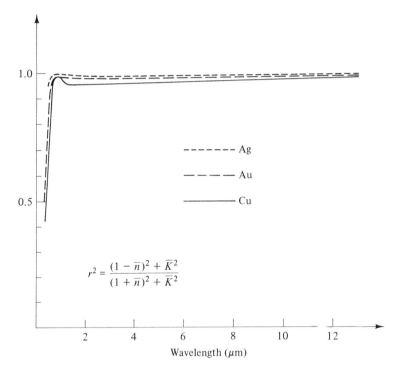

$$r^2 = \frac{(1 - \bar{n})^2 + \bar{K}^2}{(1 + \bar{n})^2 + \bar{K}^2}$$

Fig. 7.10
Calculated reflectivity for Ag, Au, and Cu. [Data from M. A. Ordal, L. L. Long,
R. J. Bell, S. E. Bell, R. R. Bell, W. Alexander, Jr., and C. A. Ward, *Applied Optics* **21**,
1099 (1983).]

near infrared is shown in Figure 7.10. Note that in the visible spectral region,
considerable changes are observed in the reflectance of metals. Toward the
infrared, the reflectance approaches a value of 90 percent and more.

The determination of the optical constants of metals through the use of
reflection methods brings us back to equations 7.28 and 7.29, where we discussed
absorbing dielectrics:

$$|r_{E_\perp}| e^{i\delta_\perp} = f(\theta, \bar{n}, \bar{K})$$
$$|r_{E_\parallel}| e^{i\delta_\parallel} = g(\theta, \bar{n}, \bar{K})$$

While the functions f and g are true for any absorbing medium, a bar over n and
K indicates that we would apply them to metals.

Methods have been developed to employ the phase angles $\delta_\perp$ and $\delta_\parallel$. To
get an idea of how this is done, first we compare reflectances $r_{E_\perp}^2$ and $r_{E_\parallel}^2$ for an
unspecified metal with what we have discussed for a dielectric (Figure 7.11). The
wavelength of light is assumed to be in the visible spectrum.

For metals, the $r_\parallel$ component is not zero at a certain angle of incidence, as
it is at Brewster's angle in the case of dielectrics. The angle where the $r_\parallel$ component has its minimum is called the **principal angle**.

The phase difference of the reflected components of a metal, $\Delta = \delta_\parallel - \delta_\perp$,
is given in Figure 7.12b and compared with the phase difference for dielectrics
(Figure 7.12a).

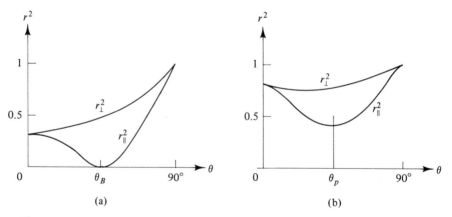

(a)   (b)

Fig. 7.11
Comparison of (a) reflection at a lossless medium (dielectric) and (b) reflection at a medium with losses (metal). Brewster's angle θ_B is replaced by the principal angle θ_P.

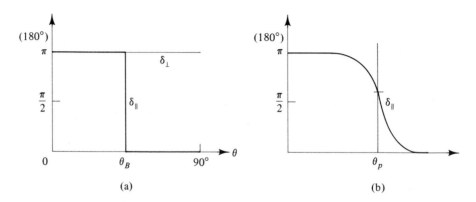

(a)   (b)

Fig. 7.12
(a) The phase angles $\delta_\parallel$ and $\delta_\perp$ for the $r_\parallel$ and $r_\perp$ components of a dielectric.
(b) The phase angle $\delta_\parallel$ for a metal.

Let us now reconsider the results for a dielectric that we obtained in Chapter 4:

$$r_{E_\parallel} = |r_\parallel| e^{i\delta_\parallel} \quad \text{and} \quad r_{E_\perp} = |r_\perp| e^{i\delta_\perp}$$

and for $\theta = 0°$, we have (see Table 4.1)

$$r_{E_\perp} = -\frac{N_2 - n_1}{N_2 + n_1} = \left(\frac{N_2 - n_1}{N_2 + n_1}\right) e^{i\delta_\perp} \to \delta_\perp = \pi$$

$$r_{E_\parallel} = -\frac{N_2 - n_1}{N_2 + n_1} = \left(\frac{N_2 - n_1}{N_2 + n_1}\right) e^{i\delta_\parallel} \to \delta_\parallel = \pi \tag{7.49}$$

and for $\theta = 90°$, we have

$$r_{E_\perp} = -1 = e^{i\delta_\perp} \rightarrow \delta_\perp = \pi$$
$$r_{E_\parallel} = 1 \quad = e^{i\delta_\parallel} \rightarrow \delta_\parallel = 0 \tag{7.50}$$

For a metal, the phase difference deviates from the rectangular curve of the dielectric. At the principal angle, the phase difference is 90°.

Since the two mutually perpendicular components of the reflected light, $r_\parallel$ and $r_\perp$, have a phase difference between them, elliptically polarized light is observed. Methods of determining $\bar{n}$ and $\bar{K}$ by employing measurements of the phase angles are called **ellipsometry**. We can qualitatively obtain the degree of elliptical polarization of the reflected waves by using the results of Chapter 6. These are tabulated in Table 7.1 (from Table 6.2) as the resulting elliptically polarized light, depending on their phase difference, for the two mutually perpendicular polarized light components having the same absolute value. As we mentioned in Chapter 6, the two reflected components $|r_\parallel|$ and $|r_\perp|$ do not have the same value for metals that they have for dielectrics. Therefore, we expect not to observe circular polarized light. The connection between these two cases is given in Table 7.1.

C. Ellipsometry and Optical Constants of Metals

To determine the optical constants of metals, we must once again consider equations 7.28 and 7.29. Two quantities of $r_{E_\perp}, r_{E_\parallel}, \delta_\perp$, and $\delta_\parallel$, or any combination of these, must be measured to obtain $\bar{n}$ and $\bar{K}$. In Figure 7.13, an optical setup is illustrated for the measurement of $\Delta = \delta_\parallel - \delta_\perp$ and $|r_\perp|/|r_\parallel| = \tan\phi$, where ϕ is the azimuthal angle, that is, the angle between the direction of the resultant of $|r_\perp|$ and $|r_\parallel|$.

For example, if we want to measure Δ, the analyzer and polarizer are set in "crossed" configuration (45° with respect to the plane of incidence). Since p and s components of the reflected light have the phase shift Δ between them, some of the reflected light will pass the analyzer. A compensator is introduced before the analyzer. With the compensator, we can compensate for the phase shift Δ, and no light will pass the analyzer.* In Figure 7.14, we show the results of measurements of ϕ and Δ as functions of θ from 0° to 90°.

After determining $\tan\phi$ and Δ as functions of θ, equations 7.28 and 7.29 are used to determine $\bar{n}$ and $\bar{K}$. The choice of measuring ϕ and Δ follows from experimental convenience. The calculations of $\bar{n}$ and $\bar{K}$ were not so simple before the introduction of desk-top computers, and various approximations had been used.

In Figure 7.15, we give a simplified presentation of the optical constants of copper. The data are from a compilation of optical constants of metals by M. A. Ordal et al.[†]

---

*For more details, see J. W. Hilton, *American Journal of Physics* **41**, 702 (1973).

[†]M. A. Ordal, L. L. Long, R. J. Bell, S. E. Bell, R. R. Bell, W. Alexander, Jr., and C. A. Ward, *Applied Optics* **21**, 1099 (1983).

Table 7.1

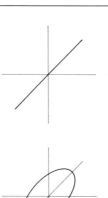

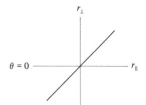
$\theta = 0$

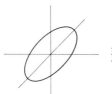

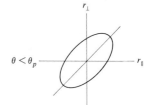
$\theta < \theta_p$

$\dfrac{\pi}{2}$

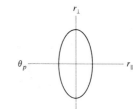
θ_p

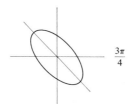
$\dfrac{3\pi}{4}$

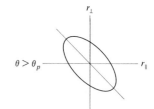
$\theta > \theta_p$

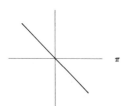
π

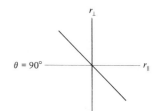
$\theta = 90°$

Addition of the real part of

$$Ae^{i(kX-\omega t)}$$

and

$$Ae^{i(kX-\omega t+\Delta)}$$

with $A = 1$. In Chapter 6, the path length X in the crystal produced the phase difference Δ.

Addition of the real part of

$$|r_\perp| e^{i(kX-\omega t)}$$

and

$$|r_\parallel| e^{i(kX-\omega t+\Delta)}$$

where $|r_\perp| \neq |r_\parallel|$. The different angles of incidence produce the different phase differences Δ.

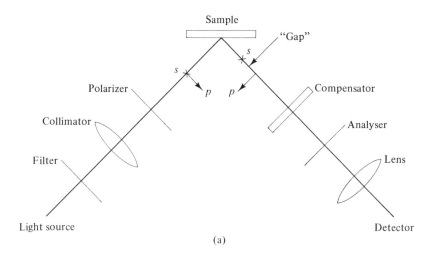

(a)

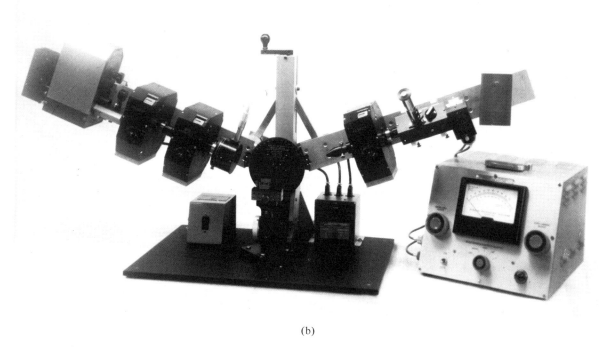

(b)

Fig. 7.13
(a) Experimental setup of an ellipsometer. The light passing the filter is considered monochromatic; it is collimated and polarized. After reflection, the parallel component p and the perpendicular component s have a phase difference, as shown by the gap in the figure. The compensator compensates for the phase difference and the analyzer will show either no light or maximum light intensity, depending on the operational mode. This is observed at the detector. (b) Photo of an ellipsometer (courtesy of Rudolph Instruments Inc., Fairfield, N.J.).

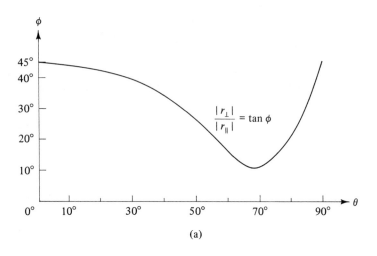

(a)

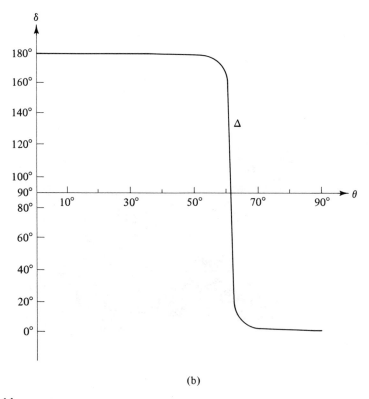

(b)

Fig. 7.14
(a) Schematic results of the measurements of ϕ as a function of θ for an aluminum film of about 30 Å thickness. (b) Schematic results of the measurement of Δ as a function of θ for an aluminum film of 20 Å.

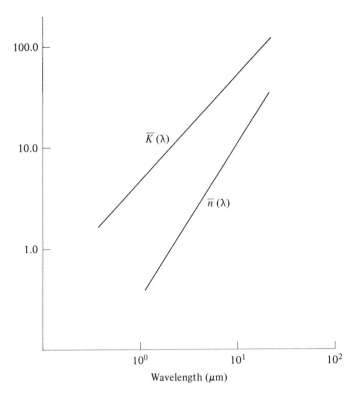

Fig. 7.15
Simplified presentation of the optical constants $\bar{n}$ and $\bar{K}$ of copper according to
M. A. Ordal, L. L. Long, R. J. Bell, S. E. Bell, R. R. Bell, W. Alexander, Jr., and
C. A. Ward, *Applied Optics* **21**, 1099 (1983).

4. FARADAY ROTATION

A. Introduction

In Chapter 6, we discussed optical activity. There, the plane of polarization of light was rotated by a certain angle α after the light passed through a certain material of thickness L. We saw that crystals made of molecules with special symmetry and in asymmetric arrangements show this phenomenon. Optical rotation can also be observed if an isotropic dielectric is placed in a magnetic field. Linearly polarized light traveling in the direction of the magnetic field will show a rotation of its plane of polarization (Figure 7.16). This is called the **Faraday effect**. The rotation angle α is proportional to the magnetic field and the length L of the material in the magnetic field. Experimentally, we find

$$\alpha = V B_X L_X \tag{7.51}$$

where V is called the **Verdet constant**. A few examples of V are given in Table 7.2. We can see from Table 7.3 that the Verdet constant depends on the wavelength.

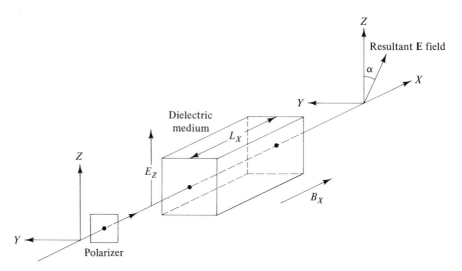

Fig. 7.16
Rotation of the plane of polarization of polarized light after passing the length L_X of a solid in the magnetic field B_X.

Table 7.2
Verdet constant for solids and liquids (Na D light)

Material	Temperature, °C	V (10^{-3} min of arc/gauss-cm)
H_2O	0	13.11
C_2H_5OH	25	11.12
n-C_3H_7OH	17.3	11.81
Acetone	15.1	11.09
CH_3Cl	18	12.9
CCl_4	15	16.03
C_6H_6	20	29.7
CS_2	0	43.41
ZnS (sphalerite)	16	225
NaCl	16	35.85
KCl	16	28.58
$PbO \cdot SiO_2$ (glass)	16	77.9

For more complete data see *International Critical Tables*, vol. 6 New York: McGraw-Hill, 1926–1930), p. 425.

Table 7.3
Dispersion of Verdet constant*

Material	Wavelength λ, μm				
	0.6	0.8	1.0	1.5	2.0
H_2O	12.6	7.0	4.4	(2.9 at $\lambda = 1.25$)	
CCl_4	16.1	8.9	5.7	2.5	1.3
C_6H_6	28.1	15.3	9.5	3.9	2.2
CS_2	39.4	21.4	13.5	5.8	3.1

*V in 10^{-3} min of arc/gauss-cm; temperature 23°C.

7. Wave Theory in Dispersive Media and Optical Constants

B. Theory

Just as we developed a theory for the refractive index using the forced oscillator model, we can give a mathematical description of the Faraday effect. The material is assumed to be represented by oscillators. These oscillators may interact with the incident electromagnetic wave. The magnetic field will enter the description in the term for the force driving the oscillator. Then we will apply Maxwell's equations to obtain the wave equation for this case. This will be done in close analogy to the method used for double refraction for uniaxial crystals (Chapter 5). The solutions of the wave equations for the Faraday rotation can also be used for the description of optical activity considered in Chapter 6.

We represent the dielectric in the magnetic field by oscillators and apply the forced oscillator equation 7.3 without a damping term. We are only interested in the Y and Z components of the oscillations, since the E field employed has no components in the X direction:

$$m\ddot{Y} + gY = F_Y$$
$$m\ddot{Z} + gZ = F_Z \qquad (7.52)$$

where g is the force constant, m is the mass of the oscillator, and F_Y and F_Z represent the **Lorentz force**, defined as

$$\mathbf{F} = -e\mathbf{E} - e[\mathbf{v} \times \mathbf{B}] = -e\mathbf{E} - e\left[\frac{d\mathbf{r}}{dt} \times \mathbf{B}\right] \qquad (7.53)$$

where $\mathbf{B}$ is the magnetic field vector. The electrical polarization vector $\mathbf{P}$ is equal to $-Ne\mathbf{r}$. Using this fact and multiplying by Ne, where N is the total number of oscillators per unit volume, we get

$$-m\ddot{P}_Y - gP_Y = NeF_Y = -e^2NE_Y + e[\dot{\mathbf{P}} \times \mathbf{B}]_Y$$
$$-m\ddot{P}_Z - gP_Z = NeF_Z = -e^2NE_Z + e[\dot{\mathbf{P}} \times \mathbf{B}]_Z \qquad (7.54)$$

or

$$m\ddot{P}_Y + gP_Y = e^2NE_Y - e(\dot{P}_Z B_X - \dot{P}_X B_Z)$$
$$m\ddot{P}_Z + gP_Z = e^2NE_Z - e(\dot{P}_X B_Y - \dot{P}_Y B_X) \qquad (7.55)$$

Note that one dot stands for d/dt, while two dots stand for d^2/dt^2.

Assuming harmonic solutions for $\mathbf{E}$ and $\mathbf{P}$, that is, $\mathbf{E} = \mathbf{E}_0 e^{i\omega t}$ and $\mathbf{P} = \mathbf{P}_0 e^{i\omega t}$, and assuming (from Figure 7.16) that $B_X \neq 0$, B_Y and $B_Z = 0$, we get

$$(-m\omega^2 + g)P_Y = +e^2NE_Y - i\omega eP_Z B_X$$
$$(-m\omega^2 + g)P_Z = +e^2NE_Z + i\omega eP_Y B_X \qquad (7.56)$$

This equation can be written in matrix form as

$$\begin{pmatrix} -\omega^2 m + g & i\omega eB_X \\ -i\omega eB_X & -\omega^2 m + g \end{pmatrix} \begin{pmatrix} P_Y \\ P_Z \end{pmatrix} = Ne^2 \begin{pmatrix} E_Y \\ E_Z \end{pmatrix} \qquad (7.57)$$

We would like to express $\mathbf{P}$ as a function of $\mathbf{E}$ as

$$\begin{pmatrix} P_Y \\ P_Z \end{pmatrix} = \varepsilon_0 \begin{pmatrix} \chi_{11} & \chi_{12} \\ \chi_{21} & \chi_{22} \end{pmatrix} \begin{pmatrix} E_Y \\ E_Z \end{pmatrix} \qquad (7.58)$$

To do this we need to calculate the inverse of the matrix in equation 7.57. This is done in problem 7 and results in

$$
(\chi_{ij}) = \begin{pmatrix} \dfrac{Ne^2}{\varepsilon_0 m}\left(\dfrac{\omega_0^2 - \omega^2}{(\omega_0^2 - \omega^2)^2 - (\omega\omega_c)^2}\right) & \dfrac{Ne^2}{\varepsilon_0 m}\left(\dfrac{(-i\omega\omega_c)}{(\omega_0^2 - \omega^2)^2 - (\omega\omega_c)^2}\right) \\ \dfrac{Ne^2}{\varepsilon_0 m}\left(\dfrac{(i\omega\omega_c)}{(\omega_0^2 - \omega^2)^2 - (\omega\omega_c)^2}\right) & \dfrac{Ne^2}{\varepsilon_0 m}\left(\dfrac{\omega_0^2 - \omega^2}{(\omega_0^2 - \omega^2)^2 - (\omega\omega_c)^2}\right) \end{pmatrix} \tag{7.59}
$$

where $\omega_0^2 = g/m$ is the resonance frequency, $\omega_p = Ne^2/\varepsilon_0 m$ is the plasma frequency, and $\omega_c = eB/m$ is the **cyclotron frequency**. Equation 7.58 will now be introduced into equation 5.22, from Chapter 5. The wave equations 5.27 and 5.34 can be obtained in a similar way, but now we use $\mathbf{D} = \mathbf{P} + \varepsilon_0 \mathbf{E}$.

The left side of equations 5.27 and 5.29 remain the same, and we must consider

$$
-\mathbf{k}(\mathbf{k} \cdot \mathbf{E}) + \mathbf{E}(\mathbf{k} \cdot \mathbf{k}) = -\mu_0 \varepsilon_0 \frac{\partial^2 \mathbf{E}}{\partial t^2} - \mu_0 \frac{\partial^2 \mathbf{P}}{\partial t^2}
$$

For our case, we use the solutions

$$
E_Y = A_Y e^{i(k_x X - \omega t)}
$$

and

$$
E_Z = A_Z e^{i(k_x X - \omega t)}
$$

where $k_X = k$, and $k_Y = k_Z = 0$, that is, $\mathbf{k} \cdot \mathbf{E} = 0$. We also set $|A_Y| = |A_Z|$. From equation 7.58, we have

$$
\mathbf{P} = \varepsilon_0(\chi)\mathbf{E}
$$

The Y and Z components are

$$
\begin{aligned}
E_Y k^2 &= -\mu_0 \varepsilon_0(-\omega^2)E_Y - \mu_0 \varepsilon_0[\chi_{11}(-\omega^2)E_Y + \chi_{12}(-\omega^2)E_Z] \\
E_Z k^2 &= -\mu_0 \varepsilon_0(-\omega^2)E_Z - \mu_0 \varepsilon_0[\chi_{21}(-\omega^2)E_Y + \chi_{22}(-\omega^2)E_Z]
\end{aligned} \tag{7.60}
$$

Setting the determinant corresponding to the homogeneous system of equations equal to zero, we get

$$
\begin{array}{cc} (E_Y) & (E_Z) \end{array}
$$
$$
\begin{vmatrix} k^2 - \mu_0\varepsilon_0\omega^2 - \mu_0\varepsilon_0\omega^2\chi_{11} & -\chi_{12}\omega^2\mu_0\varepsilon_0 \\ -\chi_{21}\omega^2\mu_0\varepsilon_0 & k^2 - \mu_0\varepsilon_0\omega^2 - \mu_0\varepsilon_0\omega^2\chi_{22} \end{vmatrix} = 0 \tag{7.61}
$$

and with $\mu_0\varepsilon_0 = 1/c^2$, we have

$$
\begin{vmatrix} k^2 - \dfrac{\omega^2}{c^2}(1 + \chi_{11}) & -\chi_{12}\dfrac{\omega^2}{c^2} \\ -\chi_{21}\dfrac{\omega^2}{c^2} & k^2 - \dfrac{\omega^2}{c^2}(1 + \chi_{22}) \end{vmatrix} = 0 \tag{7.62}
$$

where (E_Y) and (E_Z) refer to the direction of vibration of the $\mathbf{E}$ vector.

7. Wave Theory in Dispersive Media and Optical Constants

For the relation of χ_{ik} to refractive indices n_Y and n_Z, we have

$$\varepsilon = \varepsilon_0(1 + \chi), \quad 1 + \chi_{11} = n_Y^2, \quad \text{and} \quad 1 + \chi_{22} = n_Z^2 \tag{7.63}$$

similar to equation 5.38.

For the solution of the determinant in equation 7.62, we get

$$\left[k^2 - \frac{\omega^2}{c^2}(1 + \chi_{11}) \right]\left[k^2 - \frac{\omega^2}{c^2}(1 + \chi_{22}) \right] = \chi_{12}\chi_{21}\frac{\omega^4}{c^4} \tag{7.64}$$

Multiplying by c^4/ω^4 and inserting $\chi_{11} = \chi_{22}$ and $\chi_{12} = -\chi_{21}$ (equation 7.59) gives

$$\left[\frac{c^2 k^2}{\omega^2} - (1 + \chi_{11}) \right]^2 = -\chi_{12}^2 \tag{7.65}$$

As a result, we have

$$\frac{c^2 k^2}{\omega^2} - (1 + \chi_{11}) = \pm i\chi_{12} \tag{7.66}$$

or

$$k = \pm \frac{\omega}{c}\sqrt{1 + \chi_{11} \pm i\chi_{12}} \tag{7.67}$$

Before applying our result to the Faraday effect, we consider optical activity.

C. Optical Activity

If we substitute the solutions for k from equation 7.67 into equation 7.60, we have

$$\left[\frac{\omega^2}{c^2}(1 + \chi_{11} \pm i\chi_{12}) - \frac{\omega^2}{c^2} - \frac{\omega^2}{c^2}\chi_{11} \right]E_Y - E_Z\chi_{12}\frac{\omega^2}{c^2} = 0$$

which simplifies to

$$\pm \chi_{12}iE_Y = \chi_{12}E_Z$$

or

$$iE_Y = \pm E_Z \tag{7.68}$$

These solutions can be expressed as Jones vectors (Chapter 6). From equation 7.67, we have two solutions for k, which we call k_+ and k_-. As Jones vectors, they may be written as

$$E_+ = \begin{pmatrix} A_Y e^{i(k_+ X - \omega t)} \\ iA_Z e^{i(k_+ X - \omega t)} \end{pmatrix} \tag{7.69}$$

$$E_- = \begin{pmatrix} A_Y e^{i(k_- X - \omega t)} \\ -iA_Z e^{i(k_- X - \omega t)} \end{pmatrix} \tag{7.70}$$

Setting

$$|A_Y| = |A_Z| = 1, \quad k_+ = k_0 + \Delta k, \quad \text{and} \quad k_- = k_0 - \Delta k$$

and omitting the common phase factor, we have

$$E_+ = \begin{pmatrix} 1 \\ i \end{pmatrix} e^{i\Delta k}$$

$$E_- = \begin{pmatrix} 1 \\ -i \end{pmatrix} e^{i\Delta k}$$

(7.71)

In equation 7.71, one vector is interpreted as right polarized light, the other as left polarized light (see equation 6.67). The addition of both gives linearly polarized light. If linearly polarized light is incident and the medium rotates the plane of polarization in only one direction, the angle of polarization α is given as

$$\alpha = \Delta k = (n_R - n_L)\frac{\pi}{\lambda}L$$

(7.72)

(see Chapter 6, Section 4A), where we have the corresponding

$$n_R = (k_+)\frac{\lambda}{2\pi} \quad \text{and} \quad n_L = (k_-)\frac{\lambda}{2\pi}$$

D. Faraday Rotation

For the Faraday rotation, we have

$$\alpha = \tfrac{1}{2}(k_+ - k_-)$$

(7.73)

Since the Faraday rotation is a small effect compared to the permittivity of the medium produced by the applied electric field of the traveling wave, we have off-diagonal elements in the matrix of equation 7.58, small compared to diagonal elements. We therefore develop the root in equation 7.67 and have

$$k_\pm \simeq \frac{\omega}{c}\sqrt{1+\chi_{11}}\left(1 \pm \frac{1}{2}\frac{i\chi_{12}}{(1+\chi_{11})}\right)$$

(7.74)

From equation 7.63, we have

$$1 + \chi_{11} = n_Y^2 = n^2$$

and with equation 7.73 have

$$\alpha = \frac{i\chi_{12}}{n}\left(\frac{\pi}{\lambda}\right)$$

or with equation 7.59

$$\alpha \simeq \frac{\pi}{\lambda \cdot n}\left(\frac{\omega\omega_c\omega_p^2}{(\omega_0^2 - \omega^2)^2 - \omega^2\omega_c^2}\right) \qquad \text{(Note: } \alpha \text{ is } \propto \text{ to } B)$$

(7.75)

The dependence of α on the frequency is very similar to what we have seen for the refractive index n (see equations 7.15–7.17). We again have a resonance at $\omega = \omega_0$. This has been experimentally shown for F centers in KI (potassium iodide) crystals. The F centers are produced by X-ray radiation of the KI crystal, and they consist of an electron bound to a place in the crystal where a negative ion (I^-) is missing. A missing iodine ion in a perfect periodic lattice of K^+I^- has

7. Wave Theory in Dispersive Media and Optical Constants

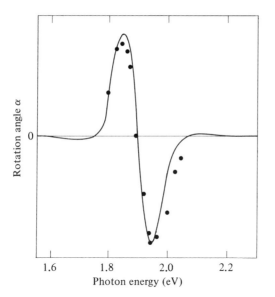

Fig. 7.17
Faraday effect. Rotation angle α in arbitrary units, depending on the frequency of the light (in photon energy). [From F. C. Brown and G. Laramore, *Applied Optics* **6**, 669 (1967).]

the effect of a positive charge; it can bind an electron. These F centers are the oscillators we described by the forced oscillator model (equation 7.52), and the magnetic field acts on them (equation 7.53).

For this case, α has been measured as a function of frequency, as shown in Figure 7.17. Using the results of such measurements and equation 7.75, the Verdet constant (see equation 7.51) can be deduced for a specific wavelength.

Problems

1. *The forced and damped oscillators.*
   A. *The forced oscillator.* Let us review the mechanical forced oscillator. We start with the free oscillator

   $$m\ddot{x} + kx = 0$$

   where m is the mass, k is the force constant, and x is the displacement, and $\ddot{x} = d^2x/dt^2$. We will call ω_0 the resonance frequency.
   a. Use the trial solution $x = A\sin(\omega_0 t + \delta)$ and show how the frequency depends on the constant k and the mass m.

   The free oscillator may be driven by a force $F_0 \sin \omega t$, and we must consider the equation

   $$m\ddot{x} + m\omega_0^2 x = F_0 \sin \omega t.$$

   b. Use the trial solution $x = A\sin \omega t$ and find the solution for the above equation.

   At the resonance frequency $\omega = \omega_0$, the amplitude is infinite. Therefore, we will now consider the more realistic damped oscillator.

B. *The damped oscillator.* To avoid the above difficulty, we introduce a damping term and consider the free damped oscillator

$$m\ddot{x} + b\dot{x} + kx = 0$$

c. Use the trial solution $x = Ae^{i\bar{\omega}_0 t}$ and give the dependence of $\bar{\omega}_0$ on the constants m, b, and k.

d. Write the solution in the form $x = Ae^{\alpha t + i\beta t}$ and discuss α and β. What is their significance?

We now turn to the damped forced oscillator equation

$$m\ddot{x} + b\dot{x} + kx = F_0 e^{i\omega t}$$

e. Use the trial solution $x = Ae^{i\omega t}$ and introduce $\omega_0^2 = k/m$ of the free oscillator. Show that we get

$$A = \frac{F_0}{m(\omega_0^2 - \omega^2) + i\omega b}$$

Note that for $b = 0$ and $\omega \to \omega_0$, we have the same problem we had for the free oscillator.

f. Rewrite the result in the form $x = \bar{A}e^{i(\omega t + \delta)}$. To determine $\bar{A}$, take the square root of the absolute value of x.

The phase angle δ is obtained by writing x (without the factor $e^{i\omega t}$) in complex form, that is, multiply numerator and denominator by

$$m(\omega_0^2 - \omega^2) - i\omega b$$

and take the ratio of the imaginary part over the real part.

g. Show that

$$\tan \delta = \frac{-\omega b}{m(\omega_0^2 - \omega^2)}$$

and give the final solution (Ans. see 2a).

h. Sketch the amplitude and phase angle as function of ω.

2. *Power considerations of the damped oscillator.* The average power may be obtained by multiplication of the (complex) force by $\dot{x}$ (where $dx/dt = \dot{x}$) and taking the time average

$$m\ddot{x} + b\dot{x} + kx = F$$
$$\overline{P_{av}} = (\overline{F\dot{x}}) = m\overline{\ddot{x}\dot{x}} + b\overline{\dot{x}^2} + k\overline{x\dot{x}}$$

The averages of $\overline{\ddot{x}\dot{x}}$ and $\overline{x\dot{x}}$ are both zero.
Therefore, we have

$$\overline{P_{av}} = b\overline{\dot{x}^2}$$

a. Calculate $\overline{\dot{x}^2}$ by using the real part of the solution above, that is,

$$x = \frac{F_0}{\sqrt{m^2(\omega_0^2 - \omega^2)^2 + \omega^2 b^2}} \cos(\omega t + \delta)$$

and observe that the time average of $\sin^2(\omega t + \delta)$ is 1/2; see problem 1, Chapter 2.

We now consider the case of different dampings. If $b = 0$, no power is taken in by the oscillator.

b. If we plot $\overline{P_{av}}$ for large and small b, we obtain two curves like this:

7. **Wave Theory in Dispersive Media and Optical Constants**

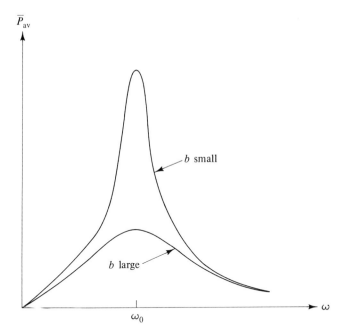

3. *Rubens-Hagen approximation for metals.* For metals, we have the approximation

$$\bar{n} \simeq \bar{K} \simeq \sqrt{\frac{\sigma}{2\varepsilon_0 \omega}}$$

a. Show that the reflectivity at normal incidence

$$R = \frac{(\bar{n} - 1)^2 + \bar{K}^2}{(\bar{n} + 1)^2 + \bar{K}^2}$$

may be written as

$$R \simeq 1 - 2\frac{1}{\sqrt{\sigma/2\varepsilon_0 \omega}}$$

if we neglect terms in

$$\left(\frac{1}{\sqrt{\sigma/2\varepsilon_0 \omega}}\right) \quad \text{with respect to} \quad \left(\frac{1}{\sqrt{\sigma/2\varepsilon_0 \omega}}\right)^2$$

This is called the *Rubens-Hagen approximation*.

b. Calculate $\sigma/2\varepsilon_0 \omega$ for infrared frequencies (use $\lambda = 1$ mm and show how close R is to the value of 1 for copper).

$$\varepsilon_0 = 8.85 \times 10^{-12} \frac{C^2}{Nm} \quad \text{and} \quad \sigma = 6 \times 10^7 \, (Am)^{-1}$$

c. What do we obtain for $\omega \to 0$?

Ans. a. $1 - (2/\bar{n})$; b. $R = 0.9985$; c. $R \to 1$

4. *Ray deviation by a prism.* In Chapter 1, we discussed the angle of deviation of a prism. Now we want to obtain the angle for minimum deviation. For the prism,

we have

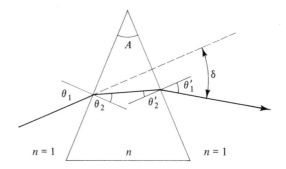

The refractive indices are indicated. We have

$$\sin \theta_1 = n \sin \theta_2$$
$$n \sin \theta_2' = \sin \theta_1'$$
$$\delta = \theta_1 + \theta_1' - A$$
$$A = \theta_2 + \theta_2'$$

These equations are invariant under the change from unprimed angles to primed angles. This suggests that we may have an optimum case for $\theta_1 = \theta_1'$, that is, a symmetric path. We now want to show that δ is minimum for the symmetric path. According to Sommerfeld we do this as follows:

a. Eliminate θ_1' and θ_2 from the four equations.
b. Differentiate with respect to θ_1 and θ_2' and show that we get

$$(\cos \theta_1) d\theta_1 + n \cos(A - \theta_2') d\theta_2' = 0$$
$$(\cos \theta_2') d\theta_2' + \frac{1}{n} \cos(\delta + A - \theta_1) d\theta_1 = 0$$

The two equation can be considered as a system of two homogeneous equations for $d\theta_1$ and $d\theta_2'$. To obtain a nontrivial solution, the determinant must vanish?

c. Show that the condition for the determinant to vanish is

$$\cos \theta_1 \cos \theta_2' = \cos(A - \theta_2') \cos(\delta + A - \theta_1)$$

d. Show that one possible solution is

$$\theta_1 = \theta_1' = \frac{\delta + A}{2}, \quad \theta_2 = \theta_2' = \frac{A}{2}$$

and that the other possible solution is not useful to determine the significance of this symmetric path.

e. Calculate δ for $n = 1.5$, and $A = 30°$ and determine $\theta_1 = \theta_1'$.
f. Calculate δ for several values of θ_1 near the symmetric position $\theta_1 = 22.8°$ by filling in the following table:

θ_1	θ_2	θ_2'	θ_1'	δ
26°				
24°				
22°				
20°				
18°				

7. **Wave Theory in Dispersive Media and Optical Constants**

Note that δ goes through a minimum. The fact that δ has its minimum value for the symmetric path may be used in the determination of the refractive index.

g. Show that the refractive index at minimum derivation is

$$n = \frac{\sin\frac{1}{2}(\delta + A)}{\sin\frac{1}{2}A}$$

5. *Prism dispersion.* The prism is a useful dispersion device. If polychromatic light is incident upon a prism, the components of different wavelength will each be deviated, depending on the refractive index of each wavelength. In ultraviolet, visible, and infrared spectrometers, prisms are used to disperse radiation from a source and produce, with the use of slits, quasi-monochromatic bands of light. In this problem, we will be concerned with those properties of prism dispersion of interest to instrument designers.

A quantity of interest in the design of monochromators is the angular dispersion $d\delta/d\lambda$, which is the rate of change of deviation δ with wavelength. By writing

$$\frac{d\delta}{d\lambda} = \frac{d\delta}{dn}\left(\frac{dn}{d\lambda}\right) = G\frac{dn}{d\lambda}$$

we see that the angular dispersion is a product of a geometric factor ($G = d\delta/dn$) and the dispersion $dn/d\lambda$, a factor dependent only on the prism characteristics.

a. Consider an equilateral prism of side B and angle A. Show that for the special case of minimum deviation, the geometric factor becomes

$$G = \frac{d\delta}{dn} = \frac{B}{b}$$

where b is the beam diameter.

b. Show that the angular dispersion is

$$\frac{d\delta}{d\lambda} = \frac{B}{b}\left(\frac{dn}{d\lambda}\right)$$

Now consider the dispersion factor. In regions where no resonances occur, the index of refraction may be represented empirically by an expression of the form

$$n(\lambda) = c_1 + \frac{c_2}{\lambda^2} + \frac{c_3}{\lambda^4}$$

where c_1, c_2, and c_3 are constants.

c. Calculate the constants c_1, c_2, and c_3 for fused quartz. Take the values

$$n = 1.486 \quad \text{for} \quad \lambda = 3034 \text{ Å}$$
$$n = 1.463 \quad \text{for} \quad \lambda = 4800 \text{ Å}$$
$$n = 1.455 \quad \text{for} \quad \lambda = 7065 \text{ Å}$$

Part. Ans. $C_1 = 1.448$, $C_2 = 3.3 \times 10^5 \text{ Å}^2$, $C_3 = 1.23 \times 10^{11} \text{ Å}^4$

6. *Comparison of dielectrics and metals.* For the dielectric case, we have

$$n^2 - K^2 = 1 + \frac{\omega_p^2(\omega_0^2 - \omega^2)}{(\omega_0^2 - \omega^2)^2 + \gamma^2\omega^2} \tag{7.76}$$

$$2nK = \frac{\gamma\omega\omega_p^2}{(\omega_0^2 - \omega^2)^2 + \gamma^2\omega^2} \tag{7.77}$$

For the conducting case, we have

$$\bar{n}^2 - \bar{K}^2 = 1 - \frac{\omega_p^2}{\omega^2 + \gamma^2} \tag{7.78}$$

$$2\bar{n}\bar{K} = \frac{\gamma}{\omega}\left(\frac{\omega_p^2}{\omega^2 + \gamma^2}\right) \tag{7.79}$$

a. In which particular situation will the formula for $2nK$ approach the formula for $2\bar{n}\bar{K}$?

b. Show that for this same case, $n^2 - K^2$ approaches $\bar{n}^2 - \bar{K}^2$.

c. Show that (7.78) and (7.79) may be satisfied by a common value of $\bar{n} = \bar{K}$. Show that we get $\bar{K} = \bar{n} = \sqrt{\gamma/2\omega}$.

d. Show that $\bar{n} = \sqrt{\gamma/2\omega_p}$ if we assume that $\omega = \omega_p$ and $\bar{n} = \bar{K}$.

7. *Inverse matrix.* Calculate the inverse matrix of

$$\begin{pmatrix} -\omega^2 m + g & i\omega eB_X \\ -i\omega eB_X & -\omega^2 m + g \end{pmatrix}$$

Frequency Resolution and Fourier Transform Spectroscopy

8

1. INTRODUCTION

In the first six chapters, we restricted most of our discussion to only one wavelength. In Chapter 7 we studied the dependence of the optical constants on frequency or wavelength.

In this chapter, we will see how spectroscopic devices are used for the separation of different wavelengths of incident light. If, for example, absorption is being studied, then the absorption of the different components of the incident light is desired. If emission is being studied, then it is of interest to find how intense and how much separated in frequency the emitted spectral lines are. The separation of spectral lines is called resolution, and different spectroscopic devices have different resolving powers, depending in part on the amount of illumination. In the visible region, enough light is incident on the spectroscopic device to consider only the diffraction limitations. In other spectral regions, such as the infrared, the limitations of the resolving power are determined by the aperture admitting the light for analysis.

We will discuss the resolving power of diffraction gratings, prisms, the Fabry-Perot spectrometer, and the Michelson interferometer. We will also discuss a different spectroscopic method, Fourier transform spectroscopy, which will also serve as an introduction to Fourier transformations.

2. RESOLVING POWER

A. Grating Spectrometers

Spectrometers that make use of diffraction gratings are discussed in most physics texts, and many students have already done experiments with them in introductory laboratory courses. Figure 8.1 shows a simple experimental setup.

Diffraction at a grating was discussed in Chapter 3. A schematic of a grating spectrometer is shown in Figure 8.2. The light from the source S_1 is concentrated

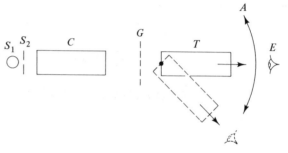

Fig. 8.1
Setup for simple experiments with a diffraction grating. S_1 is the source, S_2 the slit, C a collimator, G the grating, T a telescope, A an angle scale for positioning of T, and E the eye.

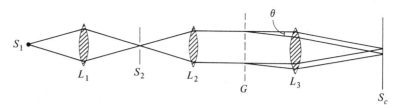

Fig. 8.2
Schematic of a grating spectrometer. S_1 is the source, L_1 a condensing lens, S_2 the slit, L_2 a collimating lens, G the grating, and L_3 a lens to produce the diffraction pattern on the screen, S_c.

by the condenser lens L_1 onto the slit S_2. The slit serves as a well-defined source opening. Lens L_2 produces parallel light, which is incident on the grating. Fraunhofer diffraction applies here, and lens L_3 is used to focus the diffraction pattern on the screen. An image may also be produced at an exit slit, which would replace the screen. In the latter case, a lens or mirror system condenses the light from the exit slit onto the detector.

The square of the amplitude, which we take as the intensity of the diffraction pattern appearing on the screen (from Chapter 3, Section 2), is given as

$$u^2 = u_0^2 \frac{\sin^2[k(d/2)\sin\theta]}{[k(d/2)\sin\theta]^2} \frac{\sin^2[Nk(a/2)\sin\theta]}{\sin^2[k(a/2)\sin\theta]} \tag{8.1}$$

where N is the number of openings in the grating, d is the width of the opening, a is the periodicity constant (grating constant), and θ is the angle of diffraction with respect to the normal. For the small-angle approximation, we have $\sin\theta = Y/X$ (Figure 8.3).

The lens in Figure 8.3 shortens the distances X and Y, as was discussed in Chapters 2 and 3. The angle θ is the angle between two central lines of two emerging parallel light bundles presenting the interfering waves for the production of the diffraction pattern.

For higher-order diffraction, θ could become quite large, and then lens L_3

8. Frequency Resolution and Fourier Transform Spectroscopy

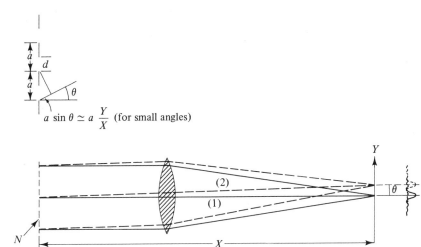

$a \sin \theta \simeq a \dfrac{Y}{X}$ (for small angles)

Fig. 8.3
Geometry for diffraction at a grating. N is the number of openings in the grating, θ is the angle of diffraction, d is the width of the openings, a is the grating constant, X is the distance of the screen from the grating, Y is the coordinate of the diffraction pattern and (1) and (2) are the central lines of the parallel light bundles emerging from the grating for two spectral lines.

would have to be reoriented. In that case, the small-angle approximation cannot be used.

The condition for the principal maxima is

$$a \sin \theta = m\lambda \qquad (8.2)$$

where m is an integer representing the order of the diffraction pattern.

From equation 8.2, we see that the wavelength is determined by a length or angle measurement. If we know the angle θ of the mth maximum, then we can obtain the wavelength from equation 8.2 if we also know the grating constant a. This is in contrast to the determination of frequencies in the microwave region. There we have a precise frequency standard for comparison, and we can determine the frequency with high accuracy. In the optical case, the frequency determination ($v = c/\lambda$) is only as good as the accuracy of the length measurement.

Resolution of two spectral lines. Now let us consider two different wavelengths, λ_1 and λ_2, producing two overlapping diffraction patterns. We will now develop a criterion to distinguish between the two diffraction patterns due to λ_1 and λ_2. The two waves may appear as two spectral lines. First we will illustrate the main maximum and side maxima and minima of the diffraction pattern due to λ_1 (Figure 8.4).

For λ_2, we obtain essentially the same result. Superposition of the two patterns reveals the wavelength shift. The two main maxima are centered at λ_1/a and λ_2/a, and these two points may be very close together (Figure 8.5).

We will consider two spectral lines resolved if they are positioned such that approximately half of the height on the right side of one spectral line is equal to

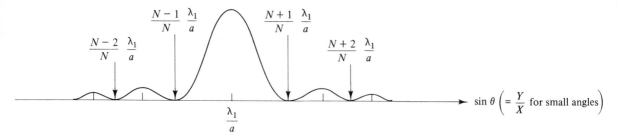

Fig. 8.4
Main maximum and side maxima and minima of the diffraction pattern obtained with a grating having N lines for one wavelength λ_1.

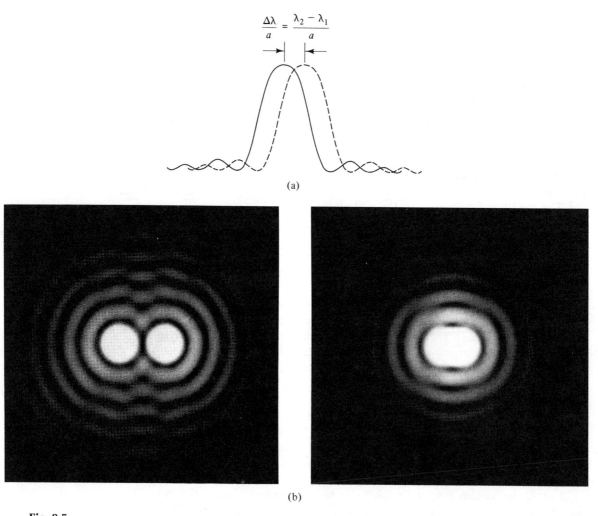

(a)

(b)

Fig. 8.5
(a) Closely positioned main maxima of the two diffraction patterns produced by λ_1 and λ_2 with $\Delta\lambda = \lambda_2 - \lambda_1$.
(b) (Left) images of two point sources sufficiently separated; (right) image of two point sources at the limit of resolution (from Cagnet, Françon, and Thrierr, *Atlas of Optical Phenomena*).

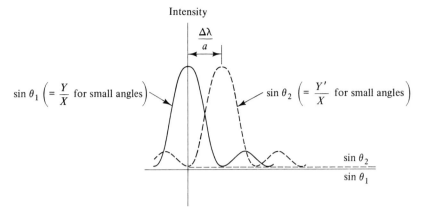

Fig. 8.6
Two spectral lines of wavelengths λ_1 and λ_2 are positioned in such a way that the main maximum of one spectral line is at the first minimum of the other. According to the Rayleigh criterion, the two spectral lines are considered resolved.

half of the height of the left side of the other (Figure 8.6). This is equivalent to the **Rayleigh criterion**, which states that two spectral lines are considered resolved if the main maximum of one spectral line is positioned at the first minimum of the other spectral line. Note our use of $\sin\theta$ for the abscissa (and Y/X for the small-angle approximation).

From Figure 8.4, the distance from the center to the first minimum is

$$\frac{N+1}{N}\left(\frac{\lambda_1}{a}\right) - \frac{\lambda_1}{a} = \frac{\lambda_1}{Na} \tag{8.3}$$

The spectral line produced by λ_2 has its maximum at the first minimum of the spectral line of λ_1, and the separation between the two can be written as

$$\frac{\Delta\lambda}{a} = \frac{\lambda_2 - \lambda_1}{a} = \frac{\lambda_1}{Na} \tag{8.4}$$

or

$$\frac{\lambda_1}{\lambda_2 - \lambda_1} = N \tag{8.5}$$

Since the difference between λ_1 and λ_2 is small but N is usually large, we can also write

$$\frac{\lambda}{\Delta\lambda} = N \tag{8.6}$$

If we repeat this procedure for the mth order, we obtain

$$m \cdot \Delta\lambda = \frac{\lambda}{N}$$

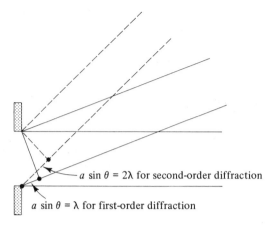

Fig. 8.7
Angles for the first- and second-order diffraction. The difference in path length is also indicated.

which can also be written as

$$\frac{\lambda}{\Delta\lambda} = mN \qquad (8.7)$$

We see that the resolving power $\lambda/\Delta\lambda$ depends on the order of diffraction. This is certainly understandable, since the path difference increases with m, as shown in Figure 8.7. Consequently, the angular difference between two spectral lines will get larger with higher-order spectra (Figure 8.8).

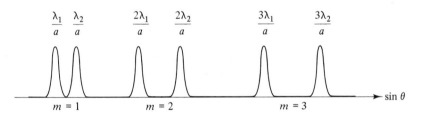

Fig. 8.8
The angular distance (depending on $\sin\theta$) between the diffraction maxima of wavelengths λ_1 and λ_2 becomes larger for higher orders. Its appearance on the screen becomes more and more distinct.

Moreover, $\lambda/\Delta\lambda$ depends only on N, and not on the spacing a of the grating. But we can only consider diffraction patterns for $\lambda \leq a$; therefore, we cannot make a as small as we like. Under these limitations, larger gratings will give us higher resolution. To derive a quantitative expression, we consider all higher orders.

From equation 8.2, the elementary formula for the position of the maxima, we find that the highest order is given by the largest angle θ we can employ, that is, $\theta = 90°$, and with $\sin 90° = 1$, we have

$$a = m_{max}\lambda \qquad (8.8)$$

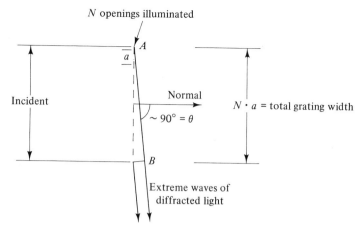

N openings illuminated

Incident

Normal

$\sim 90° = \theta$

$N \cdot a$ = total grating width

Extreme waves of diffracted light

Fig. 8.9
The total grating width $N \cdot a$ is equal to the path difference AB of the extreme waves for $\theta = 90°$.

Substitution into equation (8.7) gives us

$$\frac{\lambda}{\Delta\lambda} = \frac{a}{\lambda}N = \frac{\text{grating width}}{\lambda} \tag{8.9}$$

We find that $\lambda/\Delta\lambda$ depends on the total width of the grating or, correspondingly, on the path difference AB of the extreme waves on the left and right sides, as shown in Figure 8.9.

Later in this chapter, we will see that similar results (equation 8.9) are obtained for the relation of the maximum resolution and the path difference of the extreme waves for optical dispersion devices and methods.

Example

Consider a grating with 60 lines per millimeter and a length of 10 cm. The Wavelength of light incident on the grating is

$$\lambda = 6000 \text{ Å} = 600 \text{ nm} = 0.6 \times 10^{-6}\text{m}$$

and the order is $m = 2$:

$$N = (60 \text{ lines/mm})(100 \text{ mm}) = 6000$$

Since $\lambda/\Delta\lambda = mN$, the **resolving power** is

$$\frac{\lambda}{\Delta\lambda} = (2)(6000) = 12,000$$

$$\Delta\lambda = 0.5 \text{ Å} \quad \text{for} \quad \lambda = 6000 \text{ Å}$$

If we can take $m = 20$, then

$$\frac{\lambda}{\Delta\lambda} = 120,000$$

$$\Delta\lambda = 0.05 \text{ Å}$$

As a comparison, the Bohr radius of the hydrogen atom is 0.5 Å.

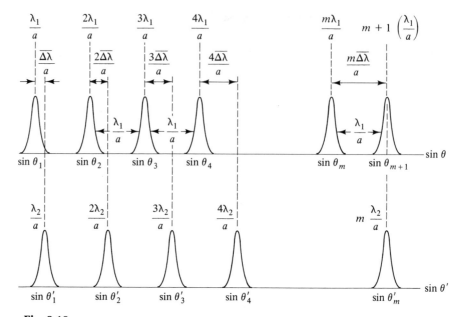

Fig. 8.10
Positions of the higher orders of spectral lines with wavelengths λ_1 and λ_2 depending on the diffraction angle θ. For λ_1, the position is indicated by $\sin \theta_m = m(\lambda_1/a)$; for λ_2, by $\sin \theta'_m = m(\lambda_2/a)$. The distance between the lines of λ_1 is λ_1/a; for λ_2 it is λ_2/a; and the distances between the same orders of λ_1 and λ_2 are indicated as multiples of $\overline{\Delta\lambda}/a$.

Free spectral range. Figure 8.8 showed that the separation of two spectral lines λ_1 and λ_2 ($\lambda_1 < \lambda_2$) increases with higher orders. But there is a limit to the highest order that can be used. When the $(m + 1)$th order of λ_1 and the mth order of λ_2 have the same value for $\sin \theta$, we speak of overlapping orders, and the observer can no longer distinguish between light having wavelength λ_1 and light having wavelength λ_2.

In Figure 8.10, we have plotted schematically the positions of all orders of the spectral lines of λ_1 and λ_2 depending on $\sin \theta$ according to the grating equation $m\lambda = a \sin \theta$, (one might think about all the higher orders of a doublet line). The separation of the lines of wavelength λ_1 is the same for all: λ_1/a; and for the lines of wavelength λ_2, we have λ_2/a.

Defining

$$\frac{\overline{\Delta\lambda}}{a} = \frac{\lambda_2}{a} - \frac{\lambda_1}{a}$$

we see that if the value of $\overline{\Delta\lambda}$ is such that the $(m + 1)$th order of λ_1 and the mth order of λ_2 overlap, we have (see Figure 8.10)

$$\frac{\overline{\Delta\lambda}}{a} m = \frac{\lambda_1}{a} \tag{8.10}$$

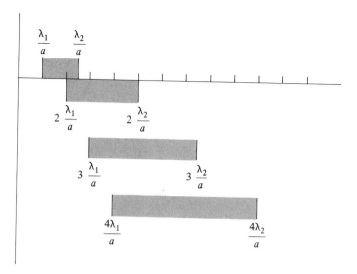

Fig. 8.11
First- to fourth-order diffraction spectra of the interval $\lambda_2 - \lambda_1$. The angle is expressed as $\sin \theta = m\lambda/a$ for the limits of each spectrum of order m.

The highest order useful is limited by the value of $\overline{\Delta\lambda}$, and we have

$$m_{\text{highest}} = \frac{\lambda_1}{\overline{\Delta\lambda}} \tag{8.11}$$

The useful spectral range is given by $\lambda/\overline{\Delta\lambda}$.

Now let us consider the following example for the resolving power of a grating.

The resolving power as expressed in equation 8.7 applies to all situations where the entrance slit can be made very narrow, for example, in the visible spectral region. If the slit width needs to be increased to provide enough energy for the detector to give a sufficient response, the spectrometer is called energy-limited. Spectrometers that operate within the infrared are in this category. The resolving power is much lower for these systems and depends on the width of the slit.

Now let us assume that a broad wavelength range $\lambda_2 - \lambda_1$ is incident on the grating. This would be the situation if we were to examine the emission of the sun with a grating spectrometer. As in acoustics, the range from λ_1 to twice the wavelength, that is, $2\lambda_1$, is called an **octave**. The visible spectrum of the sun stretches over a range just larger than an octave.

Figure 8.11 shows the overlapping of the first- to fourth-order spectra. Overlapping of the mth order with respect to the $(m-1)$th order occurs at $m(\lambda_1/a)$, and in each order, the next higher order overlaps at a wavelength distance $\sin \theta = \lambda_1/a$. Without filtering, we can only use the first order from λ_1/a to $2(\lambda_1/a)$. If we apply a filter of bandwidth smaller than one octave, we can apply the considerations leading to equation 8.11 and determine the useful spectral range.

2. Resolving Power   **319**

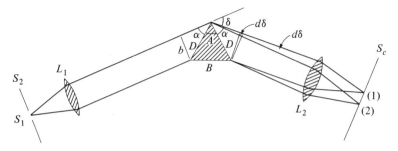

Fig. 8.12
Schematic of a prism spectrometer. S_1 is the source, L_1 a collimator lens, B the base of the prism, A the apex angle, δ the angle of deviation, D the width of the prism, and L_2 the lens forming the image of the slit on the screen S_c.

B. Prism Spectrometers

Although primarily known for his fundamental work on gravity, Isaac Newton also spent a considerable amount of time studying optics. He is the founder of the color theory of white light. To prove his theory, he used a prism to separate the colors. Then he applied a second prism to demonstrate that the separated colors could not be decomposed any further.

An experiment with a prism spectrometer is similar to what we described for a grating spectrometer. This experiment, done in introductory laboratory courses, is often applied to the observation of atomic emission lines from gas discharge tubes.

The resolving power of a prism is more complicated to explain than that of a grating, since we must now add the effects of dispersion and diffraction. Figure 8.12 illustrates the setup of a prism spectrometer.

Consider the special case in which the light passes the prism in a symmetric path, δ being the angle of deviation. Two different wavelengths emitted by the source are imaged at positions 1 and 2, on the screen. The prism is of finite width D. The projection of D perpendicular to the incident beam is b. This width b is the effective aperture opening for diffraction of each spectral line at the prism. The two lines are considered resolved if the main maximum of the pattern produced by one line is positioned at the first minimum of the pattern of the other line. The distance between the two maxima is then λ/b (Figure 8.13). The diffraction angles for the two maxima of the pattern are θ and θ'. Their difference is $d\delta$, corresponding to λ/b, and we have

$$d\delta = \frac{\lambda}{b} \tag{8.12}$$

The resolving power $\lambda/\Delta\lambda$ can be expressed in parameters that depend on the prism. For a symmetric pass (our special case), we have (from Chapters 1 and 7)

$$n = \frac{\sin\left(\dfrac{A + \delta}{2}\right)}{\sin(A/2)} \tag{8.13}$$

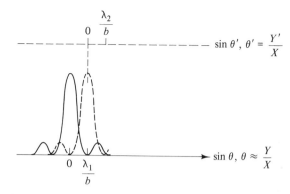

Fig. 8.13
Two coordinate systems are shown for presentation of the diffraction patterns
produced with λ_1 and λ_2. The angle θ is the angle of diffraction.

Differentiation results in

$$\frac{dn}{d\lambda} = \frac{\cos\left(\dfrac{A+\delta}{2}\right)}{\sin(A/2)}\left(\frac{1}{2}\right)\frac{d\delta}{d\lambda} \tag{8.14}$$

To express $\cos\left(\dfrac{A+\delta}{2}\right)$ in parameters of the prism, we note from Figure 8.12
that

$$\alpha = 90° - \frac{A+\delta}{2} \quad \text{and} \quad \sin\alpha = \frac{b}{D} = \cos\frac{A+\delta}{2}$$

In a similar fashion, we find that

$$\sin\frac{A}{2} = \frac{B/2}{D}$$

Introducing these relations into equation 8.14 yields

$$\frac{dn}{d\lambda} = \frac{b/D}{(B/2)/D}\left(\frac{1}{2}\right)\frac{d\delta}{d\lambda} = \frac{b}{B}\frac{d\delta}{d\lambda} \tag{8.15}$$

Combining with equation 8.12 results in

$$\frac{\lambda}{\Delta\lambda} = B\frac{\Delta n}{\Delta\lambda} \tag{8.16}$$

where Δ replaced the differentiation or a variation in the parameters. Thus, the
resolving power depends only on the base length of the prism and the change in
the refractive index with the wavelength. Consequently, different materials, such
as different glasses, may be used to obtain different resolving powers.

For example, if we choose a material with $\Delta n/\Delta\lambda = 1000(1/\text{cm})$ at $\lambda = 6000$
Å and a base $B = 6$ cm, we then have

$$\frac{\lambda}{\Delta\lambda} = 6(1000) = 6000$$

Therefore, at 6000 Å, $\Delta\lambda = 1$ Å. Here we have used a relatively large prism, whereas in Section A, we used a not very large grating. In general, prisms are inferior to gratings when it comes to resolving power. The grating can be used in higher orders, improving the resolution. This cannot be done for prisms; but as a result, there are also no overlapping orders. That is, the prism has a large free spectral range.

In high-resolution grating spectrometers, a prism can be used to do a rough separation of the incoming light; that is, it acts as a filter (and provides the free spectral range for investigation). Its filter action is convenient, since turning the prism will provide the high-resolution grating spectrometer with another band of light for dispersive analysis. In a similar way, the prism may be used with a Fabry-Perot spectrometer.

C. The Fabry-Perot Spectrometer

In Chapter 2, we saw how a plane parallel plate can produce fringes. We assume that light of wavelengths λ_1 and $\lambda_2 = \lambda_1 + \Delta\lambda$ is incident on the plate. The angle with the normal of the plate is θ, the refractive index is n, and the thickness is D. We assume that the refractive index of the plate is the same for λ_1 and λ_2. If we want to observe maxima of interference for the same angle θ of λ_1 and λ_2, we must change D. This can be accomplished by using two parallel plates and changing the gap between them (Figure 8.14a). The interference in the two plates is not considered. We use only the change in the distance between the two inner interfaces (Figure 8.14b). By changing D, the two interference patterns will appear one after the other, and the resolving power can be discussed as before.

We can assume that the two plates act as an optically dense medium, resulting in a phase shift of π in reflection but not in transmission. We will then have the same phase relation between the incident and transmitted waves as discussed in Chapter 2. Absorption will be neglected.

For the reflected and transmitted intensities, we derived the following formulas in Chapter 2 (equation 2.65):

$$I_r = \frac{g^2 \sin^2(\Delta/2)}{1 + g^2 \sin^2(\Delta/2)} \quad \text{and} \quad I_t = \frac{1}{1 + g^2 \sin^2(\Delta/2)} \tag{8.17}$$

where

$$g^2 = \frac{4r^2}{(1 - r^2)^2} \tag{8.18}$$

and r is the reflection coefficient for unpolarized light, that is, when the light is at close to normal incidence. For $\Delta/2$, we derived the following expression (equation 2.66):

$$\frac{\Delta}{2} = \frac{2\pi n D \cos\theta_2}{\lambda} \tag{8.19}$$

where D is the distance between the plates, n is the refractive index between the plates, θ_2 is the angle of refraction, and λ is the wavelength outside the plates. Equation 8.19 applies to the plane parallel plate as well as to a Fabry-Perot etalon, which, for example, can be made of two half-silvered plates at a distance D apart.

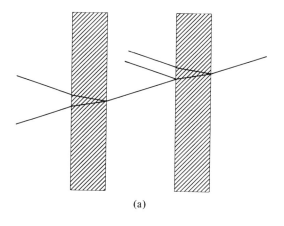

(a)

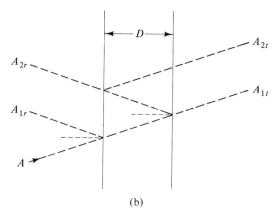

(b)

Fig. 8.14
(a) Two plane parallel plates with a distance D between them in a Fabry-Perot configuration. (b) Schematic of two reflecting and transmitting planes for the summation of multiple reflection and transmission without refraction.

Resonance distance. In the following, we assume that the refractive index between the reflecting plates is $n \neq 1$, the wavelength between the reflecting plates is λ, and the light is at normal incidence ($\theta_2 = 0$). We then have

$$\frac{\Delta}{2} = \frac{2\pi}{\lambda} D \tag{8.20}$$

for equation 8.19, and equations 8.17 bcome

$$I_r = \frac{g^2 \sin^2[2\pi(D/\lambda)]}{1 + g^2 \sin^2[2\pi(D/\lambda)]} \tag{8.21a}$$

and

$$I_t = \frac{1}{1 + g^2 \sin^2[2\pi(D/\lambda)]} \tag{8.21b}$$

2. Resolving Power 323

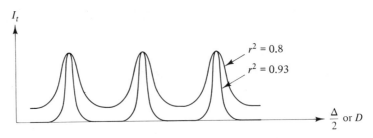

Fig. 8.15
Plots of I_t for $r^2 = 0.8$ and $r^2 = 0.93$ as functions of $(\Delta/2)$ or D.

Since r^2 is a constant, if we assume that there is only one wavelength, then I_r and I_t depend only on D. In Figure 8.15, plots of I_t are shown where $r^2 = 0.8$ and $r^2 = 0.93$. From the plots, we can see that the peaks are narrower for larger values of r^2. This was discussed in Chapter 2 for the plane parallel plate. However, there is another important point to observe. While each plate has a low transmission for large values of r^2 (that is, $r^2 \approx 1$), the assembly of two plates at a certain distance D apart has a high transmission. This particular distance D is called the **resonance distance**, and it corresponds to the case of constructive interference. It is given by

$$\frac{2\pi}{\lambda} D = m\pi \quad \text{or} \quad D = \frac{m\lambda}{2} \tag{8.22}$$

where m is an integer. From the condition of equation 8.22, we find that an integer number of half-wavelengths fit between the two plates. We recognize this as a **standing-wave pattern**, as illustrated in Figure 8.16. These different standing-wave patterns are called **modes**. If we go to the unrealistic case where $r^2 = 1$ exactly, then $g^2 = \infty$ and $I_t = 0$. In practical situations, because of absorption, $r^2 \neq 1$ and $g^2 \neq \infty$.

Resolving power. A Fabry-Perot etalon can be used for spectroscopic purposes if the transmitted light is recorded as a function of D, as shown in Figure 8.17. For slightly different wavelengths, a slightly shifted pattern is obtained. Since the resonance condition depends on the wavelength, the patterns will be displaced with respect to each other. To study this more closely, consider the wavelengths λ_1 and λ_2 when $m = 1$ (Figure 8.17).

We consider two spectral lines distinguishable if they overlap at half-height (assume $I = 1$). Then we have

$$\frac{1}{2} = \frac{1}{1 + g^2 \sin^2[(2\pi/\lambda_1)(D + \varepsilon)]} \quad \text{and} \quad \frac{1}{2} = \frac{1}{1 + g^2 \sin^2[(2\pi/\lambda_2)(D - \varepsilon)]} \tag{8.23}$$

Equating these two equations results in

$$\frac{D + \varepsilon}{\lambda_1} = \frac{D - \varepsilon}{\lambda_2} \tag{8.24}$$

8. Frequency Resolution and Fourier Transform Spectroscopy

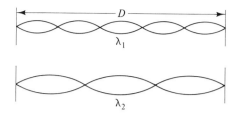

Fig. 8.16
Standing waves of two different wavelengths λ_1 and λ_2 in the resonance cavity of distance D.

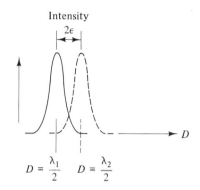

Fig. 8.17
Two spectral lines at resonance for λ_1 and λ_2 at $m = 1$. The separation of the two spectral lines is 2ε.

If we set $\lambda_1 = \lambda_2 - \Delta\lambda$, we have

$$\lambda_2(D + \varepsilon) = (\lambda_2 - \Delta\lambda)(D - \varepsilon)$$

or

$$2\lambda_2\varepsilon = (\varepsilon - D)\Delta\lambda \tag{8.25}$$

Renaming λ_2 as λ, we get

$$\frac{\lambda}{\Delta\lambda} = \frac{\varepsilon - D}{2\varepsilon} \tag{8.26}$$

To obtain ε, we consider the half-height of one of the spectral lines:

$$\frac{1}{2} = \frac{1}{1 + g^2 \sin^2[(2\pi/\lambda)(D + \varepsilon)]} = \frac{1}{1 + g^2 \sin^2[(2\pi/\lambda)\varepsilon]} \tag{8.27}$$

The last equality is true, since

$$\left| \sin \frac{2\pi}{\lambda} D \right| = 0 \quad \text{and} \quad \left| \cos\left(\frac{2\pi}{\lambda} D\right) \right| = 1$$

Since ε is small, we can approximate $\sin x \simeq x$ to get

$$1 + g^2 \left(\frac{2\pi\varepsilon}{\lambda}\right)^2 = 2 \quad \text{or} \quad 2\varepsilon = \lambda \frac{1}{g\pi} \tag{8.28}$$

We can neglect ε with respect to D in equation 8.26, and with equation 8.28, we have

$$\left|\frac{\lambda}{\Delta\lambda}\right| = \frac{g\pi D}{\lambda} \tag{8.29}$$

Introduction of $D = \lambda/2$ for the spectral line position results in

$$\left|\frac{\lambda}{\Delta\lambda}\right| = \frac{\pi g}{2} \quad \text{or} \quad \left|\frac{\lambda}{\Delta\lambda}\right| = \frac{\pi r}{1 - r^2}$$

(which shows exclusive dependence on r). (*Note*: $\lambda/\Delta\lambda$ is independent of n.)

For higher orders, we have (absolute value bars omitted)

$$\frac{\lambda}{\Delta\lambda} = m\frac{\pi g}{2} \quad \text{or} \quad \frac{\lambda}{\Delta\lambda} = m\mathscr{F} \tag{8.30}$$

where we introduce the finesse $\mathscr{F}$, defined as $\mathscr{F} = \pi g/2$. As we saw for the grating, we also have here improved resolution with higher orders of m, and we have only a certain useful free spectral range (see next section).

The resolution of the Fabry-Perot etalon does not depend on the size of the plates, but only on their reflectivity. We can obtain high resolution in a small free spectral range by using a large m, that is, a large distance between the plates.

Free spectral range. To discuss the free spectral range, let us reconsider Figure 8.17 and plot, in Figure 8.18a, all the lines for λ_1 and λ_2 ($\lambda_1 < \lambda_2$), including higher orders. One might think that all the higher orders appear as doublet lines. We assume that $\theta_2 = 0$, corresponding to normal incidence, that $n \neq 1$, and that λ is the wavelength between the plates. The spectral lines plotted in Figure 8.18a resemble those depicted in Figure 8.10 for the grating, and there is some similarity in the discussion. From equation 8.22 ($D = m\lambda/2$), the lines of λ_1 are at positions $m(\lambda_1/2)$, and the lines of λ_2 are at positions $m(\lambda_2/2)$. In first order, the lines λ_1 and λ_2 are separated by

$$\frac{\lambda_2 - \lambda_1}{2} = \frac{\overline{\Delta\lambda}}{2}$$

When the $(m + 1)$th order of λ_1 and the mth order of λ_2 overlap, we have

$$m\frac{\overline{\Delta\lambda}}{2} = \frac{\lambda_1}{2} \tag{8.31}$$

one calls $\overline{\Delta\lambda}$ the spectral range (of the FP), which may be expressed as

$$\overline{\Delta\lambda}_{SR} = \frac{\lambda}{m_n} \quad \text{or} \quad \overline{\Delta\lambda}_{SR} = \frac{\lambda^2}{2D} \tag{8.32}$$

If we go to higher orders, overlapping occurs, and the observer cannot distinguish between lines from the λ_1 or λ_2 higher-order series.

As discussed in Chapter 2, a ring pattern may be observed with a plane parallel plate or a Fabry-Perot etalon. To observe this pattern, we need a light source emitting into a solid angle. The distance between the plates is fixed; that

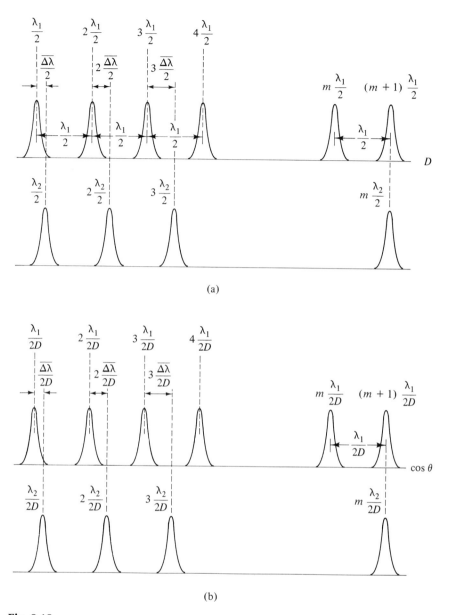

Fig. 8.18

(a) Positions of the series of lines $m(\lambda_1/2)$ and $m(\lambda_2/2)$ depending on D as the variable. The distance between the lines $m(\lambda_1/2)$ is indicated, as well as the distance between a few lines of λ_1 and λ_2. Overlapping occurs for $(m+1)(\lambda_1/2)$ and $m(\lambda_2/2)$. (b) Positions of the series of lines $m(\lambda_1/2D)$ and $m(\lambda_2/2D)$ depending on $\cos\theta$ as the variable. The distance between the lines $m(\lambda_1/2D)$ and $(m+1)(\lambda_1/2D)$ is indicated, as well as the distance between the lines $\lambda_1/2D$ and $\lambda_2/2D$.

is, D is constant. From equation 8.19 and the condition for constructive interference (equation 8.22), the position of the rings described by the angle θ_2 is

$$\frac{2\pi}{\lambda} D \cos \theta_2 = m\pi \tag{8.33}$$

With $\theta_2 = \theta$, we have

$$2D \cos \theta = m\lambda \tag{8.34}$$

In analogy to what has been discussed for the grating, we plot the series of lines for λ_1 and λ_2 as functions of $\cos \theta$ (see Figure 8.18b). Overlapping occurs when

$$m\frac{\overline{\Delta\lambda}}{2D} = \frac{\lambda_1}{2D} \quad \text{or} \quad m = \frac{\lambda_1}{\overline{\Delta\lambda}} \tag{8.35}$$

If we insert m from equation 8.34 and write λ for λ_1, we have

$$\overline{\Delta\lambda} = \frac{\lambda^2}{2D \cos \theta}$$

If θ is small, we can approximate $\cos \theta \simeq 1$ and have (see equation 8.32)

$$\overline{\Delta\lambda} = \frac{\lambda^2}{2D} \tag{8.36}$$

This is the free spectral range for a ring pattern. The range is small for a large separation of the plates, that is, for large D, which corresponds to large m.

To make use of the high resolving power at small free spectral ranges, we must use a filter. Such a filter may be a multiple-layer dielectric filter or the output of a prism.

Fabry-Perot etalons have been used for the resolution of the hyperfine structures of atomic emission lines. Analysis of the fine structures yields the nuclear moment of the atom.

Example

Let $r = 0.99$ and $m = 2$. Using

$$\frac{\lambda}{\Delta\lambda} = m\frac{r\pi}{1 - r^2}$$

results in

$$\frac{\lambda}{\Delta\lambda} = 2\pi\frac{(0.99)}{1 - 0.99^2} \simeq 300$$

This resolving power is 20 times smaller than the resolving power found for the grating in the example at the end of Section 2A. However, using a plate of improved reflectivity, where $r = 0.995$ and $m = 2$, we get $\lambda/\Delta\lambda \simeq 600$. And if we had used plates of $r^2 = 0.995$ but observed at the order of $m = 20$, we would again have $\lambda/\Delta\lambda \simeq 6000$.

As we saw from the discussion of the free spectral range, if the two lines to be investigated are very narrowly positioned, high orders can be used, up to the 100,000 range for m.

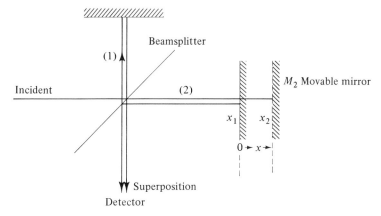

Fig. 8.19
The superimposed light traveling to the detector of the Michelson interferometer is made up of two components. The component from arm 1 travels a constant path. The component from arm 2 may travel various path lengths, depending on the movement of mirror M_2.

In comparing the resolving power of the grating with that of the Fabry-Perot etalon, we saw that large values are obtained for the grating if the path difference (counted in wavelengths) between the extreme waves is large. Similarly for the Fabry-Perot etalon, high resolution is obtained if the two traveling waves producing the standing-wave patterns travel large distances.

D. The Michelson Interferometer

The Michelson interferometer, as discussed in Chapter 2, can be employed for spectroscopic purposes. The parameter here for changing the appearance of the spectrum is the displacement of the movable Michelson mirror.

Let us consider the superposition of two beams in a Michelson interferometer (Figure 8.19). From Chapter 2, the intensity of the two superimposed components is given as

$$u^2 = 4A^2 \cos^2[k(x_2 - x_1)] \tag{8.37}$$

The movable mirror, M_2, is displaced a distance x with respect to the symmetric positions of the two mirrors at x_1. The positions of minima and maxima depending on the distance $x = x_2 - x_1$ are shown in Figure 8.20.

Resolving power.   The separation of the fringes of two waves with wavelengths λ_a and λ_b is obtained from the overlapping fringe pattern of the two different waves, with wave vectors $2\pi/\lambda_a$ and $2\pi/\lambda_b$, respectively (Figure 8.21). The maximum of the mth order of the wave with wavelength λ_a is at $2x_a = m\lambda_a$, and for the wave with wavelength λ_b, it is at $2x_b = m\lambda_b$. We consider the two waves resolved if they overlap at about half-height. In a simplified way, we assume that the maximum of one pattern is at the minimum position of the other pattern. (No fringes are observed.)

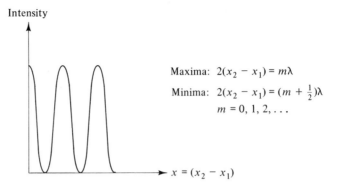

Intensity

Maxima: $2(x_2 - x_1) = m\lambda$

Minima: $2(x_2 - x_1) = (m + \frac{1}{2})\lambda$

$m = 0, 1, 2, \ldots$

$x = (x_2 - x_1)$

Fig. 8.20
Intensity minima and maxima as functions of the displacement x of the movable mirror.

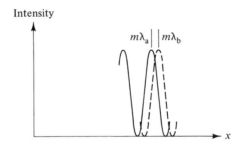

Intensity

$m\lambda_a$ $m\lambda_b$

x

Fig. 8.21
Maxima depending on the displacement x of the movable mirror for two different wavelengths, λ_a and λ_b, with order m.

Then we have

$$m\lambda_b - m\lambda_a = \frac{\lambda}{2} \tag{8.38}$$

where $\lambda/2$ is the distance from maximum to minimum expressed as the path difference. If we use $\Delta\lambda$ to represent the difference $\lambda_b - \lambda_a$, then we have

$$\frac{\lambda}{\Delta\lambda} = 2m \tag{8.39}$$

If we call $\lambda/\Delta\lambda$ the resolving power, we see that it does not depend on the parameters of the spectroscopic device. In comparison, we found that the resolving power of the grating depends on the order m and on the total number of lines of the grating; and for the Fabry-Perot etalon, it depends on the order m and on the reflection coefficient r.

Visibility. Equation 8.39 may be interpreted as saying that we must go to the order m so that the line of wavelength λ_b of the mth order is halfway between the lines of λ_a of orders m and $m + 1$. Measuring the intensity, we will observe no fringes. This situation is similar to the one discussed in Chapter 2 for the experiment of spatial coherence for an extended source.

8. Frequency Resolution and Fourier Transform Spectroscopy

Following Michelson, we can analyze the situation by using the visibility of fringes defined as

$$V(x) = \frac{I_{max} - I_{min}}{I_{max} + I_{min}} \qquad (8.40)$$

where I_{max} and I_{min} are the intensities in the immediate neighborhood of the distance x of the displacement of one mirror in the Michelson interferometer with respect to the other. To see how this works, let us consider the two narrow lines of the sodium D line. The visibility $V(x)$ becomes sharp and diffuse for increasing x. We find about a thousand fringes between two positions of maximum V. We can conclude that the sodium D line consists of two spectral lines with a wavelength difference of 1/1000 of λ. This phenomenon is analogous to the "beat note" of two frequencies.

The situation is not changed if we use a ring pattern for constant mirror distance x. Such a ring pattern has been discussed as problem 9 in Chapter 2.

For directly observing the resolution of two spectral lines, the Michelson interferometer is not as useful as the grating, the prism, or the Fabry-Perot etalon. However, a modified procedure of first observing the visibility curve $V(x)$ depending on x (called an **interferogram**) and then analyzing it with a computer (Fourier transformation) has been developed into a superior spectroscopic method.

3. FOURIER TRANSFORM SPECTROSCOPY

A. Multiplex Advantage

So far, we have discussed various spectroscopic devices and their resolving powers. In the visible region, these devices are used for two different spectroscopic methods. Let us consider a grating, such as the one shown in Figure 8.2. If we place a photographic plate on the screen S_c, we will get the total spectrum diffracted into the range of angles θ that are covered by the plate. If we wish to cover the range from $0°$ to $\pm 90°$, a circular curved photographic film arrangement is employed. However, there is another way to record the spectrum. A small detector element may be swung in a semicircle ($180°$), recording the intensity at each angle. In the first method, all the spectral details are observed in a total time T of observation. In the second method, the spectral details are observed one at a time. The first method has a multiplex advantage over the second. For low-power sources, a long observation time is needed to get appreciable signals. In **Raman spectroscopy**, before the laser was invented, a total spectrum was sometimes observed with a photographic plate in a total time of one week. Long observation time result in a a better signal-to-noise ratio. The signals can accumulate, and the fluctuations tend to average. The improvement of the signal-to-noise ratio is proportional to $\sqrt{T}$, where T is total observation time.

In the visible spectral region, a photographic plate can be conveniently used to observe all the spectral details during a time T. This is not the case for the infrared spectral region, for which other methods must be considered. One of these methods is to observe the interferogram of the spectral elements produced in a Michelson interferometer. The interferogram is recorded by displacing the movable mirror over a given range. This interferogram contains all wavelengths (spectral elements) incident onto the interferometer, and consequently, all are observed in the total time the interferometer produces the interferogram. This interferogram cannot be used directly to obtain the spectral details, since all the spectral elements are "mixed up." However, a Fourier transformation of the interferogram can be used to untangle the spectral elements. When this is done, the spectral elements available for analysis are equivalent to the results obtained with a photographic plate. Also, the advantage of $\sqrt{T}$ for the signal-to-noise ratio has been attained.

B. Interferogram Function

Let us start with the output intensity of the Michelson interferometer for one wavelength, as discussed in Chapter 2. There we found that depending on the displacement of one of the mirrors

$$u^2 = 4A^2 \cos^2[k(x_2 - x_1)] = 2A^2\{1 + \cos[2k(x_2 - x_1)]\} \quad (8.41)$$

The optical path difference between the two beams is $y = 2(x_2 - x_1)$ and we have

$$J(y) = u^2 = 2A^2[1 + \cos(ky)] \quad (8.42)$$

We introduce $v = 1/\lambda$ as the frequency (in cm^{-1}), have y in centimeters, and substitute $2G_v$ for $2A^2$ to get

$$J(y) = 2G_v[1 + \cos(2\pi vy)] \quad (8.43)$$

where G_v is the intensity of the wave with frequency v. This is the expression for monochromatic light. For polychromatic light, we have the superposition of all the intensities, each for one wavelength (or spectral element). They are summed, and we have

$$J(y) = \sum_{v_i = v_1}^{v_n} 2G_{v_i}[1 + \cos(2\pi v_i y)] \quad (8.44)$$

For the case of a continuous distribution of frequencies, we replace the sum by an integral:

$$J(y) = 2\int_0^\infty G(v)[1 + \cos(2\pi vy)]\, dv \quad (8.45)$$

For the specific case where $y = 0$, we get

$$J(0) = 4\int_0^\infty G(v)\, dv \quad (8.46)$$

$G(v)$ is a function of v representing the intensity of the spectral elements with frequency v. This is the quantity we wish to compute.

To introduce a convenient mathematical notation, we formally assume that $G(v)$ is symmetric around $v = 0$; that is, $G(v) = G(-v)$. Then we can write equation 8.46 as

$$J(0) = 4\int_0^\infty G(v)\, dv = 2\int_{-\infty}^{+\infty} G(v)\, dv \quad (8.47)$$

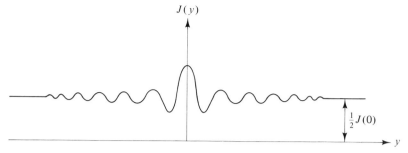

Fig. 8.22
Schematic diagram of an interferogram $J(y)$ obtained from a Michelson interferometer as a function of the displacement of one mirror.

From equation 8.45, $J(y)$ can be expressed as

$$J(y) = 2\int_0^\infty G(v)\,dv + 2\int_0^\infty G(v)\frac{e^{i2\pi vy} + e^{-i2\pi vy}}{2}\,dv \qquad (8.48)$$

where the cosine function has been written in exponential form. Since $G(v) = G(-v)$, and using equation 8.46 and 8.47 we can express equation 8.48 as

$$J(y) = \frac{1}{2}J(0) + \int_{-\infty}^{+\infty} G(v)e^{i2\pi vy}\,dv \qquad (8.49)$$

For mathematical purposes, we have extended the frequency scale to negative values; but at the end, we will consider only positive values. Using equation 8.49, we will now introduce $S(y)$, where

$$S(y) = J(y) - \frac{1}{2}J(0) = \int_{-\infty}^{+\infty} G(v)e^{i2\pi vy}\,dv \qquad (8.50)$$

$S(y)$ can be obtained experimentally, (see Figure 8.22) by moving the mirror M_2 in a Michelson interferometer from $-y$ to $+y$.

For large displacements of y, we obtain

$$J(y) \stackrel{y\to\infty}{\to} \tfrac{1}{2}J(0) \qquad (8.51)$$

If we subtract $\frac{1}{2}J(0)$ from $J(y)$, we obtain the interferogram function $S(y)$ (see equation 8.50).

C. Fourier Transformation

In Chapter 3, we came across a similar integration as in equation 8.50 for the diffraction at a slit. There we had identified the integral as a Fourier integral and spoke of a Fourier transformation.

What we want to find here is $G(v)$. If the Fourier transformation of $G(v)$ is $S(y)$, that is,

$$S(y) = \int_{-\infty}^{+\infty} G(v)e^{i2\pi vy}\,dv \qquad (8.52)$$

3. Fourier Transform Spectroscopy

Fig. 8.23
Fourier transformation of the interferogram function $S(y) \propto \cos^2 ky$, resulting in one value for the frequency.

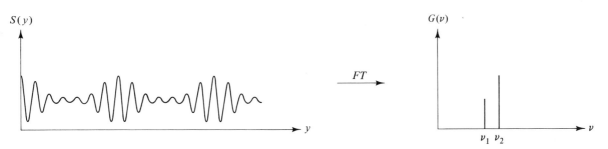

Fig. 8.24
The superposition of two waves with frequencies v_1 and v_2 results in the well-known beat phenomenon for the interferogram function $S(y)$. The Fourier transformation gives two frequencies in the frequency space (domain).

then the inverse is

$$G(v) = \int_{-\infty}^{+\infty} S(y)e^{-i2\pi vy}\,dy \tag{8.53}$$

where $S(y)$ and $G(v)$ and the dependent variables have been interchanged and -1 appears in the exponential of the inverse equation 8.53. The integration is now over the space coordinate, and the frequency appears as a parameter.

To become more familiar with the Fourier transformation and the pair of Fourier integrals $S(y)$ and $G(v)$, we will consider some examples. We will demonstrate the relationship between the spectral function $G(v)$ and its Fourier transform, the interferogram function $S(y)$.

For a monochromatic input, we obtain an interferogram function $S(y)$ corresponding to $\cos^2 ky$; $G(v)$ is just a delta function at frequency v_0 (Figure 8.23). Superposition of two waves with frequencies v_1 and v_2 results in the "beat phenomenon" in the y coordinate (Figure 8.24).

Now we ask the question, What would the interferogram function look like if there were a frequency distribution having a nonzero value between 0 and v_m and otherwise the value 0? If we extend this to $-v_m$, our question becomes, What would $S(y)$ look like to be the Fourier transformation of $G(v)$, as shown in Figure 8.25?

We can learn the result from what was done in Fraunhofer diffraction. There we found that the amplitude of the diffraction pattern of a slit was represented

8. **Frequency Resolution and Fourier Transform Spectroscopy**

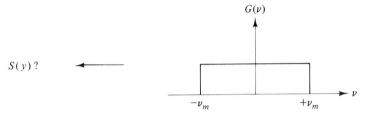

$S(y)$?

Fig. 8.25
If the function $G(v)$ is given in frequency space, what does the Fourier transform $S(y)$ look like?

by the integral

$$u(\bar{Y}) = \bar{C} \int_{-\infty}^{+\infty} P(\bar{y})e^{i\bar{y}\bar{Y}} \, d\bar{y} \tag{8.54}$$

where $P(\bar{y})$ was the slit function (step function) given as

$$P(\bar{y}) = \begin{cases} 1 & \text{for} \quad -a \le \bar{y} \le a \\ 0 & \text{otherwise} \end{cases} \tag{8.55}$$

The slit function $P(\bar{y})$ is shown in Figure 8.26. The result of the integration was

$$u(\bar{Y}) = \bar{C}' \frac{\sin(2\pi a\bar{Y})}{2\pi a\bar{Y}} \tag{8.56}$$

We find that the Fourier transform of the slit function is the function $\sin(b\bar{Y})/b\bar{Y}$ (Figure 8.27).

The answer to our question may be obtained by substituting the function $G(v)$ (equal to 1 from $-v_m \le v \le v_m$ and other wise zero) in equation 8.52:

$$S(y) = \int_{-v_m}^{+v_m} e^{i2\pi vy} \, dv = \frac{2v_m \sin(2\pi yv_m)}{2\pi yv_m} \tag{8.57}$$

The result is the familiar $\sin x/x$ function with scaling constants.

The frequencies that were used had units of 1/cm. This is convenient when the space coordinate is in centimeters. The product $v \cdot y$ must be dimensionless and appears in the exponential term of the Fourier transform integral. The two domains of the v and y coordinates are reciprocal. We saw this in the diffraction problem, where narrow obstacles result in wide diffraction patterns, and vice versa.

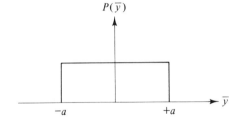

Fig. 8.26
Slit function in $\bar{y}$ space.

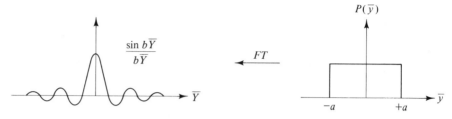

Fig. 8.27
Fourier transformation as performed in Chapter 3 for the diffraction at a slit.

In our present study, the goal is to understand the importance of the Fourier transformation process through some simple examples. Various mathematics texts are available in which rigorous developments and proofs are provided.

From what we have discussed up to this point, the Fourier transformation pair for the interferogram function $S(y)$ and the spectrum function $G(v)$ may be expressed as

$$S(y) = \int_{-\infty}^{+\infty} G(v)e^{i2\pi vy}\,dv \tag{8.58}$$

$$G(v) = \int_{-\infty}^{+\infty} S(y)e^{-i2\pi vy}\,dy \tag{8.59}$$

Now we may ask the question, Is the Fourier transform of a Fourier transform the original function? We can check this for a simple Gaussian-type function,

$$G(v) = \exp\left(-\frac{a^2v^2}{2}\right)$$

Since this function is symmetric for $v = 0$, we can use the cosine function instead of the exponential and calculate

$$S(y) = 2\int_{0}^{+\infty} e^{-a^2v^2/2}\cos(2\pi vy)\,dv \tag{8.60}$$

Using the integral formula

$$\int_{0}^{+\infty} e^{-c^2x^2}\cos(bx)\,dx = \frac{\sqrt{\pi}}{2c}e^{-b^2/4c^2} \tag{8.61}$$

we obtain

$$S(y) = \frac{\sqrt{2\pi}}{a}e^{-2\pi^2y^2/a^2} \tag{8.62}$$

Inserting $S(y)$ into equation 8.59 and applying the same formula (equation 8.61) yields

$$G(v) = e^{-a^2v^2/2} \tag{8.63}$$

We find that for this special example of a Gaussian function, the Fourier transform of a Fourier transform is the original function. But this is also true in

8. **Frequency Resolution and Fourier Transform Spectroscopy**

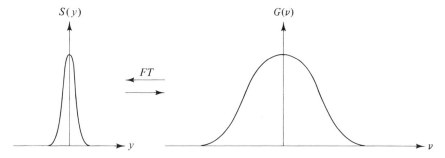

Fig. 8.28
The Fourier pair $S(y)$ and $G(v)$ are Fourier transforms of each other.

general. We can also see the reciprocity in the width of the Gaussian functions $S(y)$ in equation 8.62 and $G(v)$ in equation 8.63. A small value of a would make $G(v)$ wide and $S(y)$ narrow (Figure 8.28). It is now possible to construct a table of Fourier transformations for some of the cases we discussed. See Table 8.1 on page 338.

D. Spectrum and Resolution

Now we can go back to our spectroscopy problem. We can experimentally determine $S(y)$ if we want to get $G(v)$. The integral to be evaluated is

$$G(v) = \int_{-\infty}^{+\infty} S(y)\cos(2\pi vy)\,dy = 2\int_{0}^{\infty} S(y)\cos(2\pi vy)\,dy \qquad (8.64)$$

Since $S(y)$ is real and assumed to be symmetric, we need only the real part of $e^{i2\pi vy}$. First, the interferogram is measured; that is, we must determine $J(y)$ (see equation 8.45). This is the intensity output of the Michelson interferometer. It is measured for a series of equally spaced displacements y_i of the movable mirror. Then we determine $J(0)$ (see equation 8.46). At this point, it is more convenient to measure $J(\infty)$, corresponding to $J(y)$ for large values of y. It can be shown that $2J(\infty) = J(0)$. The next step is to subtract $\frac{1}{2}J(0)$ from $J(y)$ to get $S(y)$ (see equation 8.50) and substitute in equation 8.64. The integral in equation 8.64 is then approximated for computation purposes by a sum.

A computer program using the algorithm of the fast Fourier transform FFT calculates the function $G(v)$ for equally spaced frequency points. This is the desired spectrum, as illustrated in Figure 8.29.

Let us determine what the smallest spectral detail is that can be observed with Fourier transform spectroscopy. In our discussions of gratings, the Fabry-Perot spectrometer, and the Michelson interferometer, we saw that the most important parameter is the path difference between the two interfering waves. A similar relationship holds for the case of Fourier transform spectroscopy. It can be shown that the smallest spectral interval Δv is inversely proportional to the maximum displacement y_{max} of the movable Michelson mirror:

$$\Delta v = \frac{1}{2y_{max}} \qquad (8.65)$$

3. Fourier Transform Spectroscopy

Table 8.1
Fourier transformation pairs.

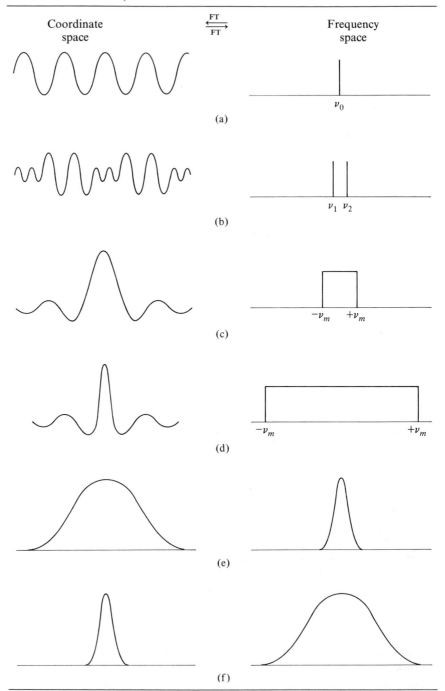

(a) Single-frequency oscillation; (b) superposition of two oscillations with frequencies v_1 and v_2; (c, d) $(\sin y)/y$ type functions with two different scaling parameters; (e, f) two Gaussian functions with two different scaling parameters.

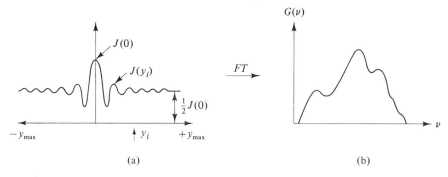

(a)   (b)

Fig. 8.29
(a) The sampled interferogram in the interval $y = -y_{max} = -L$ to $y = y_{max} = L$, where y is the displacement of the movable Michelson mirror. (b) The calculated spectrum $G(v)$.

Is it possible to make y_{max} as large as we want? In fact, there must be a limiting factor.

Let us assume that we have blackbody radiation. The incident waves have a continuous frequency distribution, and a frequency interval Δv carries a certain intensity. If the interval is made smaller, the intensity decreases. If the intensity of the interval Δv approaches the noise level of the measuring instruments, the process of making Δv smaller and y_{max} larger comes to a limit. For example, y_{max} can be of the order of meters when in the visible or infrared spectral region.

Problems

1. *Reflection grating.* The diffracted intensity for an amplitude grating at angle θ is given by

$$u^2 = u_0^2 \frac{\sin^2 \gamma(d/2)}{[\gamma(d/2)]^2} \cdot \frac{\sin^2 N\gamma(a/2)}{\sin^2 \gamma(a/2)}, \quad \gamma = \frac{2\pi}{\lambda} \sin \theta$$

If we want to use such a grating in reflection, we obtain the same formula for the diffracted intensity if the light is incident normal to the plane of the reflecting surfaces of width d.

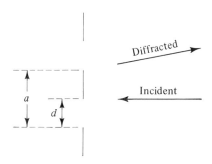

The maxima are obtained at

$$\frac{2\pi}{\lambda}\left(\frac{a}{2}\right)\sin\theta = m\pi$$

or

$$a\sin\theta = m\lambda$$

a. If $a = 10\ \mu m$, find the position of maxima for wavelength $\lambda = 2, \ldots, 7\ \mu m$ in first and second orders.
b. At what angles θ do we observe overlapping of the second-order spectrum with the first-order spectrum?
c. What is the resolution of the grating if the length is 6 cm?
d. Give the value of the resolution in fifth and tenth orders.
Ans. b. $24°$ and $37°$ c. $m \times 6 \times 10^3$ d. 30×10^3 and 60×10^3

2. *Blazed grating in reflection.* A profile of the grating with steps as shown will direct the main intensity from the zero order to the first order.

As in the transmission echelette grating, the interference factor remains the same, and the maxima are given by

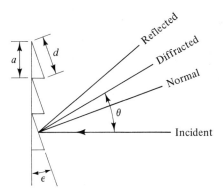

$$a\sin\theta = m\lambda$$

The diffraction factor now has the main maximum at $\theta = 2\varepsilon$:

$$\frac{\sin^2[2\pi/\lambda(d/2)\sin(\theta - 2\varepsilon)]}{[2\pi/\lambda(d/2)\sin(\theta - 2\varepsilon)]^2}$$

If we choose $\theta = 2\varepsilon$ in the interference factor, we obtain the location of the maximum

$$a\sin 2\varepsilon = m\lambda$$

a. For $a \simeq d = 10\ \mu m$ and $\varepsilon = 15°$, find all the wavelengths for orders $m = 1, \ldots, 5$ that appear at the $30°$ angle.
b. Find the wavelength range for the maxima in first order for $15°$ around the $30°$ angle.
c. The intensity distribution in the range $15°$ around the $30°$ angle is given by the envelope obtained from the diffraction factor. Plot the envelope for the range $\lambda = 3, \ldots, 7\ \mu m$ in steps of $1\ \mu m$.

8. Frequency Resolution and Fourier Transform Spectroscopy

3. *Fabry-Perot etalon.* The transmitted intensity of a Fabry-Perot spectrometer is

$$I_t = \frac{1}{1 + g^2 \sin^2(2\pi D/\lambda)}, \quad g = \frac{2r}{1 - r^2}$$

To use the Fabry-Perot etalon as a spectrometer, we must vary the distance D. The wavelength λ for which resonance occurs is the transmitted wavelength. Assume that a filter passes wavelengths from 2 to 10 μm only.

a. What is the variation of D needed to cover this region?
b. At what wavelength do we observe an overlapping of the second order (of the shorter wavelength) on the first order? The region of no overlapping is called the free spectral range.
c. Assume that $r = 0.99$ and $r = 0.999$, and calculate $\Delta\lambda$ for $\lambda = 5$ μm for the tenth and one hundredth orders.

Ans. a. $D_1 = 1$ μm, $D_2 = 5$ μm b. $D = 2$ μm

4. *Higher orders of a Fabry-Perot etalon.* Light of continuous wavelength of $\lambda = 4500$ Å to $\lambda = 5500$ Å is incident on a Fabry-Perot etalon of thickness $D = 5 \cdot 10^{-4}$ cm. At what wavelength and what order do we observe resonance, that is, do we see an interference fringe?

Par. Ans. $m = 22, 4545$ Å to $m = 19, 5263$ Å.

5. *Michelson interferometer.* The intensity of a Michelson Interferometer is given as

$$u^2 = 4A^2 \cos^2\left(\frac{2\pi}{\lambda}(x_2 - x_1)\right)$$

where $x = x_2 - x_1$ is the distance of the Michelson mirror from its symmetric position. The first-order maximum is obtained for wavelength λ at $x = \lambda/2$. By varying x, we may get first-order maxima for different λ.

a. Consider the wavelength range 2–10 μm. What variation of x is necessary to cover the range?
b. Find the free spectral range for the first order.
c. Give the resolution for $\lambda = 5$ μm and $m = 10$; for $\lambda = 5$ μm and $m = 100$.

Ans. a. x from 1 μm to 5 μm b. $\lambda_{1 \text{ highest}} = 4$ μm c. 0.25 μm and 0.025 μm

6. *Sodium D lines.* The two sodium lines, called D lines, are at $\lambda = 5896$ Å and $\lambda = 5890$ Å.

a. Apply the formula for the resolution of a grating, $\lambda/\Delta\lambda = mN$, and determine the number of lines N to be used for the resolution of the sodium lines if $m = 3$.
b. What value B for the base of a prism must we use to accomplish this resolution of the sodium D lines? For glass, at $\lambda = 5461$ Å, $n = 1.460$, and at $\lambda = 5893$ Å, $n = 1.458$.
c. If we use a Fabry-Perot etalon, the resolving power depends on the reflectivity of the plates. Assume that $r = 0.9$. What order of the Fabry-Perot etalon must be used to resolve the sodium D lines?

Ans. a. $N = 327$ b. $B = 2$ cm, c. $m = 65$

7. *Free spectral range, overlapping orders.*

a. Consider a grating of 60 lines per millimeter and two spectral lines of 4000 Å and 5000 Å. Calculate m_{highest} and the corresponding angle θ (if $m = 0$ corresponds to $\theta = 0$).
b. For the Fabry-Perot etalon, we have the same formula for the highest order, $m_h = \lambda_1/\Delta\overline{\lambda}$, which in our case equals 4. For overlapping in the ring pattern,

we have

$$\overline{\Delta\lambda} = \frac{\lambda_1^2}{2D\cos\theta}$$

We set $\cos\theta = 1$ and calculate $D = 8\cdot 10^3$ Å. This value is for a Fabry-Perot unreasonably small. A Fabry-Perot etalon is used for small $\overline{\Delta\lambda}$ and large values of D. Calculate m_h and D for $\overline{\Delta\lambda} = 1$ Å and $\lambda_1 = 4000$ Å.

Ans. a. $m_h = 4$, $\theta = 7.2°$ b. $m_h = 4000$, $D = 800$ μm

8. *Fourier transformation of a Fourier transformation.* For the Gaussian function $e^{-(a^2/2)v^2}$, show that the Fourier transformation of the Fourier transformation is the function itself. Use

$$\int_0^\infty e^{-c^2x^2}\cos bx\,dx = \frac{\sqrt{\pi}}{2c}e^{-b^2/4c^2}$$

9. *Resolving power of a grating, Fourier transform spectrometer, and Fabry-Perot spectrometer.* In the section on Fourier transform spectroscopy, we mentioned that the resolution depends on the optical path difference of the two superimposed beams. For a grating, we can use an angle of incidence close to 90° and an angle of diffraction close to 90°.

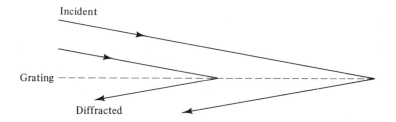

The path difference between the first and last wave is on the side of incidence equal to the length of the grating. For diffraction of the first order the path difference of the first and last wave is as well equal to the length of the grating. We have

$$L = Na$$

and

$$a(\sin\alpha_1 + \sin\alpha_2) = m\lambda$$

where a is the grating constant. The resolving power of the grating was given as

$$\frac{\lambda}{\Delta\lambda} = mN \quad \text{with} \quad N = \frac{L}{a} \quad \text{and} \quad 2a = m\lambda$$

and we have

$$\frac{\lambda}{\Delta\lambda} = m\frac{L}{a} = \frac{2a}{\lambda}\left(\frac{L}{a}\right) = \frac{2L}{\lambda}$$

$$\Delta\lambda = \frac{\lambda^2}{2L}$$

For the Fourier transform spectrometer, we have $\Delta v = 1/2L$ in 1/cm, where L is the maximum displacement of the Michelson mirror:

$$\lambda v = 1$$

$$\Delta v = \Delta \frac{1}{\lambda} = \left| \frac{\Delta \lambda}{\lambda^2} \right|, \quad \frac{\Delta \lambda}{\lambda^2} = \frac{1}{2L}, \quad \text{or} \quad \Delta \lambda = \frac{\lambda^2}{2L}$$

For the Fabry-Perot etalon with $m\lambda/2 = L$, where L is the thickness of the etalon,

$$\frac{\Delta \lambda}{\lambda} = \frac{2}{m\pi g} = \frac{2}{L(2/\lambda)\pi g} = \frac{\lambda}{L\pi g}$$

$$\Delta \lambda = \frac{\lambda^2}{\pi g L}$$

Whereas $\Delta \lambda / \lambda^2$ depends only on L for grating and Fourier transform spectra, we have the additional dependence of $\Delta \lambda / \lambda^2$ on $1/\pi g$ for the Fabry-Perot spectrum. The quantity $2/\pi g$ can be much smaller than 1. Therefore, the Fabry-Perot etalon has an advantage for resolution (in a limited small spectral region).

a. Calculate the resolution for $\lambda = 1$ μm and $L = 10$ cm for a grating.

b. For $\pi g = 2$, we would obtain the same result as in part (a) for the Fabry-Perot etalon. What is the resolution if we have $\pi g = 100$?

Ans. a. 5×10^{-6} μm b. 1×10^{-7} μm

Spatial Resolution   9

1. INTRODUCTION

Since optical instruments are capable of enlarging an object, it is of interest to know just how small an object can be and still be recognized. We know from Chapter 3 that each small object will produce a diffraction pattern. If we are considering two closely spaced points, then the phenomenon to study is the overlapping of two diffraction patterns in the image plane. We did this for two stars in Chapter 2, where we discussed spatial coherence. When we are comparing intensity patterns, it is customary to say that the objects producing those patterns are *incoherently illuminated*. (Coherently illuminated objects are discussed in Chapter 10.) The question we want to answer is, When can we just distinguish the two diffraction patterns?

We can see the analogy to frequency resolution and the Rayleigh criterion may be introduced. The Rayleigh criterion says that we can distinguish between two diffraction patterns if the maximum of one is positioned at the first minimum of the other. This concept applies to such instruments as the telescope, eye, and microscope. We will study all of these and come up with some quantitative results to the question: When can we consider two object points distinguishable?

2. SPATIAL RESOLUTION AT A SLIT

In Chapter 8, we saw the overlapping of the diffraction patterns of two waves with different wavelengths λ_1 and λ_2. We also discussed the question of how the separation $\Delta\lambda$ depends on the specific diffraction device or interferometric method.

Now we will only consider a single wavelength. We again consider two diffraction patterns, but this time the two patterns are produced by two different sources using the same slit or lens (Figure 9.1). The two diffraction patterns are generated by two sets of parallel waves having an angle γ between their center lines. The diffraction patterns are superimposed and appear as shown in Figure 9.2.

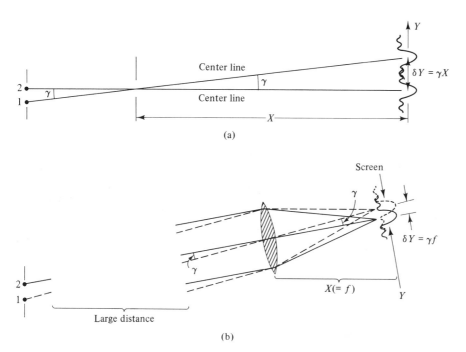

Fig. 9.1
(a) The parallel light from two sources is incident on a slit. The diffracted light produces two diffraction patterns at a faraway screen. The angle γ is the angle between the center lines of the diffraction patterns. (b) the same as (a), but a lens is used to focus the light from the two sources on a screen. The diffraction pattern of the round opening of the lens is observed in the focal plane of the lens.

The separation of these two diffraction patterns is similar to what we have discussed for the diffraction patterns of two different frequencies. We can again adopt the convention that we can distinguish between two points if the corresponding diffraction patterns appear so that the maximum of one is located at the minimum of the other (the Raleigh criterion).

3. THE TELESCOPE

When the optical instrument is a telescope, the question becomes how to distinguish the images of two narrowly spaced objects that are located far away. For the astronomical telescope, we can consider the separation of the images of a double star. The diffraction aperture here is the opening of the telescope. The Fraunhofer diffraction pattern for a round opening was calculated in Chapter 3, problem 7. In Figure 9.3, we present some results of this calculation important to our discussion.

According to the Rayleigh criterion, the separation of the peaks is equal to the distance between the first minimum and the center of the diffraction pattern, that is, the first minimum of the Bessel function at $q = 3.83$,

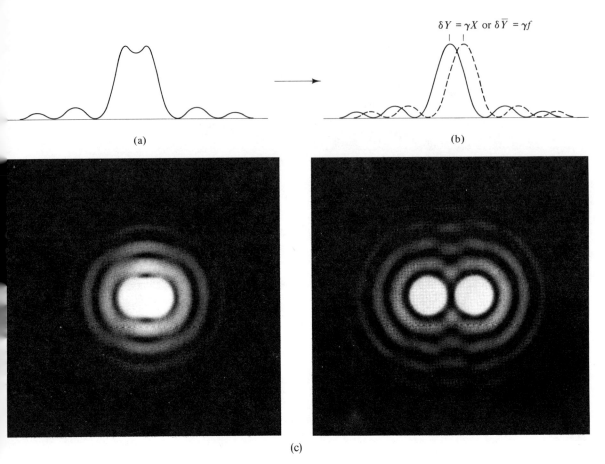

$$\delta Y = \gamma X \text{ or } \delta \bar{Y} = \gamma f$$

(a)

(b)

(c)

Fig. 9.2
(a) Appearance of the diffraction pattern of a slit of two closely spaced point sources. (b) The diffraction patterns are shown as the overlapping of individual patterns. The separation $\delta \bar{Y}$ corresponds to the separation of the two maxima in the focal plane of the lens. (c) (Left) diffraction pattern of a round aperture at limit of resolution; (right) sufficiently separated diffraction pattern (from Cagnet, Françon, and Thrierr, *Atlas of Optical Phenomena*).

and we have

$$q = 3.83 = \frac{2\pi a}{\lambda}\left(\frac{R_1}{X}\right) \quad \text{or} \quad \gamma = \frac{R_1}{X} = 0.61\frac{\lambda}{a} = 1.22\frac{\lambda}{2a} \tag{9.1}$$

where $2a$ is the diameter of the lens and γ is the angle between the two stars (Figure 9.4).

The fact that the diameter of the telescope mirror enters the expression for the resolution explains why we have such large telescope mirrors for astronomical observations; for example, the Mount Palomar mirror is about 5 m in diameter (see problem 1). If a telescope mirror has a diameter of 30 cm and the wavelength of the light is $\lambda = 5 \times 10^{-4}$ mm, then we have for the angle γ

$$\gamma = 1.22\frac{\lambda}{2a} = \frac{1.22 \times 5 \times 10^{-4} \text{ mm}}{300 \text{ mm}} \simeq 2 \times 10^{-6} \text{ radians} \tag{9.2}$$

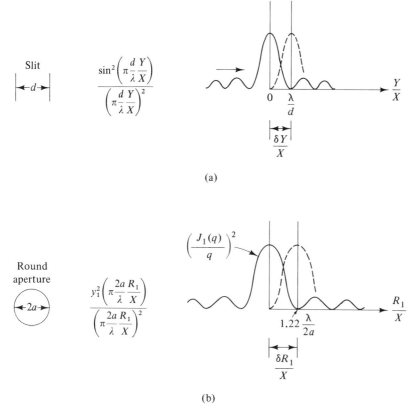

(a)

(b)

Fig. 9.3
(a) Slit of width d and diffraction pattern with first minimum at λ/d, plotted as a function of Y/X. (b) Round aperture of diameter $2a$, plotted as function of R_1/X. The diffraction pattern is described by a Bessel function (see problem 7, Chapter 3). The first minimum is at $q = 3.83$, where $q = (2\pi a/\lambda)(R_1/X)$, corresponding to $R_1/X = 1.22(\lambda/2a)$.

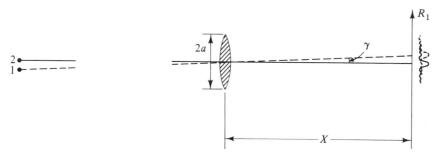

Fig. 9.4
The angle between the two stars is γ, the diameter of the telescope lens (objective) is $2a$, the distance between the observation screen and the lens is X (equal to focal length f of the lens), and the position of the details of the diffraction pattern is R_1. See also Figure 9.3.

If we convert to seconds, we get

$$\frac{(2 \times 10^{-6})(360)(60)(60)}{2\pi} = 0.4 \text{ second} \tag{9.3}$$

4. THE EYE

We observe a similar situation for the human eye. However, the pupil (i.e., the opening that governs the diffraction pattern) has a large range of openings. Depending on the amount of light, the diameter of the opening may change from 8 mm (for darkness) to 1 mm (for bright light). Again, we assume the wavelength in the middle of the visible spectrum, $\lambda = 5 \times 10^{-4}$ mm. But we must take into account that the wavelength of the light is reduced in the eye by the refractive index of the fluid in the eye. Assuming this refractive index to be $n = 1.33$, we now have

$$\gamma = 0.61 \frac{\lambda/n}{a} \tag{9.4}$$

Assuming that $a = 4$ mm, we get

$$\gamma = 0.61 \frac{5 \times 10^{-4} \text{ mm}}{1.33 \times 4 \text{ mm}} \simeq 0.6 \times 10^{-4} \text{ degrees} \tag{9.5}$$

corresponding to about 12.6 seconds. For a pupil with a radius of 0.5 mm, γ is about 1.7 min. This indicates that the best that the human eye can do is about a thousand times less resolving power than the Mount Palomar telescope (see also problem 2). Also, in the dark, the resolution is almost an order of magnitude worse.

5. THE MICROSCOPE

When the instrument involved is a microscope, then we want to know just how small an object can be seen—a problem very similar to the one we faced for the telescope. However, there are some important differences, since the object is so close to the objective lens in a microscope. Fraunhofer diffraction is actually no longer applicable, but is still approximately used. A more important consideration is that we may no longer use the paraxial theory for image formation. Some rays used for the formation of the image make large angles with the axis. The application of the diffraction limitations is similar to what has been discussed before.

For the angle γ, the angle between the minima and maxima of the diffraction pattern (see Figure 9.5), we have

$$\gamma \simeq 0.61 \frac{\lambda}{a} \simeq \frac{Y'}{b} \tag{9.6}$$

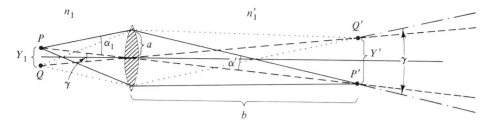

Fig. 9.5
P and Q are the object points separated by Y_1 and angle γ; a is the radius of the objective lens; P' and Q' are the maxima of the diffraction pattern separated by Y' and γ; b is the distance between the lens and the plane of the diffraction pattern; α_1 is half the angle of the light cone emanating from the object; and α' is half the angle of the light cone arriving at the diffraction maxima.

or

$$Y' \simeq 0.61\lambda\frac{b}{a} \simeq 0.61\frac{\lambda}{\alpha'} \tag{9.7}$$

where a is the radius of the lens, b is the distance from the lens to the image plane, and α' is the angle between the chief ray and the outermost ray of the light arriving at P'.

To express equation 9.6 in terms of Y_1 and α_1, we must use the imaging relations between image and object. These were introduced and developed in Chapter 1 for the paraxial theory. But since α_1 is not small, we must go through this derivation again without using the paraxial approximation.

We first assume only one spherical surface (Figure 9.6). We apply the sine theorem:

$$\frac{|s_1| + |r|}{\sin\phi} = \frac{|r|}{\sin\alpha_1}, \quad \frac{|s_2| - |r|}{\sin\theta_2} = \frac{|r|}{\sin\alpha_2} \tag{9.8}$$

Applying Snell's law,

$$n_1\sin(180 - \phi) = n_1\sin\phi = n_2\sin\theta_2$$

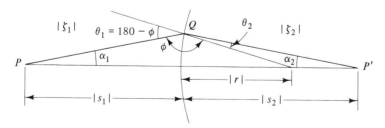

Fig. 9.6
Coordinates for imaging of points P and P'. $|s_1|$ and $|s_2|$ are the distances of P and P' from the refracting surface. Angles α_1 and α_2 are the angles of the light ray with the axis. $|r|$ is the radius of curvature of the refracting surface (see also Figure 1.10).

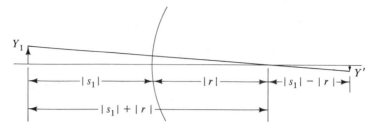

Fig. 9.7
Magnification for refraction at a single surface. The axis and the ray through the
center of curvature are employed.

we can write

$$\frac{|s_1| + |r|}{|s_2| - |r|} \cdot \frac{\sin \theta_2}{\sin \phi} = \frac{(|s_1| + |r|)n_1}{(|s_2| - |r|)n_2} = \frac{\sin \alpha_2}{\sin \alpha_1} \tag{9.9}$$

The magnification, as it applies to a single refracting surface, is illustrated in
Figure 9.7, and we have

$$\frac{Y_1}{Y'} - \frac{n_2 \sin \alpha_2}{n_1 \sin \alpha_1} \tag{9.10}$$

which might also be written so that all the quantities belonging to one side of
the interface are on one side of the equation:

$$Y_1 n_1 \sin \alpha_1 = Y' n_2 \sin \alpha_2 \tag{9.11}$$

This is the sine condition, which will be derived in Chapter 13 and also used in
Chapter 16. Note that no (paraxial) approximations have been made so far.

We have derived the sine condition for one spherical refracting surface only.
Extending our calculations to several surfaces is accomplished by finding the
equation for each surface:

$$Y_1 n_1 \sin \alpha_1 = Y_2 n_2 \sin \alpha_2 = Y' n' \sin \alpha' \tag{9.12}$$

For the lens in Figure 9.5, this may be written as

$$Y_1 n_1 \sin \alpha_1 = Y' n' \sin \alpha' \tag{9.13}$$

and it is shown in Figure 9.8. Two rays coming from extreme points of the object
must have the same optical path length to form a proper image.

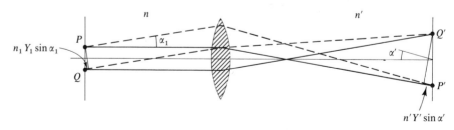

Fig. 9.8
Coordinates for demonstration of the sine condition.

9. Spatial Resolution

If we assume that $n' = 1$, we have

$$Y_1 n_1 \sin \alpha_1 = Y' \sin \alpha' \qquad (9.14)$$

We can now introduce equation 9.14 into equation 9.7 and approximate $\sin \alpha'$ with α'. Such an approximation cannot be used for $\sin \alpha_1$:

$$Y' = 0.61 \frac{\lambda}{\alpha'} = \frac{0.61\lambda}{(Y_1/Y')n_1 \sin \alpha_1} \qquad (9.15)$$

For the size of the object, we get

$$Y_1 = 0.61 \frac{\lambda}{n_1 \sin \alpha_1} \qquad (9.16)$$

The expression $n_1 \sin \alpha_1$ is called the **numerical aperture** (N.A.) (see also Chapter 12, Section 4C). Increasing n_1 makes the numerical aperture larger and decreases the value of Y_1; which otherwise depends on λ. If we use an oil with index of refraction $n_1 = 1.5$, we do better than with air, where $n_1 = 1$. For shorter wavelengths λ, we obtain smaller values for Y_1. An X-ray microscope would be of interest for making Y_1 very small.

For example, suppose that we place an object in an immersion fluid with index of refraction $n_1 = 1.5$. At the objective lens, we have $\alpha_1 = 70°$, and $n_1 \sin \alpha_1 = 1.4$. The wavelength is again taken as $\lambda = 5 \times 10^{-4}$ mm, and we get

$$Y_1 = 0.61 \frac{\lambda}{1.4} = 0.61 \frac{5 \times 10^{-4}}{1.4} = 2 \times 10^{-4} \text{ mm}$$

This is about one-half of a wavelength.

Problems

1. *Mount Palomar telescope.* Calculate the resolution of a telescope with a mirror diameter of 5.1 m, assuming light with a wavelength of 5500 Å. Give the result in radians, degrees, and seconds.
   Ans.
   1.3×10^{-7} rad, 0.74×10^{-5} degrees, or 0.0268 s

2. *Resolution of the eye.* Calculate the resolution for the pupil of the eye, given radii of $a = 0.25, 0.5, 1, 2$, and 4 mm, and assume a wavelength of $\lambda = 5500$ Å. The formula to use is

   $$\gamma = 1.22 \frac{\lambda/n}{2a}$$

   where n is the refractive index of the eye. Assume that $n = 1.33$, which is close to the refractive index of water in the visible region. Give the results in radians and minutes.

3. *Resolution of the microscope*
   a. Calculate the numerical aperture N.A. for $n = 1.55$ and $\alpha = 75°$.
   b. Calculate the size of an object in millimeters resolved by a microscope using the N.A. of part (a) and $\lambda = 5500$ Å.

c. If we could use light of 3000 Å, how much would the resolution improve?

d. The rule of thumb for the microscope is

$$Y_{min} = 0.61 \cdot \frac{\lambda}{\text{N.A.}} = \frac{0.3}{\text{N.A.}} \mu m$$

for λ approximately in the middle of the visible region. The N.A. is given on the objective for nonimmersing liquids. Calculate Y_{min} for typical cases, such as N.A. = 0.25 and N.A. = 0.45.

Ans. a. N.A. = 1.5; b. Y_{min} = 0.22 μm; c. twice as good;
  d. Y_{min} = 1.2 μm and Y_{mm} = 0.61 μm

Phase-Sensitive Imaging | **10**

1. INTRODUCTION

In Chapter 1, we discussed image formation using the law of refraction. In Chapters 2 and 3, we dealt with the interference and diffraction patterns of objects. The objects were assumed to be illuminated by incoherent light in Chapter 1 and coherent light in Chapter 2 and 3.

We will now see how introducing small changes in the diffraction pattern can result in dramatic changes in the image.

Images can also be formed by using principles of interference and diffraction only, without geometrical optics. This will be discussed for the Fresnel zone plate and holography.

The theory for these imaging processes have been developed by Ernst Abbe and Fritz Zernike for the purpose of understanding and improving the microscope. Dennis Gabor developed the theory for holography.

2. DIFFRACTION PATTERN IN THE FOCAL PLANE OF A LENS AND IMAGE FORMATION

In Chapter 1, we discussed in detail the imaging properties of a lens according to refraction principles. In Chapter 2, we saw how an interference pattern appears in the focal plane of a lens. And in Chapter 3, we discussed the formation of various diffraction patterns. In problem 7 of Chapter 3, we studied the Fraunhofer diffraction on a round aperture, and in Chapter 3, Section 3, we discussed Fresnel diffraction on a round aperture for a symmetric position of object and image points. Now we will consider the combined processes of forming an image and forming a diffraction pattern of an object and the role the lens plays with respect to refraction and diffraction. Figure 10.1 shows an experiment in which the diffraction pattern and the image of an object can be observed.

We can divide this double action into two familiar processes. The first is

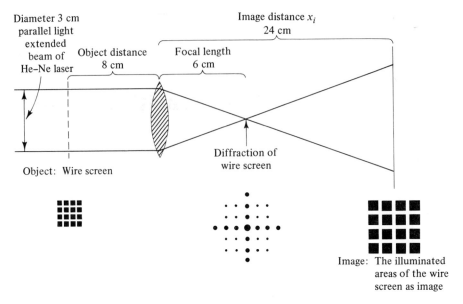

Fig. 10.1
A lens presents the diffraction pattern at its focal plane at 6 cm and the image at image distance 24 cm. Dark spots indicate light at the object plane, focal plane, and image plane.

imaging according to geometrical optics. We can consider two illuminated squares of the object as examples. The images are produced at the observation screen as shown in Figure 10.2. The second process is the transfer of the diffraction pattern into the focal plane of the lens. The various plane parallel waves (Huygens wavelets) diffracted from all the object points are concentrated by the lens on its focal plane (Figure 10.3).

For purposes of explanation, we have depicted the transfer of the diffraction pattern and the formation of the image as two separate steps; in actuality,

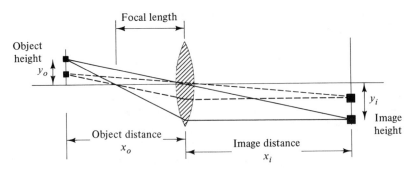

Fig. 10.2
The geometrical optical imaging process of two illuminated sample squares of the object.

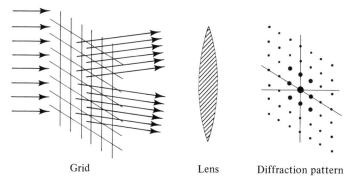

Grid   Lens   Diffraction pattern

Fig. 10.3
Various parallel waves diffracted by all the openings of the grid appear in the focal plane of the lens as a diffraction pattern of the grid.

however, they are intimately connected. A small disturbance of the diffraction pattern may seriously change the image. We can change the diffraction pattern by removing certain spots with a mask, that is, by changing the intensity in parts of the pattern. We can also change the phase of some parts of the amplitude diffraction pattern. These two possibilities will be discussed in Sections 4 and 5.

3. IMAGE FORMATION AND FOURIER TRANSFORMATION

We have seen how a lens produces the diffraction pattern of the object in its focal plane and an image at the geometrical optical image distance. The appearance of the diffraction pattern at the focal plane and the image at the image distance follows from geometrical optics.

We obtained the diffraction pattern in Chapter 3 by a summation process over the aperture. Using integration and the Fraunhofer observation, we summed up all the Huygens wavelets emanating from all points of the aperture, taking into account the amplitudes and relative phases. We showed mathematically that the diffraction at a slit is obtained by the Fourier transformation of the slit function, as illustrated in Figure 10.4.

In Chapter 8, we saw that the Fourier transformation of the Fourier transfor-

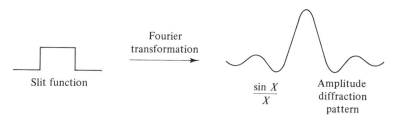

Slit function   Fourier transformation $\dfrac{\sin X}{X}$ Amplitude diffraction pattern

Fig. 10.4
The Fourier transformation of the slit function is the function $\sin X/X$.

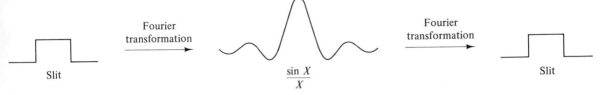

Fig. 10.5
Schematic representation of the Fourier transformation of the Fourier transformation of a slit, resulting in a slit pattern.

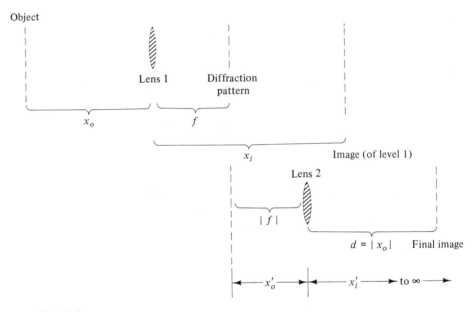

Fig. 10.6
Image formation by two lenses used as Fourier lenses. Lens 1 produces the diffraction pattern of the object at distance f. Lens 2 transforms the diffraction pattern into the final image of the object, at a distance d determined by symmetry.

mation of a function results in the original function. We show this schematically for a slit in Figure 10.5 and by calculation for a Gaussian function in problem 1. We can demonstrate it by using two lenses of focal length f, as shown in Figure 10.6. The first lens is used as in Figure 10.1. It produces first the diffraction pattern at distance f to the right and then the image at image distance x_i. The second lens is placed at distance $|f|$ to the right of the diffraction pattern. It produces the final image of the object at distance $d = |x_o|$, which is unequal to $x_i' = \infty$. This would be the image distance if the diffraction pattern were the object for the second lens. Note the symmetric arrangement: object–lens 1–diffraction pattern –lens 2–final image. Lenses used to produce the diffraction pattern of the object and not to form the geometrical image are frequently called **Fourier lenses**.

10. Phase-Sensitive Imaging

4. OPTICAL FILTERING AND INTENSITY CHANGES, SPATIAL FREQUENCIES AND SPATIAL WAVES

A. Intensity Changes of the Diffraction Pattern

At this point, let us return to the experiment depicted in Figure 10.1 and place two different masks at the focal plane of the lens. Each mask blocks off certain parts of the diffraction pattern; the resulting changes in the image are shown in Figure 10.7. If we block off all spots except the ones belonging to the diffraction pattern of the vertical lines, the image shows only the vertical lines. If we let pass only the spots on the diagonals at 45°, we see a grid rotated by 45°. The spots passing are the most important spots of the diffraction pattern corresponding to an object made of lines rotated at 45°, although such lines do not exist in the original object screen.

Now let us consider a plane grating with periodicity constant a, as shown in Figure 10.8. The diffraction pattern is observed in the focal plane of the lens. If we block off every second maximum of the diffraction pattern, we are left with a diffraction pattern with a spacing of $2\lambda/a$ between the maxima. The image we observe is that of a grating structure with a periodicity constant of $a/2$ instead of a (see Figure 10.9).

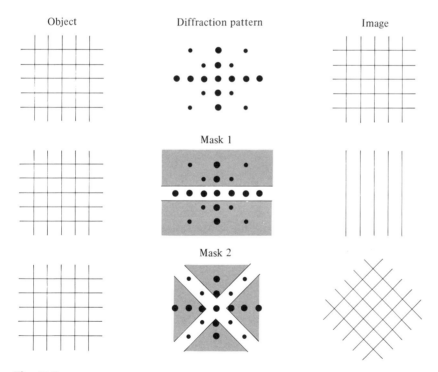

| Object | Diffraction pattern | Image |

Mask 1

Mask 2

Fig. 10.7
In a variation of the experiment of Figure 10.1, masks 1 and 2 are placed at the focal plane of the lens, blocking off certain parts of the diffraction pattern. The resulting images are shown on the right.

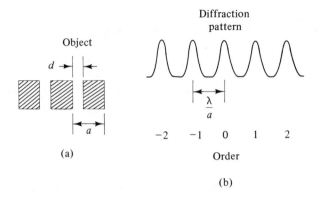

Fig. 10.8
(a) Object of a plane grating with periodicity constant a. (b) Diffraction pattern with a spacing of λ/a between the maxima.

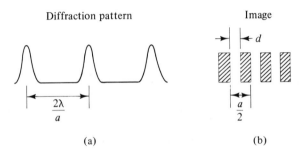

Fig. 10.9
(a) The disturbed diffraction pattern with a spacing of $2\lambda/a$ between maxima leads to (b) an image of periodicity $a/2$.

B. Spatial Frequencies and Spatial Waves

By making intensity changes in the diffraction pattern, we obtained a change in the image. In the example of the grating, we accomplished this by removing every second peak from the diffraction pattern of the object. The diffraction pattern of the grating normally has peaks at $m\lambda$, where $m = 0, 1, 2, 3, 4, 5, \ldots$. Removing every second peak results in a diffraction pattern with peaks at $m\lambda$, where $m = 1, 3, 5, \ldots$, corresponding to an object grating with only half the spacing of the original grating. This is seen as the image (see Figure 10.9).

The diffraction pattern may be interpreted as a recording of frequencies corresponding to waves used to make up the periodic structure of the object by superposition (see also problem 4). These frequencies are called **spatial frequencies**, and the waves corresponding to them are called **spatial waves**. The spatial waves can be visualized by considering periodically darker and lighter sections on a gray background (Figure 10.10b, c, and d). If we represent such a wave by $A \sin[2\pi(x/\lambda)]$, we can associate the darkest part with $+A$, the lightest

10. Phase-Sensitive Imaging

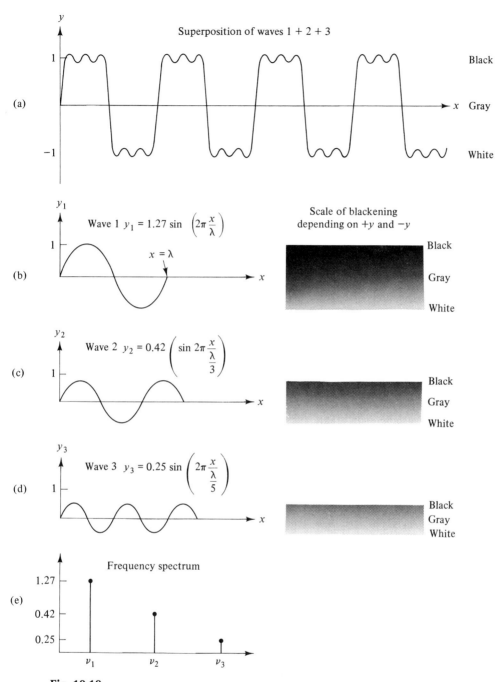

Fig. 10.10
Periodic structure and its decomposition in spatial waves and frequencies.
(a) Superposition of the three spatial waves shown in (b), (c), and (d). (b) Spatial
wave of frequency ν_1 and amplitude 1.27. (c) Spatial wave of frequency ν_2 and
amplitude 0.42. (d) Spatial wave of frequency ν_3 and amplitude 0.25. (e) Frequency
spectrum of the three superimposed waves as shown in (a).

4. Optical Filtering and Intensity Changes, Spatial Frequencies and Spatial Waves   359

part with $-A$, and a gray tone with zero, depending on the phase of $\sin[2\pi(x/\lambda)]$, which depends on the distance x. By taking the gray tone to be average, we can record phase information.

The superposition of several spatial waves again results in a periodic structure (see Figure 10.10a). The corresponding frequency spectrum, shown in Figure 10.10e, is the spectrum of the spatial frequencies. If the object pattern is narrow, the diffraction pattern is wide, and vice versa, because they are Fourier transformations of each other. As a consequence, high frequencies of the spatial waves of the object are recorded in the diffraction pattern far away from the center. The process of changing the diffraction pattern to influence the image is called **optical filtering**.

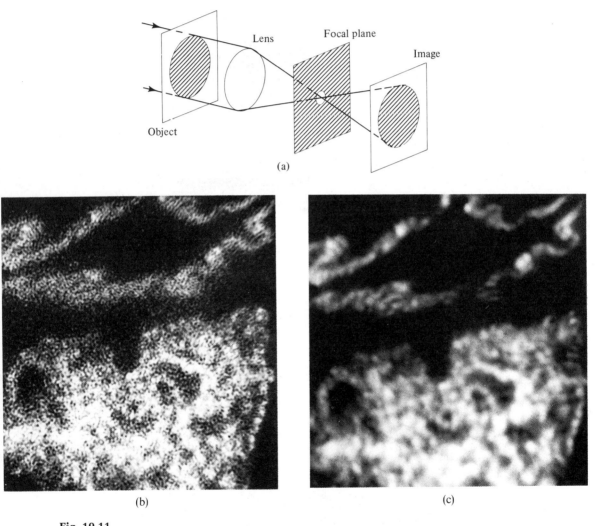

Fig. 10.11
(a) A single hole is used as a filter in the focal plane of the lens to block off most of the diffraction pattern. (b) Photomicrograph of the intestine of a snail, before filtering; (c) photo after filtering (from Cagnet, Françon, and Thrierr, *Atlas of Optical Phenomena*).

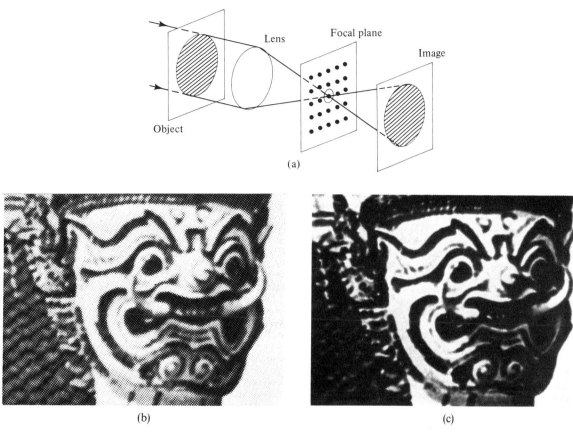

Fig. 10.12
(a) A screen with many holes is used as a filter in the focal plane of the lens to block off certain orders of the diffraction pattern. (b) Halftone photo before filtering; (c) halftone photo after filtering (from Cagnet, Françon, and Thrierr, *Atlas of Optical Phenomena*).

C. Optical Filtering and Photography

One application of optical filtering is the elimination of the graininess of a photograph. Figure 10.11 shows a filter with a single hole placed in the focal plane of the lens. Figure. 10.12 shows filtering by a screen. The filter removes certain parts of the diffraction pattern. In the case shown in Figure 10.11, only the central part of the diffraction pattern passes. All higher orders are removed. The higher orders of the spatial frequencies represent spatial waves of higher frequencies. Because these spatial waves are now missing in the formation of the image, fewer details can be observed. In the case shown in Figure 10.12, a periodic structure is used as a filter to eliminate the dot pattern from halftone photographs.

    The theory of image formation as it relates to the diffraction pattern of the object was developed by F. Abbe in 1873. Abbe observed that the image quality of a simple object in the microscope was better when the aperture opening at the objective lens was made larger (and not smaller, as one might think according to the theory of geometrical optics and aberration). We can see this in Figure 10.11, where the fine structure of the image was removed by bringing a smaller aperture into the beam (at the plane of the diffraction pattern).

5. OPTICAL FILTERING AND PHASE CHANGES

In Section 4, we saw how blocking off certain parts of the diffraction pattern affects the image. We will now see how a phase change in part of the diffraction pattern in the focal plane of the lens may influence the image.

When using a microscope to investigate biological samples, we must deal with the problem of contrast. The change in refractive index and the change in thickness across the sample area are usually too small to result in appreciable differences in the transmitted light. One way to introduce contrast is to use organic compounds that color some parts of the sample but not others. Another way is the **phase-contrast method** developed by Fritz Zernike (Nobel Prize in physics, 1953).

The phase of the diffraction pattern can be changed by a quarter-wave plate. To discuss this, we follow A. Sommerfeld for a model case, using a phase grating as the object. The phase grating is a transparent plate of refractive index n with regular grooves of height h and periodicity constant d. We will calculate the diffraction pattern and then show that a phase change in the diffraction pattern leads to an image of a grating with black and white lines. The grating we begin with is called a *phase grating*, since only the phase of the transmitted light is changed when the light passes through the grating. After the phase change is

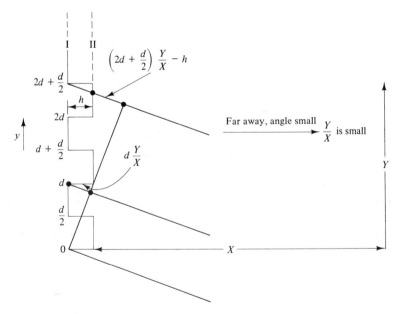

Fig. 10.13
Coordinate system for the calculation of the diffraction pattern of a transparent step phase grating. The step height is h and the periodicity of the step is d.

10. Phase-Sensitive Imaging

introduced in the diffraction pattern, we get an *amplitude grating*—that is, periodic dark and bright sections in the image.

Figure 10.13 shows a transparent medium with steps of height h. Huygens' wavelets originating in plane II of Figure 10.13 have a phase difference compared to waves originating in plane I, corresponding to the length h. We neglected the contribution to the phase difference from the refractive index of the material of the step of height h. Thus, we must sum up the Huygens wavelets generated in planes I and II by the incident plane wave (see Figure 10.13). We have indicated the phase difference for various rays at large angles; however, we only want to discuss small angles, in order to be able to assume that the intensity of the light is constant over the area.

The diffraction pattern is obtained much the way it was in Chapter 3. We must sum up the integrals, each extending over a section of width $d/2$, (see problem 6) as we did for the sequence of N slits in Chapter 3:

$$u_p = \int_0^{d/2} \exp\left[ik\left(y\frac{Y}{X} - h\right)\right]dy + \int_d^{d+d/2} \exp\left[ik\left(y\frac{Y}{X} - h\right)\right]dy$$

$$+ \text{ all integrals of plane II} \tag{10.1}$$

$$+ \int_{d/2}^{d} \exp\left[ik\left(y\frac{Y}{X}\right)\right]dy + \int_{d+d/2}^{2d} \exp\left[ik\left(y\frac{Y}{X}\right)\right]$$

$$+ \text{ all integrals of plane I} \tag{10.2}$$

For this phase grating, we have

$$u_p = \left[2\exp\left(-ik\frac{h}{2}\right)\right]\left(\frac{-\sin[k(h/2)] + \sin[k(d/2)(Y/X) + k(h/2)]}{k(Y/X)}\right)$$

$$\times \left(\frac{\sin[kN(d/2)(Y/X)]}{\sin[k(d/2)(Y/X)]}\right)\exp\left[ikN\left(\frac{d}{2}\right)\left(\frac{Y}{X}\right)\right] \tag{10.3}$$

We recognize the second factor as the diffraction factor, and the third factor as the interference factor. The other two are phase factors. Note that superimposing the diffraction factor onto the interference factor may suppress certain orders of interference. This was shown for the example of four slits in Chapter 3.

In equation 10.3, we want to consider only the maxima. They are given as the maxima of the interference factor, for which we have

$$k\left(\frac{d}{2}\right)\left(\frac{Y}{X}\right) = 0, \pi, 2\pi, \ldots \tag{10.4}$$

As a result, the third factor in equation 10.3 will be N, and the fourth factor will be equal to 1. Introducing $k(d/2)(Y/X)$ from equation 10.4 into the second factor (diffraction factor), we obtain the following result for m odd, m even but $\neq 0$, and $m = 0$:

m odd: $\dfrac{-\sin[k(h/2)] + \sin[\pi m + k(h/2)]}{(2\pi m/d)} = \dfrac{-d\sin[k(h/2)]}{\pi m}$

m even ($\neq 0$): $= 0$ (10.5)

$m = 0$: we obtain $\dfrac{0}{0}$ and find $= \dfrac{d}{2}\cos\left(k\dfrac{h}{2}\right)$

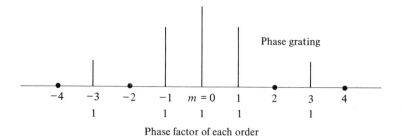

Fig. 10.14
Diffraction pattern of a phase grating according to equations 10.3–10.5. The absolute values of the amplitudes and the total phase factors are indicated for each order with intensity $\neq 0$.

The phase factor $e^{-ik(h/2)}$ is the same for all orders and is of no interest. Using the results of equations 10.4 and 10.5, we can find the diffraction pattern of equation 10.3, as shown in Figure 10.14.

To convert a phase grating into an amplitude grating, the material of the steps of height h is changed to absorb light. To express the absorption, we must make h a complex number; that is, $h = a - ib$. We know that in such a case an attenuation factor can factored from the wave equation (see Chapter 7). In Figure 10.15, we see the conversion from phase grating to amplitude grating.

The contribution of $e^{-ik(a/2)}$ to the phase factor is of no importance, since the light is suppressed in that section of the grating. Therefore, we can set $a = 0$. This makes it easier to find the transition from phase grating to amplitude grating. To implement the transition, we substitute $-ib$ for h in equation 10.3 (maxima only; see equation 10.4). Then for m odd we have

$$\frac{-\sin[k(-ib/2)] + \sin[\pi m + k(-ib/2)]}{2\pi m/d} \tag{10.6}$$

or

$$\frac{-2\sin[k(-ib/2)]}{2\pi m/d} = \frac{d}{\pi m}\sin\left(i\frac{kb}{2}\right) \tag{10.7}$$

For $m = 0$, we get

$$\frac{d}{2}\cos\left(-i\frac{kb}{2}\right) \tag{10.7}$$

We can now summarize as we did in equations 10.5:

$$
\left.
\begin{array}{ll}
m\text{ odd:} & \dfrac{d}{\pi m}\sin\left(i\dfrac{kb}{2}\right) = \dfrac{id}{\pi m}\sinh\left(\dfrac{kb}{2}\right) \\[2mm]
m\text{ even }(\neq 0) & = 0 \\[2mm]
m = 0: & \dfrac{d}{2}\cos\left(-i\dfrac{kb}{2}\right) = \dfrac{d}{2}\cosh\left(\dfrac{kb}{2}\right)
\end{array}
\right\} \tag{10.8}
$$

where $\sin[i(kb/2)]$ and $\cos[i(kb/2)]$ have been rewritten as the hyperbolic sine

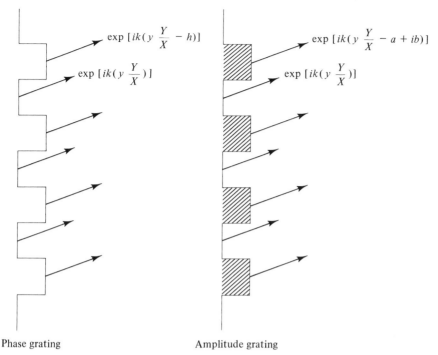

Phase grating   Amplitude grating

Fig. 10.15
The conversion from (a) phase grating to (b) amplitude grating is accomplished by introducing the absorption expressed as an imaginary part of h in the phase factor.

and cosine functions:

$$\sin\left(i\frac{kb}{2}\right) = i\sinh\left(\frac{kb}{2}\right) \tag{10.9}$$

$$\cos\left(i\frac{kb}{2}\right) = \cosh\left(\frac{kb}{2}\right) \tag{10.10}$$

To find the approximate heights of the lines as discussed previously, we assume that b is chosen so that

$$\frac{\sin[k(h/2)]}{\cos[k(h/2)]} \simeq \frac{\sin[k(b/2)]}{\cos[k(b/2)]} \tag{10.11}$$

Plotting the diffraction pattern by considering equations 10.8–10.11, we get Figure 10.16. The diffraction pattern in Figure 10.16 is very similar to the one in Figure 10.14, except that the phase is different for all orders of $m \neq 0$. This is easily seen from equation 10.8, where we have a factor i for odd orders; this translates into a phase factor of $e^{i(\pi/2)}$.

This important difference between the two diffraction patterns seems to be a very small one. For the phase grating, there is no phase difference between the zero order and all the other higher orders. In contrast, in the amplitude grating, the zero order has a phase difference of $\pi/2$ with respect to all other orders.

We have already seen that a change in the diffraction pattern may change the image. If we make a phase change of $\pi/2$ in the zero order part of the

5. Optical Filtering and Phase Changes   365

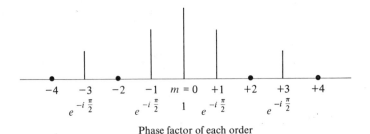

Phase factor of each order

Fig. 10.16
Diffraction pattern of the amplitude grating according to equations 10.8–10.11 (instead of equations 10.3–10.5). The absolute values of the amplitudes and the total phase factors are indicated for each order with intensity $\neq 0$.

diffraction pattern of the phase grating, the image will appear as if it came from an amplitude grating.

Let us now apply this concept to the microscope. The zero order of the object appears in the center of the microscope in the focal plane of the objective lens. There we place a small quarter-wave plate, changing the phase of the zero order to $\pi/2$. Now there is a phase difference between the zero order and all the other orders. What was a phase grating now appears as an amplitude grating. With this "trick," a hard-to-recognize structure will now appear as a well-contrasted structure. This phase-contrast method for the improvement of the image contrast in the microscope was published by Zernike in 1935.

6. IMAGE FORMATION BY INTERFERENCE AND DIFFRACTION

We have seen how light incident on a periodic structure produces a diffraction pattern in the focal plane of a lens. Amplitude or phase changes of this diffraction pattern change the final image we observe.

Now we want to discuss how images are formed when light is incident on periodic or nonperiodic structures imposing phase relations on the incident light.

A. The Fresnel Zone Plate

The **Fresnel zone plate** is made of black and white rings (Figure 10.17). A spherical wave is produced by a point source and is assumed to be incident on the plate. The ring pattern produces phase relations between the light beams passing the plate. All these beams are superimposed and form a point as image. We will regard this procedure as an extention of the diffraction on a circular opening or circular stop (presented in Chapter 3). For the diffraction pattern of the circular opening, we saw that the intensity at the image point has maxima or minima depending on the radius a of the opening. From equation 3.58, we have

$$\text{min: } a^2 = m\lambda\rho_0$$
$$\text{max: } a^2 = (m + \tfrac{1}{2})\lambda\rho_0 \tag{10.12}$$

where λ is the wavelength, ρ_0 is the distance from the aperture to the observation screen, and m is an integer.

10. Phase-Sensitive Imaging

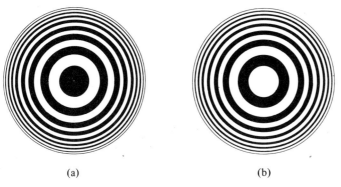

(a)   (b)

Fig. 10.17
Fresnel zone plates. Parts (a) and (b) are complementary screens.

This result was obtained by integration (summing up the phase factors) over all waves passing through the circular opening. In the Fresnel zone plate, we produce a maximum at the observation point with a plate divided in circular zones. The width and radius of the zones are chosen so that the light passing the open zones have a path difference of λ with their neighboring open zones and a path difference of $\lambda/2$ with their neighboring closed zones. The light from the open zones all superimpose. The closed zones are blocking off the light that would cancel the light from the open zones.

To calculate the radii of the zones, see Figure 10.18. The path difference of the light from S to O through the open zones is

$$(R + \rho) - (R_0 + \rho_0) = m\lambda \tag{10.13}$$

and through the blocking zones is

$$(R + \rho) - (R_0 + \rho_0) = m\frac{\lambda}{2} \tag{10.14}$$

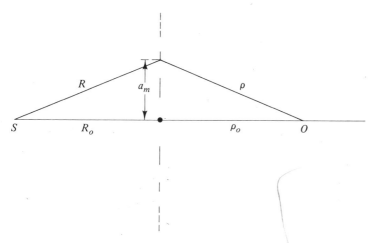

Fig. 10.18
Coordinate system for calculation of the mth radius a_m of a Fresnel zone plate.

For the radius a_m of the mth blocking zone, we have

$$a_m^2 = \rho^2 - \rho_0^2 = R^2 - R_0^2 \qquad (10.15)$$

Since a_m is small compared to R and ρ, we expand:

$$\rho = \sqrt{\rho_0^2 + a_m^2} \simeq \rho_0 \left[1 + \frac{1}{2}\left(\frac{a_m^2}{\rho_0^2}\right) \right]$$
$$R = \sqrt{R_0^2 + a_m^2} \simeq R_0 \left[1 + \frac{1}{2}\left(\frac{a_m^2}{R_0^2}\right) \right] \qquad (10.16)$$

Introducing equations 10.16 into equation 10.14 yields

$$R_0 + \frac{1}{2}\left(\frac{a_m^2}{R_0}\right) + \rho_0 + \frac{1}{2}\left(\frac{a_m^2}{\rho_0}\right) - R_0 - \rho_0 = m\frac{\lambda}{2} \qquad (10.17)$$

or

$$\frac{1}{R_0} + \frac{1}{\rho_0} = \frac{m\lambda}{a_m^2} \qquad (10.18)$$

The right side of equation 10.18 is a constant and is interpreted as the inverse focal length of the plate. We then obtain an imaging equation that resembles the thin-lens equation:

$$\frac{1}{R_0} + \frac{1}{\rho_0} = \frac{1}{f} \qquad (10.19)$$

where $f = a_m^2/m\lambda$. For the order of magnitude of the radii a_m, consider the following numerical example. For $R_0 = \rho_0$, we have

$$a_m^2 = \frac{m\lambda\rho_0}{2}$$

Assuming that $\rho_0 = 1000$ mm and $\lambda = 0.5 \times 10^{-3}$ mm, we get

$$a_m^2 = 0.25m \text{ mm}^2$$

or

$$a_m = 0.5\sqrt{m} \text{ mm}$$

The focal length f for the symmetric case of Figure 10.18 is obtained from $R_0 = \rho_0$, and we have $f = \rho_0/2$. (See also problems 7 and 8 for an asymmetric light pass and achromatism.)

B. Image Formation by a Steel Ball

Now let us consider image formation by a steel ball. An experimental setup, as used by R. Pohl,* is shown in Figure 10.19. We can think of the steel ball as a

---

*R. W. Pohl, *Einführung in die Optik* (Berlin: Springer-Verlag, 1948), p. 83.

10. Phase-Sensitive Imaging

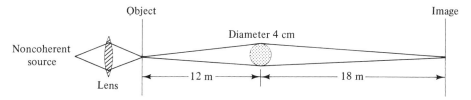

Fig. 10.19
Schematic demonstration of image formation with a 4 cm diameter steel ball, according to the experiment by R. Pohl.

degenerate Fresnel plate composed of only one section. It would then show the diffraction pattern of a stop with a bright spot in the middle. In Chapter 3, we discussed the theoretical consideration leading to a bright spot (Poisson spot) in the diffraction pattern of a spherical obstacle, located in the shadow region of the obstacle.

For the imaging process, we divide the object into bright and dark spots. The steel ball forms of each bright spot of the object a bright spot on the image plane, and dark spots of the object remain dark spots. The image is upside down. The dimensions shown in Figure 10.19 indicate that the image and object distances are large compared to the diameter of the steel ball. A. Sommerfeld[†] notes that the surface of the steel ball must be very smooth. Deviations must be small compared to the wavelength of the light.

C. Holography

We saw that the Fresnel zone plate introduces phase changes in the incident light. The transmitted light forms an image point. As an extension of this procedure, one might think that we can produce a periodic or nonperiodic structure of an object. Illumination of this structure will then produce an image that is similar to the object. First let us consider a point object illuminated by light from an expanded laser beam, as shown in Figure 10.20. The parallel coherent light and the light diffracted from the point object are superimposed and recorded on the photographic plate. The exposure of the plate is held to a gray level in order to be able to record phase information, see also Figure 10.10 and a ring pattern is visible. This photographic plate is called a **hologram**. If it is illuminated with laser light, real and virtual images result (Figure 10.21).

The hologram records the amplitude and phase information at the first diffraction process, depicted in Figure 10.20. The diffracted light in the second diffraction process consists of converging and diverging waves. The converging waves form the real image, and the diverging waves may be traced back to form a virtual image (see Figure 10.21). Our eye then sees the virtual image at the original object point. It appears three dimensional, since the hologram has recorded all the necessary phase and amplitude information.

[†]A. Sommerfeld, *Vorlesungen über Theoretische Physik*, Band IV, Optik (Wiesbaden: Dieterich'sche Verlagsbuchhandlung, 1950), p. 220.

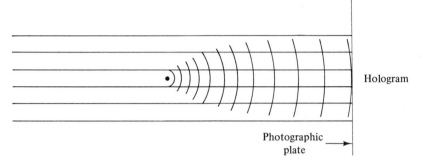

Fig. 10.20
Hologram of a pointlike object. Parallel coherent light is incident on a photographic plate. The diffracted light from the point obstacle is superimposed on the incident light.

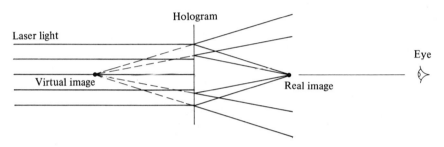

Fig. 10.21
Illumination of a hologram with laser light results in real and virtual images. The eye looks along the optic axis of the experimental arrangement.

From Figure 10.21, we can see that the real image is in the same direction as the virtual image. Clearly, slightly different "off-axis" arrangements would make it possible to see the virtual image without having to look at the real image at the same time.

Figure 10.22 shows the setup for recording a hologram. The reference beam is directed to the photographic plate without having any interaction with the object. The signal beam hits the object, and the light from the object is directed to the photographic plate. The superposition of the two beams forms the more or less gray structure of the hologram.

Let

$$u_r = u_{r_0} e^{i(kx+\phi_r)} \quad \text{and} \quad u_s = u_{s_0} e^{i(kx+\phi_s)}$$

be the reference and signal beams, respectively. For the absolute value squared of the superposition of u_r and u_s, as they appear at the hologram, we get

$$|u_r + u_s|^2 = u_{r_0}^2 + u_{s_0}^2 + 2u_{r_0} u_{s_0} \cos(\phi_r - \phi_s) \tag{10.20}$$

Depending on the distance each beam travels and the influence the object has on the signal beam, we get negative or positive values for $\cos(\phi_r - \phi_s)$. Therefore, the constant value $u_{r_0}^2 + u_{s_0}^2$ may be enlarged or reduced, resulting in more gray or less gray levels on the photographic plate.

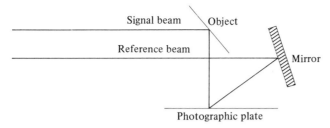

Fig. 10.22
Signal and reference beams for recording a hologram.

Using the right amount of light for the reference beam, we can record the phase information on the photographic plate in a manner similar to the way we record the modulation of a carrier wave in communication theory. Illumination of the hologram will result in the image.

Consider a simplified example of producing a hologram: a diffraction process and reconstruction of the image by a second diffraction process. The object will be a grating.

The diffraction amplitude of a grating that has periodicity constant a and M lines is given as

$$u_s = u_0 \frac{\sin \gamma(d/2)}{\gamma(d/2)} \left(\frac{\sin M\gamma(a/2)}{\sin \gamma(a/2)} \right) e^{i\phi_s} \tag{10.21}$$

where $\gamma = (2\pi/\lambda)\sin\theta$ or $\gamma = (2\pi/\lambda)(Y/X)$ (small angle approximation); see also equations 3.35 and 3.36. A phase factor occurs in the expression because it is the amplitude pattern. The interference factor gives maxima at distances λ/a, and the diffraction pattern makes the maxima slowly decrease at large values of Y/X (or $\sin\theta$ if d is small compared to a).

These diffracted waves undergo interference with the reference waves $u_r e^{i\phi_r}$, and we observe a grating like structure on top of the background on the hologram (Figure 10.23). The pattern looks somewhat similar to the intensity

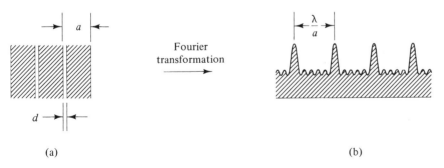

(a)   (b)

Fig. 10.23
(a) The object is a grating structure with periodicity constant a. (b) The diffraction pattern (hologram) has a spacing between its maxima equal to λ/a. It would be shaded white-gray-black.

6. Image Formation by Interference and Diffraction

patterns we have discussed before, but the side minima and maxima are now recorded differently on top of the background. These side maxima and minima are important if we try to make a diffraction pattern of the hologram. Now we can illuminate the diffraction pattern (hologram) with laser light and regain the image of the original grating (Figure 10.24).

In all these considerations where there is a diffraction pattern of a diffraction pattern, we must realize that it is the amplitude diffraction pattern (containing the phase information) that is important. Holography is based on recording the available phase information and on the second process using it for reproducing the image. The image cannot be reproduced if we work only with intensity patterns. (The observation of intensity patterns was the subject of Chapter 3.)

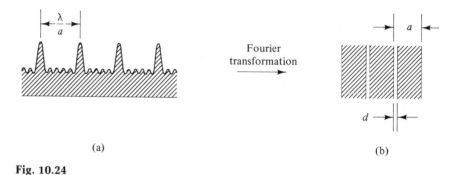

Fourier transformation

(a)

(b)

Fig. 10.24
(a) Structure of the diffraction pattern (hologram). It would be shaded white-gray-black.
(b) The reproduced original grating structure with periodicity constant a.

The off-axis hologram is illustrated in Figure 10.25. A reflection hologram is shown at the beginning of the book.

It is possible to use X rays to obtain a hologram. Figure 10.26 shows a hologram obtained with X rays of $\lambda = 60$ Å. The object is composed of three slits, each 3 μm wide and 9μm apart. The reference beam is produced by a another slit. Huygens' principle is used to obtain coherent X rays. The hologram is recorded on a film. A helium-neon laser is used to make the object "visible." The wavelength of the helium-neon laser is 6328 Å, a hundred times larger than the wavelength of the X rays.

Fig. 10.25
Viewing the off-axis hologram.

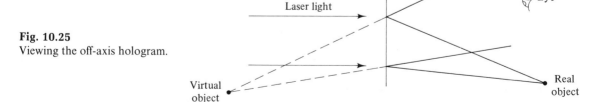

10. Phase-Sensitive Imaging

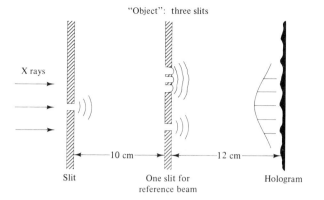

"Object": three slits

X rays

Slit

One slit for
reference beam

←——10 cm——→ ←——12 cm——→

Hologram

Fig. 10.26
Hologram produced by X rays of $\lambda = 60$ Å.
Object slits each 3 μm wide and 9 μm apart.

Problems

1. *Fourier transformation, Gaussian function.* Functions for which $f(x) = f(-x)$ are even functions and symmetric with respect to $x = 0$. For such functions, the cosine-Fourier transform integral is all we need to use. We see this from

$$\int_{-\infty}^{+\infty} f(x)[\cos(xy) - i\sin(xy)]\,dy$$

$$= \int_{-\infty}^{+\infty} f(x)\cos(xy) - i\int_{-\infty}^{+\infty} f(x)\sin(xy)\,dy$$

   a. Show that the second integral is equal to zero.
   b. The Gaussian function is defined as $f(x) = e^{-a^2x^2}$. Calculate the Fourier transform of $f(x)$. Does it suffice to use only the cosine transformation?
   c. The inverse Fourier transform was defined as

$$f(x) = \frac{1}{2\pi}\int_{-\infty}^{+\infty} F(y)e^{ixy}\,dy$$

   Take the result from part (b), put it into the inverse Fourier transformation integral, and obtain back $f(x) = e^{-a^2x^2}$. We see for the Gaussian function, we can show in a simple way that the Fourier transformation of the Fourier transform is the original function.

2. *Fourier transform: box.* In Chapter 3, we calculated the amplitude diffraction pattern of a slit for Fraunhofer diffraction. The diffraction pattern may also be obtained as the Fourier transformation of the slit function $P(x)$. We may use

$$f(x) = \frac{1}{2\pi}\int_{-\infty}^{+\infty} F(y)e^{ixy}\,dy, \quad F(y) = \int_{-\infty}^{+\infty} f(x)e^{-iyx}\,dx$$

where

$$f(x) = \begin{cases} 1 & \text{for} \quad -a \text{ to } +a \\ \text{otherwise } 0 \end{cases}$$

and the Fourier transformation yields

$$F(y) = 2a\frac{\sin(ay)}{ay}$$

Problems   373

Calculate the Fourier transform of the function

$$\overline{f(x)} = \begin{cases} 0 & \text{for} \quad -\infty \text{ to } -d \\ -1 & \text{for} \quad -d \text{ to } 0 \\ +1 & \text{for} \quad 0 \text{ to } +d \\ 0 & \text{for} \quad d \text{ to } \infty \end{cases}$$

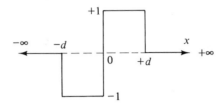

Make a sketch of the intensity diffraction pattern if $\overline{f(x)}$ is the aperture.

3. *Fourier transform: periodic function.* Consider the periodic function

$$f(x) = \begin{cases} \cos(y_0 x) & \text{for} \quad -a \text{ to } +a \\ \text{otherwise } 0 \end{cases}$$

Calculate the Fourier transform and make a sketch. The Fourier transformation may be interpreted as the distribution of spatial frequencies of the periodic pulse function.

Ans. $F(y) = a \left[\dfrac{\sin(y_0 + y)a}{(y_0 + y)a} + \dfrac{\sin(y_0 - y)a}{(y_0 - y)a} \right]$

4. *Fourier series.* A periodic function can be represented by the superposition of cosine and sine functions with different frequencies and amplitudes. For simplicity, we consider only odd functions, that is, $f(x) = -f(-x)$ with respect to $x = 0$. For example, consider the following periodic function defined in the interval 0 to λ:

$$f(x) = \begin{cases} +1 & \text{for} \quad 0 < x < \dfrac{\lambda}{2} \\ -1 & \text{for} \quad \dfrac{\lambda}{2} < x < \lambda \end{cases}$$

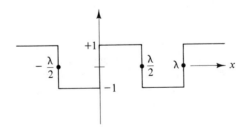

We start by determining the sum of sine functions that can represent this function:

$$f(x) = \sum_{m=1}^{\infty} B_m \sin\left(2\pi m \frac{x}{\lambda}\right)$$

$$= \sum_{m=1}^{\infty} B_m \sin(mkx) \quad \text{where} \quad k = \frac{2\pi}{\lambda}$$

a. Calculate

$$\frac{2}{\lambda} \int_0^\lambda f(x) \sin(nkx)\, dx$$

by introducing

$$f(x) = \sum_{m=1}^\infty B_m \sin mkx$$

into the integral and show that only the term with $n = m$ is not equal to zero and that we get

$$B_m = \frac{2}{\lambda} \int_0^\lambda f(x) \sin mkx\, dx$$

b. Introduce the definition of $f(x)$ given in the preceding information and calculate B_m.
c. Calculate B_1 to B_5 and give the Fourier series.
d. Add the first three terms to get an idea of how fast the approximation works.
e. In the Fourier transformation integral, we obtained the spatial frequencies from the Fourier transformation of the space functions. Here we can read off the frequencies and their weights (amplitudes) from the series representation. Draw the frequency spectrum.

5. *The Dirac delta function and Fourier transformation.* The Dirac delta function is defined by

$$\delta(x) = \begin{cases} 0 & \text{for} \quad x \neq 0 \\ \infty & \text{for} \quad x = 0 \end{cases} \quad \text{and} \quad \int_{-\infty}^{+\infty} \delta(x)\, dx = 1$$

If it is used as a product with another function $f(x)$ in an integral from $-\infty$ to $+\infty$, one can show

$$\int_{-\infty}^{+\infty} \delta(x - x_0) f(x)\, dx = f(x_0)$$

a. Calculate the integral

$$\int_{-\infty}^{+\infty} \delta(x - x_0) f(x)\, dx \quad \text{for} \quad f(x) = e^{-ixy}$$

b. Can we say that the Fourier transform of $\delta(x - x_0)$ is $e^{-ix_0 y}$?
c. Calculate the Fourier transform of

$$A[\delta(x - x_0) + \delta(x + x_0)]$$

Show that we get as result $2A \cos(x_0 y)$. Compare with the interference from two narrow slits in chapter 2.

In Chapter 2, we found that interference patterns were obtained by a summation process. In Chapter 3, we used an integration process to derive the diffraction pattern. In part (c) of this problem we obtained formally the interference pattern of two source points by integration. We calculated the Fourier transform of two Dirac delta functions.

d. Calculate the Fourier transformation of

$$\sum_{n=0}^{n=+N-1} A\delta(x - nx_0)$$

In Chapter 2, we summed up

$$\sum_{n=0}^{n=N-1} e^{inx_0 y}$$

and obtained

$$\frac{\sin[N(x_0 y/2)]}{\sin(x_0 y/2)} e^{i(N-1)(x_0 y/2)}$$

If we square this expression, we have the following interference pattern observed from N (narrow sources):

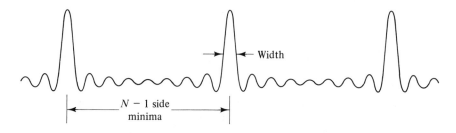

We found that for large values of N, the width of the main maxima becomes narrower, the number of side maxima increases, and their individual heights decrease. For $n \to \infty$, we expect that the Fourier transform of

$$f(x) = \sum_{n=-\infty}^{n=+\infty} A\delta(x - nx_0)$$

is a sum of delta functions, and one can show that the Fourier transform is

$$F(y) = \sum_{n=-\infty}^{+\infty} \frac{2\pi A}{x_0} \delta\left(y - n\frac{2\pi}{x_0}\right)$$

e. If $f(x)$ is graphically expressed as

$$> |x_0| <$$

$$|\quad|\quad|\quad|\quad|\quad|\quad|\quad\}\ A$$

make a sketch of $F(y)$ and give the separation and heights of the "bars."

6. *Phase grating*
   a. Fill in all steps in the calculation of the phase grating and arrive at

$$u_p = 2e^{-ik(h/2)} \frac{-\sin[k(h/2)] + \sin[k(d/2)(Y/X) - k(h/2)]}{k(Y/X)}$$

$$\times \frac{\sin[kN(d/2)(Y/x)]}{\sin[k(d/2)(Y/X)]} e^{ikN(d/2)(Y/X)}$$

   b. Show that we have

$$\frac{-\sin(kh/2) + \sin[\pi m + (kh/2)]}{(2m\pi/d)} = \begin{cases} \text{for} \quad m \text{ odd} & -\dfrac{d}{\pi m}\sin\left(\dfrac{kh}{2}\right) \\[2ex] \text{for} \quad m \text{ even but} \neq 0 & 0 \\[2ex] \text{for} \quad m = 0 & \dfrac{d}{2}\cos\left(\dfrac{kh}{2}\right) \end{cases}$$

7. *Fresnel zone plate.* Derive the focal length of a Fresnel zone plate by assuming

parallel light incident from the left. Check your result by comparison with the thin-lens equation for the zone plate when $R_0 \to \infty$.

Ans: $\dfrac{1}{\rho_0} = \dfrac{m\lambda}{a_m^2}$

8. *Fresnel zone plate and chromatic aberration.* The focal length of the "thin-lens equation" for the zone plate (equations 10.18 and 10.19) depends on the wavelength. Parallel light incident on the plate will be focused to the focal point $f = \rho_0$. If the incident light has a different wavelength, we will obtain a different focal length. This corresponds to chromatic aberration of lenses (see Chapter 13). Calculate the difference in the focal length for light of wavelength $\lambda_1 = 0.5 \times 10^{-3}$ mm and $\lambda_2 = 0.55 \times 10^{-3}$ mm, where $\rho_0 = 1000$ mm.

Ans. $\Delta f = \Delta \rho = 91$ mm

Radiometry and Photometry | **11**

1. INTRODUCTION*

In Chapter 2, we studied the superposition of light waves. We defined the intensity of a light wave as proportional to the square of the amplitude. Since the amplitude depends on space and time, we discussed the averaging process depending on the time coordinate. A factor resulting from this averaging process is usually ignored when interference problems are discussed. We were able to associate the intensity of a wave with the energy or power of the wave without having to define a quantitative relationship.

In Chapter 4, we learned that the power per unit area is given by the Poynting vector. By calculating the time average of the Poynting vector, we established the numerical proportionality factor for the average power per unit area.

In our study of the photoelectric effect (Chapter 14) and blackbody radiation (Chapter 15), we will discuss the energy of a photon and the energy density of radiation.

In this chapter, we will deal with radiometric and photometric quantities. In radiometry, we have specific names for the quantities corresponding to energy, energy density (per area), power, and power (per area): They are **radiant energy**, **radiant energy density**, **radiant flux**, **radiant exitance**, and **irradiance** (see Table 11.1). Using these quantities, we are only interested in the energy and power as they are emitted and detected, independent of the wavelength dependence of the light involved. As we will discover in the discussion of blackbody radiation (Chapter 15), the energy density depends on temperature and wavelength through **Planck's law**. The total emitted energy is the integral over all the wavelengths given by the **Stefan-Boltzmann law**. In photometry, however, we are interested in the visible spectral region only, and new units are introduced for the radiometric quantities.

---

*The definitions in this chapter follow R. W. Boyd, *Radiometry and Detection of Radiation* (New York: Wiley, 1983).

Table 11.1
Quantities used in radiometry

Radiant energy	J (joules)
Radiant energy density	$\dfrac{J}{m^2}$ (joules per square meter)
Radiant flux	$\dfrac{J}{s}$ (watts)
Radiant exitance (emitter)	$\dfrac{W}{m^2}$ (watts per square meter)
Irradiance (receiver)	$\dfrac{W}{m^2}$ (watts per square meter)

2. IRRADIANCE, RADIANCE, AND SMALL SOLID ANGLES

A. Irradiance

Suppose that we want to measure the power produced by the sun in a solar cell. Let the detector have an area equal to da' (in square meters) and be placed perpendicular to the light from the source (Figure 11.1a). The **irradiance** E is the radiant flux (power in watts) per unit area (square meter) incident on the detector:

$$E = \frac{\text{radiant flux}}{\text{area}} \left[\frac{W}{m^2}\right] \tag{11.1}$$

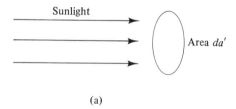

(a)

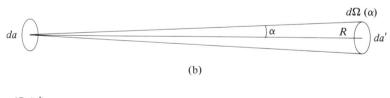

(b)

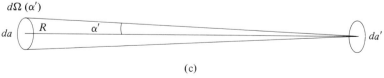

(c)

Fig. 11.1
(a) Sunlight arriving at the detector area da'. (b) The solid angle $d\Omega(\alpha) = da'/R^2$.
(c) The solid angle $d\Omega(\alpha') = da/R^2$.

For radiation from the sun, we get $E = 1.35 \times 10^3$ W/m². This numerical value is called the **solar constant**.

We now assume that the emitter has a flat circular area with a small diameter compared to the distance R to the receiver. We call the area of the emitter da. From each "point" of this area emerges a cone of light. The opening of this cone is determined by the area of the receiver da' (see Figure 11.1b). The solid angle of the cone, $d\Omega(\alpha)$, is given as

$$d\Omega(\alpha) = \frac{da'}{R^2} \tag{11.2}$$

where α is the half-angle of the cross section and da' relates to the receiver.

B. Radiance

The radiant flux arriving at the receiver is proportional to the solid angle and area da of the emitter, and we have

$$P = (\text{const})\, da\, d\Omega(\alpha) = (\text{const})\, da\, \frac{da'}{R^2}\,[\text{W}] \tag{11.3}$$

The constant (const) is called the **radiance** L, and it is equal to the radiant flux per unit area and solid angle:

$$\text{Radiance: } L = \frac{\text{radiant flux}}{(\text{area})(\text{solid angle})} \left[\frac{\text{W}}{\text{m}^2\,\text{sr}} \right] \tag{11.4}$$

where the nondimensional sr (**steradians**) is the unit of the solid angle (see below). Using equation 11.4, we rewrite equation 11.3 as

$$P = L\, da\, d\Omega(\alpha) = L\, da\, \frac{da'}{R^2} \tag{11.5}$$

C. Units of Solid Angles

The solid angle in steradians is defined as the area the cone cuts out from the sphere of radius $R = 1$ (Figure 11.2a). For the solid angle of the entire sphere, we have 4π sr; for a hemisphere, 2π sr. If we can assume small solid angles, the spherical section may be approximated by a flat section (see Figure 11.2b), and the solid angle is expressed as

$$d\Omega(\alpha) = \frac{\pi(R\sin\alpha)^2}{R^2} = \pi\sin^2\alpha \tag{11.6}$$

If we introduce this into equation 11.5, that is, if we assume that the emitter and receiver areas are both small compared to their distance from one another and that they are perpendicular to the connecting line, we have

$$P = L\, da\, \pi\sin^2\alpha \tag{11.7}$$

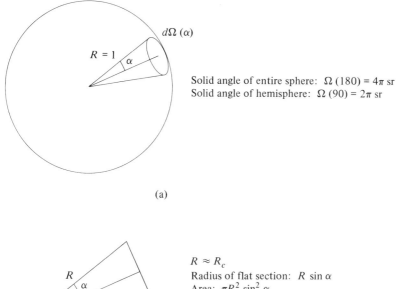

Solid angle of entire sphere: $\Omega(180) = 4\pi$ sr
Solid angle of hemisphere: $\Omega(90) = 2\pi$ sr

(a)

$R \approx R_c$
Radius of flat section: $R \sin\alpha$
Area: $\pi R^2 \sin^2\alpha$
Solid angle: $\pi \dfrac{R^2 \sin^2\alpha}{R^2} = \pi \sin^2\alpha$

(b)

Fig. 11.2
(a) Solid angle for the entire sphere and for the hemisphere. (b) Small solid angle.
The spherical section is approximated by a flat section (cross section shown here).

D. Relation Between Irradiance and Radiance

The radiant flux from the emitter arriving at the receiver may also be expressed
using equation 11.5:

$$P = L\,da\,d\Omega(\alpha) = L\,da\,\frac{da'}{R^2} = L\,da'\,\frac{da}{R^2} = L\,da'\,d\Omega(\alpha') = P' \qquad (11.8)$$

On the left we have expressed the power proportional to the area da of the emitter
multiplied by the solid angle $d\Omega(\alpha)$. On the right we have expressed the power
proportional to the area da' of the receiver multiplied by the solid angle $d\Omega(\alpha')$.
The proportionality constant L is the same, because we have related the power
to the area and the solid angle for emitter and receiver in a symmetric way
(see also Figure 11.1b and c). We can express the power arriving at the receiver
by using the irradiance:

$$E = \frac{P}{da'} \qquad (11.9)$$

and with

$$L\,da'\,d\Omega(\alpha') = E\,da' \qquad (11.10)$$

2. Irradiance, Radiance, and Small Solid Angles

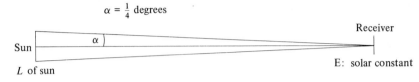

Fig. 11.3
The sun is viewed from the earth under the small solid angle $d\Omega(\alpha)$, where $\alpha = 1/4°$.

we have

$$L\,d\Omega(\alpha') = E \tag{11.11}$$

The radiance L (W/m^2 sr) multiplied by the solid angle results in the irradiance E.

For example, consider the sun as a small, flat emitter. To obtain L for the sun, we must divide E (numerically the solar constant) by the solid angle from the earth (Figure 11.3). For the sun, we have $\alpha = (1/4)°$ or $\alpha = 0.004$ rad; and for small angles,

$$d\Omega = \pi \sin^2(0.25) = \pi(0.004)^2 = 6 \times 10^{-5}\text{ sr}$$

we get

$$L = \frac{E}{d\Omega} = \frac{1.35 \times 10^3\text{ W}}{6 \times 10^{-5}\text{ m}^2\text{ sr}} = 2.25 \times 10^7\frac{\text{W}}{\text{m}^2\text{ sr}} \tag{11.12}$$

E. Radiometric Quantities

In Section 1, we introduced radiant energy, radiant energy density, radiant flux, radiant exitance and irradiance. In this section, we introduced irradiance, radiance, and units of solid angle. The last quantity to be defined is **radiant intensity** I, the power emitted into a solid angle, expressed in watts per steradians (W/sr). All radiometric quantities are summarized in Table 11.2.

Table 11.2
Radiometric quantities

	Symbol	*Unit*
Radiant energy		J
Radiant energy density	u	$\dfrac{\text{J}}{\text{m}^2}$
Radiant flux (power)	P	W
Radiant exitance (energy leaving the source)	M	$\dfrac{\text{W}}{\text{m}^2}$
Irradiance (energy arriving at the receiver)	E	$\dfrac{\text{W}}{\text{m}^2}$
Radiance (constant of equation 11.3)	L	$\dfrac{\text{W}}{\text{m}^2\text{ sr}}$
Radiant intensity	I	$\dfrac{\text{W}}{\text{sr}}$

11. Radiometry and Photometry

3. LAMBERT'S LAW: SMALL SOLID ANGLES

A. The Cosine Law and the Projected Area

We have assumed that the two small areas da and da' are parallel. If they are not, we obtain a correction factor in the expression for the emitted power. This factor can only be determined by an experiment. Consider the setup shown in Figure 11.4a. The area da is tilted through the angle θ with respect to the line connecting da and da', the areas of the emitter and the detector, respectively. The emitter may be a heated piece of metal, the receiver a thermocouple as described in Chapter 14. If we measure the power for all possible positions of da as a

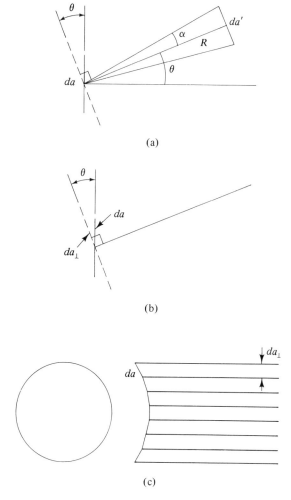

(a)

(b)

(c)

Fig. 11.4
(a) The emitter area da is tilted by the angle θ with respect to the receiver area da'.
(b) The projected area $da_\perp$. (c) Each division da has the same projected area $da_\perp$.
The emitted power from all $da_\perp$ is equal; therefore, the sun looks like an even disk, uniformly emitting.

function of the angle θ, we have

$$P(\theta) = L \frac{da'}{R^2} da\, g(\theta) \tag{11.13}$$

where $g(\theta)$ describes the dependence of P on θ as it is experimentally determined. We find that $g(\theta) = \cos\theta$, and the power is thus given as

$$P(\theta) = L \frac{da'}{R^2} da\cos\theta \tag{11.14}$$

The $\cos\theta$ factor was found by Johann Lambert in 1769 and is called the **cosine law of Lambert**.

If we take the $\cos\theta$ factor and combine it with da, we have $da_\perp = da\cos\theta$, where $da_\perp$ is the projection of the area da on the plane perpendicular to the connecting line between emitter and receiver. Lambert's law tells us that the area da radiates with the same L as the smaller area $da_\perp$ when looked at under the angle θ. Alternatively, we may have a tilted receiver, in which case a similar result is obtained.

B. Lambert's Law and the Sun

We consider the sun as a sphere emitting the power P into the solid angle $d\Omega = da'/R^2$; that is,

$$P = L\, da(\cos\theta)\frac{da'}{R^2} = L(da_\perp)\frac{da'}{R^2}$$

or

$$L = \frac{P}{d\Omega\, da_\perp} \tag{11.15}$$

Assuming Lambert's law for all equal sections da of a spherical emitting surface such as the sun, we have the same power P for each section $da_\perp$ (see Figure 11.4b and c). That is why we see the sun as a uniformly emitting disk. Lambert's law is not always fulfilled. The best Lambert source is a blackbody emitter. An extremely poor Lambert source is a laser beam; the emitted light forms a strongly directional beam.

4. LAMBERT'S LAW: LARGE SOLID ANGLES

A. Small Emitter Area and Large Receiver Area

Consider the case where the emitter has a small area da and the receiver has a large area a' (Figure 11.5a). All the power arriving at a' passes the section of the spherical surface of solid angle $\Omega(\alpha)$ and radius R. This spherical section is divided into rings of width $R\,d\gamma$ and radius $R\sin\gamma$ (see Figure 11.5b). The area of each ring is $2\pi R(\sin\gamma)R\,d\gamma$.

11. Radiometry and Photometry

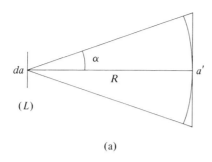

(a)

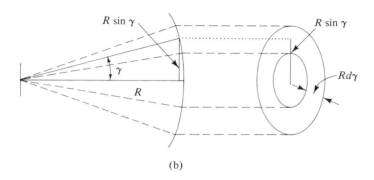

(b)

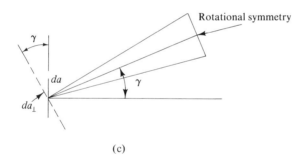

(c)

Fig. 11.5
(a) The emitter has small area da; the receiver has large area a'. (b) Ring section of the spherical surface. (c) Projection of da onto the plane of $da_\perp$.

The power arriving at a' and passing through one ring is

$$dP = L\,da\cos\gamma\,\frac{2\pi(R\sin\gamma)R\,d\gamma}{R^2} \tag{11.16}$$

where $da\cos\gamma$ may be interpreted as the projection of da onto the plane parallel to the surface of the section of the ring (see Figure 11.5c). The total power arriving at a' is now

$$P = \int_{\gamma=0}^{\gamma=\alpha} 2\pi L\,da\cos\gamma\sin\gamma\,d\gamma \tag{11.17}$$

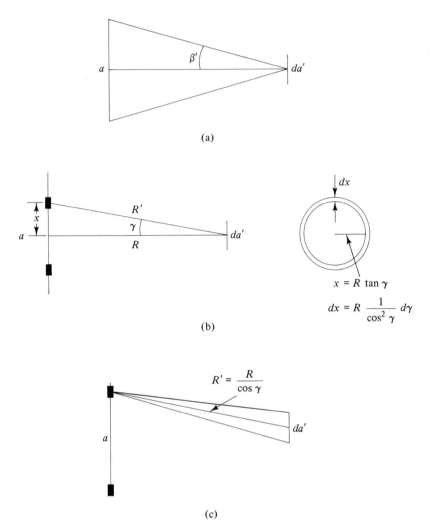

Fig. 11.6
(a) The emitter has large area a; the receiver has small area da'. (b) Division of the surface a into a ring pattern. (c) Cone to the receiver; center distance $R' = R/\cos\gamma$.

Integration yields

$$P = 2\pi L\, da \left|\tfrac{1}{2}\sin^2\gamma\right|_0^\alpha = L\, da\, \pi \sin^2\alpha \tag{11.18}$$

This is the same result obtained for the case of small emitter and receiver areas (see equation 11.7). There we assumed that the two areas were both perpendicular to the connecting line, and Lambert's law did not have to be considered.

Using equation 11.18, we find that the radiant exitance of the emitter is

$$M = \frac{P}{da} = \pi L \sin^2\alpha \tag{11.19}$$

If $\alpha = 90°$, then $M = \pi L$.

11. Radiometry and Photometry

B. Large Emitter Area and Small Receiver Area

Now let us interchange the size of emitter and receiver (Figure 11.6a). We divide the emitter into a ring pattern on the surface a with radius x (see Figure 11.6b). The power from one ring arriving at da' is given as

$$dP = L2\pi(x\,dx\cos\gamma)\frac{da'\cos\gamma}{(R')^2} \tag{11.20}$$

Introducing x, dx, and R' and integrating from $\gamma = 0$ to $\gamma = \beta'$, we get

$$P = L\,da'\,(\pi\sin^2\beta') \tag{11.21}$$

Comparing this with the result of the inverse case (equation 11.18), we see that the area involved is now that of the receiver, and the angle opens toward the emitter while L again refers to the emitter.

C. Power and Area Times Solid Angle

If we assume in equation 11.14 that da is tilted, then we can write

$$P = L\,da_\perp\frac{da'}{R^2}$$

Then we can say that the power for all these cases is obtained by multiplying the area, the solid angle, and L. But for the case of a tilted area, we must take the projected area to be the area. The area and the apex of the solid angle are always on one "side":

$$P = L(\text{area})_\perp(\text{solid angle}) \tag{11.22}$$

For equations 11.5, 11.7, and 11.8, the product of the area and the solid angle applies to the emitter and receiver in a symmetric way. Table 11.3 presents a summary of the cases discussed so far.

Table 11.3
Summary of discussed cases

Case	Equation	Equation number
da and da' small and perpendicular to the connecting line	$P = L\dfrac{da\,da'}{R^2}$	(11.5)
	$P = L\,da\,\pi\sin^2\alpha$	(11.7)
	$P' = L\,da'\,\pi\sin^2\alpha'$	(11.8)
da and da' small and tilted with respect to each other (Lambert's law)	$P = L\,da\cos\theta\left(\dfrac{da'}{R^2}\right)$	(11.14)
da small, receiver area large (Lambert's law used)	$P = L\,da\,\pi\sin^2\alpha$	(11.18)
Emitter area large, da' small (Lambert's law used)	$P = L\,da'\,\pi\sin^2\beta'$	(11.21)

5. RADIANCE, IRRADIANCE, RADIANT EXITANCE, AND IMAGING

A. Radiance and an Optical System

Consider a lens forming an image of a small area da of an object with a small area da'. The solid angles involved are assumed not to be small because the lens is not small (Figure 11.7a). The power arriving at the lens is (equation 11.18)

$$P = da\, L\pi \sin^2 \alpha \tag{11.23}$$

where L is the radiance at da. The power arriving at da' is (equation 11.19)

$$P' = da'\, L'\pi \sin^2 \beta \tag{11.24}$$

where L' is the radiance at the lens.

If we assume ideal conditions for the lens—that is, no reflection, absorption, or diffraction—then we have

$$da\, L\pi \sin^2 \alpha = da'\, L'\pi \sin^2 \beta \tag{11.25}$$

Abbe's sine condition, expressed in Chapter 9 as

$$Y_1 n_1 \sin \alpha_1 = Y_2 n_2 \sin \alpha_2$$

is equivalent to

$$\sqrt{da}\, \sin \alpha = \sqrt{da'}\, \sin \beta$$

Squaring this latter expression gives us

$$da \sin^2 \alpha = da' \sin^2 \beta \tag{11.26}$$

This tells us that the area da times the solid angle $\Omega(\alpha)$ is equal to the product da' times $\Omega(\beta)$. From equations 11.24 and 11.25,

$$L = L' \tag{11.27}$$

The radiance at da (power per area and per solid angle) is the same as the radiance at the lens. This is demonstrated in Figure 11.7b. The eye looks at a lamp covered with a plate of frosted glass. The solid angle $\Omega(\alpha)$ is small, and the eye is far away. Then a stop is introduced and reduces the area a to the much smaller area da. A lens of area a is introduced at distance f equal to its focal length. The solid angle at da is now $\Omega(\alpha')$. The faraway eye "sees" the same radiance for the lens as it saw for the frosted glass of area a.

Another demonstration is provided in Figure 11.7c. A small emitter is at the focal point of a parabolic mirror (e.g., the headlight mirror of a car). The area of the mirror viewed at a great distance seems to be as "bright" as the area of the emitter viewed at a short distance. The product of the area times solid angle is approximately the same in each case, and L does not change.

In infrared spectrometers, there is less available light than in the visible region, and we must be very economical with the light flux. As much light from the source as possible must be guided to the detector. Usually, the detector area and the largest possible solid angle at the detector determine the maximum

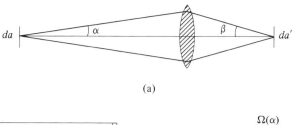

(a)

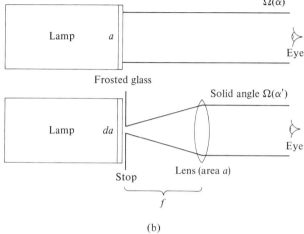

Lamp a

$\Omega(\alpha)$

Eye

Frosted glass

Lamp da

Solid angle $\Omega(\alpha')$

Eye

Stop

Lens (area a)

f

(b)

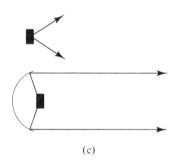

(c)

Fig. 11.7
(a) Power transfer by a lens. (b) The radiance of the frosted glass is the same as the radiance of the lens. (c) A small emitter seems to be as "bright" as the parabolic mirror if the emitter is placed at the focal point of the mirror. (d) In an infrared instrument, the product of area times solid angle at the different components must be equal to the product of area times solid angle at the detector.

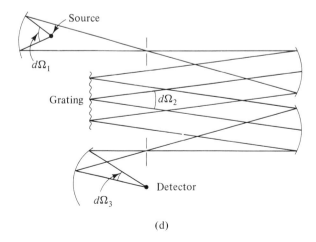

Source

$d\Omega_1$

Grating

$d\Omega_2$

Detector

$d\Omega_3$

(d)

5. **Radiance, Irradiance, Radiant Exitance, and Imaging**   **389**

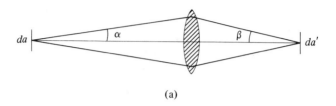

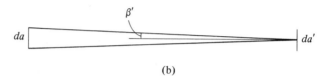

Fig. 11.8
(a) Emitted light from *da* to *da'* using a lens. (b) Emitted light from *da* to *da'* without using a lens.

value for the product of the area times solid angle to be applied to the other components. Figure 11.7d demonstrates how the product of area times solid angle at each component most be the same. The product of the source area times $d\Omega_1$, must be equal to the grating area multiplied by $d\Omega_2$ and equal to the detector area multiplied by $d\Omega_3$.

For all the systems we have discussed, the product of area and solid angle is constant. Multiplying this product with L gives us the power. We have a conservation of power at the different components of the system.

B. Irradiance Depending on the Solid Angle of an Optical System

In Chapter 1, we discussed angular magnification by a lens. We compared the magnification the eye detects with and without a lens by using the angles to the object with and without a lens. Now we will calculate the irradiance on a target with and without a lens (Figure 11.8). Since the radiant flux per area and solid angle (radiant exitance) of the source and lens is the same ($L = L'$), we have with the lens

$$P_1 = \pi L \, da' \sin^2 \beta \tag{11.28}$$

and without the lens

$$P_2 = \pi L \, da' \sin^2 \beta' \tag{11.29}$$

The irradiance at *da'* is $E_1 = P_1/da'$, and we have with the lens

$$E_1 = \pi L \sin^2 \beta \tag{11.30}$$

and without the lens

$$E_2 = \pi L \sin^2 \beta' \tag{11.31}$$

or

11. Radiometry and Photometry

$$E_1 = E_2 \frac{\sin^2 \beta}{\sin^2 \beta'} \qquad (11.32)$$

The irradiance with the lens is larger by the ratio of the solid angles. This is demonstrated when a magnifying glass is held up to the sun and a hole is burned in a sheet of paper beneath it (see Problem 7).

C. The Searchlight*

A searchlight has a light-emitting area at the focal point of a lens or parabolic mirror. The image of the object is at infinity (Figure 11.9a). The geometrical optical construction method, which has its limitations for this case, has been used in Figure 11.9a. The object is composed of small areas da, each emitting light into a solid angle $\Omega(\alpha)$. The two extreme cones from the object are indicated in Figure 9.11b. Considering the irradiance (power per area) at a plane at the right side of the lens, we see that we have several zones illuminated differently. Zone I is common to all cones from the object; zones II and III are not; and zone IV is not illuminated. To find the irradiance at a plane to the right of the lens, we consider three planes, as indicated in Figure 11.9c. To evaluate the irradiance, we take a second lens L_2 with focal length f' and bring it into planes 1, 2, and 3 at the axis (Figure 11.9d). In plane 1, the product of area times solid angle for the light leaving the object, arriving at L_1, and then arriving at L_2, is given by

$$a\Omega(\alpha) = a'\Omega(\alpha') = a''\Omega(\beta) \qquad (11.33)$$

We use R_1 to represent the distance between L_2 and L_1; with $\alpha' = \alpha$, the power arriving at L_2 is expressed by

$$P_1 = L\pi a' \sin^2 \alpha' = L\pi a' \frac{r_o^2}{r_o^2 + R_1^2} = \frac{\left(\dfrac{a''}{\pi}\right)}{\left(\dfrac{a''}{\pi}\right) + R_1^2} \qquad (11.34)$$

where r_o is the radius of the object. We have derived the power at one point on the axis in zone. I. It can be shown that this is the power at any point in zone I. We now move the lens to plane 2, and the power there is expressed by

$$P_c = L\pi a' \sin^2 \alpha' = L\pi a' \frac{r_L^2}{r_L^2 + R_c^2} \qquad (11.35)$$

(see Figure 11.9d) where R_L is the radius of L_2 and R_c is the distance from the lens to zone II. At this point, the area of L_2 fills the solid angle $\Omega(\alpha')$.

Moving the lens to plane 3, we find that the power is

$$P_{II} = L\pi a' \sin^2 \alpha'' = L\pi a' \frac{r_L^2}{r_L^2 + R_{II}^2} \qquad (11.36)$$

At this point, the solid angle $\Omega(\alpha'')$ is smaller than $\Omega(\alpha')$, and the area of lens L_2 does not fill the solid angle $\Omega(\alpha')$ and therefore the solid angle is smaller. The power P_{II} is smaller than $P_c = P_I$. Although we have derived P_{II} for a point on

*After R. W. Boyd, *Radiometry and Detection of Radiation* (New York: Wiley, 1983), p. 86.

5. Radiance, Irradiance, Radiant Exitance, and Imaging   391

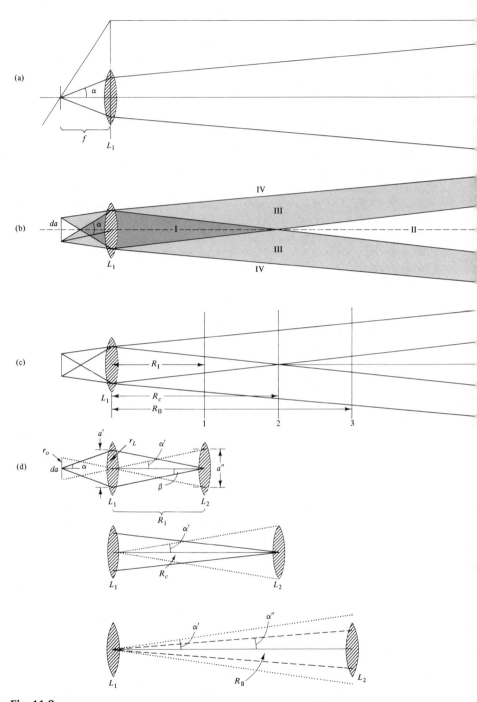

Fig. 11.9
The Searchlight. (a) The central cone from the center of the emitting area through the lens (geometrical image construction). (b) Extreme cones from the extreme points of the emitting area and formation of zones I, II, III, and IV. (c) Planes 1, 2, and 3 at different distances R_I, R_c, and from lens L_1. (d) The solid angles $\Omega(\alpha)$, $\Omega(\alpha')$, $\Omega(\beta)$, and $\Omega(\alpha'')$ indicated by the angles α, α', β, ar a' is the area of L_1; a'' is the area of L_2.

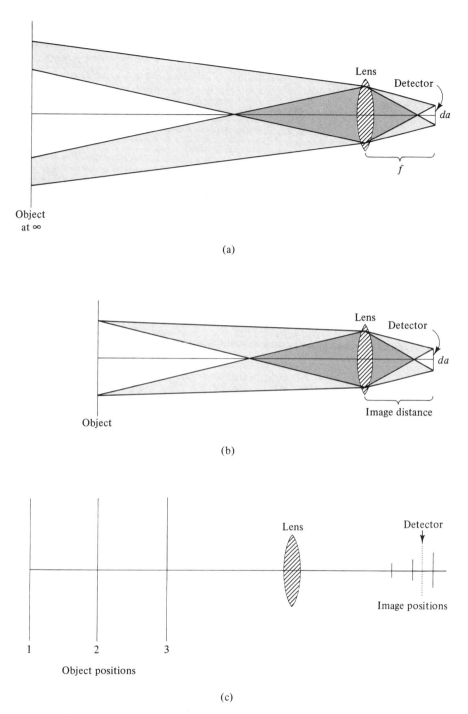

Fig. 11.10
The Radiometer. (a) Same situation as depicted in Figure 11.9b. We are now interested in the zone structure on both sides of the lens. (b) The two extreme light cones from object to image for the case of a finite object distance. (c) Different object planes 1, 2, and 3 and the corresponding image planes.

5. Radiance, Irradiance, Radiant Exitance, and Imaging

393

the axis, it holds for all points in zone II. The irradiance for each plane has a certain value at the axis. Moving from the axis to the outside, the irradiance is constant in zones I and II, decreases monotonically while moving through zone III, and is zero in zone IV. As long as we are in zone I, the solid angle $\Omega(\alpha')$ determines the irradiance; but in zone II, the solid angle $\Omega(\alpha'')$ is the limitation on the irradiance.

D. The Radiometer

If a radiometer is used to measure the emitted radiation of a large object at infinity, we have the case inverse to that discussed for the searchlight (Figure 11.10a). The object is at infinity, and the radiation detector of area da is at the focal point of the lens. Again we obtain a zone structure for the light on both sides of the lens.

If the object is not at infinity, then the detector should be placed at the image distance (see Figure 11.10b).

We have a more difficult situation if the light emitted from a large source area must be measured for various distances between the source and the lens.

If we use a detector that is larger than the image of the object at all distances, the detector may be placed at a range of distances from the lens (see Figure 11.10c).

In most radiometers, a small detector is used at a fixed distance to the first lens. Then we must analyze the light flux from the point of view of entrance and exit pupils and the product of area times solid angle. (We will discuss entrance and exit pupils in Chapter 12.)

E. Application to Astronomical Telescopes

When we use an astronomical telescope, we need to distinguish between an extended image formed by the sun, moon, or planets and a diffraction pattern produced by faraway stars. For the moon, we have the usual situation of imaging. A larger solid angle, obtained with a larger-diameter telescope, will bring more power to the photographic plate. The power $P = L\,da\,\sin^2\alpha$ depends on the solid angle we can use (Figure 11.11a). The dependence on the diameter B and focal length f is given by

$$\sin\alpha = \frac{B/2}{f} = \frac{B}{2f} \tag{11.37}$$

The power depends on the ratio $B^2/4f^2$, not on B^2 alone.

For stars, we do not observe an image, but only a diffraction pattern. The ratio of the diameter B of a large telescope to the distance R of the star from earth gets into the order of magnitude of the wavelength of the light being used. If we take for a model the diffraction of a slit, we find that the angular difference between the first two orders is $2\lambda/B$. With larger B, the diffraction ring pattern becomes narrower; that is, more power per unit area arrives at the photographic plate. In addition, the situation "area times solid angle" holds, and a large opening B of the telescope is as well desirable and must be accompanied by an appropriate focal length.

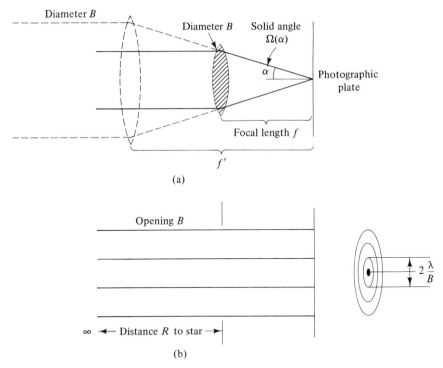

Fig. 11.11
(a) When viewing the moon, the power arriving at the photographic plate depends on the solid angle $\Omega(\alpha)$ (given by B, f or B', f'). (b) Diffraction pattern produced by a faraway star.

6. PHOTOMETRY

A. The Eye as Detector and the Lumen as Power Unit

So far, we have assumed that the dependence of the power on different wavelength distributions may be ignored. A certain value of the irradiance, (i.e., power per unit area) may be produced by more visible light and less infrared light, and vice versa. In radiometry, only the final value of the power is considered. In Chapter 14, we will discuss the wavelength-dependent sensitivity of detectors. If the detector is insensitive to a certain wavelength range, the measured power cannot be assumed to be the incident power.

Photometry is concerned with the spectral region for which the eye is sensitive (Figure 11.12). The eye is the important detector. Light waves having wavelengths too large or too small to be detected by the eye do not play a role in photometry. The sensitivity of the eye to different wavelengths also must be taken into account. This makes the measurement of radiation from the point of view of photometry more specific. Therefore, different names and units are used for photometric quantities, as shown in Table 11.4 with their radiometric counterparts. Note that the symbols of photometry carry the subscript v to indicate that only the visible spectrum is being considered. The units used are the **lumen** (lm),

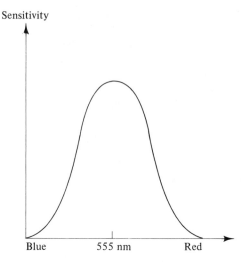

Fig. 11.12
Schematic presentation of the sensitivity of the eye depending on the wavelength. The central wavelength is about 555 nm.

the second (s), the meter (m), and the steradian (sr). The unit of power is defined as the **luminous flux** of monochromatic radiation of wavelength 555 nm, whose flux is (1/683) W. The **illuminance** corresponds to the irradiance. The solar constant of the sun has an irradiance of 1.3×10^3 W/m². The photometric illuminance of the sun is 1.2×10^5 lm/m². If we divide this by 683, we obtain 0.17×10^3 W/m². This number is smaller than 1.3×10^3 W/m², indicating that the wavelength range for the photometric consideration is smaller than the range in radiometry.

Table 11.4

Radiometry	Symbol	Unit	Photometry	Symbol	Unit
Radiant energy		J	Luminous energy		lm-s, or talbot
Radiant energy density	u	$\dfrac{J}{m^3}$	Luminous density	u_v	$\dfrac{lm\text{-}s}{m^3}$
Radiant flux (power)	P	W	Luminous flux	P_v	lm
Radiant exitance	M	$\dfrac{W}{m^2}$	Luminous exitance	M_v	$\dfrac{lm}{m^2}$, or lux
Irradiance	E	$\dfrac{W}{m^2}$	Illuminance	E_v	$\dfrac{lm}{m^2}$, or lux
Radiance	L	$\dfrac{W}{m^2\,sr}$	Luminance	L_v	$\dfrac{lm}{m^2\,sr}$, or nit
Radiant intensity	I	$\dfrac{W}{sr}$	Luminance intensity	I_v	$\dfrac{lm}{sr}$, or candela

1 lm = 0.00146 W

11. Radiometry and Photometry

Table 11.5
Luminance of sources

	$\dfrac{\text{lm}}{\text{m}^2\,\text{sr}}$	$\dfrac{\text{cd}}{\text{cm}^2}$
Surface of sun	2×10^9	2×10^5
Carbon arc lamp	1×10^8	1×10^4
750 W projection lamp filament	2×10^7	2×10^3
Tungsten at 2800 K	1×10^7	1000
Electric lamp filament	5×10^6	500
Laboratory mercury lamp	1×10^5	10
60 W frosted incandescent lamp	9×10^4	9
40 W fluorescent lamp	0.5×10^4	0.5
Overcast sky	0.2×10^4	0.2
Self-luminous paint	3×10^{-2}	3×10^{-6}

The eye has its peak sensitivity at about the same wavelength that the sun (as a blackbody radiator) has its peak energy density emission. The sensitivity curve of the eye falls off at higher and lower wavelengths. Luminance values of various sources are given in Table 11.5.

B. Application to Point and Area Emitters and Receivers

We will now consider illumination by several sources that have different geometrical shapes. In all these applications, we follow the definitions and apply the results developed in the preceding sections.

Point source, area receiver. A pointlike source of area da and luminance (radiance) L_v may be used to illuminate an area of radius r (Figure 11.13). The luminous flux (radiant flux) P_v at the receiver is (see also Section 4A)

$$P_v = L_v \pi \, da \sin^2 \beta$$

or   (11.38)

$$P_v = L_v \pi \, da \frac{r^2}{r^2 + R^2}$$

Where R is the distance from the source. The area of the receiver is πr^2. For r small compared to R, we have the familiar $1/R^2$ law. Division of P_v by the area

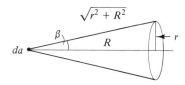

Fig. 11.13
Illumination of the area πr^2 by the area da having the luminance L_v.

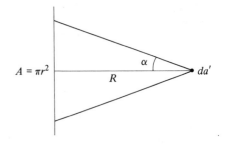

Fig. 11.14
Illumination of the area da' by the area $A = \pi r^2$ having the luminance L_v.

$A = \pi r^2$

R

α

da'

of the receiver results in the illuminance (irradiance) E_v:

$$E_v = \frac{P_v}{\pi r^2} = \frac{L_v \, da}{(r^2 + R^2)} \tag{11.39}$$

E_v also depends on $1/R^2$ when r is small compared to R. If we enlarge the area in such a way that the solid angle $\Omega(\beta)$ is a constant, then the constant luminous flux (radiant flux) P_v (see equation 11.38) falls onto the larger and larger area with larger and larger distance. The same is not true for the illuminance (irradiance) E_v (equation 11.36) because we divided by the area πr^2.

Since a point source has no area, we combine da with L_v for the **luminance intensity** I_v in units of lm/sr, or candelas, and the luminous flux is given as

$$P_v = I_v \pi \sin^2 \beta$$

or $\tag{11.40}$

$$P_v = \frac{I_v \pi r^2}{r^2 + R^2} \simeq \frac{I_v \pi r^2}{R^2}$$

For an example, see problem 8.

Point receiver, area emitter. Now let us consider the inverse case. A large area $A = \pi r^2$ illuminates the pointlike area da' in Figure 11.14 (see also Section 4B). The luminous flux (radiant flux) P_v is given as

$$P_v = L_v \pi \sin^2 \alpha \, da' \tag{11.41}$$

As long as we have a constant angle α, we have constant luminous flux (radiant flux) P_v incident on the area da' for different distances R. This also holds for the illuminance (irradiance) E_v at the receiver. We have

$$E_v = \frac{P_v}{da'} = L_v \pi \sin^2 \alpha \tag{11.42}$$

(da' is constant). We can keep the angle α constant for all distances R if the area A fills the solid angle $\Omega(\alpha)$ for all distances R. The angle is usually established at the receiver.

C. Line Sources

Let us consider a line source, such as a fluorescent lamp. To obtain the luminous flux (radiant flux) at a point receiver, we must sum up all the contributions from the pointlike elements of the line source (Figure 11.15). The mathematical prob-

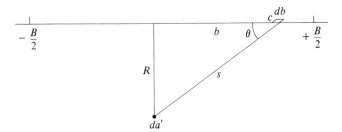

Fig. 11.15
Calculation of the irradiance at da' from a line source of length B and width c.

lem is similar to the derivation of the Biot-Savart law for magnetism. The power incident on the receiver of area da' and emitted from the element $da = c\,db$ of the line source is

$$dP_v = L_v\,da'\,\frac{c\,db\,\sin\theta}{s^2} \tag{11.43}$$

where c is the width of the line source, db is the small length element of the line source of total length B, and $\sin\theta$ gives the projection. For the total power at the receiver, we have

$$P_v = L_v\,da'\,c\int_{b=-B/2}^{b=B/2}\frac{db\,\sin\theta}{s^2} \tag{11.44}$$

From Figure 11.15,

$$b = R\cot\theta, \quad db = -\frac{R}{\sin^2\theta}d\theta, \quad \sin\theta = \frac{R}{s}. \tag{11.45}$$

So

$$P_v = -L_v\,da'\,c\frac{1}{R}\int_{\theta_1}^{\theta_2}\sin\theta\,d\theta \tag{11.46}$$

where θ_1 and θ_2 correspond to the values of b for $B/2$ and $-B/2$, respectively.

To solve the integral, we add the small contributions we would have if the line source were infinitely long; that is, we extend the integral to the limits $\theta_1 = 0°$ to $\theta_2 = 180°$ and have

$$P_v = L_v\,da'\left(c\frac{2}{R}\right) \tag{11.47}$$

If the area of the point receiver is not specified, we can give the illuminance (irradiance) at the receiver as

$$E_v = L_v\left(c\frac{2}{R}\right) \tag{11.48}$$

where the expression in parentheses, $(c2/R)$, gives the solid angle in steradians.

We have a $1/R$ law of attenuation of the emitted light (irradiance) toward a point receiver.

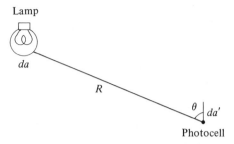

Fig. 11.16
A point source (lamp) illuminates a point receiver (photocell) under the angle θ.

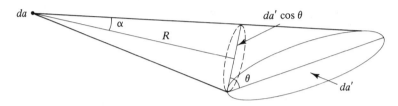

Fig. 11.17
Oblique illumination of an area by a point source.

D. Oblique Illumination: Point Source to Point Receiver

Consider a point source, such as a lamp, illuminating a point receiver such as a photocell. The light from the lamp is incident obliquely on the photocell. (Figure 11.16). The luminous flux (radiant flux) incident on the photocell is

$$P_v = L_v \, da \frac{da'}{R^2} \cos \theta \tag{11.49}$$

E. Oblique Illumination of an Area by a Point Source

If a point source illuminates an area, and the area is tilted by the angle θ with respect to the normal of the source area, the projected area must be used (Figure 11.17). For a small area, the luminous flux (radiant flux) at the receiver is expressed by

$$P_v = L_v \, da \frac{da' \cos \theta}{R^2} \tag{11.50}$$

For a large area a', we have

$$P_v = L_v \, da \, \pi \sin^2 \alpha \cos \theta \tag{11.51}$$

The illuminance (irradiance) is given as

$$E_v = L_v \pi \sin^2 \alpha \cos \theta \tag{11.52}$$

7. THE f NUMBER OF A CAMERA

In a camera with a lens of focal length f, the distance from the lens to the film is f for objects far away. For objects that are closer, the lens must be focused. For a compound lens system, some parts of the lens system can be moved relative to other parts.

The shutter regulates the amount of light that will pass the lens during a specific period of time. The aperture stop regulates the amount of light entering the lens over the extension of the opening.

Using the concept of area times solid angle, where the area a' is the area of the film and the solid angle Ω extends from the film to the lens, we can express the luminous flux (radiant flux) passing the lens as

$$P = L_v \, da \, a' \Omega \tag{11.53}$$

where da is the area of the object.

Since energy is power multiplied by time, the shutter can reduce the energy by having the opening of the lens effective only during a very short time.

For a faraway object, the solid angle is

$$\Omega = \frac{\pi (d/2)^2}{f^2} = \frac{\pi}{4} \left(\frac{d}{f} \right)^2 \tag{11.54}$$

where d is the diameter of the aperture. The ratio f/d is called the f **number** and is given for various values at the rim of the camera's lens. The values used are 2.8, 4, 5.6, 8, 11, 16, and 22. The numbers from 2.8 to 22 characterize larger to smaller openings. Since the f of a camera lens is indicated on the lens, the aperture opening $\pi (d/2)^2$ can be calculated if needed.

The shutter speed is given in multiples of 2. Table 11.6 compares some values of shutter speed and aperture opening, showing how much the values change from one number to the next. To get the same energy incident on the film, we can trade off shutter speed for a larger aperture opening. In Table 11.6, we find the change of shutter speed by a factor of 2 and the change in aperture opening by a factor of 2.

If there is enough light, we use short shutter speeds for fast-moving objects.

Table 11.6

Relative shutter speed changes by		Relative aperture area changes by		
		f/d	Square of f/d	
1/1000 ⎫		2	4	⎫
1/500 ⎬ Factor of 2		2.8	7.84	⎪
1/250 ⎭		4	16	⎪
		5.6	31.36	⎬ About factor of 2
1/100 ⎫		8	64	⎪
1/50 ⎬ Factor of 2		11	121	⎪
1/25 ⎭		16	256	⎭
		22	484	

Small aperture openings will result in a better image quality. We operate closer to the paraxial theory, reducing aberrations; that is, we operate closer to the pinhole camera concept, which shows no aberration at all.

Problems

1. *Radiance of a star.* Calculate the radiance L for a star that has the irradiance of the sun and is so far away that its angular diameter could be determined by the Michelson stellar interferometer to be $\alpha = 0.05$ s.
   Ans. $L = 7.5 \times 10^{15}$ W/m^2 sr

2. *Solid angle.* Show that the solid angle (in units of steradians) subtended by the right circular cone of half-angle α' is

$$\Omega = 4\pi \sin^2 \frac{\alpha'}{2}$$

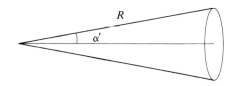

3. *Radian and steradian for small angles*

   a.
$$1 \text{ rad} = \frac{360°}{2\pi} = 57.3°$$

   Calculate the number of radians for 1°; for 1 min; for 1 s.
   b. How many degrees is 20 mrad if 1 mrad equals 10^{-3} rad?
   c. For small angles, we have $d\Omega = \pi \sin^2 \alpha$ in steradians. Since α is small, we have $\sin \alpha \simeq \alpha$. If α is in degrees for $\sin \alpha$ (left side), it must be in radians for the right side. Check this out with a calculator for 1°, 1 min, and 1 s. Radians and steradians are dimensionless quantities. However, the "dimension" of the solid angle is $[\text{rad}]^2 = [\text{sr}]$.

4. *Large emitter and small receiver area.* For the case of a large emitter area and a receiver with a small area da', do the inegration leading to

$$P = L \, da' \, \pi \sin^2 \beta'$$

5. *Power for Lambert and non-Lambert sources.* Compare the power P for Lambert and non-Lambert sources, assuming a small-area emitter and a large-area receiver, where the emitter area is da and the half-angle is α.

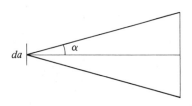

a. Show that the formulas are the same for small angles.
b. Give the ratio P_L/P_{NL} for $\alpha = 30°$.
c. Give the ratio P_L/P_{NL} for $\alpha = 60°$.
Ans. a. $P_{NL} = L \, da \, \pi 4(\alpha/2)^2$; b. $P_L/P_{NL} = 0.933$; c. $P_L/P_{NL} = 0.75$

6. *Irradiance of the sun with and without a magnifier.* For the sun, $E = 1.35 \times 10^3$ W/m². What is the irradiance if we use a magnifier of diameter 5 cm and focal length 5 cm?
Ans. $E_1 = 1.42 \times 10^7$ W/m²

7. *Sunlight and magnifier.* Suppose that sunlight is falling at normal incidence onto an area with a radius of 0.5 cm.
a. Calculate the incident power on the area.
b. A lens is placed before the area. Its radius is 0.5 cm and its focal length is 5 cm. Calculate the area da' of the image of the sun. Compare your answer with the one the paraxial theory would give for the area of the spot from chief rays from the sun.

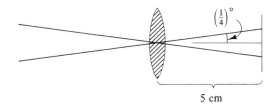

5 cm

8. *Point source illuminating a screen.* A point source is emitting $I_v = 10$ lm/sr or 10 candelas. How much light (in lumens) is arriving at a screen with radius $r = 1$ cm if the screen is a distance $R = 30$ cm away. Convert lumens to watts.
Ans. $P_v = 0.035$ lm $= 5.1 \times 10^{-5}$ W

9. *A 40 W light bulb.* A 40 W light bulb considered as a point source has approximately 40 lm/sr, or 40 candelas. How large are the luminous flux P_v and radiant flux P into a solid angle having an area 10 cm × 10 cm and a distance of 2 m?
Ans. $P_v = 0.1$ lm, $P = 1.46 \times 10^{-4}$ W

10. *A light bulb emitting different values of I_v in different directions.* We approximate a light bulb as a point source. The bulb emits 6 lm/sr in a direction perpendicular to the axis and 10 lm/sr in a direction parallel to the axis. How many lumens are incident on a screen with radius $r = 1$ cm, placed 30 cm away in both directions? Convert lumens to watts.

10 lm/sr

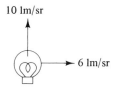

6 lm/sr

Ans. $P_v(10) = 0.035$ lm $= 5.1 \times 10^{-5}$ W, $P_v(6) = 0.021$ lm $= 3 \times 10^{-5}$ W

11. *Area illuminating a point receiver.* In most cases, the area of the receiver (detector) is known. If not, we can consider the illuminance $E_v = L_v \pi \sin^2 \alpha$. Consider a cathode-ray tube (CRT) with area 20 cm × 20 cm and luminance $L_v = 5 \times 10^4$ lm/m²sr placed

a distance of 1 m from a detector with area 1 mm². Calculate the luminous flux P_v and the illuminance E_v at the detector.
Ans. $P_v = 0.157 \times 10^{-2}$ lm, $E_v = 0.157 \times 10^4$ lm/m²

12. *Infinite line source.* Show that for an infinite line source, the luminous flux

$$P_v = L_v \, da' \left(c \int_{-\infty}^{+\infty} \frac{db \sin \theta}{s^2} \right)$$

results in

$$P_v = L_v \, da' \, c \frac{2}{R}$$

where R is the shortest distance from da' to the line source. The solid angle is $c(2/R)$ in steradians.

13. *Finite line source.* A line source is symmetrically placed with respect to the receiver area da' (with shortest distance R to the receiver) and has an extension $-B/2$ to $+B/2$ so that θ has the corresponding values θ_1 and θ_2. Calculate P_v.
Ans. $P_v = L_v \, da' (c/R) 2 \cos \theta_2$

14. *Fluorescent lamp and point receiver.* Consider a fluorescent lamp of luminance 0.5×10^4 lm/m² sr having a width of 3 cm and a length B of 2 m. Calculate the luminous flux P_v received by a detector with area 1 mm² symmetrically placed at a distance R of 1 m.
Ans. $P_v = 2 \times 10^{-4}$ lm

15. *Solar panel.* A solar panel with dimensions 1 m × 2 m is illuminated by the sun. Assume that the solar detectors are sensitive only in the wavelength region where the eye is sensitive.
   a. Calculate the luminous flux for the case when the sun is at its zenith position and for the later cases when the angle with the zenith position is 30° and 60°.
   b. Calculate the total power arriving at the panel when the sun is at its zenith position and when the sun is 30° and 60° off.

Image Formation and Light Throughput

1. INTRODUCTION

In Chapter 1, we studied image formation in the paraxial approximation. The image of an object was described by its position and size. For the construction of the image, only two rays were used (Figure 12.1a). The size of the lens did not enter the procedure.

In Chapter 11, we found that the total power of the light passing through an optical system is proportional to the product of the area the light passes through and the solid angle into which it travels (see Figure 12.1b).

We will now discuss image formation in optical instruments using both aspects and study the effect of stops and field lenses on the image formation process.

2. PUPILS

A **pupil** is usually a round aperture that is smaller than the image-forming lens. It restricts the amount of light passing from the object through the lens to the image. Its area is the smallest area common to all light cones passing through the optical system on object and image side, respectively.

We next treat cases where the entrance pupil is located before the lens to the left; where the lens itself is the entrance pupil; and where the entrance pupil is to the right of the lens.

A. Entrance Pupil Left, Exit Pupil Right of Lens

In Figure 12.1b, all the light passes through the lens within the solid angle $\Omega(u)$, where u indicates the half-angle of the cross section. (For units of the solid angle, see Chapter 11.) If we place a stop between the lens and object, as shown in Figure 12.1c, then less light passes through the system. The intensity passing through the system is now proportional to the product of the area of the hole in the stop (diaphragm) and the solid angle $\Omega(\omega')$. The lens produces an image of the

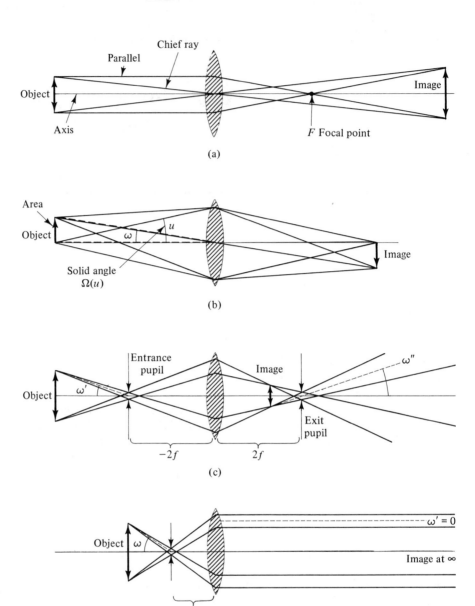

Fig. 12.1
(a) Image construction using a parallel ray from each edge of the image and the chief ray. (b) Two light cones from two different points of the object passing the total area of the lens and converging to two image points. The chief ray of one cone is the axis. The chief ray of the other cone has the angle ω with the axis. The total light passing the opening of the lens is proportional to the area of the lens times the solid angle $\Omega(\omega)$. (c) A stop between the object and the lens makes the cones of light smaller leaving the points on the object and passing through the lens. The image of the entrance pupil by the lens is called the exit pupil. (d) The pupil is at the focal point on the left. The light cones are parallel on the right. The angle ω' is zero. (In actual applications, however, the angle is not zero and the light either slightly diverges or converges.)

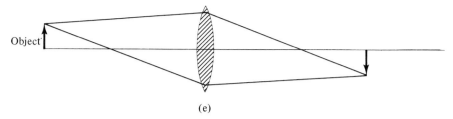

(e)

Fig. 12.1 (continued)
(e) The lens as both entrance and exit pupil.

diaphragm. For simplicity, we have placed the stop at a distance $2f$ from the lens; therefore, the image is of equal size at a distance $2f$ to the right of the lens. Many small light cones diverge from the object, filling the diaphragm and traveling through part of the lens, converging toward the image. From there they travel on and diverge. They fill the image of the diaphragm. The diaphragm is called the **entrance pupil**, and its image by the lens is called the **exit pupil**. The product of the area and the solid angle, formed at the entrance and exit pupils, is the same:

$$a_{\mathrm{ent}}\Omega(\omega') = a_{\mathrm{exit}}\Omega(\omega'') \tag{12.1}$$

In our simple case,

$$a_{\mathrm{ent}} = a_{\mathrm{exit}} \quad \text{and} \quad \omega' = \omega'' \tag{12.2}$$

where the half-angles ω' and ω'' are determined by the chief rays of the small cones passing through the diaphragm (see Figure 12.1c). All such chief rays pass through the center of entrance and exit pupils.

An application is shown in Figure 12.2a. A small mirror is introduced into an optical system (similar to Figure 12.1a–c) to measure part of the light emitted from the object. This mirror acts as an entrance pupil to the system employing the lens L_2 and the detector having area a_D. The power of the measured light is proportional to the area of the mirror (45° projection) times the solid angle opened toward lens L_2.

B. Entrance Pupil Before the Lens at the Focal Point

In Figure 12.1d, we show a system where the entrance pupil is placed at the focal point of the lens. An application is shown in Figure 12.2b. The entrance pupil is a small mirror at the focal point of lens L_2. The exit pupil is at infinity, and the chief rays of the small cones leaving lens L_2 are almost parallel. Such a system is an example of a **telecentric system**.

Assume that the mirror is rotating about an axis perpendicular to the page. The reflected light from a bar-code pattern placed at the image plane of the object plane (source area) can pass back through the optical system. A semi-transparent plate (beam splitter) reflects part of the light to a detector. At the detector, the bar-code pattern can be observed as a periodic "light" and "no light" pattern.

C. Entrance Pupil Within the Focal Length of the Lens on the Left

This is the reverse of the case discussed in Section 2E and will be discussed there.

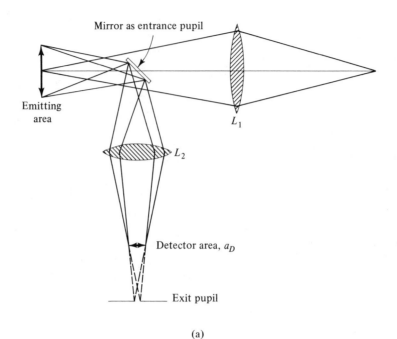

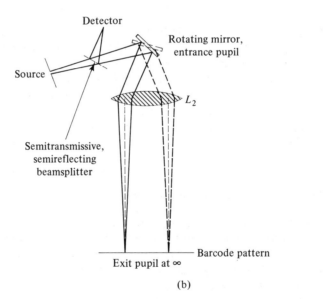

(b)

Fig. 12.2
(a) A small mirror M is the entrance pupil to the imaging process of lens L_2.
(b) The entrance pupil is at the focal point of the lens; the exit pupil is at infinity.
This is called a telecentric system. Such a system is used in bar-code reading.

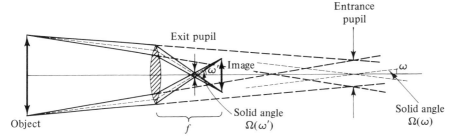

Fig. 12.3
The exit pupil is within the focal length of the lens. The entrance pupil is a virtual image to the right.

D. Lens as Entrance and Exit Pupil

If we move the entrance pupil, the image of it (which is the exit pupil) also moves. An extreme case is when entrance and exit pupils are together at the rim of the image-forming lens (see Figure 12.1e).

E. Pupil on the Right Within the Focal Length of the Lens

In Figure 12.3, a diaphragm is placed on the right side of the lens between lens and image. It is the exit pupil of the system; the entrance pupil is the image of the exit pupil. Since the entrance pupil is within the focal length of the lens, the exit pupil is farther to the right. The area of the entrance pupil is not equal to the area of the exit pupil, nor is the solid angle $\Omega(\omega')$ equal to the solid angle $\Omega(\omega)$. But the respective products of area and solid angle are equal.

F. Pupil on the Right of the Lens at Distances of the Focal Length or Farther Away

These are the reverse cases of those discussed in Sections 2A and 2B.

G. Source as Entrance Pupil

This case actually belongs before the case described in Section 2A, but it is discussed here because it leads into our discussion of condenser lenses, covered in Section 3. Figure 12.4a, shows the illumination of a transparent object by a source and the projection of the image of the object on a screen. The entrance pupil is the rim of the source opening. The exit pupil is the image of the source positioned between lens and image. An application of such a system is the projector, where we want to illuminate a slide and project it onto a screen. Note the problem with the arrangement shown in Figure 12.4a. Only a certain area of the slide is illuminated by the source. We could place the slide close to a source with similar dimensions, but this is not a practical solution. How we resolve this difficulty by using a condenser lens (Figure 12.4b) will be discussed in the next section.

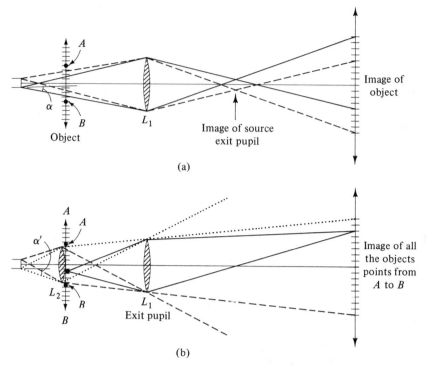

Fig. 12.4

(a) Wrongly assembled projector. There is no light scattered from certain parts of the slide. (b) Rightly assembled projector. A condenser lens illuminates the slide from A to B and bends the light toward the screen. The light scattered from all parts of the slide is imaged onto the screen. (After Pohl, *Einführung in die Optik*.)

3. CONDENSER LENSES

The difficulty in Figure 12.4a can be eliminated by using a **condenser lens** before the object (see Figure 12.4b). The condenser lens L_2 produces an image of the source as exit pupil at lens L_1. For the imaging process of L_1, the exit pupil of L_2 serves as entrance and exit pupil of L_1 (see Figure 12.1e). The angle α' in Figure 12.4b is larger than α in Figure 12.4a. Consequently, the light flux through the object is ensured to cover a larger area. The chief rays of the small cones coming from the edges of the object now form a larger angle at the image-forming lens L_1, and the image appears larger and evenly illuminated at the screen.

4. FIELD LENSES

A. Lens at the Position of the Image

Field lenses are used to improve the flow of power through an optical system. Their function is similar to that of the condenser lens. They are positioned at the location of the images. In Figure 12.5a, we see a lens L_1 forming an image of

12. **Image Formation and Light Throughput**

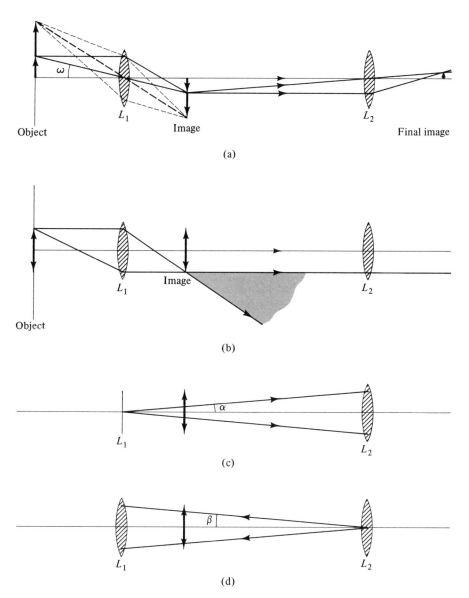

Fig. 12.5
(a) Image construction with two rays, using the paraxial method. The chief ray of a small cone of light coming from the edge of the object determines the image size. (b) A cone of light from an object point. The light never arrives at L_2. (c) The power from L_1 to L_2 is proportional to the area of L_1 times the solid angle $\Omega(\alpha)$. (d) The power from L_2 to L_1 is proportional to the area of L_2 times the solid angle $\Omega(\beta)$.

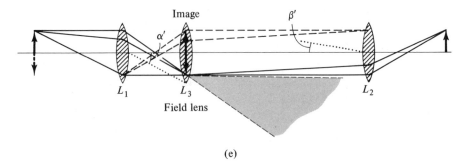

(e)

Fig. 12.5 (continued)
(e) A field lens at the position of the image bends the light so that the area of L_1 times $\Omega(\alpha')$ is equal to the area of L_2 times $\Omega(\beta')$. The transfer of light from L_1 to L_2 is indicated by one small cone of light.

a large object. The image is viewed with the eye presented by L_2. We use the simple construction method, developed in Chapter 1 for the paraxial case, to find the position of the final image. The chief ray of a cone of light from the edge determines the size of the image. The angle the chief ray makes with the axis is labcled ω. In Figure 12.5b, we see the cone of light from the edge will never get to lens L_2.

The power of the light passing through L_1 and L_2 is proportional to the area of lens L_1 times the solid angle $\Omega(\alpha)$ (see Figure 12.5c). Reversing the light path, we can take the area of lens L_2 times the solid angle $\Omega(\beta)$ (see Figure 12.5d).

How can we get all the light of the image formed by L_1 to pass through lens L_2? We need a light-bending process such that all the light leaving L_1 passes L_2 without affecting the image-forming process. This can be accomplished by introducing a lens L_3 at the position of the image of L_1 having a focal length such that the plane of L_1 is imaged into the plane of L_2 (see Figure 12.5e). Lens L_3, which is called a field lens, does not affect image formation, but transfers the light leaving lens L_1 with the solid angle $\Omega(\alpha')$ into the light passing the area of L_2 with solid angle $\Omega(\beta')$. Such a power transfer was detailed in Chapter 11.

B. Pipes: Field of View

The **field of view** is the cross section angle 2α, obtained when viewing a faraway objects. If we look through a window at a landscape, the field of view depends on the size of the window (Figure 12.6a). If we look through a pipe, the field of view is very limited, depending on the length of the pipe (Figure 12.6b). The principles involved in looking through a pipe were once employed in the **cystoscope**, an optical instrument used to examine certain organs of the human body, including the urinary bladder. (Today, fiber optics have replaced the cystoscope.) And probably the best-known example of a pipe as a visual instrument is a submarine **periscope**.

The limited field of view can be enlarged with the introduction of field lenses. In Figure 12.7a, we show the field of view as it is transferred through a pipe with image-forming lenses. The first lens produces an image of the outside object, and the next lens transfers the image farther to the right. The area of each lens is the

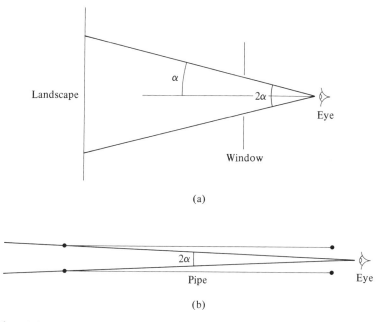

(a)

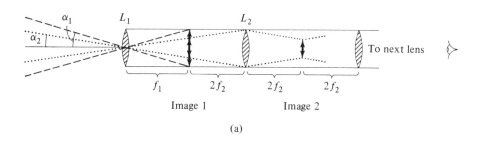

(b)

Fig. 12.6
(a) The field of view is characterized by the angle 2α. (b) Field of view of a pipe.

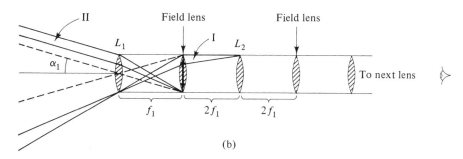

Fig. 12.7
Field of view through a pipe with and without field lenses. (a) only image transfer lenses are used. (b) Use of field lenses. The angle α' is larger than α. The light cone I shows the transfer of light from L_1 to L_2; cone II shows the limit of the field of view.

same. The solid angle $\Omega(\alpha_2)$ at lens L_2 is determined by the opening of the tube. Because of the (arbitrarily chosen) equal focal length of L_2 and the following lenses, the solid angle is the same at all lenses to the right.

We now introduce field lenses at the image positions 1 and 2 (see Figure 12.7b). The field lens transfers the light used for image formation from the plane of L_1 to the plane of L_2. This is indicated by the small cone I. The small cone II indicates the limit of the field of view which is now given by $2\alpha_1$. The field of view through the pipe is enlarged by using field lenses.

C. The Microscope: Numerical aperture

Figure 12.8a shows the image formation in a microscope, as discussed in Chapter 1. Lens L_1 is the objective lens, lens L_2 is the ocular, and lens L_3 is the lens of the eye. An actual microscope contains many more lenses (see Figure 12.8b, c). The objective lens consists of a system of lenses to reduce **aberration** (Chapter 13). It forms a magnified image of the object and utilizes a large solid angle on the object side. The ocular consist of two lenses: One is the actual image-forming lens, and the other is a field lens at image position 1 (of the objective lens) (see Figure 12.8b).

The object is illuminated by a lamp. A condenser lens fills the total solid angle of the objective lens. It forms an image (C_2) of the image (C_1) of the filament (C) of the lamp at the objective lens (see Figure 12.8d). The lamp also has a condenser lens forming the image (C_1) of the actual filament of the lamp. The large angle u that the objective lens accepts from the object is important for the characterization of the resolution of the microscope (Chapter 9). It is related to the numerical aperture, $n \sin \alpha$. In Chapter 9, we found that the smallest distance that can be resolved is

$$Y = \frac{\lambda}{2n \sin \alpha} \tag{12.3}$$

If the angle α is not filled with light, it has the same effect as putting an angle-reducing aperture between object and objective lens. Such a smaller aperture increases the diffraction effect and makes Y larger. To fill such a large angle with light, a condenser lens is employed.

The numerical aperture contains the refractive index n. Using an immersion liquid in which the object and the objective lens are immersed, we can make $n \sin \alpha$ even larger and Y even smaller. Immersion liquids have a refractive index of about $n = 1.4$. The angle α can be about $70°$.

D. Telescopic Systems

Image formation in the telescope was discussed in Chapter 1; it is depicted again in Figure 12.9a. The angular magnification was found to be $M = -f_1/f_2$. We now want to look at this system from the point of view of power flow. The light enters the entrance pupil, that is, the rim of lens L_1, almost as parallel light. The angle of view is determined by the size of the image of the object in the focal plane (see Figure 12.9b). The exit pupil is the image of the entrance pupil (diameter of L_1) formed by L_2. The solid angle of the light entering the telescope

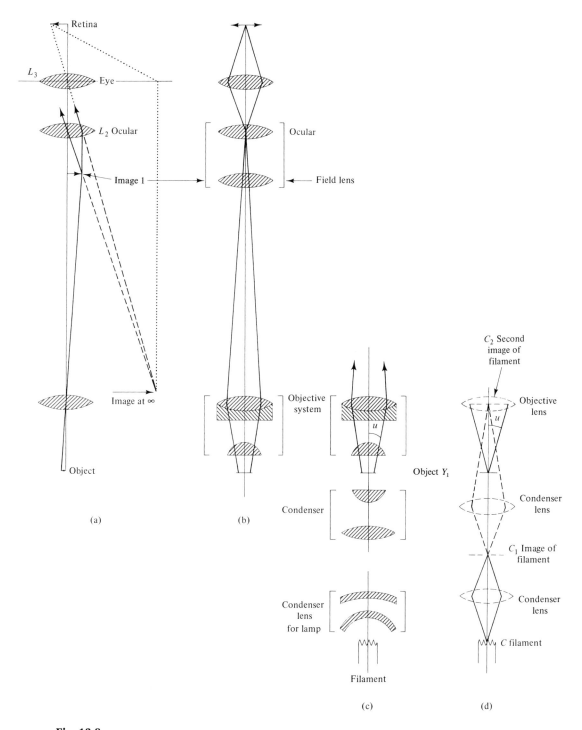

Fig. 12.8
The microscope. (a) Image formation, paraxial theory of Chapter 1. (b) A field lens is used in the ocular. (c, d) The condenser of the microscope and the condenser of the lamp conduct light under large angles through the object.

4. Field Lenses   **415**

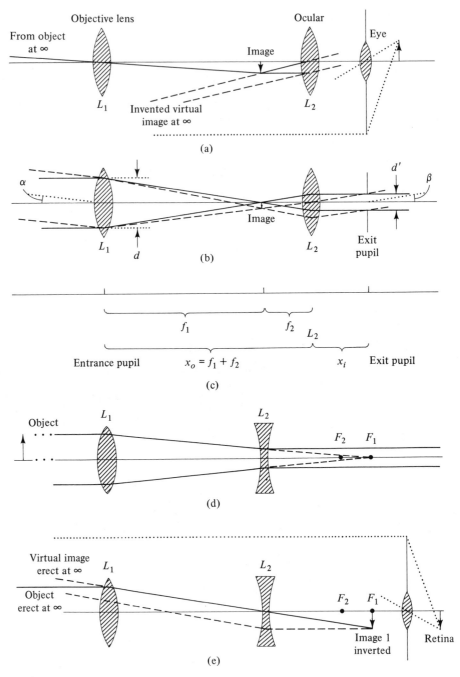

Fig. 12.9
Telescope and beam expander. (a) Image formation in a telescope. The object is erect on the retina, but one "sees" it inverted. (b) Solid angles at entrance and exit pupils of the telescope. (c) Image and object distance using the thin-lens equation. (d) Beam contraction in the Galilean telescope. (e) Image formation in the Galilean telescope. As in the magnifier, the virtual image is erect. On the retina it is inverted, but one "sees" it erect.

416   12. **Image Formation and Light Throughput**

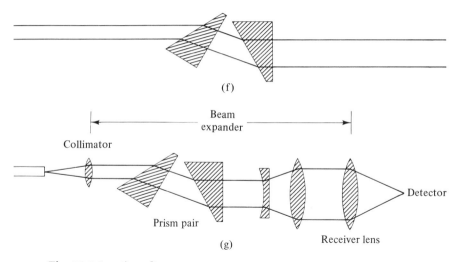

Fig. 12.9 (continued)
(f) Beam contraction using two prisms. (g) Two prisms as beam expander.

is $\Omega(\alpha)$; the solid angle of the light leaving the telescope is $\Omega(\beta)$. The power flux into the telescope is proportional to the area of lens L_1 times the solid angle $\Omega(\alpha)$, that is, $a(L_1) \cdot \Omega(\alpha)$. Assuming no losses, this product is equal to the area of the exit pupil a (exit pupil) times $\Omega(\beta)$ (see Figure 12.9b, where two light cones are shown). We have

$$a(L_1)\Omega(\alpha) = a(\text{exit pupil})\Omega(\beta) \qquad (12.4)$$

We call the diameter of the incident parallel light beam d, and we call the diameter of the leaving light beam d'. Taking the cross-sectional angles and diameters of the product (area times solid angle) for incident and leaving beams, we have

$$\frac{\beta}{\alpha} = \frac{d}{d'} \qquad (12.5)$$

Since d corresponds to object size and d' is the image of d, we have (from Chapter 1) a magnification of

$$m = \frac{x_i}{x_0} = \frac{y_i}{y_0} = \frac{d'}{d} \qquad (12.6)$$

Using the thin-lens equation and applying it to Figure 12.9c, we get

$$\left|\frac{\beta}{\alpha}\right| = \left|\frac{x_0}{x_i}\right| = \left|(f_1 + f_2)\left(\frac{1}{f_2} - \frac{1}{f_1 + f_2}\right)\right| = \left|\frac{f_1}{f_2}\right| \qquad (12.7)$$

This is the same formula (absolute values) for angular magnification of a telescopic two-lens system as derived in Chapter 1. Considering this point of view for the construction of a telescope, we need only produce a telescopic system with a parallel exit beam that is smaller than the parallel entrance beam (see Figure 12.9b).

---

*R. W. Pohl, *Einführung in die Optik* (Berlin: Springer-Verlag, 1948), p. 48.

The Galilean telescope is shown in Figure 12.9d. It consists of a positive lens and a negative lens. The negative lens is positioned so that the focal point F_1 of lens L_1 is slightly to the right of the focal point F_2 of the negative lens. The object is considered at infinity, and the first image is at F_1. The first image serves as a virtual object for L_2. We discussed the graphical construction for this case in Chapter 1 and show it again in Figure 12.9e. Lens L_2 forms a virtual upright image at infinity. As a result, the observer sees the object upright. This is in contrast to the telescopes discussed so far, where the virtual (second) image was upside down and the observer sees the object inverted (see Figure 12.9a).

We saw that if a parallel beam enters lens L_1, then a contracted beam leaves lens L_2. The ratio of their diameters gives the magnification. Clearly, the magnification is small and the light throughput large. Such telescopes are quite useful in dim areas; they are used as night glasses on ships and in theaters. A contraction of the entrance beam can also be obtained using two prisms, as proposed by Pohl (see Figure 12.9f). We obtain the same telescopic effect, but now the image is distorted. To eliminate the distortion, we employ two pairs of prisms, one pair rotated against the other by 90°. While this system is not actually used for a telescope, it demonstrates the telescopic system and the "area times solid angle" approach in a telescope.

The enlargement of the diameter of a beam by two prisms is used as a beam expander (see Figure 12.9g). The diameter of the output beam of a laser is often very small. If directly applied for an absorption measurement, the laser light can heat the sample too much and possibly damage it or change the material properties. To avoid this, we expand the laser beam to a larger diameter. This is accomplished by using two prisms and a system of positive and negative lenses. The beam is again contracted toward the detector (see Figure 12.9g). In contrast, we discussed beam expanders using two positive lenses and a positive and a negative lens in problem 9 of Chapter 1.

E. The Eyepiece

The second lens in the simple microscope and telescope may actually be composed of more than one lens and is called the **eyepiece** or ocular. We show some simplified examples in Figure 12.10. It produces a virtual image at infinity of the (first) image of the object. This first image is produced by the objective (first) lens. The eye forms the final image on the retina.

In Figure 12.10a, we see light passing through an eyepiece made of one lens. The additional lens introduced in the eyepiece shown in Figure 12.10b and c is a field lens. The other lens is called the *eye lens*. The field lens is located close to the first image and enlarges the solid angle (field of view) so that a larger area of the object can be seen. The combination of two lenses, given their appropriate choice and mount, may be utilized to reduce aberration (Chapter 13). The field lens is not positioned in the plane of the first image (contrast what has been discussed in section 4.A) thereby enabling a **cross wire** to be placed in that plane. This also prevents any dust on the surface of the field lens from being in the plane of the first image. The combination shown in Figure 12.10b is called the **Ramsden eyepiece**. Most of the spherical aberration can be corrected, and the cross wires are outside the two lenses, that is, outside of the eyepiece. In Figure 12.10c, the

12. Image Formation and Light Throughput

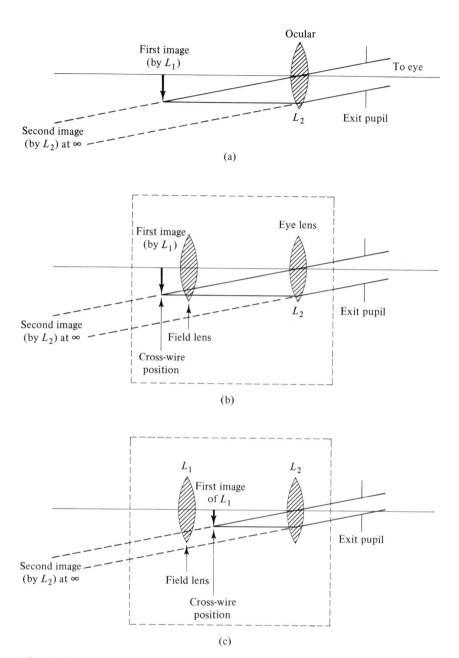

Fig. 12.10
Eyepieces, or oculars. (a) Eyepiece (L_2), or ocular, for the simple microscope or telescope. (b) Ramsden eyepiece. The field lens increases the light throughput. The cross wires may be mounted outside the eyepiece. (c) Huygens eyepiece. Cross wires would be inside the eyepiece. The field lens increases the light throughput.

Huygens eyepiece is shown; it may be corrected for chromatic aberration and off-axis coma, the cross wires are inside the eyepiece.

5. THE ANGLE OF VIEW AND TWO EYES

In Chapter 1 and the preceding section, we used a lens and a sensitive light-detecting area in place of the eye. This is equivalent to using a camera and a photographic plate. In reality, we move our eye suddenly by focusing from one spot to another. In using the microscope, we move the eye to different parts of the object, taking advantage of the large field of view.

In Chapter 1, we indicated the direction an object appears to be in when it is under water when we look at it from on top of the water. We also discussed the virtual image at a mirror and mentioned that we need a cone of light if we want to judge how far away an object appears to be.

The object under water and the image in the mirror are usually viewed not with one eye, but with both eyes. The eyes have a certain distance between them, and rays from the object to both eyes form a triangle (Figure 12.11a). It is this fact that makes it possible to judge the distance of the object.

Let us discuss the position that an object under water seems to be located if viewed at different angles by an observer above the water. In Figure 12.11b, we repeat Figure 1.3 (from Chapter 1). The direction of the apparent object is the back-tracing of the refracted ray from the object, as shown. But at what

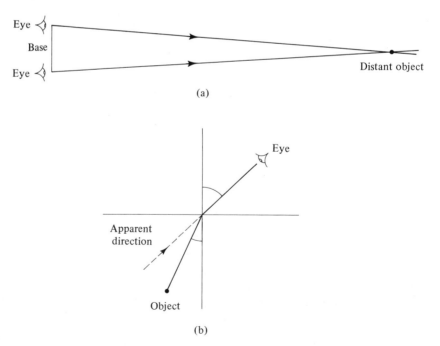

Fig. 12.11
(a) Two eyes looking at an object. A triangle is formed. (b) The direction of the apparent object as an extension of the refracted beam.

12. **Image Formation and Light Throughput**

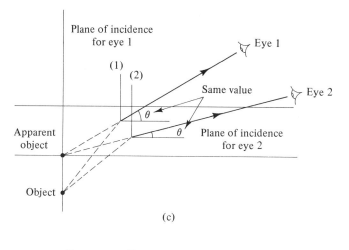

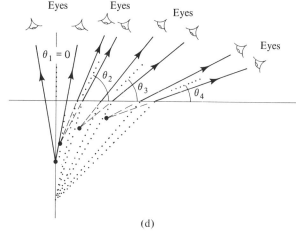

(d)

Fig. 12.11 (continued)
(c) The location of the apparent object on the straight line to the surface when viewed with two "horizontal" eyes. Different angles will move it up and down along a straight line. (d) The location of the apparent object on a curved line before the straight line [see part (c)] in direction to the observer when viewed with two "vertical" eyes. Different angles will move the apparent location on the curved line as indicated.

location on this line will the object appear? In textbooks on optics, two explanations are offered.

1. Pohl states that a distance cannot be judged with one eye; we need both.* For each eye, the ray from the object and the refracted ray form a plane of incidence and refraction. These two planes of incidence are not parallel and meet at the perpendicular line from the object to the surface of the water. The object appears on this line where the back-tracing of each refracted ray

*Pohl, p. 42.

meet (see Figure 12.11c). If we enlarge the angle θ, the apparent object moves down on a straight line.

2. A different description using only one eye and a cone of light is given by Jenkins and White.* The position of the apparent object is the meeting point of the back-tracing of the refracted rays of the outermost rays of the cone, as shown in Figure 12.11d. For different angles θ_i, the apparent object moves toward or away from the observer on a curved line.

We will not try to judge if and how close one may see a distance with only one eye. Both explanations use two refracted rays and discuss their meeting point when traced back. In the first explanation, the two rays are "horizontal"; in the second explanation, they are "vertical." To enlarge the angle for the second explanation, we can also use two eyes; we just turn our head by 90°. Now let us do a simple experiment. We let a small object hang from a string in a bucket of water. Moving our eyes in "horizontal" position up and down moves the apparent object on a straight line up and down. Turning our head by 90° and looking with our eyes in "vertical" position, we see the object between the straight line and us. Moving our eyes up and down (more difficult to do), we see the apparent object moving away from us or toward us, respectively.

6. ORIENTATIONS OF IMAGES: INTERCHANGES OF LEFT-RIGHT, UP-DOWN, AND ROTATIONS

A. Orientation of Images with Lens, Mirror, and Prism

In Chapter 1, we saw that the real image of an object is upside down. If we extend the object to the left and right, we see that the image is upside down (u-d) and that left and right (l-r) are interchanged (Figure 12.12a). However, the image remains right-handed (rh). Looking toward the light, right-handed means that we may turn the solid arrow to the broken arrow by 90° in the mathematically positive sense. Left-handed means we must turn it in the opposite direction.

We now compare the real image formed by a positive lens with the virtual image formed by a mirror (see Figure 12.12b). In the virtual image, the solid arrow is in the direction opposite the object, (looking against the light). The broken arrow remains perpendicular to the plane of incidence. As a result, a right-handed object is transformed into a left-handed image. The terminology comes from the fact that a mirror turns our left hand into an image similar to our right hand. Ambulances have the word "AMBULANCE" spelled out on the front of the vehicle with left and right (l-r) interchanged. Looking into the rearview mirror, the driver of the car ahead can read the word "AMBULANCE" in the normal manner. Look at this page using a mirror.

At the **Porro prism**, the light is reflected twice and the angle between the two reflecting surfaces is 90° (see Figure 12.12c). Looking always against the direction of the light, we have an interchange of right to left for the solid arrow and up to

---

*F. A. Jenkins and H. E. White, *Fundamentals of Optics*, 4th ed. (New York: McDraw-Hill, 1965), p. 37.

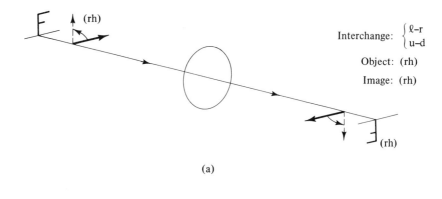

(a)

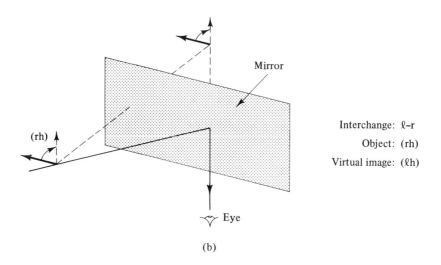

(b)

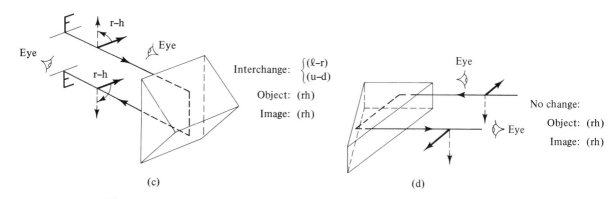

(c)   (d)

Fig. 12.12
(a) Image of a positive lens; real object, real image; interchanges; l-r and u-d (rh).
(b) The virtual image of a mirror; interchange: l-r; the object is right-handed (rh), the
image is left-handed (lh). (c) Porro prism in vertical position. Looking against the
light, the object is changed l-r and u-d, but remains rh. (d) Porro prism in horizontal
position. Looking against the light, there is no change.

6. Orientations of Images: Interchanges of Left-Right, Up-Down, and Rotations   423

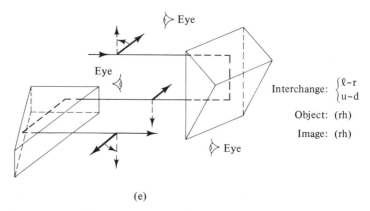

Interchange: $\begin{cases} \ell-r \\ u-d \end{cases}$

Object: (rh)

Image: (rh)

(e)

Fig. 12.12 (continued)
(e) Combination of two Porro prisms, one in vertical position and the other in horizontal position. Looking against the light, we have interchange l-r and u-d (rh) and incident and emerging light traveling in the same direction, all as observed for a positive lens [see part (a)].

down for the broken arrow. If we use a second Porro prism, but now rotated by 90°, and again looking against the direction of the light, the solid arrow is to the left for the incident light and to the left for the light leaving the prism; the broken arrow is down for incident light and down for the light leaving the prism (see Figure 12.12d).

Figure 12.12e shows the combination of the two Porro prisms. Comparing the final image with the object, we see that left and right as well as up and down are interchanged. Considering the fact that the incident light and emerging light are traveling in the same direction, we see that the action of the double Porro prism is equivalent to the action of a positive lens; see Figure 12.12a. It is used in a terrestial telescope.

Figure 12.13 illustrates the **Dove prism**. This prism rotates the image by the angle 2α if it is rotated itself by α, assuming that we look at the object against the light and at the image with the light (see problem 13).

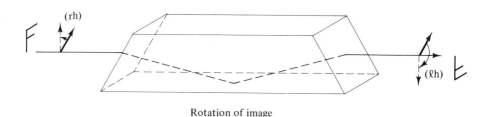

Rotation of image

Fig. 12.13
The Dove prism. The image is rotated by 2α if the prism is rotated by α. However, this simple description is only true if we look at the object against the light and at the image with the light.

12. Image Formation and Light Throughput

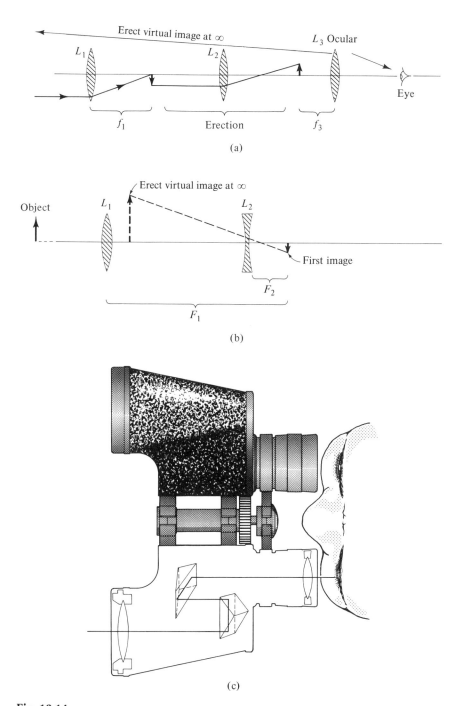

Fig. 12.14
Erecting the image in a telescope. (a) An additional lens is used in the two-positive-lens telescope. The image of the first lens is now erect. (b) In the Galilean telescope, there is already an erect image. (c) In binoculars, the image is made erect with a double Porro prism.

6. Orientations of Images: Interchanges of Left-Right, Up-Down, and Rotations

B. Application to Telescopes and Binoculars

In preceding chapters, we have discussed the astronomical telescope and the microscope and have mentioned that both systems are in principle simple. Both use two lenses (or lens systems). The image produced by the first lens is upside down and left and right are interchanged. The second lens forms a virtual image with no changes. We have an interchange of up-down, left-right. The eye lens produces another interchange of up-down, left-right, and we have the final image on the retina up, but we "see" it down. If we apply up-down and left-right considerations to the Galilean telescope, we find that the image traveling to the eye is erect (no up-down and left-right interchange). The eye brings it down on the retina, and we "see" it up.

When we use an astronomical telescope, our observation of the moon and stars is not really affected if the image is up-down and left-right interchanged. But when we use a terrestrial telescope or binoculars we want to have a magnification of the object without any accompanying up-down or left-right interchange. To accomplish this, we can add an additional lens for the sole purpose of erecting the first image (Figure 12.14a). Another possibility is already realized in the Galilean telescope. There the ocular is replaced by a negative lens. It is positioned with respect to the focal point of the first lens so that the image toward the observer is upright (see Figure 12.14b and Section 4 of this Chapter).

The additional lens introduced into the astronomical telescope to have it used as a terrestrial telescope makes the telescope longer. In a short telescopic system, we use the double Porro prism, as in binoculars (see Figure 12.14c).

Problems

1. *Power through lens and entrance and exit pupils.* A lens of radius $r' = 5$ cm and focal length $f = 20$ cm produces an image of a self-emitting object of radius $r = 1$ cm, and the lens is placed at a distance of 60 cm. Assume $L = 1$ (W/m^2 sr).

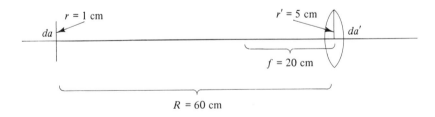

a. Calculate the power incident on the lens.
b. A round aperture of radius 0.8 cm is placed 40 cm to the left of the lens. Calculate the power P_1 passing the entrance pupil.

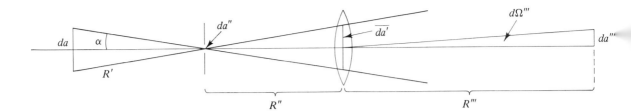

c. Express the power P_1 using da'' and the solid angle da/R'^2 and calculate the half-angle α.

d. Using the half-angle α, calculate the solid angle $d\Omega$ extending from the entrance pupil to the lens. Express the solid angle as $d\Omega = \overline{da'}/R''^2$ and calculate $\overline{da'}(R'' = 40$ cm$)$.

e. Calculate the product of the area $\overline{da'}$ and the solid angle from the lens to the exit aperture. If multiplied by L, we must get P_1.
Ans. $P_1 = 1.56 \times 10^{-6}$ W

2. *Entrance and exit pupils.* Consider a self-emitting object of radius 1 cm placed 60 cm to the left of a lens of radius 5 cm and focal length 20 cm. A round aperture of radius 0.8 cm is introduced at a distance 30 cm to the left and serves as the entrance pupil. Calculate the solid angle of the light leaving the exit pupil.
Ans. $d\Omega_{\text{exit}} = 5.47 \times 10^{-4}$ sr

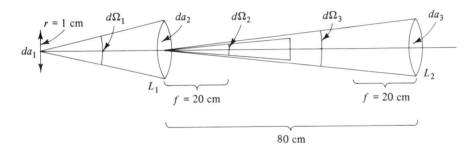

3. *Field lens and condenser lens.* Consider two lenses of radius 5 cm and focal length 20 cm. An object of radius 1 cm is placed 40 cm to the left of the first lens. The distance between the lenses is 80 cm. The image after the second lens is 40 cm to the right of it.

a. Calculate the product $da_1\, d\Omega_1$ for the power incident onto the first lens.

b. Calculate the solid angle into which the light travels from lens L_1.

c. How large is the solid angle $d\Omega_3$ from lens L_1 to lens L_2? Does all the light from the object arrive at lens L_2?

d. The object now has a 5 cm radius. How large is the solid angle into which the light travels from lens L_1? How much light does not arrive at lens L_2?

e. A field lens is placed at the image position. Show that all the light arrives at L_2.
Ans. a. $da_1\, d\Omega_1 = 15.4 \times 10^{-6}$ sr m^2; b. 0.196×10^{-2} sr; c. 1.22×10^{-2} sr

4. *Numerical aperture of the microscope.* In Chapter 9, we found that the smallest distance the microscope can resolve is

$$Y = \frac{\lambda}{2n \sin \alpha}$$

where α is the half-angle of the solid angle of the light from the object to the objective lens. The solid angle enters the expression through the usefully illuminated diameter of the objective lens. The index n is the refractive index of the immersion oil. Compare the resolvable distance for the case where there is no condenser ($\alpha = 10°$) and the case where there is a condenser and immersion oil ($\alpha = 70°$ and $n = 1.5$).
Ans. $Y_{\text{no cond}} = \lambda/0.35$; $Y_{\text{cond}} = \lambda/2.8$

5. *Telescope.* Consider a telescopic system with an objective lens of diameter $d = 5$ cm and an ocular of $d' = 1$ cm.
   a. Show that

$$\frac{d}{d'} = \left|\frac{f_1}{f_2}\right| = \frac{\beta}{\alpha}$$

   and give the angular magnification β/α, where f_1 is the focal length of the objective, f_2 that of the ocular, α is the angular size seen on the retina without the telescope, and β the angular size seen by the ocular of the first image (produced by the objective lens).
   b. If $f_1 = 10$ cm and $f_2 = 2$ cm, then we satisfy the above conditions. What happens if $f_1 = 20$ cm and $f_2 = 2$ cm? Give d/d''. Since d corresponds to the entrance pupil, what happens to the exit beam and the exit solid angle?

6. *Orientation of the image of a lens.* Consider the real image of a real object of a positive lens.
   a. Give the transformation in coordinates from object to image if a coordinate system is used as shown in the following figure.

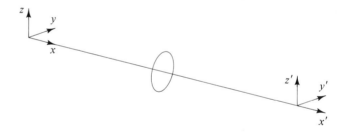

   b. If we have an object of the type

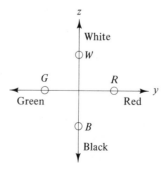

what is the configuration of the image?

12. **Image Formation and Light Throughput**

c. Can we obtain the image configuration from the object configuration by rotation of the object?

d. The object contains a right-handed rotation configuration B-R-W-G. Does the image also have a right-handed configuration?

7. *Orientation of the image produced by a mirror*

   a. Give the transformation of the coordinates from a real object and the virtual image produced by a mirror. If we trace back the image to the object with parallel rays, reflected at the mirror, we find the same transformation. (Take the angle of incidence at 45° to the normal.) make a sketch.

   b. If x, y, z is a right-handed coordinate system, (i) what is the coordinate system at the virtual image and (ii) what is the coordinate system of the reflection projection (right- or left-handed)?

   c. If we place the object

$$\begin{array}{ccc} & \uparrow z & \\ & W & \\ G & & R \rightarrow y \\ & B & \end{array}$$

   in the right-handed coordinate system of the y, z plane, how does the image configuration appear when we look into the light from the source?

   d. Can we rotate the object configuration to get the image configuration of (c)?

8. *One Porro prism, vertical extension*

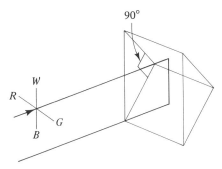

   a. Consider the Porro prism shown in the figure, where two reflecting surfaces make a 90° angle. If we look against the travel direction of the light, the object has the configuration

$$\begin{array}{ccc} & W & \\ G & & R \\ & B & \end{array}$$

   Give the image configuration looking against the travel direction of the light and looking with the light. (Left and right in reference to: perpendicular to the plane of incidence.)

   b. Which of the two cases in (a) can be obtained by rotating the object configuration by 180°?

   c. How is the direction of the light changed with respect to the vertical plane (plane of incidence)?

9. *One Porro prism, horizontal extension*

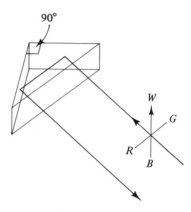

90°

W
G
R
B

a. Consider the Porro prism shown in the figure, where two reflecting surfaces make an angle of 90°. If we look against the light, the object has the configuration

$$W$$
$$G \quad R$$
$$B$$

Give the image configuration looking against and with the direction of the travel of the light. (Left and right in the plane of incidence when *W* is up and *B* is down.)

b. What is the rotation of the object configuration to obtain the configuration "with" and "against" the direction of the light?

c. What does the prism do to the light if it is not changing the configuration of the object? [Case: against the direction of the light; see part (b).]

10. *The double Porro prism, one vertical, one horizontal*

a. Use the two Porro prisms and give the configuration of the object after one prism; after two prisms. Always look against the direction of the light.

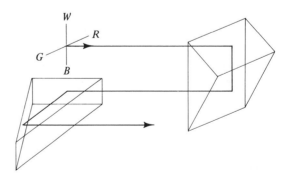

W
R
G
B

b. Can we obtain the image configuration from the object configuration by rotation?

c. What does the first prism do to the direction of the light compared to what the second prism does?

d. Compare the action of the double Porro prism to the action of the positive lens, with respect to the imaging of real object to real image.

11. *Configuration changes and group theory.* We can make a list of the action of a lens, a mirror, and various Porro prisms. Looking always into the direction of the light, we have the following:

Object Configuration	Operator	Image Configuration
$\begin{matrix} & W & \\ G & & R \\ & B & \end{matrix}$	L (lens)	$\begin{matrix} & B & \\ R & & G \\ & W & \end{matrix}$
$\begin{matrix} & W & \\ G & & R \\ & B & \end{matrix}$	M (mirror)	$\begin{matrix} & W & \\ R & & G \\ & B & \end{matrix}$
$\begin{matrix} & W & \\ G & & R \\ & B & \end{matrix}$	PV (Porro vertical)	$\begin{matrix} & B & \\ R & & G \\ & W & \end{matrix}$
$\begin{matrix} & W & \\ G & & R \\ & B & \end{matrix}$	PH (Porro horizontal)	$\begin{matrix} & W & \\ G & & R \\ & B & \end{matrix}$
$\begin{matrix} & W & \\ G & & R \\ & B & \end{matrix}$	DP (double Porro)	$\begin{matrix} & B & \\ R & & G \\ & W & \end{matrix}$

There is no interchange of B or W with R or G. Therefore, we can characterize the operations as follows:

$$L: (B \leftrightarrow W)(R \leftrightarrow G)$$
$$M: (R \leftrightarrow G)$$
$$PV: (B \leftrightarrow W)(R \leftrightarrow G)$$
$$PH: \text{No change}$$
$$DP: (B \leftrightarrow W)(R \leftrightarrow G)$$

We have three different operations:

$$L: (B \leftrightarrow W)(R \leftrightarrow G)$$
$$M: (R \leftrightarrow G)$$
$$PHE: (E) \text{ (no action)}$$

If we use these operations one after another (a process called multiplication), we find that all products, regardless of how often we multiply the operations with one another, are

$$L: (B \leftrightarrow W)(R \leftrightarrow G)$$
$$M: (R \leftrightarrow G)$$
$$K = L \cdot M: (B \leftrightarrow W)$$
$$E: (E) \text{ (no action)}$$

The operation K is the product of L and M, because twice the operation $(R \leftrightarrow G)$ results in "doing nothing," equivalent to operation (E), and E times E is E.

a. Fill in the multiplication table, where the elements across the upper row come first, the elements in the left column come second:

	E	K	L	M
E				
K				
L				
M				

b. The elements E, K, L, and M form a finite group (called a Klein four-group). To form a group, we must have
   (1) A multiplication rule: $A \cdot B = C$ for all elements in the group
   (2) A distribution law: $(A \cdot B)C = A(B \cdot C)$
   (3) A unit element and an inverse element
   If $A \cdot B = B \cdot A$, the group is called an Abelian group. Show that all this is fulfilled by the Klein four-group. (Note: It was not necessary to say which of the elements should be multiplied first.)

c. A subgroup contains fewer elements than the full group. How many subgroups does the Klein four group have?

12. *Homomorphism of two groups.* With respect to the property "left-handed coordinate system" or "right-handed coordinate system," we have the following properties of the operations:

$$L: \text{No change } (e)$$
$$M: l \leftrightarrow r \ (a)$$
$$K: l \leftrightarrow r \ (a)$$
$$E: \text{No change } (e)$$

We have the relation that L and E act as a unit operation (e) with respect to direction of coordinate system changes, and M and K act as an operation of change denoted by (a).

Show that (e), (a) form a group. The group (e, a) has two elements; its order is 2. The Klein four group has four elements; its order is 4. The relation between the Klein four-group and the two-group (e, a) is a process of mapping. We have mapped the four-group into the two-group, and more than one element of the four-group corresponds to one element of the two group. Properties of group theory are useful to bring order into configurations and to answer the question, How many different image configurations can be obtained with a certain number of optical (operator) elements?

13. *The Dove prism.* We will replace the light pass in the Dove prism by three reflections; see figure (B is below the page, W above; the mirror is perpendicular to the page).

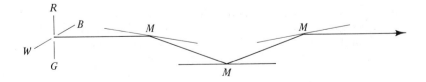

12. Image Formation and Light Throughput

We place an object

W

G   R

B

to the left. The configuration is for looking against the direction of travel of the light.

a. Using what we have learned about group theory, show that the effect of three reflections is equivalent to one reflection.

b. Give the image configuration.

c. We now turn the prism by 90° into the following position:

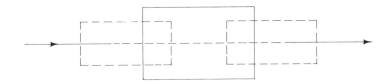

Give the image formation.

d. How does the image look if we look at it with the light (not against the light)?

e. Compare the result from part (d) with the object looked at against the light, and indicate the rotation of 180° by the object to get this image.

Aberration | **13**

1. INTRODUCTION

In Chapter 1, we used the paraxial approximation to describe image formation. All rays from the object to the image were assumed to make small angles with the axis of the system. The object and image points were assumed to be on the axis. Off-axis points were assumed to be on the axis of a system rotated by a small angle (Figure 13.1). Now we want to study the image formation of on-axis and off-axis points employing rays making small angles and large angles with respect to the axis (Figure 13.2). In this chapter, we are concerned with the image quality of an extended object. We will use nonparaxial theory for the image formation.

To get an idea of what "small" and "large" angles may be, let us go back to equation 1.7 (Chapter 1). We write this equation as

$$\frac{|s_1| + |r|}{|s_2| - |r|} = \frac{|\zeta_1|\, n_1}{|\zeta_2|\, n_2} \tag{13.1}$$

The paraxial theory was introduced by setting

$$|\zeta_1| = |s_1| \quad \text{and} \quad |\zeta_2| = |s_2|$$

(Figure 13.3). The exact relation between these quantities can be expressed as

$$|\zeta_1| \cos \alpha_1 = |s_1| + |\varepsilon|$$
$$|\zeta_2| \cos \alpha_2 = |s_2| - |\varepsilon| \tag{13.2}$$

where α_1, α_2, and ε are indicated in Figure 13.3. We expand $\cos \alpha_1$ and $\cos \alpha_2$ into a power series

$$\cos \alpha = 1 - \frac{\alpha^2}{2!} + \frac{\alpha^4}{4!} \tag{13.3}$$

Using the paraxial theory, we neglect all higher terms in the cosine expansion, as well as ε.

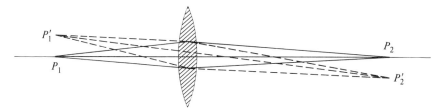

Fig. 13.1
Image formation in the paraxial approximation for points on the axis and off the axis.

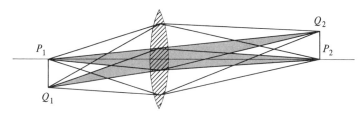

Fig. 13.2
Image formation for on-axis and off-axis points in the paraxial approximation (shaded areas) and for "large" angles.

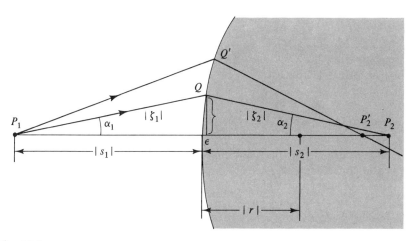

Fig. 13.3
Paraxial ray P_1Q and marginal ray P_1Q' forming image points P_2 and P_2', respectively.

More exact treatments take higher-order terms into account. In the discussion of aberration, theoretical calculations using only the first term are called **first-order theory**; those using the first and second terms are called **third-order theory**; and those using the first, second, and third terms are called **fifth-order theory**. These names are adopted from the Taylor expansion of $\sin \alpha$, which in first order is proportional to the angle and has third- and fifth-power terms in the expansion

$$\sin \alpha = \alpha - \frac{\alpha^3}{3!} + \frac{\alpha^5}{5!} + \cdots \tag{13.4}$$

Table 13.1

Expansion of $\sin \alpha$ into a power series for various angles, with errors for first- and second-order terms compared with $\sin \alpha$

α	$0°$	$5°$	$10°$	$20°$	$30°$	$40°$
α (rad)	0	0.087222	0.174444	0.348888	0.523333	0.697777
$\sin \alpha$	0	0.087111	0.173561	0.341853	0.499770	0.642516
% error	0	0.126907	0.508987	2.057908	4.714813	8.600764
$\alpha - \dfrac{\alpha^3}{3!}$	0	0.087111	0.173559	0.341810	0.499445	0.641153
% error	0	-0.00004	-0.00077	-0.01256	-0.06502	-0.21207

Table 13.1 lists numerical values for $\sin \alpha$, α, and the first two terms in the $\sin \alpha$ expansion. We see that for $\alpha = 20°$, the error introduced by approximating $\sin \alpha$ by α is 2.06 percent, while it is reduced to 0.01 percent if we use the first and second terms of equation 13.4.

In this chapter, we will describe different types of aberrations. The mathematical treatment is, for the most part, covered in the chapter problems, where we use the so-called Coddington shape and position factors. Since the expressions to be calculated in the problems are very long, we have adopted the same notation as that in J. Morgan's book, for a convenient comparison by the reader.[*] A different notation for Coddington's factors is employed in the book by F. A. Jenkins and H. E. White.[†]

2. SPHERICAL ABERRATION

First, let us consider only on-axis points for imaging. In Figure 13.3, we see that the paraxial and marginal rays from the object point form different image points. In Figure 13.4, we show this for a single refracting surface for a convex and concave thin lens and for a plane parallel plate. The plane parallel plate may seem out of place in this series because it is not an image-forming device. However, it may act as an element in a compound lens system, and its **spherical aberration** must be taken into account. For the single refracting surface, we will discuss in problem 1 the position of P'_2 as a function of the position of the object point P_1, the radius of curvature r, the refractive index n, and the distance d of point Q from the axis. For small d, we are back to the paraxial approximation.

A. Eliminating Spherical Aberration, Single Lens

For elliptical and parabolic mirrors, all rays converging to the focal point meet at one point. For lenses we may use aspherical surfaces for special purposes, such

---

[*] J. Morgan, *Introduction to Geometrical and Physical Optics* (New York: McGraw-Hill, 1953).
[†] F. A. Jenkins and H. E. White, *Fundamentals of Optics*, 4th ed. (New York: McGraw-Hill, 1965).

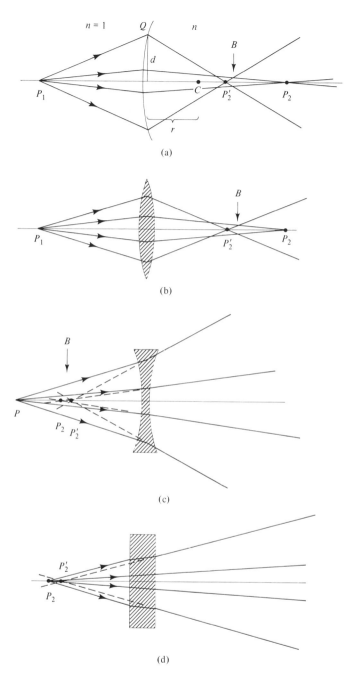

Fig. 13.4
Marginal and paraxial image points. (a) Single spherical surface, real image points. (b) Convex lens, real image points. (c) Concave lens, virtual image points. (d) Plane parallel plate. The marginal image point is always closer to the lens. Measured in direction of the propagation of light (left to right), P comes after P' in (a) and (b). This situation is called undercorrected or positive spherical aberration. In (c) and (d), P comes before P', and we have overcorrected or negative spherical aberration. The arrow at B indicates the circle of least confusion.

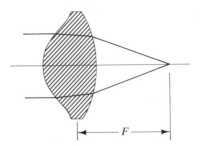

Fig. 13.5
Aspherical lens surfaces are used for the construction of a condenser lens.

as the construction of certain condenser lenses (Figure 13.5). It is possible to construct special refracting surfaces that will not show spherical aberration. However, the common procedure for obtaining quality images is to use lenses with spherical surfaces to form lens systems with them. The combination of lenses may be so chosen that aberrations are reduced or eliminated.

For a single spherical lens, spherical aberration cannot be eliminated if the object and image are both real. But for a real object and a virtual image, it can be accomplished. An example is the **aplanatic lens** (Figure 13.6). The refractive index is n, and the two radii of curvature of the lens are r_1 and r_2, where

$$r_2 = \frac{n}{n+1} r_1$$

Object and image distances are $x_0 = r_1$ and $x_i = nr_1$. We see from Figure 13.6 that the emerging rays from x_0 are not refracted at the surface of radius r_1. Therefore, we can regard point P at x_0 as being in a medium of refractive index n and serving as an object point with distance x_0' from the spherical surface of radius r_2. The image point P' at x_i is not affected by this consideration because P' is a virtual image point. It is obtained by tracing back the emerging ray at the surface of radius r_2. Points P and P' are called **aplanatic points** with respect to the spherical surface with radius r_2 and the refractive index n. The quantities

$$x_0, x_i, \quad \text{and} \quad f = \frac{1+n}{1-n} r_2 = \frac{n}{1-n} r_1 \tag{13.5}$$

are connected by the thin-lens equation; distances are measured from the surface with radius r_1.

This arrangement of aplanatic points is used in the microscope so that large-angle rays can enter the objective lens without producing spherical aberration. The object point is embedded in an immersion oil having the same index of refraction as the objective lens. Together they form a medium with a spherical surface containing the object point (see Figure 13.6c).

B. Reducing Spherical Aberration, Single Lens

If we introduce an aperture stop close to the lens, we limit the angles that the rays, extending from object to image, can form with the axis. In this way, spherical

13. Aberration

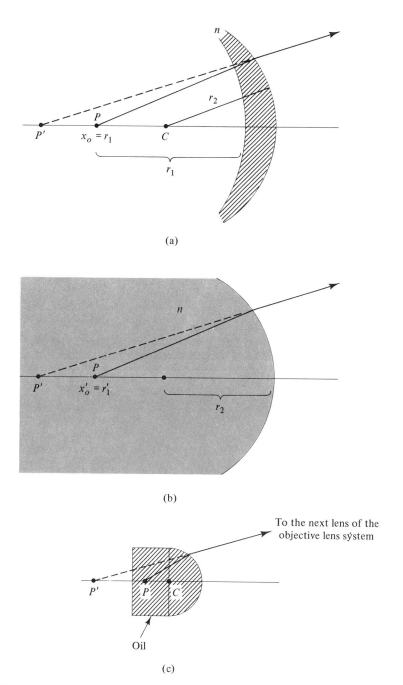

Fig. 13.6
Aplanatic lens. (a) P and P' for the aplanatic lens. (b) P and P' for the aplanatic sphere. (c) Immersion oil and planoconvex thick lens, both with the same index of refraction, as used in the microscope for reduction of spherical aberration.

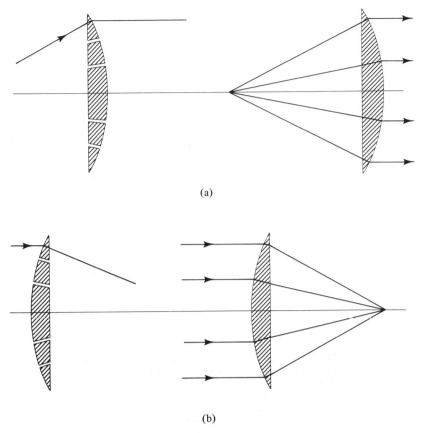

(a)

(b)

Fig. 13.7
To reduce spherical aberration using a planoconvex lens, the rays making large angles with the axis are at the flat side of the lens. (a) As used in the microscope.
(b) As used in the telescope. Regarding the lens as a set of prisms, and knowing that a symmetric path results in minimum deviation, we can understand that the image points of the marginal ray and the paraxial ray are closer together compared to the case where the lens faces the opposite direction.

aberration is reduced by getting closer to the paraxial approximation. (As we will see below, an aperture stop in a lens system can do more than simply reduce the angle size to the paraxial approximation.) A reduction in spherical aberration results if we use a planoconvex lens in such a way that the rays making the larger angles with the axis are on the flat side. In the microscope, the lens is used as shown in Figure 13.7a; in the telescope, the arrangement is reversed (see Figure 13.7b).

We can understand the reduction of spherical aberration in this manner if we think of the lens as a set of prisms. We know that the minimum deviation for a prism is obtained for a symmetric path. A symmetric path means that the ray enters the prism under the same angle that it leaves it. Minimum deviation of the marginal rays will result in a smaller deviation of P_2' from the paraxial image point P_2 (see Figures 13.3, 13.4, and 13.7).

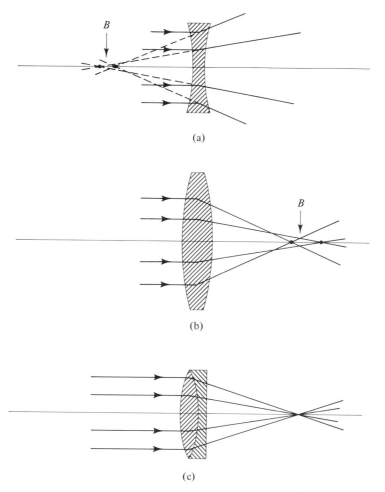

(a)

(b)

(c)

Fig. 13.8
Compensation of spherical aberration using a biconvex lens and a biconcave lens in combination. B is the circle of least confusion.

C. Eliminating and Reducing Spherical Aberration for a System of Lenses

The combination of a positive and a negative lens can eliminate spherical aberration, as schmatically shown in Figure 13.8. A suitable combination of a concave lens and a convex lens can result in the compensation of the "extra" deviation of the marginal rays. The compensation can only be done for two zones of the lenses and for a certain object and image distance. In principle, the two indices of refraction do not have to be different. However, we usually try to correct for **chromatic aberration** at the same time and for that the indices must be different (see Section 9).

For example, consider the **Brouwer lens**, composed of a biconvex lens and a **meniscus lens**, as shown in Figure 13.9. The first element has radii of curvature $r_1 = 61.070$ L and $r_2 = -47.107$ L, thickness $t_1 = 4.044$ L, a refractive index of 1.56178, and a focal length of 51.0 L, where L is a length unit (i.e., L may stand

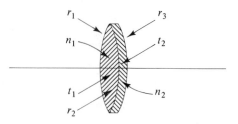

Fig. 13.9
The Brouwer lens. Example: $r_1 = 61.070L$, $r_2 = -47.107L$, $t_1 = 4.044L$, $n_1 = 1.56178$, $r_3 = -127.098L$, $t_2 = 2.022L$, $n_2 = 1.70100$ (where L is a length unit).

for mm, cm, m, etc.). To reduce spherical aberration, a miniscus lens is added having radii $r_2 = -47.107$ L and $r_3 = -127.098$ L, thickness $t_2 = 2.022$ L, and a refractive index of 1.70100.

The matrix method discussed in Chapter 1 was used for the computer calculation of this example. In that chapter, the matrices we used were obtained for the paraxial approximation. For the present example, more general matrices were used, taking into account spherical aberration; see the book by A. Nussbaum and R. A. Phillips.*

3. ABBE'S SINE CONDITION

So far, spherical aberration has been discussed for on-axis points only; the means considered to correct such aberration are only valid for on-axis points. To correct spherical aberration for both on- and off-axis points, Abbe's sine condition must be obeyed. The sine condition was derived and used in Chapter 9. We will also derive it here by using two waves emerging from an illuminated aperture (Figure 13.10).

We assume that a plane wave is incident on the aperture. One part of the wave travels along the axis and forms the image. The second part of the wave travels through the outer part of the lens. Marginal parts (rays) of the first wave are symmetric and equally long. Marginal parts (rays) of the second wave must

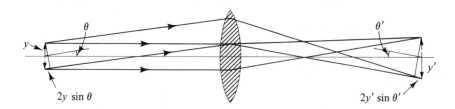

Fig. 13.10
Derivation of the sine condition for a special case. The projections of object and image width on the two light rays indicated must be equal for proper image formation.

*A. Nussbaum and R. A. Phillips, *Contemporary Optics for Scientists and Engineers* (Englewood Cliffs, N. J.: Prentice-Hall, 1976).

have the same length as the rays of the first wave for proper image formation. Therefore, the projection of the object width on one ray must be equal to the projection of the image width on the other ray. So

$$2y \sin \theta = 2y' \sin \theta' \quad \text{or} \quad \frac{\sin \theta}{\sin \theta'} = \frac{y'}{y} \tag{13.6}$$

If this sine condition is obeyed, the lens is aplanatic. An aplanatic lens is free of spherical aberration and coma. (This latter aberration is discussed in Section 5.)

4. MONOCHROMATIC ABERRATIONS

In Section 2, we discussed spherical aberration, which served as an introduction to monochromatic aberrations. **Monochromatic aberrations** constitute aberrations involving light having only one wavelength. Chromatic aberration involves a range of wavelengths and will be discussed in Section 9.

To get an idea of the mathematical distinctions and parameters involved in the calculation of aberrations, let us consider Figure 13.11. An image is formed by a single refracting surface. A stop is introduced between the refracting surface and the image. Three rays are shown. Neither the ray (1) traveling along the axis nor the ray (2) through the center of curvature is refracted. The third ray is a general ray passing through a point Q. In Figure 13.12, we see the position of these three rays at points in the plane of the aperture. Their positions are

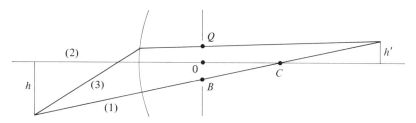

Fig. 13.11
Ray 1 passes through the center of curvature C and is not refracted. Ray 2, which passes along the axis is also not refracted. Ray 3 is a general ray through point Q in the aperture.

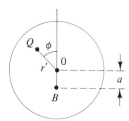

Fig. 13.12
The points where the three rays of Figure 13.11 pass through the aperture. Parameters r', ϕ, and a are indicated.

Table 13.2

Terms considered for elimination	Aberration eliminated
S_1	Spherical aberration
$S_1 + S_2$	Coma
$S_1 + S_2 + S_3 + S_4$	Astigmatism and curvature
$S_1 + S_2 + S_3 + S_4 + S_5$	Distortion

described by three parameters, r', ϕ, and a. One can show that a is related to the length h' of the image. The optical path difference between the ray through the center of curvature and the general ray (through Q) may be expressed as

$$C[r'^4 + 4ar'^3 \cos\phi + 4a^3 r' \cos\phi + 2a^2 r'^2 + 4a^2 r'^2 \cos^2\phi] \tag{13.7}$$
$$\uparrow \qquad \uparrow \qquad \uparrow \qquad \uparrow \qquad \uparrow$$
$$S_1 \qquad S_2 \qquad S_3 \qquad S_4 \qquad S_5$$

The sequence of the terms indicates both the order of importance and the order in which we discuss them. The elimination of some aberrations assumes that certain other aberrations have already been eliminated (Table 13.2).

5. COMA

The discussion of spherical aberration dealt only with on-axis points. Now we will discuss aberration when off axis points are taken into account. The corresponding aberration is called **coma**. For its elimination, we take into account the two terms S_1 and S_2 (see Table 13.2); that is, we assume that spherical aberration is eliminated. The rays from an off-axis point passing through different zones of the lens do not meet at one image point (Figure 13.13). If the rays through the outer zone of the lens are closer to the axis than the chief ray, we speak of **negative coma** (see Figure 13.13a); otherwise, we are dealing with **positive coma** (see Figure 13.13b). If the lens shows no coma, we have the situation shown in Figure 13.13c.

If coma is present, we have a cometlike structure of image points in the plane of the image (Figure 13.14c). This cometlike structure is the superposition of circles formed by light rays passing through different zones of the lens. In Figure 13.14a, the zones are numbered; and in Figure 13.14b, we see various points in each zone, where a, b, c, and d cover only half a circle. Figure 13.14c indicates where the rays passing through a, b, c, and d on each circle of the lens arrive at the image plane. Their arrangement shows, for example, that the two rays going through the two a points on the lens arrive at one point a' in the image plane (and similarly for b, c, and d for each zone 1, 2, 3, and 4). The opening of the cometlike arrangement of circles is opposite for negative and positive comas.

A. Eliminating Coma

The aplanatic lens discussed in Section 2 is free of coma. The image and object distances are fixed, and our discussion of spherical aberration was limited to on-axis points. The advantageous application to off-axis points may be imagined

13. Aberration

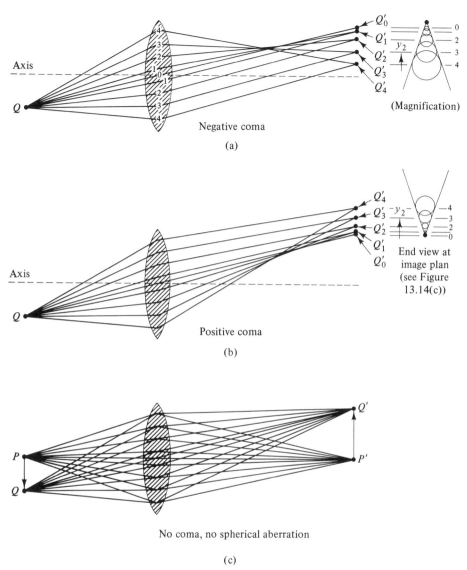

Fig. 13.13
(a) Negative coma. (b) Positive coma. (c) No coma.

from the spherical symmetry of the aplanatic arrangement. This arrangement can be found in the microscope, as seen by the use of the aplanatic sphere (Figure 13.6b) and the spherical arrangement made with immersion oil (Figure 13.6c).

B. Reducing Coma

To reduce coma for arbitrary object and image positions, we need a system of lenses. A single planoconvex lens shows less coma if the curved surface faces the incident light. This is in agreement with the use of such a lens to reduce spherical

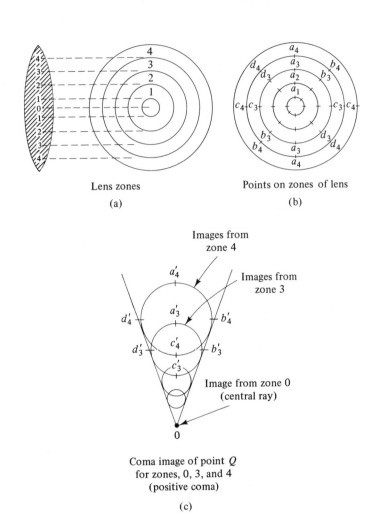

Lens zones

(a)

Points on zones of lens

(b)

Coma image of point Q
for zones, 0, 3, and 4
(positive coma)

(c)

Fig. 13.14
Zone structure of lens for the understanding of coma. (a) Lens zones. (b) Points on
zones 1–4 of the lens. (c) Coma image of point Q for zones 0, 3, and 4 (positive coma).

aberration in the telescope (see Figure 13.7b). In the microscope, however, we
are close to the aplanatic arrangement when the lens is used as in Figure 13.7a.

A lens that has minimum spherical aberration will also have only a small
amount of coma (see problems 5 to 7).

C. The Czerny-Turner Mount

Spherical mirrors often must be used in grating spectrometers intended for use
in the infrared region, because no suitable lens material is available. Figure 13.15a
shows how we would like to employ lenses at a transmission grating. Figure
13.15b and 13.15c show two arrangements using mirrors. The use of the two
spherical mirrors as shown in Figure 13.15c compensates for coma by its symmetric arrangement. The gratings in the infrared are usually reflection gratings.

13. Aberration

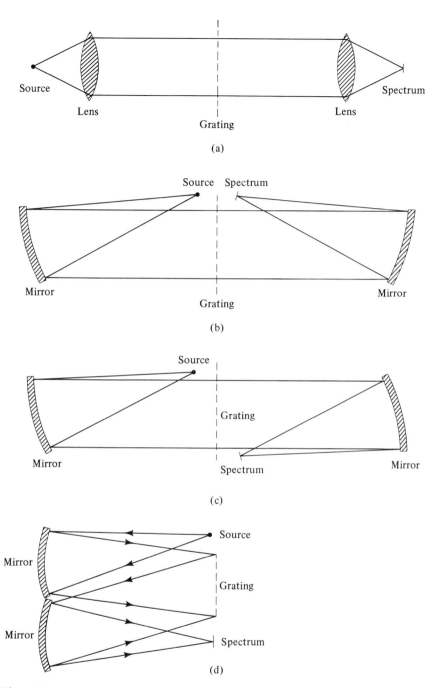

Fig. 13.15

Collimating lenses and mirrors for use with gratings; the Czerny-Turner mount.
(a) Lenses used so that parallel light will be incident on and emerging from a transmission grating. (b) Mirrors using a transmission grating so that parallel light will be used with a grating in a grating spectrometer. (c) Mirrors used in a spectrometer using a transmission grating to avoid coma. (d) Mirrors used with a reflection grating to avoid coma. This arrangement is called the Czerny-Turner mount.

Figure 13.15d shows how mirrors with a reflection grating are used in an arrangement compensating for coma. This is called the **Czerny-Turner mount**.

6. ASTIGMATISM

Recalling our list of monochromatic aberrations (Table 13.2), we now assume that the two terms S_1 and S_2 corresponding to spherical aberration and coma are zero and that these aberrations are eliminated. We now turn to S_3 and S_4, corresponding to astigmatism and curvature of field. We will discuss these two aberrations together. For spherical aberration, we discussed only on-axis points; for coma, off-axis points. However, these off-axis points could not be placed too far away from the axis. When coma is eliminated, rays farther away from the axis do not meet at the image off-axis point. They form different image points, depending on their path through the lens in the vertical and horizontal planes. This is called **astigmatism** and is shown in Figure 13.16.

All the rays in the horizontal plane coming from object positions on line A meet at image points on line H, which is parallel to A and the vertical plane (Figure 13.17b). These object points are called the **sagittal focal points**.

All the rays in the vertical plane coming from object points on line D meet at a line V, which is parallel to D and the horizontal plane (see Figure 13.17a). These points are called the **meridional focal points**. In Figure 13.17c, the object is a spoked wheel with its center on the axis. The image at the plane containing V shows the circle of the wheel (D is a section of it), and the spokes are fuzzy. The image in the plane containing H shows the spokes (A is one of them), and the circles are fuzzy. Between these two planes is the "compromise" plane containing the **circle of least confusion** (see Figure 13.16).

The off-axis object point Q has two image points Q_V and Q_H (see Figure

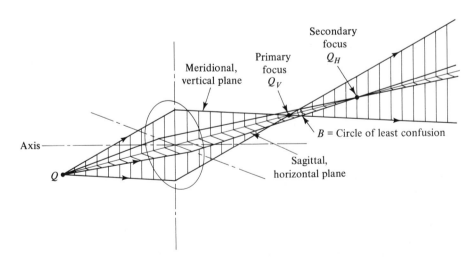

Fig. 13.16
Astigmatism. Two rays from Q in the vertical plane meet at Q_V. Two rays from Q in the horizontal plane meet at Q_H. Circle of least confusion: B.

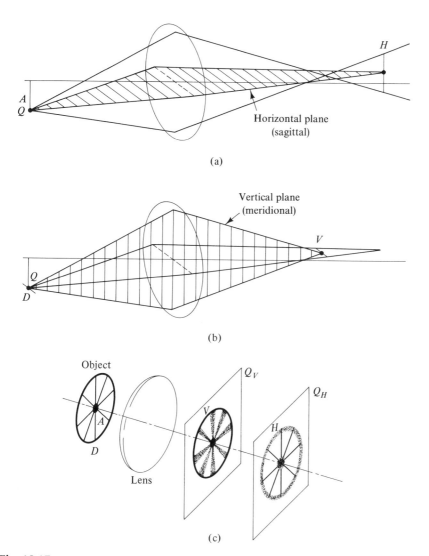

Fig. 13.17
Astigmatism demonstrated with a spoked wheel. (a) All rays from points on line A meet at points on line H, called sagittal focal points. H is $\parallel$ to A and $\perp$ to the horizontal plane. (b) All rays from points on line $D \perp$ to line A meet at points on line V, called meridional focal points. V is $\parallel$ to D and $\perp$ to the vertical plane. (c) The image points Q_V are in a plane containing line V. The image points Q_H are in a plane containing line H.

13.16). If Q is located on one plane, Q_V and Q_H are located on different paraboloidal surfaces (Figure 13.18).

A. Eliminating and Reducing Astigmatism

Astigmatism cannot be eliminated for a single thin lens. We can, however, use a stop and reduce it. Lens systems can be designed which have no astigmatism.

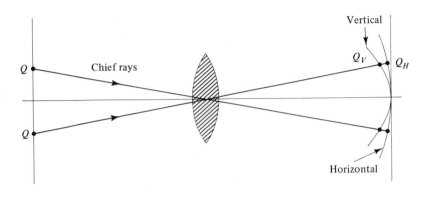

Fig. 13.18
The image points Q_V and Q_H are on paraboloidal surfaces when Q is on one plane.

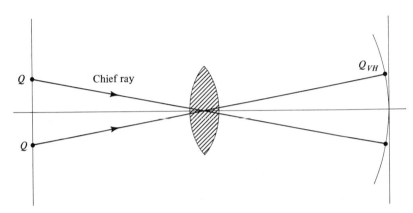

Fig. 13.19
The Petzval paraboidal surface.

The surfaces of Q_V and Q_H of the system can be so altered that both surfaces coincide to form one paraboloidal surface, called the **Petzval surface** (Figure 13.19). Here all rays from the off-axis object point Q arrive at one image point Q_{VH}. Such a lens system is called an **anastigmat**.

7. CURVATURE OF FIELD

Because the Petzval surface is curved, a flat object will have a curved image. This is called **curvature of field**. For a thin lens, it cannot be eliminated. A stop in a lens system can be used in connection with the other elements to reduce curvature of field. A famous example is the lens and stop combination used in the **Zeiss Sonnar Anastigmat** (Figure 13.20).

8. DISTORTION

For the discussion of this last monochromatic aberration, we assume that all other aberrations have been corrected; that is, we assume that

$$S_1 + S_2 + S_3 + S_4 = 0$$

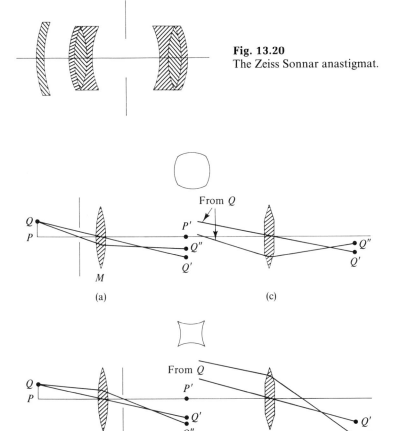

Fig. 13.20
The Zeiss Sonnar anastigmat.

(a)

(c)

(b)

(d)

Fig. 13.21
(a) Barrel-shaped, or negative, distortion. (b) Pincushion, or positive, distortion. In (c) and (d) prisms are used schematically for lenses to demonstrate the distortion, as in (a) and (b).

Distortion of the image is said to be the effect of nonuniform lateral magnification. To demonstrate this effect, we use a stop before or after the lens. Then the cone of light from an off-axis point Q passes through only a small portion of the lens. This portion is not in the center of the lens; the cone is different for the two cases where the stop comes before or after the lens.

In Figure 13.21a, the stop is placed before the lens. The paraxial image points of P and Q are P' and Q', respectively. The ray passing through the center of the stop from Q to M forms an image point Q''. The sequence of image points in direction to the axis is Q', Q'', P'. In Figure 13.21b, the stop is placed after the lens. A similar consideration of rays leads to a sequence of image points Q'', Q', P' in direction to the axis. Different portions of the lens are used in these two cases. This is shown in a simplified way by replacing the lens by a prism (see Figure 13.21c and d).

In the case of Figure 13.21a, the nonuniform magnification of the extended image leads to a distortion called **barrel-shaped** or **negative distortion**. In Figure 13.20b, the distortion is called **pincushion** or **positive distortion**.

9. CHROMATIC ABERRATIONS

So far, we have discussed monochromatic aberrations; that is, we have assumed that only one wavelength is involved. Since the law of refraction depends on the wavelength, it might be expected that all aberrations would have to be corrected for each wavelength employed in the imaging process. However, the effect of the different wavelengths is usually minimal. It is customary to account for aberrations that are due to the presence of different wavelengths for spherical aberration only.

An on-axis point shows different image points because of the different refractive indices at different wavelengths. If the image point of the shorter-wavelength light (blue) is closer to the lens than the image point of the longer-wavelength light (red), then we speak of **positive chromatic aberration** (Figure 13.22). The reverse case is called **negative chromatic aberration**. Chromatic aberration can be demonstrated as shown in Figure 13.23. A slit is illuminated with red and blue light. A screen shows the image of the slit in blue at a closer position to the lens than the image of the slit in red. On a tilted screen farther away, the red area is smaller than the blue area.

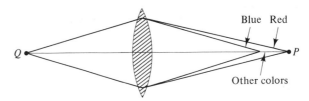

Fig. 13.22
Positive chromatic aberration.

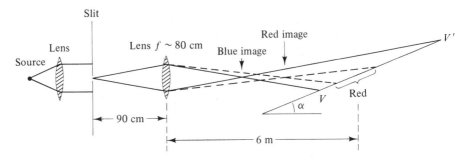

Fig. 13.23
Chromatic aberration demonstrated according to Pohl. The red light appears within the bracket, the violet light from V to V'. (After Pohl, *Einführung in die Optik*.)

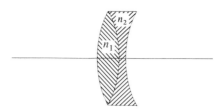

Fig. 13.24
Correction for chromatic aberration. The refractive indices of the two lenses are different.

A. Eliminating Chromatic Aberration

Chromatic aberration can be compensated for by using a positive lens and a negative lens of different refractive indices (see Figure 13.24 and problem 9). We can also eliminate chromatic aberration using two lenses. In problem 10, we show that two lenses having focal length f_1 and f_2 and made of the same material will not show chromatic aberration if placed with the special separation $t = (f_1 + f_2)/2$ between them.

10. ABERRATION AND THE PINHOLE CAMERA

Now that we have discussed monochromatic and chromatic aberrations, let us consider the possible aberrations encountered with a pinhole camera. Actually, there are no aberrations. The light throughput, however, of the pinhole camera is so small that its usefulness is very limited. To enlarge the light throughput, we would have to make the pinhole larger, a lens would be necessary for image formation, and there would be aberrations.

11. NONSPHERICAL MIRRORS

At the beginning of this chapter, we discussed the fact that spherical lenses will show aberration. Although nonspherical lenses could be designed that are free of aberration, it is the usual procedure to use systems of spherical lenses for the reduction of aberration. Spherical mirrors also show aberration, since the paraxial and nonparaxial rays are focused at different points on the axis (Figure 13.25). It is customary to use nonspherical mirrors to reduce aberration. Among those frequently used are the parabolic and the elliptical mirror.

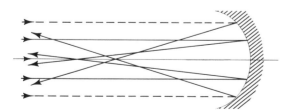

Fig. 13.25
Spherical aberration shown for a spherical mirror.

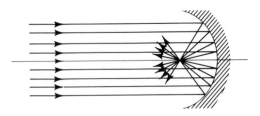

Fig. 13.26
A parabolic mirror: All incident rays parallel to the axis are reflected through the same point.

A. The Parabolic Mirror

In the **parabolic mirror**, all rays parallel to the axis are reflected through the same point, thereby introducing no spherical aberration (Figure 13.26). Parabolic reflectors are used to concentrate incident parallel light onto a single focal point (as in astronomical and radio telescopes) or to have the light from a pointlike source emerge parallel from the mirror (as in the headlights of a car).

B. The Elliptical Mirror

If a reflector is shaped like an ellipse, then light emerging from one focus and reflected at the inner surface will be focused on the other focus. In applications, we usually use only a certain section of the mirror (Figure 13.27b and c). **Elliptical off-axis mirrors** are used in infrared spectrometers for the concentration of light onto the small detector window. Mirrors are used in the infrared spectral region because often no suitable lens material is available.

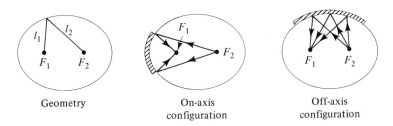

Geometry   On-axis configuration   Off-axis configuration

Fig. 13.27
An elliptical mirror (a) geometry of the ellipse; (b) on-axis section of the ellipse; (c) off-axis section of the ellipse.

Problems

1. *Spherical aberration of a single refracting surface.* Consider the rays from P_1 to P_2 passing through the refracting surface at distance ρ (see figure at top of page 455). In Chapter 1,

13. Aberration

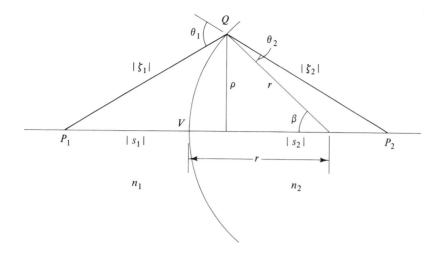

$$\frac{|s_1| + |r|}{|s_2| - |r|} = n \frac{|\zeta_1|}{|\zeta_2|} \qquad \text{(equivalent to equation 1.7 with } n_1 = 1, n_2 = n)$$

Now we do not set $s_1 \simeq \zeta_1$ and $s_2 \simeq \zeta_2$, but introduce in the following a better approximation. We neglect the absolute bars. Using the law of cosines, we get

$$\zeta_1^2 = r^2 + (s_1 + r)^2 - 2r(s_1 + r)\cos \beta$$

Using

$$\cos \beta = 1 - \frac{\beta^2}{2}$$

and setting $\beta = \rho/r$ for small β, we get

$$\zeta_1^2 = r^2 + (s_1 + r)^2 - 2r(s_1 + r)\left(1 - \frac{\rho^2}{2r^2}\right)$$

a. Show that we obtain

$$\zeta_1^2 = s_1^2 + (s_1 + r)\frac{\rho^2}{r}$$

and

$$\zeta_2^2 = s_2^2 - (s_1 - r)\frac{\rho^2}{r}$$

(*Note:* For $\rho \to 0$, we are back to the approximation used in Chapter 1.)

b. Expand the root

$$\zeta_1 = \sqrt{s_1^2 + (s_1 + r)\frac{\rho^2}{r}}$$

according to

$$y = (1 + x^2)^{1/2} = 1 + \frac{x^2}{2}$$

and get

$$\zeta_1 = s_1 + \left(\frac{1}{s_2} + \frac{1}{r}\right)\frac{\rho^2}{2}$$

$$\zeta_2 = s_2 + \left(\frac{1}{s_2} - \frac{1}{r}\right)\frac{\rho^2}{2}$$

We introduce ζ_1 and ζ_2 to the first equation.

c. Show that

$$\frac{s_1 + r}{s_2 - r} = n\frac{s_1 + [(1/s_1) + (1/r)](\rho^2/2)}{s_2 + [(1/s_2) - (1/r)](\rho^2/2)}$$

can be written as

$$\frac{1}{s_1} + \frac{n}{s_2} + \frac{1-n}{r} = \left(\frac{1}{r} + \frac{1}{s_1}\right)\left(\frac{1}{r} - \frac{1}{s_2}\right)\left(\frac{n}{s_1} + \frac{1}{s_2}\right)\frac{\rho^2}{2}$$

d. To get to the usual expression for the image formation of a single refracting surface, introduce $s_1 \to -x_o$ and $s_2 \to x_i$ and get

$$-\frac{1}{x_o} + \frac{n}{x_i} = \frac{n-1}{r} + \left(\frac{1}{r} - \frac{1}{x_o}\right)\left(\frac{1}{r} - \frac{1}{x_i}\right)\left(\frac{1}{x_i} - \frac{n}{x_o}\right)\frac{\rho^2}{2}$$

Note that for $\rho \to 0$, we are back to our original approximation. The coefficient of $\rho^2/2$ depends on x_o and x_i. We would like to have it depend only on x_o. Then we could judge how, for the same x_o, x_i varies depending on ρ. We introduce into this coefficient our original (zero-order) approximation:

$$\frac{1}{x_i} = \frac{n-1}{nr} + \frac{1}{nx_o}$$

e. Show that we obtain the following as our first-order approximation:

$$-\frac{1}{x_o} + \frac{n}{x_i} = \frac{n-1}{r} + \frac{n-1}{n^2}\left(\frac{1}{r} - \frac{1}{x_o}\right)^2\left(\frac{1}{r} - \frac{n+1}{x_o}\right)\frac{\rho^2}{2}$$

f. Give the position x_i for the marginal ray and the paraxial ray if $x_o \to \infty$. Use $n = 1.5$.

g. Show in a sketch the location of x_i (first order) with respect to x_i (zero order) for:

(1) $n_1 = 1 \quad \big(\quad n_2 = n$

Interface

(2) $n_1 = 1 \quad \big) \quad n_2 = n$

Interface

(3) $n_1 = n \quad \big(\quad n_2 = 1$

Interface

(4) $n_1 = n \quad \big) \quad n_2 = 1$

Interface

2. *Spherical aberration of a thin lens.* Let us consider the results for one interface (problem 1) where we had the refractive index $n = 1$ on the left and the refractive index n on the right. We rewrite the equations for the case where the index n is on

the left of the interface and the index 1 is on the right. This means that the light is now going from the medium of index n to the medium of index 1. The result can be obtained by substituting $x_o \to x_i'$ and $x_i \to x_o'$. Calling the radius of curvature of the second surface r_2, we have

$$-\frac{1}{x_o} + \frac{n}{x_i} = \frac{n-1}{r_1} + \frac{n-1}{n^2}\left(\frac{1}{r_1} - \frac{1}{x_o}\right)^2\left(\frac{1}{r_1} - \frac{(n+1)}{x_o}\right)\frac{\rho^2}{2}$$

and

$$-\frac{n}{x_o'} + \frac{1}{x_i'} = -\frac{n-1}{r_2} - \frac{n-1}{n^2}\left(\frac{1}{r_2} - \frac{1}{x_i'}\right)^2\left(\frac{1}{r_2} - \frac{n+1}{x_i'}\right)\frac{\rho^2}{2}$$

a. Subtract these two equations.

b. Substitute the zero order into the first-order terms to eliminate $1/x_i'$:

$$\frac{1}{x_i'} = \frac{1}{x_o} + (n-1)\left(\frac{1}{r_1} - \frac{1}{r_2}\right)$$

c. Finally, obtain

$$-\frac{1}{x_o} + \frac{1}{x_i'} = (n-1)\left(\frac{1}{r_1} - \frac{1}{r_2}\right) + \frac{n-1}{n^2}\left[\left(\frac{1}{r_1} - \frac{1}{x_o}\right)^2\left(\frac{1}{r_1} - \frac{n+1}{x_o}\right)\right.$$
$$\left. -\left(\frac{1}{x_o} + \frac{n-1}{r_1} - \frac{n}{r_2}\right)^2\left(\frac{n^2}{r_2} - \frac{n+1}{x_o} - \frac{n^2-1}{r_1}\right)\right]\frac{\rho^2}{2}$$

Note: Longitudinal spherical aberration (L.S.A.) is defined as the distance differential for $\rho \neq 0$ and $\rho = 0$; that is,

$$\text{L.S.A.} = x_i(\rho \neq 0) - x_i'(\rho = 0)$$

and may be expressed as:

$$\text{L.S.A.} = \left(\frac{1}{x_i'(\rho = 0)} - \frac{1}{x_i(\rho \neq 0)}\right)x_i(\rho \neq 0)x_i'(\rho = 0)$$

3. *The parameters π and σ.* We introduce the parameters

$$\pi = \frac{x_i' + x_o}{x_i' - x_o}$$

and

$$\sigma = \frac{r_2 + r_1}{r_2 - r_1}$$

a. Using

$$\frac{1}{x_i'} - \frac{1}{x_o} = \frac{1}{f} = (n-1)\left(\frac{1}{r_1} - \frac{1}{r_2}\right)$$

find the expressions

$$\frac{1}{x_o} = -\frac{\pi + 1}{2f}, \quad \frac{1}{x_i} = \frac{1 - \pi}{2f}$$
$$\frac{1}{r_1} = \frac{\sigma + 1}{2f(n-1)}, \quad \frac{1}{r_2} = \frac{\sigma - 1}{2f(n-1)}$$

b. Introducing the parameters π and σ into the result of problem 2, show that we finally get

$$-\frac{1}{x_o} + \frac{1}{x_i'} = (n-1)\left(\frac{1}{r_1} - \frac{1}{r_2}\right) + \frac{\rho^2}{f^3}(A\sigma^2 + B\sigma\pi + C\pi^2 + D)$$

where

$$A = \frac{n+2}{8n(n-1)^2}, \quad B = \frac{n+1}{2n(n-1)}$$

$$C = \frac{3n+2}{8n}, \quad D = \frac{n^2}{8(n-1)^2}$$

4. *Thin lens and spherical aberration.* Can the spherical aberration be corrected for a thin lens? If the expression

$$Y = A\sigma^2 + B\sigma\pi + C\pi^2 + D$$

could be equal to zero for specific values of σ, π, and n, then we would have no spherical aberration. Y can be plotted as a function of σ and will give a family of parabolas depending on n and π as parameters.

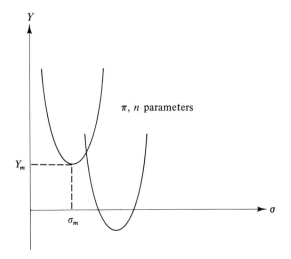

Now we want to show that $Y = 0$ is not possible for any value of σ if $n = 1.5$.

a. Calculate A, B, C, and D for $n = 1.5$.

b. Show that the minimum value of $Y(\sigma)$ is for

$$\sigma_m = -\frac{B\pi}{2A} = -0.71\pi$$

c. Show that

$$Y(\sigma_m) = 4.464\sigma_m^2 + 1.125$$

and that, consequently, there is no real solution for σ_m.

We see that for a single thin lens, spherical aberration cannot be eliminated for any variation of x_o, x_i, r_1, or r_2, given $n = 1.5$. Nor is it possible to eliminate aberration for other practical values of n.

5. *Coma.* In a similar way as discussed for spherical aberration, we can obtain an expression for the distance Q_0' to Q_4' (see Figure 13.13). Q_0' refers to the image point of the chief ray; Q_4' refers to the image point of the two outermost rays. This expression is

$$Q_0'Q_4' = \frac{3\rho^2 \tan \beta}{2fn(n-1)}[(n+1)\sigma + (2n+1)(n-1)\pi]\frac{1}{1-\sigma}$$

where

$$\pi = \frac{x_i + x_o}{x_i - x_o} \quad \text{and} \quad \sigma = \frac{r_2 + r_1}{r_2 - r_1}$$

and where β is the angle made by the chief ray with the axis, and where ρ is the distance of the zone being considered from the axis, and f is the focal length of the lens.

a. Show that coma is eliminated if the condition

$$\sigma = -(2n+1)\frac{n-1}{n+1}\pi$$

is met.

b. Calculate σ for the case where the object is at infinite distance and $n = 1.5$.

c. What is the percentage difference of the values of σ for no coma and σ for minimum spherical aberration for $\pi = -1$?

Ans. b. $\sigma = 0.8$; c. 11.25%

6. *Aplanatic lens.* A lens free of spherical aberration and coma is called an aplanatic lens. For such a lens, the sine condition is fulfilled. We know that spherical aberration can only be eliminated if object and image are not both real. For the elimination of coma, we found that

$$\sigma = -\frac{(2n+1)(n-1)}{n+1}\pi$$

a. Show that the two values

$$\sigma = -(2n+1) \quad \text{and} \quad \pi = \frac{n+1}{n-1}$$

make the expressions for spherical aberration and coma both equal to zero.

b. Introduce these values of σ and π into the following expressions discussed in problem 3:

$$\frac{1}{x_o} = -\frac{\pi+1}{2f}, \quad \frac{1}{x_i} = \frac{1-\pi}{2f}$$

$$\frac{1}{r_2} = \frac{\sigma-1}{2(n-1)f}, \quad \frac{1}{r_1} = \frac{\sigma+1}{2(n-1)f}$$

Derive the following relations:

$$f = \frac{n+1}{1-n}r_2, \quad x_o = \frac{n+1}{n}r_2, \quad x_i = (n+1)r_2, \quad r_1 = \frac{n+1}{n}r_2$$

from which we get

$$r_2 = \frac{n}{n+1}r_1, \quad x_o = r_1, \quad x_i = nr_1.$$

These are the formulas used in the discussion of the aplanatic lens (Section 2 of this chapter).

c. Consider an aplanatic lens of refractive index $n = 1.5$ and radius of curvature $r_1 = -5$ cm. Calculate r_2, x_o, x_i, and f, and confirm that the thin-lens equation is fulfilled.

d. Make a sketch of the positions of x_o, x_i, and r_2.

7. *Aplanatic sphere.* Let us now treat the aplanatic sphere with the imaging equation at one surface. We have from Chapter 1:

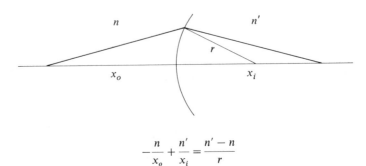

$$-\frac{n}{x_o} + \frac{n'}{x_i} = \frac{n' - n}{r}$$

a. Write the imaging equation for one surface of a glass sphere, where $n = 1.5$ and $n' = 1$.

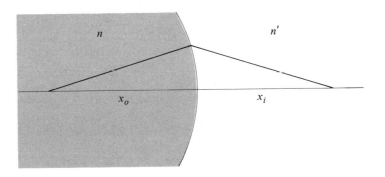

b. Assume for

$$x_0 = -\frac{n+1}{n}r$$

(we set $r_2 = -r$) that is the aplanatic distance of point P at x_o from the curved surface. Show that we get $x_i = nx_o$, that is the aplanatic distance for $x_i = P'$ from the curved surface.

8. *Astigmatism at a single surface.* We first consider the focal points in the horizontal plane (sagittal) (see Figure 13.17 and following figure). We measure x_o along the line

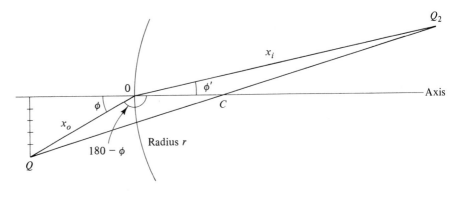

$Q0$ and x_i along the line $0Q_2$ (as in the paraxial case). For the triangles, we get

$$Q0C = \tfrac{1}{2}r|x_o|\sin\phi$$
$$0CQ_2 = \tfrac{1}{2}rx_i\sin\phi'$$
$$Q0C + 0CQ_2 = Q0Q_2 = \tfrac{1}{2}|x_o|x_i\sin(180 - \phi + \phi')$$

or

$$r|x_o|\sin\phi + rx_i\sin\phi' = |x_o|x_i\sin(\phi - \phi')$$

a. Apply

$$\sin(\alpha + \beta) = \sin\alpha\cos\beta + \cos\alpha\sin\beta$$

and

$$\sin\phi = n\sin\phi'$$

and obtain

$$-\frac{1}{x_o} + \frac{n}{x_i} = \frac{n\cos\phi' - \cos\phi}{r}$$

where at the end we set $|x_o| = -x_o$ according to our sign convention. We now consider the vertical (or meridional) focal points (see Figure 13.17 and following figure).

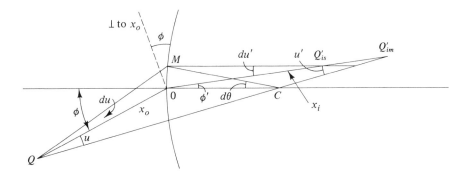

For small angles, we have

$$du = \frac{OM\cos\phi}{|x_o|}, \quad |d\theta| = \frac{OM}{r}, \quad |du'| = \frac{OM\cos\phi'}{x_i}$$

We differentiate $\sin\phi = n\sin\phi'$; that is,

$$\cos\phi\, d\phi = n\cos\phi'\, d\phi'$$

We have

$$\phi = u + |\theta|, \quad \phi' = |\theta| - |u'|$$

that is,

$$(du + |d\theta|)\cos\phi = n(|d\theta| - du')\cos\phi'$$

b. Combine the preceding relations and arrive at

$$-\frac{\cos^2\phi}{x_o} + \frac{n\cos^2\phi'}{x_i} = \frac{n\cos\phi' - \cos\phi}{r}$$

When $\phi = \phi' = 0$, both focal points Q_1 and Q_2 are determined by the same equation

$$-\frac{1}{x_o} + \frac{n}{x_i} = \frac{n-1}{r}$$

We choose $x_o = -40.000$ cm, $r = 10.000$ cm, and $n = 1.5000$

c. Find x_i.

d. For $r = 10.000$ cm, $x_o = -40.000$ cm, $n = 1.5000$, and $\phi = 3°$, calculate the values of x_{is} from (a) and x_{im} from (b), where s and m stand for sagittal and meridional, respectively, and give the difference $x_{is} - x_{im}$ in centimeters.

Ans. $x_{is} - x_{im} = .417$ cm

9. *Achromatic doublet with $t = 0$.* In Chapter 1, we obtained the following focal lengths for two thin lenses in contact:

$$\frac{1}{f} = \frac{1}{f_1} + \frac{1}{f_2} = (n_1 - 1)\left(\frac{1}{r_1} - \frac{1}{r_2}\right) + (n_2 - 1)\left(\frac{1}{r_1'} - \frac{1}{r_2'}\right)$$

where n_1, r_1, r_2 and n_2, r_1', r_2' are as indicated:

$$\begin{pmatrix} & \text{Medium} & \\ & n_1 & \\ r_1 & & r_2 \\ & \text{Lens 1} & \end{pmatrix} \qquad \begin{pmatrix} & \text{Medium} & \\ & n_2 & \\ r_1' & & r_2' \\ & \text{Lens 2} & \end{pmatrix}$$

We abbreviate

$$\frac{1}{r_1} - \frac{1}{r_2} = k_1 \quad \text{and} \quad \frac{1}{r_1'} - \frac{1}{r_2'} = k_2$$

then we have

$$\frac{1}{f} = (n_1 - 1)k_1 + (n_2 - 1)k_2$$

The focal length should be independent of the refractive index in the wavelength region λ_{blue} to λ_{red}. The condition for $1/f$ is

$$\frac{d}{d\lambda}\left(\frac{1}{f}\right) = \frac{d}{d\lambda}[(n_1 - 1)k_1 + (n_2 - 1)k_2] = 0$$

or

$$\frac{dn_1}{d\lambda}k_1 + \frac{dn_2}{d\lambda}k_2 = 0$$

or

$$\frac{k_1}{k_2} = -\frac{dn_2}{dn_1}$$

We approximate

$$dn_1 = n_{1B} - n_{1R} \quad \text{and} \quad dn_2 = n_{2B} - n_{2R}$$

where n_B stands for the refractive index of blue light and n_R for the refractive index of red light. We also introduce the medium refractive index n_D, where D stands for the sodium line. We use the medium refractive index to relate k_1 and f_1 and k_2 and f_2

$$k_1 = \frac{1}{f_1(n_1 - 1)} = \frac{1}{f_1(n_{1D} - 1)}$$

$$k_2 = \frac{1}{f_2(n_2 - 1)} = \frac{1}{f_2(n_{2D} - 1)}$$

We define

$$V_1 = \frac{n_{1B} - n_{1R}}{n_{1D} - 1} \quad \text{and} \quad V_2 = \frac{n_{2B} - n_{2R}}{n_{2D} - 1}$$

a. Show that we have

$$\frac{f_2}{f_1} = -\frac{V_2}{V_1}$$

b. Assume the following values:

$$n_{1D} = 1.700 \quad n_{2D} = 1.500$$
$$n_{1B} = 1.735 \quad n_{2B} = 1.525$$
$$n_{1R} = 1.665 \quad n_{2R} = 1.475$$

What follows for f_1 and f_2, and what type of lenses must be used?

c. The resulting lens combination of problem (b) does not act as a lens. What condition must be placed on V_1 and V_2 to have a converging lens?

10. *Achromatic doublet with $t \neq 0$.* For two thin lenses at a distance t apart, we have the following focal length according to Chapter 1:

$$\frac{1}{f} = \frac{1}{f_1} + \frac{1}{f_2} - t\left(\frac{1}{f_1 f_2}\right)$$

For one lens, we have

$$\frac{1}{f} = (n - 1)\left(\frac{1}{r_1} - \frac{1}{r_2}\right)$$

Differentiation with respect to n yields

$$\Delta\left(\frac{1}{f}\right) = \Delta n\left(\frac{1}{r_1} - \frac{1}{r_2}\right) = \frac{\Delta n}{f(n - 1)}$$

In problem 9, we defined

$$V_1 = \frac{n_{1B} - n_{1R}}{n_{1D} - 1}$$

which is now written as

$$V = \frac{\Delta n}{n - 1}$$

and we have

$$\Delta\left(\frac{1}{f}\right) = \frac{V}{f}$$

Differentiation of $1/f$ for the two-lens system gives us

$$\Delta\left(\frac{1}{f}\right) = \frac{V_1}{f_1} + \frac{V_2}{f_2} - t\left[\frac{1}{f_2}\Delta\left(\frac{1}{f_1}\right) + \frac{1}{f_1}\Delta\left(\frac{1}{f_2}\right)\right]$$

$$= \frac{V_1}{f_1} + \frac{V_2}{f_2} - t\left(\frac{V_1}{f_1 f_2} + \frac{V_2}{f_1 f_2}\right)$$

a. Set

$$\Delta\left(\frac{1}{f}\right) = 0$$

and show that we require the following condition for no chromatic aberration:

$$t = \frac{V_1 f_2 + V_2 f_1}{V_2 + V_1}$$

b. Show that the condition for no chromatic aberration with two lenses of equal material is that they are separated by a distance of

$$t = \frac{f_1 + f_2}{2}$$

Detectors | **14**

1. INTRODUCTION

We all use our eyes to detect light. While the range of orders of magnitude of intensities the eye can detect is very large, the spectral range is relatively narrow. It covers a wavelength range of about 0.4 μm (400 nm) to 0.7 μm (700 nm). If high-intensity light from the sun or lasers directly enters the eye, the eye can get damaged. For this reason, we must ensure a proper attenuation with filters.

The photographic plate is sensitive to the spectral range detected by the eye, but in addition stretches to both longer and shorter wavelengths. In the shorter-wavelength region, it covers quite a large range and can be used for the detection of X rays. The photographic plate detects light quanta and integrates the incoming light. The response is not linear, and photometric calibration curves are used for the quantitative comparison of measured light. For wavelengths longer than the visible range, we presently use semiconductor detectors and thermal detectors. Semiconductor detectors are also photon detectors.

In this chapter, we will discuss thermal detectors and photon detectors for a frequency range extending from the visible to the far infrared. Although photon detectors are more useful in the shorter-wavelength region and thermal detectors in the longer-wavelength region, some detectors cover a large overlapping range. Each detector has its specific frequency range of response, speed of response characterized by the time constant, and its specific sensitivity, mostly characterized by **D***, inversely related to the **NEP**. This abbreviation stands for **noise equivalent power** and corresponds to the smallest signal detectable. If the signal is equal to the noise, it cannot be detected, and we are operating at the limit of the noise equivalent power of the detector.

First we will consider noise for a thermal detector and an electrical resistor. Then we will discuss the bolometer as a representative thermal detector, as well as the noise involved. Other thermal detectors will also be presented. Finally, we will discuss photoconductive detectors as an example of photon detectors and discuss their physical principles and characterization. Photovoltaic and other related detectors will be considered as well.

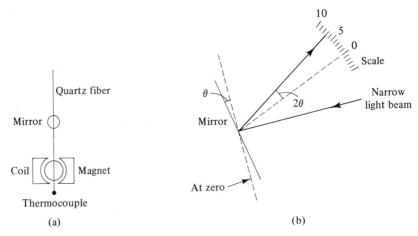

Fig. 14.1
Galvanometer. (a) Schematic of a galvanometer. (b) Reading the deflection with a light beam.

2. THE GALVANOMETER AND NOISE

The **galvanometer** was an important radiation detector many years ago. By examining this detector, we can understand the influence of thermal motion on the production of random fluctuations of the final reading. This is called the **noise** of the instrument. In the case of the galvanometer, noise is produced by Brownian motion of the molecules around the movable parts of the instrument. In Figure 14.1, the working principle of the galvanometer is explained. The radiation is absorbed by a thermocouple, and a current is produced in a coil hanging on a quartz fiber for possible torsional motion. The magnetic moment of the coil interacts with the magnetic field of a permanent magnet. The torsional motion is detected by a light beam reflected from a small mirror attached to the torsional fiber.

A. Noise from Brownian Motion

If no light is falling onto the thermocouple, we might expect that the reading of the angle 2θ is zero. This is not the case. The random movement of the air molecules hitting the mirror from both sides causes random deviations of the mirror from the zero position.

If θ is the angle of deviation of the mirror from the zero position, then the time average is $\bar{\theta} = 0$. A measure of the fluctuation is obtained through the average of $\overline{\theta^2}$, since here the random distribution of negative and positive deviations do not cancel each other.

We can obtain a value of $\overline{\theta^2}$ from statistical considerations. The fluctuating motion of the mirror can be described by an oscillator. The moment of inertia of the movable parts is called M, and the force constant of the torsional motion is G. We then have

$$\tfrac{1}{2}M\dot{\theta}^2 \quad \text{for kinetic energy} \tag{14.1}$$

and

$$\tfrac{1}{2}G\theta^2 \quad \text{for potential energy} \tag{14.2}$$

where $\dot{\theta}$ is $d\theta/dt$. From statistical mechanics, we know that for a microoscillator in thermal equilibrium, the kinetic energy and potential energy each carry on the average an amount of energy equal to $\tfrac{1}{2}kT$, where T is the absolute temperature and k the Boltzmann constant.*

Taking the time average, we have

$$\begin{aligned}
\tfrac{1}{2}M\overline{\dot{\theta}^2} &= \tfrac{1}{2}kT \\
\tfrac{1}{2}G\overline{\theta^2} &= \tfrac{1}{2}kT
\end{aligned} \tag{14.3}$$

The small deviations of the mirror are now regarded as the oscillations of a microoscillator. The average fluctuations of the galvanometer depend on the force constant of the fiber and the absolute temperature. No damping has been considered.

B. The Forced, Damped Oscillator and Noise Spectrum

To consider damping of the instrument, we describe the galvanometer as a forced, damped oscillator,

$$M\frac{d^2\theta}{dt^2} + D\frac{d\theta}{dt} + G\theta = \text{driving force} \tag{14.4}$$

is the driving force. This equation has been discussed in detail in Chapter 7. The damping constant is called D. Since no radiation is falling onto the detector, the oscillator is driven by random short pulses hitting, e.g., the mirror on both sides. The frequency spectrum of each short pulse is large, as seen from the time-dependent Fourier transformation of Gaussian functions; see Figure 14.2. In Chapters 8 and 10 we discussed Fourier transformation between the frequency and space domain. Here the Fourier transformation is between the frequency and time domain.

We approximate the frequency spectrum by a rectangle, as shown in Figure 14.2. The waves making up the pulses have the same amplitude A_f and frequencies between 0 and f_h. Considering one component of these waves as the driving force, we have

$$M\frac{d^2\theta}{dt^2} + D\frac{d\theta}{dt} + G\theta = A_f e^{i2\pi ft} \tag{14.5}$$

Using a trial solution of

$$\theta = \theta_f e^{i2\pi ft}$$

where $2\pi f = \omega$ is the angular frequency,

---

*Paul Tipler, *Modern Physics* (New York: Worth, 1969).

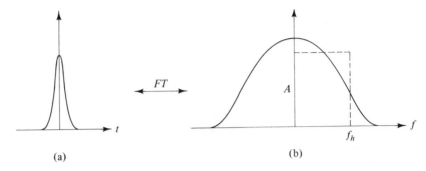

(a)   (b)

Fig. 14.2
The Fourier transformation of Gaussian functions and noise spectrum. A short pulse in the time domain (a) corresponds to a wide pulse in the frequency domain (b). In the frequency domain, the frequency distribution is approximated by a rectangle with amplitude A_f, extending to the highest frequency f_h.

we obtain

$$\theta_f = \frac{A_f}{(G - M\omega^2) + i\omega D} \tag{14.6}$$

(see problem 1). We are interested in $\overline{\theta^2}$, and for the absolute value of equation 14.6, we have

$$\overline{\theta_f^2} = \frac{\overline{A_f^2}}{(G - \omega^2 M)^2 + \omega^2 D^2} \tag{14.7}$$

(see problem 2). $\overline{A_f^2}$ is the time average of the square of the amplitude of each noise wave of frequency f. We have assumed that all the A_f's have the same absolute value. Therefore, we can express the summation of all the $\overline{A_f^2}$ values in the frequency interval Δf as

$$\sum \overline{A_f^2} = F \Delta f \tag{14.8}$$

where F is a constant. We may express one of the $\overline{A_f^2}$ values as

$$\overline{A_f^2} = F \, df \tag{14.9}$$

From equation 14.3, the result of thermodynamic considerations, we obtained $\overline{\theta^2} = kT/G$ as an average over all frequencies. Therefore, we average equation 14.7 over all frequencies. Since we assumed that there are no wave components in the pulse spectrum for frequencies higher than a so-far unspecified frequency f_h, we can integrate from 0 to ∞ and obtain according to R. A. Smith et. al*

$$\overline{\theta^2} = \frac{1}{2\pi} \int_0^\infty \frac{F \, d\omega}{(G - \omega^2 M)^2 + \omega^2 D^2} = \frac{F}{4GD} \tag{14.10}$$

Combining equations 14.10 and 14.3 gives us

$$F = 4DkT \tag{14.11}$$

---

*R. A. Smith, F. E. Jones, and R. P. Chasmar, *The Detection and Measurement of Infrared Radiation* (Oxford, Eng.: The Clarendon Press, 1957).

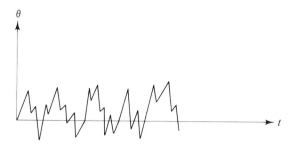

Fig. 14.3
Fluctuations of θ depending on time for the case of a small damping constant.

Introducing F from equation 14.11 into equation 14.9 yields

$$\overline{A_f^2} = 4Dk\,Tdf \tag{14.12}$$

The result is a value for $\overline{A_f^2}$ depending on the damping constant D and the temperature T.

We introduce equation 14.12 into equation 14.7 and get the time average of the square of the deviation θ corresponding to frequency f:

$$\overline{\theta_f^2} = \frac{4k\,TD\,df}{(G - \omega^2 M)^2 + \omega^2 D^2} \tag{14.13}$$

Of interest is the case where the damping constant is small. In that case, the oscillator responds to only a small frequency interval of the pulse spectrum close to the resonance frequency of the oscillator, $\omega_0^2 = G/M$. The appearance of the deviation from zero by the oscillator is sinusoidal. But since we are not dealing with a single frequency, it appears distorted (Figure 14.3).

3. THE NOISE OF A RESISTOR: JOHNSON NOISE

If we measure the voltage across a resistor using a very sensitive method, we will find voltage fluctuations. As in the case of the galvanometer, the fluctuations are of thermal origin. We want to apply the model of the forced damped oscillator to the noise problem of a resistor. Let us consider the electrical circuit shown in Figure 14.4. The fluctuations in the voltage are considered to be generated by thermal fluctuations of the electrons in the resistor and are described by a fictitious emf (electromotive force) of voltage u. For the energy balance of the circuit, we have

$$Iu = \frac{d}{dt}\left(\frac{1}{2}LI^2\right) + \frac{d}{dt}\left(\frac{1}{2C}q^2\right) + RI^2 \tag{14.14}$$

where I is the current, q is the charge on the capacitor, L is the inductance, C is the capacitance, and R is the resistance, as shown in Figure 14.4a (see problem 3). After differentiation, cancellation of I, and a second differentiation, we have

$$\dot{u} = L\frac{d^2I}{dt^2} + \frac{1}{C}I + R\frac{dI}{dt} \tag{14.15}$$

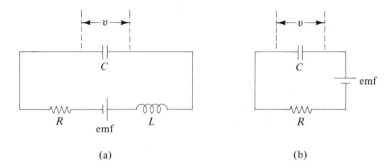

Fig. 14.4
(a) Equivalent circuit for noise generation in a resistor. R is resistance, C is capacitance, L is inductance, emf is the noise-generating source, and v is the voltage across the capacitor. (b) If L is small, we may neglect it and have the R-C circuit as shown.

Considering the trial solutions

$$u = u_f e^{i\omega t} \quad \text{and} \quad I = I_f e^{i\omega t}$$

where f stands for one frequency only, we obtain

$$\frac{I_f}{i\omega C} = \frac{u_f}{(1 - LC\omega^2) + i\omega RC} \tag{14.16}$$

(see problem 3). The left side of the equation is the voltage v_f across the capacitor (see Figure 14.4a). By squaring equation 14.16 and taking the time average, we get

$$\overline{v_f^2} = \frac{\overline{u_f^2}}{(1 - LC\omega^2)^2 + \omega^2(RC)^2} \tag{14.17}$$

This equation is the analog of equation 14.7.

We want to determine $\overline{u_f^2}$ and assume a noise frequency spectrum similar to what has been discussed for the galvanometer shown in Figure 14.2. In analogy to equations 14.8 and 14.9, we have

$$\sum \overline{u_f^2} = F' \Delta f \tag{14.18}$$

or

$$\overline{u_f^2} = F' \, df \tag{14.19}$$

where F' is a constant.

Introducing equation 14.19 into equation 14.17 and averaging over all frequencies, we get

$$\overline{v^2} = \frac{1}{2\pi} \int_0^\infty \frac{F' \, d\omega}{(1 - LC\omega^2)^2 + \omega^2(RC)^2} \tag{14.20}$$

The result of the integration is obtained by substituting the values of F', 1, LC, and RC, for F, G, M, and D, respectively, into equation 14.10. We then obtain

$$\overline{v^2} = \frac{F'}{4RC} \tag{14.21}$$

14. Detectors

Substituting equation 14.21 into equation 14.19

$$\overline{u_f^2} = \overline{v^2} 4RC \, df \tag{14.22}$$

A statistical consideration analogous to what we discussed for the galvanometer is applied to the average of the potential energy of the circuit (without damping). We have

$$\tfrac{1}{2}C\overline{v^2} = \tfrac{1}{2}kT \tag{14.23}$$

Combining equations 14.22 and 14.23 yields

$$\overline{u_f^2} = 4RkT \, df \tag{14.24}$$

This is called **Nyquist's formula**. It tells us that the time average of the square of the noise voltage (fictitious emf voltage) depends on the temperature T, the resistance R, the frequency interval df, and the Boltzmann constant k.

Introducing equation 14.24 into equation 14.17 gives us (for the frequency $f = \omega/2\pi$)

$$\overline{v_f^2} = \frac{4RkT \, df}{(1 - LC\omega^2)^2 + \omega^2(RC)^2} \tag{14.25}$$

The interpretation of this result is different from the mechanical case discussed previously. The inductance L analogous to the inertia term in the mechanical case is so small that we may write

$$\overline{v_f^2} = \frac{4RkT \, df}{1 + \omega^2(RC)^2} \tag{14.26}$$

We have no resonance. As in the mechanical case, the damping is small. Since L is small, we have the R-C circuit shown in Figure 14.4b. The differential equation 14.15 can now be written as

$$\frac{(RI)}{RC} + \frac{d(RI)}{dt} = 0 \tag{14.27}$$

where we have set $\dot{u} = 0$ because we are now interested in the frequency response of the R-C circuit. With $RI = v$, we have

$$\frac{v}{RC} + \dot{v} = 0 \tag{14.28}$$

with the solution

$$v = v_0 e^{-t/RC} \tag{14.29}$$

The product RC has the dimension of time and is called the time constant τ of the circuit. If the time constant is small, the circuit responds quickly to an applied sinusoidal voltage. We introduce the time constant τ for RC in equation 14.26 and get

$$\overline{v_f^2} = \frac{4RkT \, df}{1 + \omega^2\tau^2} \tag{14.30}$$

3. The Noise of a Resistor: Johnson Noise

For most applications, $\omega^2\tau^2 \ll 1$ and

$$\overline{v_f^2} = 4RkT\,df \qquad (14.31)$$

This is the equation for **Johnson noise**. The result is that the voltage fluctuation across the resistor is independent of the frequency spectrum of the noise-generating source. It depends only on R and T. The subscript f will sometimes be omitted.

We have expressed the frequency interval of f as df. Very often, this interval is set equal to 1 cycle per second (cps) and not explicitly written in the formula. When numerical calculations are done, we must take into account the dimension of $1/s$.

For example, suppose that $\Delta f = 1$ cps $= 1$ Hz, $R = 1 \times 10^6\ \Omega$, $T = 300$ K, and $k = 1.381 \times 10^{-23}$ J/K. Then

$$\sqrt{\overline{v^2}} = 12.8 \times 10^{-8}\ \text{V} \qquad (14.32)$$

This voltage is generated by thermal fluctuations of the electrons and appears across the resistor.

4. THE BOLOMETER

The **bolometer** is an example of a thermal detector. The helium-cooled bolometer is a very sensitive detector for the entire infrared region. It is commercially available and is used in the laboratory and for field measurements. It has been used in balloon experiments for the observation of radiation from outer space.

The incident radiation is absorbed by an absorbing layer, and a resistor is heated. The change in resistivity is a measure of the incident radiation. With the application of a bias voltage, the change in resistivity is transformed into a change in voltage. It is usually easier to measure and amplify an alternating voltage than a constant voltage. This is done by placing a chopper before the detector. During the heating cycle, the change in resistivity goes in one direction; during the dark cycle, it goes back.

The detector element is mounted in such a way that there is good heat contact to the cooling "heat bath." This bath is usually at helium temperature, that is, 4.2 K. The incident radiation is absorbed during the heating cycle, and the temperature of the detector element rises; during the dark cycle, the temperature decreases. The construction and mounting of the bolometer must be done in a sensitive way to accomplish optimum radiation and heating-cooling response. We will discuss the electrical and thermal functioning of the bolometer and obtain the responsivity. Then we will discuss the thermal and electrical noise and obtain an expression for the total noise and its characterization as noise equivalent power (NEP).

A. Electrical Functioning

A schematic of the bolometer's electric circuit is shown in Figure 14.5. The

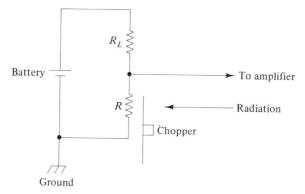

Fig. 14.5
Electric circuit for a bolometer. R_L is the load resistor, and R the detector resistor.

temperature coefficient of resistance, α, is defined as

$$\alpha = \frac{1}{R}\left(\frac{dR}{dT}\right) \quad \left[\frac{1}{K}\right] \tag{14.33}$$

where R is the resistance at temperature T. For semiconductors, we can present the resistivity as $R = R_0 e^{A/T}$. Thus, $\alpha = -A/T^2$ and α is negative. With $A = 3000$ K, we find $\alpha = -0.033$ [1/K] for $T = 300$ K.

From equation 14.33 the change in the detector resistance is given as

$$\Delta R = \alpha R\, \Delta T \tag{14.34}$$

where ΔT is the corresponding temperature change. A battery supplies the voltage V across the resistor. A load resistor R_L is used, and for the change in voltage depending on the change in resistivity, we obtain

$$\Delta v = \frac{R_L V}{(R_L + R)^2} \Delta R \tag{14.35}$$

(see problem 6). Using $V = (R_L + R)I$, equation 14.35 can be rewritten as

$$\Delta v = \frac{I R_L \Delta R}{R_L + R} \tag{14.36}$$

Since in most practical cases we have R small compared to R_L, we have approximately

$$\Delta v = I\, \Delta R \tag{14.37}$$

B. Thermal Functioning

The detector is mounted as shown in Figure 14.6. The temperature of the surroundings is T_0. The detector element absorbs the signal radiation ΔW and the background radiation B and is heated by the bias current with the power $W_0 = R I^2$. Heat is dissipated through the wires and by re-radiation. However, re-radiation is small and can be neglected.

First let us consider the balance of heat for the case where $\Delta W = 0$ and the

4. The Bolometer   473

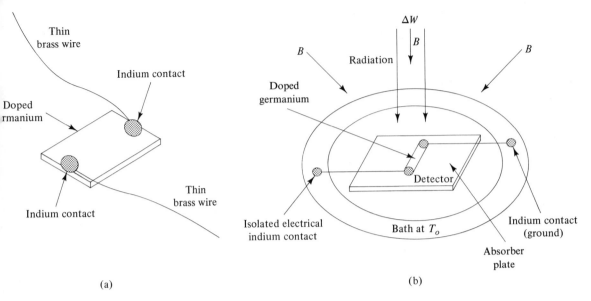

Fig. 14.6
Mounting of the bolometer with respect to thermal conduction. (a) A germanium chip detector. [F. Low, *J. Opt. Soc. Am.* **51**, 1300 (1961). (b) A compound detector. [See N. S. Nishioka, P. L. Richards, and D. P. Wood, *Appl. Opt.* **17**, 1562 (1978).]

temperature of the detector is T. We have

$$\text{Change of heat} = \text{Heat "in" minus heat "out"}$$

The change of the heat content of the detector is

$$\tilde{C}\frac{d(T - T_0)}{dt}$$

where $\tilde{C}$ is the heat capacity of the detector, and we have

$$\tilde{C}\frac{d(T - T_0)}{dt} = (W_0 + B) - \tilde{G}_0(T - T_0) \tag{14.38}$$

where $\tilde{G}_0$ $[W/K]$ is the constant of heat conduction, often expressed as heat resistance $\tilde{G}_0 = 1/\tilde{R}_0$. Now we consider the fact that the incident power $\Delta W e^{i\omega t}$ falls on the detector and changes the temperature to $T + \Delta T$. We then have

$$\tilde{C}\frac{d(T - T_0 + \Delta T)}{dt} = (W + B + \Delta W e^{i\omega t}) - \tilde{G}_0(T - T_0 + \Delta T) \tag{14.39}$$

where W is now the heat produced in the detector by the bias current. This is only slightly different from W_0, but the difference is important. We expand W in a power series with respect to the temperature:

$$W = W_0 + \frac{dW_0}{dT}\Delta T \tag{14.40}$$

With equation 14.33, we have

$$I^2 R \alpha = \frac{d(RI^2)}{dT} \rightarrow W_0 \alpha = \frac{dW_0}{dT} \tag{14.41}$$

and we can write

$$W = W_0 + \alpha W_0 \Delta T \tag{14.42}$$

For the heat balance equation for the case when radiation is falling onto the detector, we have

$$\tilde{C}\frac{d(T - T_0 + \Delta T)}{dt} = (W_0 + \alpha W_0 \Delta T + B + \Delta W e^{i\omega t}) - \tilde{G}_0(T - T_0 + \Delta T) \tag{14.43}$$

From this equation, we subtract equation 14.38 and get

$$\tilde{C}\frac{d(\Delta T)}{dt} + (\tilde{G}_0 - \alpha W_0)\Delta T = \Delta W e^{i\omega t} \tag{14.44}$$

We set

$$\tilde{G}_0 - \alpha W_0 = \tilde{G} = \frac{1}{\tilde{R}} \tag{14.45}$$

and must solve

$$\tilde{C}\frac{d(\Delta T)}{dt} + \tilde{G}\Delta T = \Delta W e^{i\omega t} \tag{14.46}$$

If the right side of equation 14.45 is zero, then it is a homogeneous equation, and we must solve an equation similar to equation 14.27. The result for the homogeneous equation is

$$\Delta T = \Delta T_0 e^{-\frac{t}{\tilde{R}\tilde{C}}} \tag{14.47}$$

The quantity $\tilde{R}\tilde{C}$ is called the thermal time constant τ_T and in analogy to equation 14.29 is larger than zero. Since $\tilde{C} > 0$, we must have $\tilde{R} > 0$, or as a consequence, $\tilde{G} > 0$.

From equation 14.45, it follows that

$$\tilde{G}_0 > \alpha W_0 \tag{14.48}$$

which states that the bias current must not overheat the bolometer. The solution for the inhomogeneous equation of equation 14.46 is analogous to the solution of equation 14.5 (see problem 1). We just have to set $M = 0$, $\tilde{C} = D$, $1/\tilde{R} = G$, $\Delta T = \theta$, and $\Delta W = F$, and we have

$$\Delta T = \frac{\Delta W}{\sqrt{(1/\tilde{R})^2 + \omega^2 \tilde{C}^2}} \tag{14.49}$$

For most applications, the product of the thermal time constant and the frequency of the incident power (chopping frequency) is small compared to 1, that is, $\omega^2 \tilde{R}^2 \tilde{C}^2 \ll 1$, and we have

$$\Delta T = \tilde{R}\Delta W \tag{14.50}$$

C. Responsivity

The **responsivity** r of the bolometer is defined as the change in voltage divided by the change in incident power:

$$r = \frac{\Delta v}{\Delta W} \tag{14.51}$$

Using equations 14.34, 14.37, and 14.50, we have

$$r = \frac{I \, \Delta R}{\Delta T} \tilde{R} = \alpha I R \tilde{R} \tag{14.52}$$

Let us consider a semiconducting bolometer for which $T = 315$ K, $T_0 = 300$ K, $\alpha = -0.03$ 1/K, $\tilde{G}_0 = 10^{-4}$ W/K, and $R = 3 \times 10^6$ Ω. In equation 14.38, we have at equilibrium that the left side is zero:

$$W_0 + B = \tilde{G}_0(T - T_0) \tag{14.53}$$

With a filter at the right temperature, we can make B so small that it can be neglected, and we get

$$W_0 = \tilde{G}_0(T - T_0) \tag{14.54}$$

Equation 14.45 therefore gives us

$$\tilde{G} = \tilde{G}_0[1 - \alpha(T - T_0)] = 1.45 \times 10^{-4} \text{ W/K} \tag{14.55}$$

From equations 14.41 and 14.54, we have

$$W_0 = I^2 R = \tilde{G}_0(T - T_0) = 15 \times 10^{-4} \text{ W} \tag{14.56}$$

and can obtain I as 2.2×10^{-5} A. For the absolute value of the responsivity, $|r| = |\alpha I R \tilde{R}|$, we then have about 1.4×10^4 V/W.

D. Bolometers and Noise

The noise of the bolometer is a combination of thermal noise and electrical noise.

Thermal noise. The temperature of the detector will fluctuate at the equilibrium condition even when radiation is not falling onto the detector. This is a similar situation as discussed for the galvanometer. But our present model for evaluation of the statistical thermodynamic variations is not an oscillator model. The fluctuations are coming from the motion of the atoms in the material. The atoms are not moving separately, but are coupled to their neighbors; therefore, we have the model of a linear chain of atoms undergoing fluctuations. The noise resulting from these fluctuations is called **phonon noise**. For such a system, thermodynamic considerations give us the following time average of the square of the energy fluctuations:

$$\overline{(\delta \tilde{E})^2} = kT^2 \tilde{C} \tag{14.57}$$

The thermal energy $\tilde{E}$ of a body fluctuates as

$$\delta \tilde{E} = \tilde{C} \, \delta T \tag{14.58}$$

Combining equations 14.57 and 14.58 gives us

$$\overline{(\delta T)^2} = \frac{kT^2}{\tilde{C}} \tag{14.59}$$

The next step is to analyze the frequency spectrum. We have a heat balance equation of the fluctuations similar to equation 14.44:

$$\tilde{C} \frac{d}{dt}(\delta T) + \frac{\delta T}{\tilde{R}} = \delta P \, e^{i\omega t} \tag{14.60}$$

where δP is the power of the noise, and ω is the frequency with which the detector is heated. We need to determine δP in the frequency range of the thermal noise frequencies. We can do this the same way we did it for equations 14.5 and 14.7 (galvanometer). The result is obtained by substituting the values δT, δP, 0, $1/\tilde{R}$, and $\tilde{C}$ for θ_f, A_f, G, M, and D, respectively, into equation 14.7. As a result, we obtain

$$\delta T = \frac{\delta P}{\sqrt{(1/\tilde{R})^2 + \omega^2 \tilde{C}^2}} \tag{14.61}$$

Integration over all noise frequencies (see also problem 4) gives us, in analogy to equation 14.10 and for the frequency interval Δf,

$$\overline{(\delta T)^2} \, \Delta f = \frac{\overline{\delta P^2}}{4\tilde{C}/\tilde{R}} \tag{14.62}$$

Combining this result with the result from statistical thermodynamics, equation 14.59 yields a noise power of

$$\overline{(\delta P)^2} = \frac{4kT^2}{\tilde{R}} \Delta f \tag{14.63}$$

Electrical noise. The electrical noise is the Johnson noise of the resistor; that is, according to equation 14.31,

$$\overline{(\delta v)^2} = 4RkT \, \Delta f \tag{14.64}$$

where we have now written $\overline{(\delta v)^2}$ for $\overline{v_f^2}$ and Δf for df.

Combination of noise and noise equivalent power (NEP). The resulting noise from several noise contributions is obtained by adding together the time averages of the square of the noise power of the individual contributions. We write equation 14.51 as

$$\overline{W^2} = \frac{\overline{(\delta v)^2}}{r^2}$$

and for the combination of thermal and electrical noise (equations 14.63 and 14.64), we have

$$\overline{W_B^2} = \overline{(\delta P)^2} + \frac{\overline{(\delta v)^2}}{r^2} \tag{14.65}$$

Using equations 14.31 and 14.52, we get

$$\overline{W_B^2} = 4\frac{kT^2}{\tilde{R}}\Delta f + \frac{4kTR\,\Delta f}{(R\tilde{R}\alpha I)^2} = \frac{4kT^2}{\tilde{R}}\left(1 + \frac{1}{TR\tilde{R}\alpha^2 I^2}\right)\Delta f \qquad (14.66)$$

With equation 14.56 and assuming $\tilde{G}_0 \simeq \tilde{G}$, we can write $RI^2 = \tilde{G}(T - T_0)$. Using $\tilde{G} = 1/\tilde{R}$, we have

$$\overline{W_B^2} = \frac{4kT^2}{\tilde{R}}\left(1 + \frac{1}{T(T - T_0)\alpha^2}\right)\Delta f \qquad (14.67)$$

Using the numbers from our example, $\tilde{R} = 1.42 \times 10^4$ K/W, $T = 315$ K, $T_0 = 300$ K, and $\alpha = -0.03$ 1/K, we obtain for the second term of expression in large parentheses in equation 14.67

$$\frac{1}{T(T - T_0)\alpha^2} = 0.22 \qquad (14.68)$$

From equation 14.67, we see that this value compared to 1 is the ratio of Johnson noise to thermal noise. We multiply $\sqrt{\overline{W_B^2}}$ by $\sqrt{\Delta f} = \sqrt{1/s}$ and obtain for the noise equivalent power (NEP)

$$\text{NEP} = \sqrt{\overline{W_B^2}} = \sqrt{\frac{4kT^2}{\tilde{R}}(1.24)} \simeq 2 \times 10^{-11}\ \text{W} \qquad (14.69)$$

(see problem 8).

5. MODULATION NOISE

So far, we have discussed Johnson noise and thermal noise (phonon noise). There is another noise source, called **modulation** or **1/f noise**. This is the noise introduced by the sudden variation of incident light, such as by chopping the radiation for amplification. This noise is proportional to $1/f^n$, where n is between 0.8 and 2.

6. NEP AND D*

We have described the noise equivalent power, NEP, as the lowest signal voltage equivalent to the noise voltage. In the literature, we find an expression that is inversely proportional to the NEP, including the detector area a_D. It is called $D*$:

$$D* = \frac{\sqrt{a_D \Delta f}}{\text{NEP}}\left[\frac{\text{cm}\sqrt{\text{Hz}}}{\text{W}}\right] \qquad (14.70)$$

The noise bandwidth Δf is sometimes included, but for all applications it is set to 1/s and usually not explicitly mentioned. To indicate the wavelength at which $D*$ has been measured, the chopping frequency, and the noise bandwidth Δf, we give $D*$(wavelength, frequency, Δf). For example, if the wavelength is $\lambda = 10\ \mu m$, the modulation frequency is f in hertz and $\Delta f = 1/s$, then $D*$ is given as $D*(10, f, 1)$.

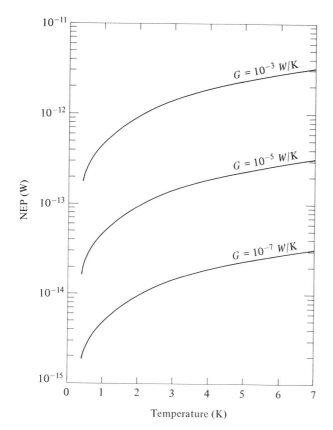

Fig. 14.7
NEP as a function of temperature and heat conduction (parameter G) as reported by
F. Low. [From F. Low, *J. Opt. Soc. Am.* **51**, 1300 (1961).]

7. HELIUM-COOLED BOLOMETERS

Bolometers used at helium-cooled temperatures are presently used for detection
over a broad wavelength range in the infrared and millimeter region. They are
used, for example, in the detection of radiation coming from stars, galaxies, and
molecular clouds, as well as the detection of cosmic background radiation. The
investigation is done with telescopes, balloons, or satellites. One such bolometer
was reported by F. Low (see Figures 14.6 and 14.7). It used doped germanium
as its radiation absorber and detector resistor. The leads for the bias voltage and
detector signal were made of thin brass wires with a diameter of about 25 μm
(see Figure 14.6a). These wires held the detector element in place and served for
the heat conduction of the absorbed radiation energy to the heat bath, which
was at helium temperature or lower. Figure 14.7 presents the dependence of the
NEP on temperature T and the heat conduction parameter G.

At helium temperature, as discussed in the last part of Section 4D for higher
temperatures, Johnson noise is about equal to thermal noise. If the temperature
is decreased below 4.2 K (e.g., by pumping on the helium), smaller NEP values

are possible. A compound bolometer using gallium-doped germanium as its resistor element is shown in Figure 14.6b. It has been reported to have an NEP of 3×10^{-15} when operated at 1.2 K. A Ge:In:Sb bolometer operated at 0.35 K is reported to have an NEP of 6×10^{-16} W.

8. THERMOCOUPLES AND THERMOPILES

The familiar **thermocouple**, used for temperature measurements in the laboratory, is also used as a radiation detector. If a number of thermocouples are attached in series, we have a **thermopile** (Figure 14.8). Each thermocouple consists of two different metals. To use a thermocouple as a simple radiation detector, we can use the absorbing area as one metal and make a junction on this area with another metal. A small piece of tellurium sheet is welded on one side to a Constantan wire (Figure 14.9). Both metals are connected to a sensitive ampere meter. To increase the absorption of radiation, we put some absorbing black on one side of the tellurium.

In Figure 14.10, we make a comparison of some thermal detectors and photodetectors, to be discussed shortly. The detectivity D^* is plotted over the wavelength. For example, a thin-film thermopile has a D^* of about 1.5×10^8. A bolometer of an NEP of 10^{-12} having a detector area of 1 mm^2 would have a

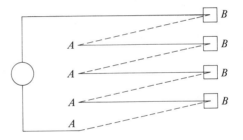

Fig. 14.8
Thermopile. Junctions A at room temperature, junctions B at higher temperature at the absorbing plate.

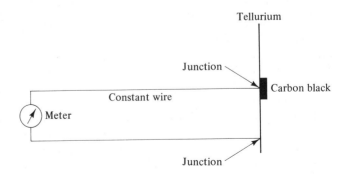

Fig. 14.9
Thermocouple for demonstration purposes. (After Pohl, *Einführung in die Optik.*)

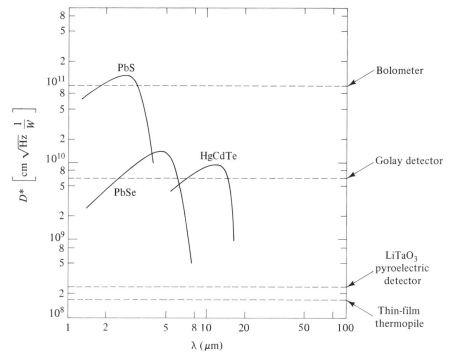

Fig. 14.10
Comparison of D^* as a function of wavelength for thermal and photoconductive detectors.

D^* of 10^{11} and is also marked on Figure 14.10. Descriptions of the pyroelectric and Golay detectors follow.

9. THE PYROELECTRIC DETECTOR

Pyroelectric detectors are made of ferroelectric crystals such as triglycine sulfate or lithium tantalate. Such crystals have a permanent polarization for temperatures below a certain temperature (called the Curie temperature). If the crystals are heated, the lattice distance changes, and a change in polarization results. In an electric circuit, this polarization change is detected as a change in the capacitance and can be translated into a current and voltage change. These changes are proportional to the change in incident radiation. If chopped radiation is used, the change in voltage can be amplified by an ac (alternating-current) amplifier.

Figure 14.11 presents some data about pyroelectric detectors. For more details, see *The Infrared Handbook*.* The detector noise and D^* depending on frequency are shown. Since the detector is a thermal detector, it has a slow response. D^* drops considerably for frequencies of $f = 1000$ Hz.

---

* W. L. Wolfe and G. J. Zissis, Office of Naval. Research, Dept. of the Navy, *The Infrared Handbook* (Washington, D.C.: U.S. Government Printing Office).

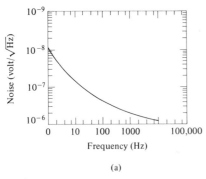

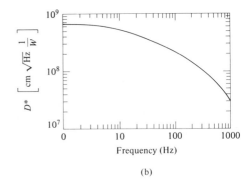

(a)

(b)

Fig. 14.11
Data on pyroelectric detectors. (a) Frequency response of detector noise. (b) D^* at λ_{max} depending on frequency.

10. THE GOLAY DETECTOR

The **Golay detector** is also called a pneumatic detector. It works on the principle that heat causes changes in pressure. The incident radiation is absorbed by a thin film in the pressure cell, and the gas in the pressure cell is heated. A small tube connects the pressure cell to a second cell sealed with a flexible film. The resulting pressure change in the first cell distorts the film in the second cell. The film is used as a deflecting mirror, as in the galvanometer. A light detection system, more sophisticated than the one used in the galvanometer (blocking fringe system), is used to measure the deflection of the film, which is proportional to the incident radiation. Figure 14.12 shows schematically the pressure cell and optical system, also indicating the flexible mirror and blocking fringe system. The light detection system conducts photons to the phototube, proportional to the amount of alternating incident light. Figure 14.13a shows the dependence of D^* on the chopping frequency; Figure 14.13b shows the dependence of the response on the wavelength. For more details, see *The Infrared Handbook*. The Golay cell is a slow detector. It responds to wavelengths from the visible to the millimeter region.

Fig. 14.12
Schematic of a Golay detector.

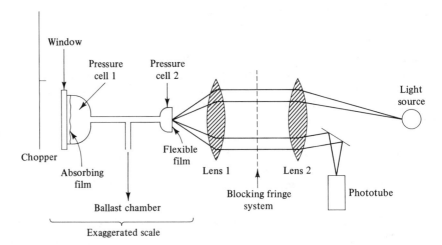

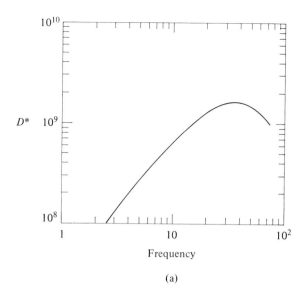

(a)

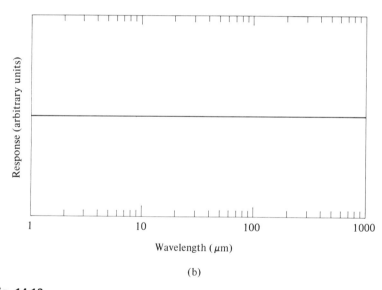

(b)

Fig. 14.13
(a) $D^*(\lambda, f, 1)$ cm$\sqrt{\text{Hz}}$/W as a function of chopping frequency for a Golay detector.
(b) Spectral response for a Golay detector.

11. PHOTON DETECTORS

In this section, we will discuss the photoemissive detector and its noise. Following sections will deal with photoconductive and photovoltaic detectors, together with some other photon detectors.

Photoemissive detectors work on the principle of the photoelectric effect. A light quantum produces an electron at the cathode. A voltage accelerates the

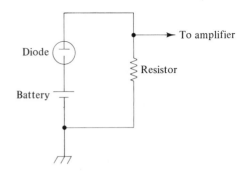

Fig. 14.14
Schematic circuit for a photoemissive
detector (photodiode).

electron to the anode, and a current can be measured. Figure 14.14 shows a
schematic of the circuit. A variation of this detection system is the **photomultiplier**.
The electron from the cathode first hits another target, where it produces a
number of electrons. Each new electron is accelerated toward yet another target,
where it, in turn, produces a number of electrons. This process can be repeated
10 to 15 times to obtain an appreciable amplification.

A. Functioning of the Photocell

We assume that monochromatic light of power W per unit area is incident on
the cathode of the photocell. The number of light quanta per second and per unit
area (of the cathode) is called N, and we have

$$N = \frac{W}{h\nu} \tag{14.71}$$

where h is Planck's constant and ν is the frequency of the light quantum. The
number of electrons produced depends on the quantum efficiency of the material.
The quantum efficiency q is the fraction of incident light quanta producing one
electron. The number of electrons produced per second is

$$n = qN = \frac{qW}{h\nu} \tag{14.72}$$

The resulting photocurrent I and signal voltage v at the load resistor become

$$I = en = eq\frac{W}{h\nu} \tag{14.73}$$

and

$$v = IR_L = eR_L q\frac{W}{h\nu} \tag{14.74}$$

B. Responsivity

The responsivity r of the photocell is given as

$$r = \frac{\Delta v}{\Delta W} = \frac{eR_L q}{h\nu} \tag{14.75}$$

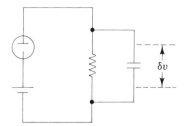

Fig. 14.15
Schematic circuit for noise consideration of photodiodes.

For example, for $\lambda = 1$ μm, $R_L = 2 \times 10^7$ Ω, and $q = 10^{-2}$ we get

$$r = 1.64 \times 10^5 \frac{V}{W} \tag{14.76}$$

C. Noise and NEP: Schottky's Formula

Let us consider the circuit shown in Figure 14.15 to discuss the photon noise of the "signal electrons." We assume that n_t electrons pass from the cathode to the anode in time interval t. The current is

$$I_t = \frac{e n_t}{t} \tag{14.77}$$

From a statistical analysis using the Poisson probability distribution, the variance of n_t from its time average $\bar{n}_t$ is given as

$$\overline{(n_t - \bar{n}_t)^2} = \bar{n}_t \tag{14.78}$$

where the overbar indicates the average of the quantities in question. Multiplication by e^2/t^2 gives the fluctuation of the current in the resistor:

$$\overline{(I_t - \bar{I}_t)^2} = \overline{(\delta I_t)^2} = \frac{e}{t}\bar{I}_t \tag{14.79}$$

The voltage across the capacitor (see Figure 14.15) and resistor is

$$\overline{(\delta v)^2} = R_L^2 \overline{(\delta I_t)^2} = \frac{R_L^2 e \bar{I}_t}{t} \tag{14.80}$$

The time t is taken representative of the frequency response of the electrical circuit. (See problem 10.) We know that after time $t = R_L C$, the voltage is decreased to $(1/e)$, or $(1/2.718)$; after $t = 2R_L C$, to about $1/10$. Taking $t = 2R_L C$, we have

$$\overline{(\delta v)^2} = R_L^2 \overline{(\delta I_t)^2} = \frac{R_L^2 e \bar{I}_t}{2 R_L C} = \frac{R_L e}{2C}\bar{I}_t \tag{14.81}$$

For the frequency dependence of $(\delta I_t)_f$, we assume a similar spectrum as we did for the galvanometer and resistor in Sections 2 and 3. We have

$$\sum R_L^2 \overline{(\delta I_t)_f^2} = F'' \Delta f \tag{14.82}$$

11. Photon Detectors 485

or

$$R_L^2 \overline{(\delta I_t)^2_f} = F'' \, df \tag{14.83}$$

Since we are considering an RC circuit, the frequency averaging is, in analogy to the integral in equation 14.20,

$$\overline{(\delta v)^2} = \frac{1}{2\pi} \int_0^\infty \frac{F'' \, d\omega}{1 + \omega^2 (R_L C)^2} \tag{14.84}$$

where we have set in equation 14.20 $L = 0$ and $F' = F''$. The result is

$$\overline{(\delta v)^2} = \frac{F''}{4 R_L C} \tag{14.85}$$

(see as well equation 14.21).

Combining equations 14.83 and 14.85 and then using equation 14.81 yields

$$\overline{(\delta v^2)} = \frac{R_L^2 \overline{(\delta I_t)^2_f}}{df \, 4 R_L C} = \frac{R_L e}{2C} \overline{I}_t \tag{14.86}$$

or

$$\overline{(\delta I_t)^2} = 2e \overline{I}_t \, df \tag{14.87}$$

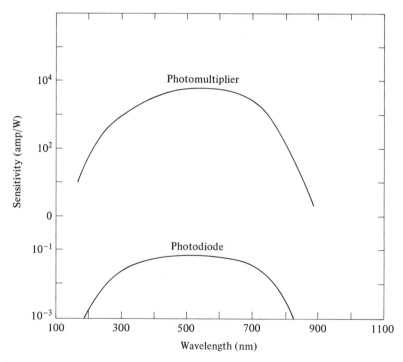

Fig. 14.16
Data on the photodiode and photomultiplier. The sensitivity is given in amperes per watt and depends on the wavelength in nanometers. (For most of these tubes, 600 V gives the highest S/N ratio.)

14. Detectors

where we ignored the f on $\overline{(\delta I_t)^2_f}$ because at the end it does not depend on the frequency.

The current fluctuation of the quantized charges is proportional to the time average of the current. This type of fluctuation is called shot noise, and equation 14.87 is called **Schottky's formula**.

Example

For the time average of the power fluctuation, we have from equation 14.73

$$\overline{W^2} = \frac{h^2 v^2}{q^2 e^2} (\delta I)^2 \tag{14.88}$$

Introduction of Schottky's formula gives us

$$\overline{W^2} = \frac{h^2 v^2}{q^2 e^2} 2 e \overline{I_t} \, df \tag{14.89}$$

If we take $q = 10^{-2}$, $\lambda = 1 \ \mu$m, $\overline{I_t} = 1 \ \mu$A, and $df = 1/s$, we get

$$\sqrt{\overline{W^2}} = 7 \times 10^{-11} \ \text{W} \tag{14.90}$$

Electrons are also produced by the background radiation. The background radiation is not modulated, but the fluctuation of the produced electrons enter the combined noise. A similar consideration yields a numerical value of

$$\sqrt{\overline{W_B^2}} = 10^{-11} \ \text{W} \tag{14.91}$$

Let us calculate the value of Johnson noise. To get the power fluctuation corresponding to Johnson noise, we introduce equation 14.31 into equation 14.88 and get

$$\overline{W^2} = \frac{h^2 v^2}{q^2 e^2} \left(\frac{\overline{(\delta v)^2}}{R_L^2} \right) = \frac{h^2 v^2}{q^2 e^2} \left(\frac{4k T R_L \, df}{R_L^2} \right)$$
$$= \frac{h^2 v^2}{q^2 e^2} \left(\frac{4k T}{R_L} \right) df \tag{14.92}$$

With $R_L = 10^7 \ \Omega$, $\lambda = 1 \ \mu$m, $q = 10^{-2}$, $T = 300$ K, and $\Delta f = 1/s$, we have

$$\sqrt{\overline{W^2}} = 0.5 \times 10^{-11} \ \text{W} \tag{14.93}$$

We see that the NEP is of the order of 10^{-11} and that the current noise (14.91) is of the same order of magnitude as Johnson noise. In Figure 14.16, we present some data on the photodiode and the photomultiplier.

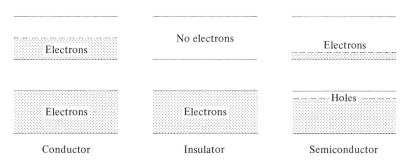

Fig. 14.17
Schematic of the energy band structure of conductors, insulators, and semiconductors.

12. PHOTOCONDUCTIVE DETECTORS

Most photoconductive detectors are semiconductors. In Figure 14.17, we show a simplified schematic of the energy band structure of conductors, insulators, and semiconductors. In conductors, the highest energy band occupied with electrons is half filled. This is called the **conduction band**. The band below is called the **valence band**. Electrons in the conduction band are free to participate in the conduction process. In an insulator, the conduction band is empty and the valence band is completely filled. The electrons cannot participate in the conduction process. There is a large gap to the valence band. In semiconductors, this gap is small, and by thermal excitations, electrons from the valence band can get into the conduction band, leaving empty places, or "holes," in the band below. Conduction is now possible. If the temperature is lowered, fewer and fewer electrons are present in the upper band.

The band structure of the energy states of the electrons may be understood if we start off from the energy schematic of one atom. There are discrete energy levels. All the atoms in a solid have the same energy levels, but the interaction displaces these levels against one another slightly, and a band of states is formed. As a result of **Pauli's principle**, each state can be occupied with two electrons. For a conductor, each atom supplies one electron corresponding to one state, and therefore in the band structure, half the band is filled. The highest occupied state is called the **Fermi level**.

Semiconductors may also contain impurities, in which case they are called **extrinsic semiconductors**; if no impurities are present, they are called **intrinsic semiconductors**. The impurities may have electrons whose energy levels are in the range of the band gap. If these impurities can give an electron away easily, they are called donors; if they are ions and attract an electron, they are acceptors. Donors supply an electron to the upper band; acceptors create a hole in the lower band. The different excitation processes are shown in Figure 14.18.

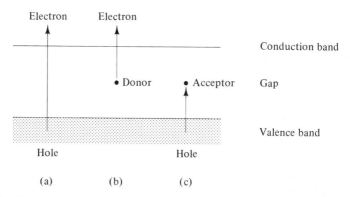

Fig. 14.18
The absorption of light, supply of electrons, and creation of holes. (a) Absorption across the band. (b) Absorption at a donor impurity. (c) Absorption at an acceptor impurity.

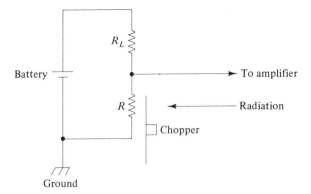

Fig. 14.19
Electric circuit for a semiconductor detector. R_L is the load resistor, and R is the detector resistor.

A. Functioning of Semiconductor Detectors

If light falls onto an intrinsic semiconductor, electrons are "lifted" from the lower band into the higher band, and more electrons and "holes" can participate in the conduction process. If the wavelength is too long, the corresponding energy of the photon is not sufficient to lift an electron from below into the conduction band. For extrinsic semiconductors, a similar condition exists, depending on the energy state of the impurities. Cooling the semiconductor can increase the sensitivity by reducing the thermally excited electrons, which are in "competition" with electrons excited by the light (photon excited). The change in conductivity is measured as the voltage change across the detector resistor R (Figure 14.19).

We assume, for simplicity, that we have only electrons as current carriers. We first deal with thermally excited electrons. We assume that the change in the number n of the electrons in the upper state is proportional to their number n. This is similar to what we find for radioactive decay, and we have

$$\frac{dn}{dt} = A - \frac{n}{\tau} \tag{14.94}$$

where τ is the time constant of the transition process to lower energy. The constant A follows from the equilibrium condition $dn/dt = 0$, and $A = n_0/\tau$, where n_0 is the number of electrons in that state. Using the equilibrium condition and the definition $\Delta n = n - n_0$, we can write equation 14.94 as

$$\frac{d\,\Delta n}{dt} = -\frac{\Delta n}{\tau} \tag{14.95}$$

We have the solution

$$\Delta n = De^{(-t/\tau)}$$

D being the value of Δn for $t = 0$.

If we now consider radiation of intensity J falling onto the semiconductor, carriers are created by the rate

$$\frac{d}{dt}\Delta n = \gamma J - \frac{\Delta n}{\tau}$$

(14.96)

where γ is a constant.

A solution of this equation is

$$\Delta n = \gamma J \tau (1 - e^{-t/\tau})$$

(see problem 13). For real devices, we must take into account not only the electrons produced by the light, but also the holes left in the lower band. The properties of the holes may be slightly different.

B. Responsivity

Consider again the electric circuit shown in Figure 14.19. For the change in voltage across the detector element, we have a similar situation as described by equation 14.36.

$$\Delta v = \frac{IR_L \Delta R}{R_L + R}$$

(14.97)

The incident power is assumed to be monochromatic with a frequency of v. The number of electrons excited by the incident light is, according to equation 14.72,

$$n' = q\frac{\Delta W}{Ahv}$$

(14.98)

Here, n' is the number of electrons produced per detector area A and per second. From experiments, we find that the change in surface resistivity $\Delta\sigma$ is proportional to n', that is,

$$\Delta\sigma = \beta n'$$

(14.99)

The resistivity decreases when more photons are incident on the detector. The change in resistivity is then

$$\Delta R = \frac{\ell\,\Delta\sigma}{b}$$

(14.100)

where ℓ is the length and b is the width of the detector element. Combining equations 14.98–14.100 yields

$$\Delta R = \frac{\beta q\ell\,\Delta W}{hvAb}$$

(14.101)

or, expressed in the voltage change with equation 14.97,

$$\Delta v = \frac{IR_L\beta q\ell\,\Delta W}{(R_L + R)hvAb}$$

(14.102)

The responsivity r is then

$$r = \frac{\Delta v}{\Delta W} = \frac{R_L I \beta q \ell}{(R_L + R) h v A b} \tag{14.103}$$

Photoconductors with a resistivity of about 1 MΩ used with a current of 100 μA have a responsivity of about $r = 10^4$ V/W (see problem 14).

C. Noise and Response Time

Photoconductive detectors have a quantum efficiency of about $q = 1$ at their response maximum, depending on wavelength.

For low chopping frequencies, the $1/f$ noise is dominant. Choosing modulation frequencies of about 1000 Hz is, in most cases, sufficient to reduce this noise contribution.

Photoconductive detectors show a particular type of noise originating from the statistical fluctuations in generation of the carriers and the recombination rate of electrons and holes. This is called **generation recombination noise (G-R noise)**.

Johnson noise is due to the temperature and resistivity, according to equation 14.31, as $\overline{v^2} = 4RkT\,df$.

The noise components must be added as the square of the average:

$$\overline{W^2} = + \overline{W_M^2} \quad \text{(Modulation noise)}$$
$$+ \overline{W_{G\text{-}R}^2} \quad \text{(G-R noise)}$$
$$+ \overline{W_J^2} \quad \text{(Johnson noise)}$$

The first two types of noise and the square of the responsivity r (see equation 14.103) are proportional to I^2. Johnson noise can be neglected if we work with currents that do not heat up the detector element. Modulation noise can be reduced by using high chopping frequencies with narrow-band amplifiers. The remaining noise is G-R noise (see problem 15).

Compared to thermal detectors, semiconductor detectors have a short response time, in the range 10^{-3}–10^{-7} s.

D. Commercially Available Photoconductive Detectors

For an example of a photoconductive detector, consider Figure 14.20, which shows the frequency response and D^* of a PbS detector at 193 K. In Figure 14.21, we give D^* (for bandwidth 1) depending on the wavelength for a large number of commercially available photoconductive and photovoltaic detectors.

13. PHOTOVOLTAIC DETECTORS AND SOLAR CELLS

Photovoltaic detectors are made of semiconductor *p-n* junctions. For example, let us consider a solar cell. In Figure 14.22a, we see a simplified schematic of a *p-n* junction; Figure 14.22b shows the junction with radiation falling onto it.

Without illumination, the Fermi levels in the *p* and *n* regions are equal. If

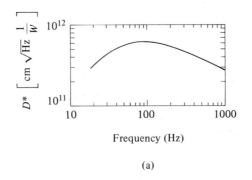

(a)

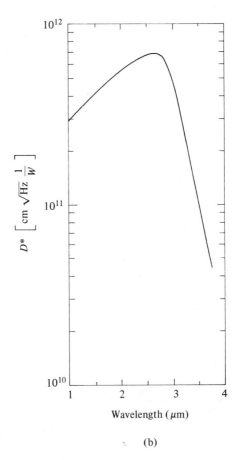

(b)

Fig. 14.20
PhS detector. (a) Frequency response and (b) spectral response of a PbS detector at 193 K. (For more detail, see *The Infrared Handbook*.)

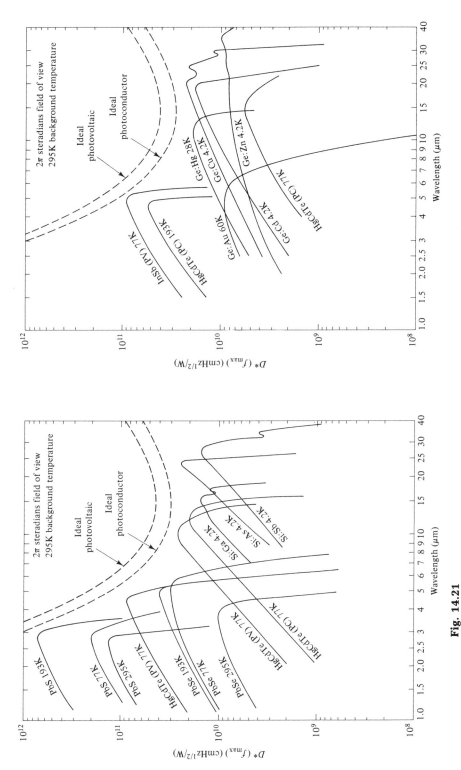

Fig. 14.21

Spectral response D^* depending on wavelength for a number of commercially available detectors. [From R. W. Boyd, *Radiometry and Detection of Optical Radiation* (New York: John Wiley & Sons, 1983); Courtesy of Santa Barbara Research Center, Goleta, Calif.]

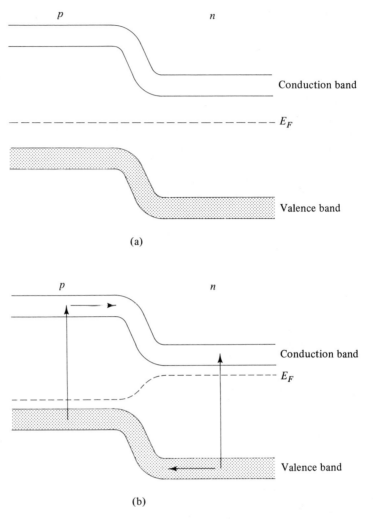

Fig. 14.22
Schematic of photovoltaic effect and *p-n* junction (E_F is Fermi level). (a) Conduction and valence band of a *p-n* junction. (b) Movement of electrons and holes when the *p-n* junction is illuminated.

light is absorbed in both regions, electrons are lifted up in the *p* region and holes are created. Both electrons and holes have a tendency to move "to the other side." This changes the Fermi level, and a voltage difference is created. This voltage difference is detected with an electric circuit and may result in the detection of radiation.

A silicon solar cell is shown in Figure 14.23. The silicon crystal contains layers of *n*-type and *p*-type silicon, forming a *p-n* junction. The incident photons create hole-electron pairs, and a current flows in the cell and through attached wires to the load resistor. The relative spectral response is shown in Figure 14.24. It has its peak in the infrared region. The peak of the emission of the sun is about 480 nm.

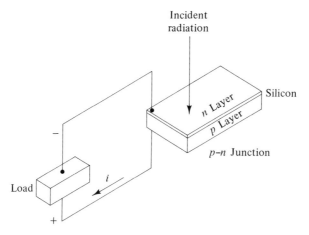

Fig. 14.23
Schematic of silicon solar cell.

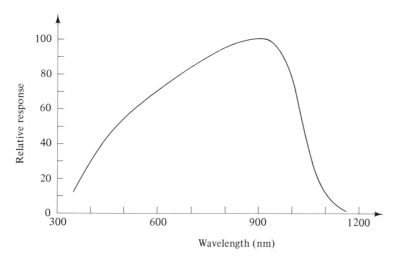

Fig. 14.24
Relative spectral response for a silicon cell operated at short circuit.

Panels made from solar cells of 40–10,000 m^2 may produce 15–10,000 KWh. NASA uses such panels to supply electricity to space vehicles.

In the laboratory, the solar cell is a handy, inexpensive radiation detector for laser light, and it can be used with an inexpensive ampere meter. For example, cells with an area of about 1 cm^2 have sensitivities of 0.1–0.2 μA/μW.

14. THE EYE AND THE PHOTOGRAPHIC PLATE

There are certain similarities between the human eye and a camera. From the point of view of detection of radiation, both the eye and the photographic plate are photon detectors, and both employ chemical processes for the detection

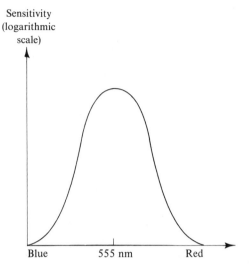

Fig. 14.25
Sensitivity of the eye to different wavelengths
(arbitrary logarithmic scale).

Sensitivity
(logarithmic
scale)

Blue   555 nm   Red

of light. The spectral sensitivity of the eye is highest at about $\lambda = 555$ nm (Figure 14.25). Close to this wavelength, at 480 nm, the spectrum of the sun has its maximum. The sensitivity of the eye to different wavelengths (colors) is shown in Figure 14.25 on a logarithmic scale. The ultimate sensitivity appears only for black and white and is given as 6×10^{-13} W/m², corresponding to 100 light quanta in a second.

The photographic plate uses silver bromide crystals as grain particles. The light quanta incident on the plate are absorbed at impurities in the crystals. Development of the photographic plate will produce about 10^8 silver atoms for each absorbed light quantum. The photographic plate can easily be used as an integrator of incident radiation. The amount of blackening of the plate can be related to the total number of incident light quanta arriving in a certain time interval. For low-level radiation, long exposure times can be employed. This was done in observation of the Raman effect before the laser was used. Exposure times of a week were not unusual.

The photographic plate can also be used for the simultaneous detection of several wavelengths. The emerging light from a grating may be focused into a plane where a photographic plate is positioned. To use other types of detectors, it is customary to scan or use an array of detectors. For more information on photographic plates see *Kodak Plates and Films for Scientific Photography.**

Problems

1. *Finding the amplitude of the trial solution* θ_f *(equation 14.5).* The differential equation

$$M\ddot{\theta} + D\dot{\theta} + G\theta = A_f e^{i2\pi ft}$$

Kodak Plates and Films for Scientific Photography, Kodak Publication No. P-315 (Rochester, N.Y.: Eastman Kodak Co., 1973).

is solved with the trial solution

$$\theta = \theta_f e^{i2\pi ft}$$

Find θ_f as a function of A_f, G, M, D, and ω.

2. *Damped oscillator.* Consider the value obtained in problem 1,

$$\frac{A_f}{(G - M\omega^2) + i\omega D}$$

Give the absolute value and the absolute value squared.

Ans.
$$|\theta_f| = \frac{A_f}{\sqrt{(G - M\omega^2)^2 + (\omega D)^2}}$$

$$\theta_f^2 = \frac{A_f^2}{(G - M\omega^2)^2 + (\omega D)^2}$$

3. *The R-L-C circuit.* The total power of the R-L-C circuit is given as

$$Iu = \frac{d}{dt}(\text{magnetic energy}) + \frac{d}{dt}(\text{electric energy}) + \text{losses}$$

We have I current, $\frac{1}{2}LI^2$ magnetic energy, L inductance, RI losses, u voltage, $\frac{1}{2}(q^2/C)$ electrical energy, C capacitance, q charge, and R resistance.

a. Get

$$\dot{u} = L\frac{d^2I}{dt^2} + \frac{1}{C}I + R\frac{dI}{dt}$$

by differentiation of the electrical and magnetical energy.

b. Use the trial solution $u = u_f e^{i\omega t}$ and $I = I_f e^{i\omega t}$ and obtain

$$\frac{I_f}{i\omega C} = \frac{u_f}{(1 - LC\omega^2) + i\omega RC}$$

4. *Average over the noise frequency spectrum of $\overline{v_f^2}$.* Go back to equation 14.17:

$$\overline{v_f^2} = \frac{\overline{u_f^2}}{(1 - LC\omega^2)^2 + \omega^2(RC)^2}$$

Before averaging over all frequencies from 0 to ∞, we introduce the approximation $L \to 0$ and have

$$\overline{v^2} = \frac{1}{2\pi}\int_0^\infty \frac{\overline{u_f^2}\, d\omega}{1 + \omega^2(RC)^2}$$

using the integral formula

$$\int_0^\infty \frac{dx}{1 + x^2} = \frac{\pi}{2}$$

Show that we have

$$\overline{v^2} = \frac{\overline{u_f^2}}{4RC}$$

Introduction of this result into

$$\frac{1}{2}C\overline{v^2} = \frac{1}{2}kT$$

yields Nyquist's formula,

$$\overline{u_f^2} = 4RkT$$

5. *Numerical calculation of $\overline{v^2}$.* Calculate

$$\sqrt{\overline{v^2}} = \sqrt{4RkT\,df}$$

for $T = 2$ K and $R = 1 \times 10^6\ \Omega$.
Ans. $\sqrt{\overline{v^2}} = 1 \times 10^{-8}$ V

6. *Calculation of Δv.* Consider the circuit in Figure 14.5. The ratio of the voltage drop across R and $R_L + R$ is

$$\frac{v_D}{V} = \frac{R}{R_L + R}$$

Calculate the change in voltage Δv depending on the change in the resistor ΔR.

7. *Responsivity of a bolometer.* Calculate the responsivity

$$r = \frac{\Delta v}{\Delta W} = \alpha I R \tilde{R}$$

for the semiconducting bolometer using $\alpha = -0.03$ 1/K and $R = 10^6\ \Omega$.
a. Calculate $\tilde{R}$ by using $\tilde{G}_0 = 10^{-4}$ W/K and $T - T_0 = 10$ K (equation 14.55).
b. Calculate I from

$$W_0 = I^2 R = \tilde{G}_0(T - T_0)$$

c. Calculate the absolute value of the responsivity using $r = \alpha I R \tilde{R}$.
Ans. a. $\tilde{R} = 0.77 \times 10^4$ K/W; b. $I = 3.16 \times 10^{-5}$ A; c. $|r| = 0.73 \times 10^4$ V/W

8. *Thermal noise and Johnson noise.* Calculate the combination of thermal noise and Johnson noise for the bolometer. Use $T = 4$ K, $T_0 = 3.999$ K, $\alpha = -(3000\ \text{K}/T^2)$ [1/K], $\tilde{G}_0 = 10^{-4}$ W/K, and $k = 1.4 \times 10^{-23}$ J/K.
a. Calculate α.
b. Calculate $\tilde{R}$ from $\tilde{G} = \tilde{G}_0[1 - \alpha(T - T_0)]$
c. Calculate

$$\text{NEP} = \sqrt{\overline{W_B^2}} = \sqrt{\frac{4kT^2}{\tilde{R}}\left(1 + \frac{1}{T(T - T_0)\alpha^2}\right)\Delta f}$$

Par. Ans. b. $\tilde{R} = 0.84 \times 10^4$ K/W; c. NEP $= 3.3 \times 10^{-13}$ W

9. *Responsivity of a photocell.* Calculate the responsivity of the photocell, using

$$r = \frac{eR_L q}{h\nu}$$

where $\lambda = 0.8\ \mu$m, $R_L = 10^6\ \Omega$, $q = 10^{-2}$, $h = 6.6 \times 10^{-34}$ J·s, and $e = 1.6 \times 10^{-19}$ C.
Ans. $r = 6.46 \times 10^3$ V/W

10. *Electrical bandwidth and frequency response.* If a system has a uniform frequency response in the interval $\Delta f = f_2 - f_1$ and no response at other frequencies, Δf is called the electrical bandwidth. For a more complicated case, where the frequency response is nonuniformly over the frequency range, we define the electrical bandwidth as

$$\Delta f = \int_0^\infty \left|\frac{r(f)}{r_{\text{max}}}\right|^2 df$$

where $r(f)$ is the responsivity of the detector as a function of frequency. For example, for the RC circuit, we have

$$r(f) = \frac{r_{max}}{\sqrt{1 + (\omega CR)^2}}$$

and we get

$$\Delta f = \int_0^\infty \frac{1}{1 + (\omega CR)^2} df$$

With our result from problem 4, we have

$$\Delta f = \frac{1}{4RC}$$

If the responsivity $r(f)$ is averaged over the sampling time T, we must calculate

$$r(f) = \frac{1}{T} \int_{-T/2}^{+T/2} e^{-i2\pi ft} dt$$

a. Show that we get (for the real part) $r(f) = \dfrac{\sin \pi f T}{\pi f T}$

b. Calculate Δf for this case.

11. *Shot noise.* Calculate the "shot noise" (equation 14.89) for the values $\lambda = 0.8$ μm, $\bar{I}_t = 2$ μA, $\Delta f = 1/s$, $q = 10^{-2}$, $e = 1.6 \times 10^{-19}$ C, and $h = 6.6 \times 10^{-34}$ J·s.
Ans. $\sqrt{\overline{W^2}} = 12 \times 10^{-11}$ W

12. *Johnson noise.* Calculate Johnson noise for the photodiode described by

$$\overline{W^2} = \frac{h^2 v^2}{q^2 e^2}\left(\frac{4kT}{R_L}\right)$$

Use $R_L = 10^7$ Ω, $\lambda = 0.8$ μm, $q = 10^{-2}$, $T = 300$ K, $e = 1.6 \times 10^{-19}$ C, $h = 6.6 \times 10^{-34}$ J·s, and k $= 1.4 \times 10^{-23}$ J/K.
Ans. $\sqrt{\overline{W^2}} = 0.63 \times 10^{-11}$ W

13. Find the solution $\Delta n(t)$ for the equation

$$\frac{d}{dt}\Delta n = \gamma J - \frac{\Delta n}{J}$$

14. *Responsivity of a photoconductive detector.* Calculate the responsivity of a photoconductive detector (equation 14.103), assuming that R_L is of the order of $R_L + R$. Use $I = 100 \cdot 10^{-6}$ A, $q = 10^{-2}$, $\ell = 10$ b, $h = 6.6 \times 10^{-34}$ J·s, $v = 3 \times 10^{14}$ 1/s ($\lambda = 1$ μm), and $A = 10$ mm^2. We take $\beta = 1 \times 10^{-14}$ Ω·s·m^2 to get r in the order of 10^4 V/W, as given in R. A. Smith, F. E. Jones, and R. P. Chasmar, *The Detection and Measurement of Infrared Radiation* (Oxford, Eng.: The Clarendon Press, 1957).
Ans. $r = 5 \cdot 10^4$ V/W

15. *Johnson noise and G-R noise.* Compare Johnson noise with G-R noise for a photoconductive detector. For Johnson noise, we get with $R = 10^6$ Ω, $T = 300$ K, and k $= 1.4 \times 10^{-23}$ J/K the value $\sqrt{\overline{v^2}} = 1.3 \times 10^{-7}$ V. Calculate G-R noise. Take

$$\overline{(\delta v)^2} = \frac{R^2 I^2 2\tau}{\bar{N}}$$

and use $I = 100 \times 10^{-6}$ A, $\tau = 10^{-4}$ 1/s, $\bar{N} = 10^{18}$ 1/s, and R from above.
Ans. $\sqrt{\overline{(\delta\sigma)^2}} = 1.4 \times 10^{-9}$ V

Light Sources | **15**

1. INTRODUCTION

The source of light originally used for the study of optical phenomena was the sun. Later, lamps were developed, and today, a very important light source is the laser.

As examples of light sources, we will discuss blackbody radiators, atomic emission lamps, and lasers. The different kinds of light emitted by these sources was discussed in Chapter 2 from the point of view of the coherence length. Now we want to study in more detail the dependence of the intensity on the frequency. The discussions in this chapter will be helpful in understanding the light emission process in other types of light sources, such as gas discharge lamps, and arc lamps.

2. BLACKBODY RADIATION

The light emitted from the sun or from a hot wire is similar to the light emitted from a blackbody source. A schematic of a blackbody source is shown in Figure 15.1. If we heat a wire, we observe that with increasing temperature, the color of the wire changes. First it gets red, and as the temperature increases it becomes white. This shows us that the frequency distribution of the emitted intensity depends on the temperature. The radiation is electromagnetic in nature and draws its energy from the heating of the wire. Thermodynamic and electrodynamic theory are both involved in finding the dependence of the intensity on the frequency and temperature. The famous law describing this result is Planck's formula, and its interpretation led to the quantum theory of light. A measurement of the dependence of the intensity on the frequency shows that we have no discontinuity in the frequency spectrum (the way we have for the emission of the sodium lamp, for example). If we decrease the frequency interval, we will observe less and less energy until the interval is so small that we can no longer observe any radiation, observing only noise. This is different from what we have referred

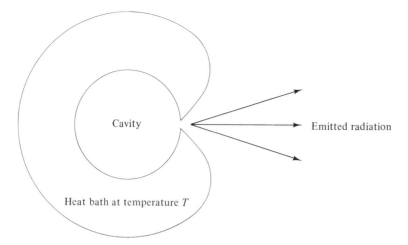

Fig. 15.1
Schematic of a blackbody radiator. The cavity is surrounded by a body of temperature T, which keeps the cavity at this temperature. Electromagnetic radiation is emitted. The temperature T characterizes the intensity distribution depending on frequency.

to in preceding chapters (in an idealized way) as monochromatic light, with one frequency and a certain amplitude.

A. Rayleigh-Jeans Law

We can describe the blackbody as a rectangular box having dimensions ℓ_X, ℓ_Y, and ℓ_Z (Figure 15.2a). The heat reservoir keeps the box at the equilibrium temperature T. We assume that electromagnetic waves are running back and forth between the walls of the box. We also assume that oscillators on the walls absorb and re-emit the electromagnetic waves, and in so doing, maintain the thermodynamic-electrodynamic equilibrium. To obtain the energy content in the box, we count all waves traveling back and forth in all directions. In the direction X, we have traveling waves

$$Ae^{i(k_X X - \omega t)} \quad \text{and} \quad Ae^{i(k_X X + \omega t)}$$

where

$$k_X = \frac{2\pi}{\lambda} \quad \text{and} \quad \omega = 2\pi v = \frac{2\pi}{T} \tag{15.1}$$

The two traveling waves can be combined into a standing wave as

$$Ae^{ik_X X} 2\cos\omega t \quad \text{or} \quad \begin{cases} 2A\cos k_X X \cos\omega t \\ 2A\sin k_X X \cos\omega t \end{cases} \tag{15.2}$$

The wave $2A\sin k_X X \cos\omega t$ is fixed with two nodes on each reflecting surface (see Figure 15.2b). The amplitude maxima vibrate between A and $-A$ during the period T. Only for certain wavelengths is such a standing wave possible for length ℓ_X; the standing-wave condition is

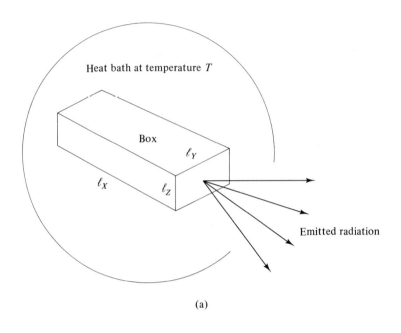

(a)

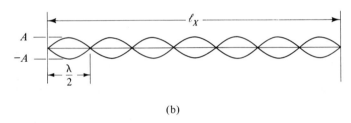

(b)

Fig. 15.2
(a) A box with dimensions ℓ_X, ℓ_Y, and ℓ_Z, as a blackbody radiator. The box is surrounded by a heat bath of temperature T. (b) Standing-wave pattern in one dimension.

$$\frac{\ell_X}{\lambda/2} = M_X, \qquad M_X \text{ an integer}$$

or

$$2k_X\ell_X = M_X 2\pi \tag{15.3}$$

We now consider two traveling waves, one for the X direction and one for the Y direction:

$$Ae^{i(k_X X \pm \omega t)}, \quad Ae^{i(k_Y Y \pm \omega t)} \tag{15.4}$$

We have $k_X = k\cos\theta$ and $k_Y = k\sin\theta$, with $k = 2\pi/\lambda$ (Figure 15.3). The standing-wave condition for each direction is

$$2k_X\ell_X = M_X 2\pi, \quad 2k_Y\ell_Y = M_Y 2\pi, \qquad M_X, M_Y \text{ integers} \tag{15.5}$$

15. Light Sources

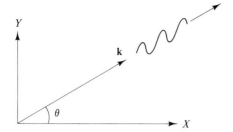

Fig. 15.3
A wave propagating in direction θ with wave vector **k**.

We can combine these conditions by using

$$k^2 = k_X^2 + k_Y^2 = \frac{\omega^2}{c^2}$$ (15.6)

where $c = \lambda v$ and have

$$k^2 = \pi^2 \left[\left(\frac{M_X}{\ell_X} \right)^2 + \left(\frac{M_Y}{\ell_Y} \right)^2 \right]$$ (15.7)

In three dimensions, we similarly have (see problem 1)

$$k^2 = \pi^2 \left[\left(\frac{M_X}{\ell_X} \right)^2 + \left(\frac{M_Y}{\ell_Y} \right)^2 + \left(\frac{M_Z}{\ell_Z} \right)^2 \right] = \frac{\omega^2}{c^2}$$ (15.8)

Depending on the actual values of k_X, k_Y, and k_Z, the standing wave may have any direction in space (X, Y, Z). Considering a cube where $\ell_X = \ell_Y = \ell_Z = \ell$ and using $\omega = 2\pi v$, we get

$$M_X^2 + M_Y^2 + M_Z^2 = \frac{4\ell^2 v^2}{c^2}$$ (15.9)

We see that the three M values originally introduced independently through the standing-wave conditions in each direction are now interconnected in the three-dimensional case. While in the one-dimensional case they were assumed larger than zero in order to have a wave and not a constant, we can now have one or two of the M's equal to zero. This corresponds to the case of a standing wave in two or one dimension, respectively.

Equation 15.9 is the condition for standing waves in directions X, Y, and Z. For each frequency, we can have several sets of the three M values satisfying equation 15.9. We can visualize these combinations for the two-dimensional case if we plot the integer numbers M_X and M_Y in the X, Y plane (Figure 15.4). If we draw a circle with radius $R = \sqrt{M_X^2 + M_Y^2}$, we see that the combinations M_X and M_Y having points close to the circle belong to the same approximate frequency. We can check this by the example shown in Table 15.1.

If we are looking at all the combinations between R and $R + dR$ (see Figure 15.4), we obtain the number of combinations belonging to the frequency intervals v and $v + dv$. From equation 15.7, we have $R = \sqrt{M_X^2 + M_Y^2} = 2\ell v/c$, and the number of combinations in the area $2\pi R\, dR$ is

$$dZ_2 = \tfrac{1}{4}(2\pi R\, dR)$$ (15.10)

2. Blackbody Radiation

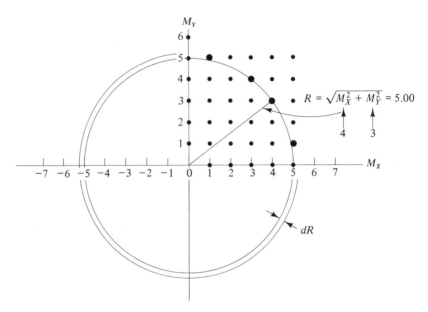

Fig. 15.4
Plot of the integer numbers M_X and M_Y in the X, Y plane.

Table 15.1
Combinations of M_X and M_Y
situated close to a circle
of radius $R = 5$

M_X	M_Y	$R = \sqrt{M_X^2 + M_Y^2}$
1	5	5.09
3	4	5.00
4	3	5.00
5	1	5.09

The factor $\frac{1}{4}$ comes from the fact that all M_X and M_Y are positive. But in the X, Y plane in Figure 15.4, we have also used negative values for M_X, M_Y in determining the area between the two circles.

With $R = 2\ell v/c$, we have

$$dZ_2 = \tfrac{1}{4}(2\pi)\left(\frac{2\ell v}{c}\right)\left(\frac{2\ell\, dv}{c}\right) \tag{15.11}$$

A similar consideration in three dimensions leads to

$$dZ = \tfrac{1}{8}(4\pi)R^2\, dR = 4\pi\frac{\ell^3}{c^3}v^2\, dv \tag{15.12}$$

This is the number of possible standing waves in the frequency intervals v and $v + dv$ in three dimensions, also called modes. In equilibrium, the number of possible modes times the average energy of each mode gives us the total radiation energy in the cavity. In this derivation, it is assumed that each mode carries the

15. Light Sources

same energy and the square of the amplitude has not been considered. The absorption and re-emission process by the oscillators is in thermal equilibrium with the radiation and heat reservoir. In thermal equilibrium, we have for each degree of freedom the energy $\frac{1}{2}kT$, where k is the Boltzmann constant and T the temperature in degrees Kelvin. The oscillator has this amount for each, its kinetic and its potential energy,* and for each mode we have the average energy kT. The average radiation energy per frequency interval is

$$\frac{dE}{dv} = kT\frac{dZ}{dv} = 4\pi kT\frac{\ell^3}{c^3}v^2 \qquad (15.13)$$

and the average energy density per frequency interval is

$$\frac{du}{dv} = \frac{1}{V}\left(\frac{dE}{dv}\right) = 4\pi kT\frac{v^2}{c^3} \qquad (15.14)$$

where $\ell^3 = V$, the volume.

So far, we have only considered one direction of polarization. Including both directions gives us an additional factor of 2 on the right side; that is,

$$\frac{du}{dv} = 8\pi kT\frac{v^2}{c^3} \qquad (15.15)$$

This formula is called the **Rayleigh-Jeans law of radiation**. It has been derived with the assumptions of classical thermodynamics and electrodynamics. Our assumption that all modes have the same average energy is only true for long wavelengths. For shorter and shorter wavelengths, (i.e., increasing frequency), equation 15.15 gives in the limit infinite energy density. This is called the **ultra-violet catastrophe**. Experimentally, on the other hand, we find that for high frequencies, the energy density goes to zero. The difficulties were solved by Planck's derivation of his famous radiation law. In his derivation, he was forced to assume that light is absorbed and emitted not continuously, as in electro-dynamics, but in quanta; that is, $E = hv$, where v is the frequency of the light and h is Planck's constant. We have, however, such light quanta for a continuous frequency spectrum.

B. Planck's Formula, Einstein's Coefficients

We will derive Planck's formula by using Einstein's idea of employing induced emission, induced absorption, and spontaneous emission. These terms will appear later in the discussion of lasers (Section 5). Figure 15.5 shows the three processes schematically. In the induced processes, an electromagnetic wave is needed to get the process going. The spontaneous emission is not initiated by an electro-magnetic wave.

The absorption process is already familiar to us. Light is absorbed and in most cases lost as thermal energy. In the induced emission process, an incoming wave picks up a light quantum and enlarges its intensity. The energy of the light quantum corresponds to the energy difference between the higher and lower

---

*See Paul Tipler, *Modern Physics* (New York: Worth, 1978).

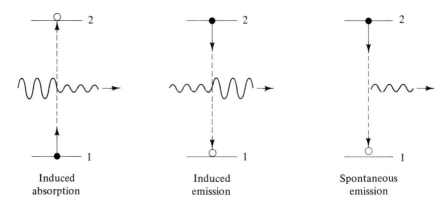

Fig. 15.5
Schematic of induced absorption and induced and spontaneous emission.

levels of the unspecified oscillator. In the spontaneous emission process, a new wave is created.

The probability of a transition of induced absorption between any levels numbered 1 and 2 is called W_{12}. It is proportional to the energy density per frequency interval du/dv,

$$W_{12} = B_{12}\frac{du}{dv} \qquad (15.16)$$

where B_{12} is a constant.

Similarly, the probability of a transition of induced emission from level 2 to 1 is called W_{21}

$$W_{21} = B_{21}\frac{du}{dv} \qquad (15.17)$$

where B_{21} is a constant.

The probability of spontaneous emission is not proportional to du/dv, and we have

$$\overline{W}_{21} = A_{21} \qquad (15.18)$$

where A_{21} is a constant. The constants B_{12}, B_{21}, and A_{21} are called Einstein's coefficients of induced and spontaneous emission and absorption. In a thermal equilibrium, we will have as many transitions up as we have down. If we call N_1 the number of possible light quanta at level 1 to be absorbed and N_2 the number at level 2 available for emission, we have

$$N_1\left(B_{12}\frac{du}{dv}\right) = N_2\left(B_{21}\frac{du}{dv} + A_{21}\right) \qquad (15.19)$$

The Boltzmann distribution for the occupation of states of energy E_1 and E_2 in thermal equilibrium tells us that

$$N_1 = N_0 e^{-(E_1/kT)} \quad \text{and} \quad N_2 = N_0 e^{-(E_2/kT)} \qquad (15.20)$$

where N_0 is a constant.

15. Light Sources

The ratio, from equations 15.19 and 15.20, is

$$\frac{N_2}{N_1} = e^{-[(E_2 - E_1)/kT]} = \frac{B_{12}(du/dv)}{B_{21}(du/dv) + A_{21}} \tag{15.21}$$

The energy density per frequency interval is

$$\frac{du}{dv} = \frac{A_{21}}{B_{12}e^{[(E_2 - E_1)/kT]} - B_{21}} \tag{15.22}$$

Einstein's argument to obtain Planck's formula is as follows: If $T \to \infty$, we have $du/dv \to \infty$ and it follows that B_{12} must be equal to B_{21}. Therefore, we can write

$$\frac{du}{dv} = \frac{A_{21}/B_{21}}{e^{[(E_2 - E_1)/kT]} - 1} \tag{15.23}$$

The remaining constant A_{21}/B_{21} is obtained quantitatively from the requirement that for long wavelengths, the Rayleigh-Jeans law is valid. We introduce $E_2 - E_1 = hv$, where h is Planck's constant, and we have in the limit

$$\lim_{v \to 0} \frac{du}{dv} = \frac{A_{21}/B_{21}}{1 + (hv/kT) - 1} = \frac{A_{21}}{B_{21}}\left(\frac{kT}{hv}\right) \tag{15.24}$$

This must be equal to equation 15.15, and we have

$$\frac{A_{21}}{B_{21}} = 8\pi \frac{hv^3}{c^3} \tag{15.25}$$

Substituting the result from equation 15.25 into equation 15.23 gives us Planck's formula,

$$\frac{du}{dv} = \frac{8\pi hv^3}{c^3} \frac{1}{e^{(hv/kT)} - 1} \tag{15.26}$$

In problem 3, we will see how the ultraviolet catastrophe is now avoided.

Equation 15.26 can be rewritten by using the radiance L_B (Chapter 11, Section 2B) of the blackbody per frequency interval or per wavelength interval, instead of the energy density per frequency interval. Planck's formula then appears as

$$\frac{dL_B}{d\lambda} = \frac{C_1}{\lambda^5} \frac{1}{e^{(C_2/\lambda T)} - 1} \quad \left[\frac{W}{(1/s) \cdot m^2 \cdot sr}\right] \tag{15.27a}$$

or

$$\frac{dL_B}{dv} = C_3 v^3 \frac{1}{e^{[C_4(v/T)]} - 1} \quad \left[\frac{W}{m \cdot m^2 \cdot sr}\right] \tag{15.27b}$$

where $C_1 = 2hc^2 = 1.176 \times 10^{-16}$ W·m^2, $C_2 = (hc/k) = 1.432 \times 10^{-2}$ m·K, $C_3 = (2h/c^2) = 1.47 \times 10^{-50}$ (W·s^4/m^2), and $C_4 = (h/k) = 4.78 \times 10^{-11}$ s·K. In Figure 15.6, we show a plot of Planck's formulas.

The derivation of Planck's formulas leading to the units of radiance/wavelength interval and radiance/frequency interval is similar to the derivation leading to units of energy density/frequency interval as presented above. In Section C, we will show the proportionality of the radiance to the energy density.

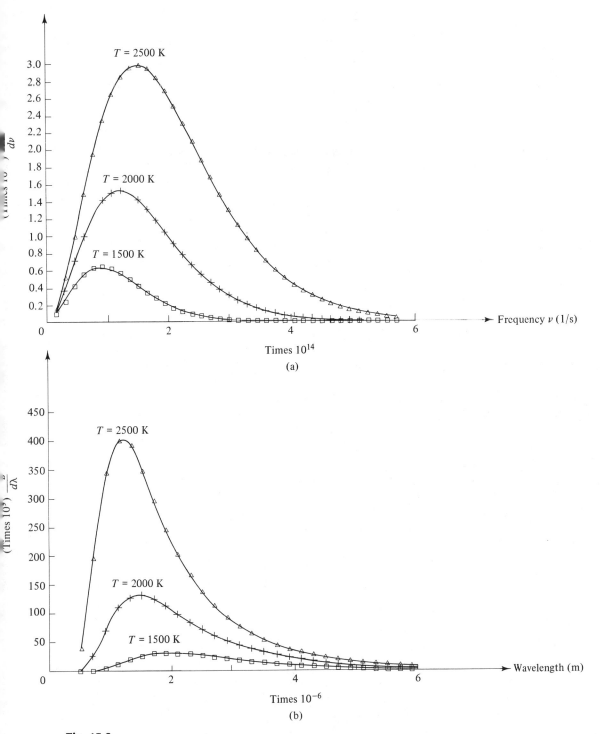

Fig. 15.6
Planck's radiation law for different temperatures. (a) Plot of radiance dL_B/dv as a function of frequency. (b) Plot of radiance $dL_B/d\lambda$ as a function of wavelength.

15. Light Sources

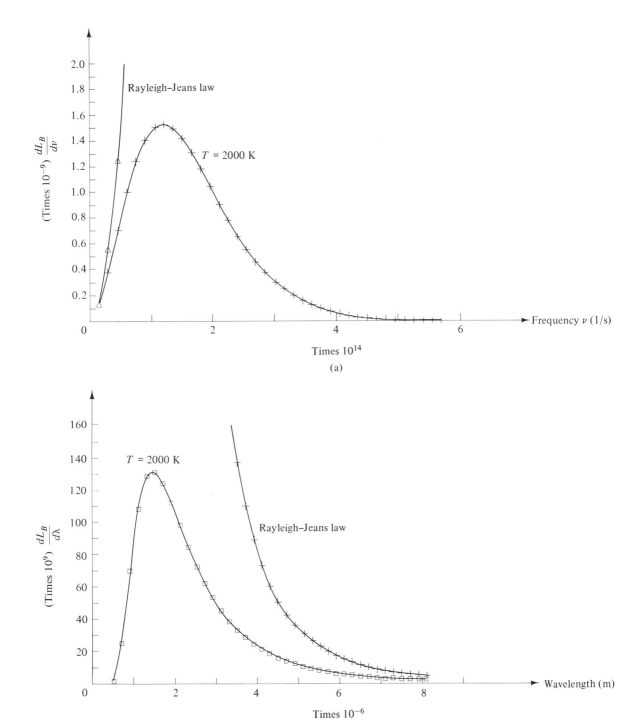

Fig. 15.7
Planck's law and the Rayleigh-Jeans law. (a) As a function of frequency. (b) As a function of wavelength.

In Figure 15.7a and b, we plot Planck's law depending on frequencies and wavelengths for one temperature. Using the frequency-dependent Planck's formula, we can expand the exponential and retain only the first two terms to obtain the Rayleigh-Jeans law showing the frequency dependence:

$$\frac{dL_B}{dv} = \frac{8\pi h v^3}{c^3}\left(\frac{1}{1 + (hv/kT) - 1}\right)$$
$$= \frac{8\pi kT}{c^3}v^2 \tag{15.28}$$

The wavelength dependence formulas are obtained by differentiation of $\lambda v = c$; that is,

$$\frac{d\lambda}{\lambda} = -\frac{dv}{v} \quad \text{or} \quad \frac{dv}{c} = \left|\frac{d\lambda}{\lambda^2}\right| \tag{15.29}$$

Introducing equation 15.29 into equation 15.28 results in the Rayleigh-Jeans law in terms of wavelength:

$$\frac{dL_B}{dv} = \frac{dL_B}{(d\lambda/\lambda^2)c} = \frac{8\pi kT}{c^3}\left(\frac{c^2}{\lambda^2}\right) \quad \text{or} \quad \frac{dL_B}{d\lambda} = \frac{8\pi kT}{\lambda^4} \tag{15.30}$$

(see problem 4). The Rayleigh-Jeans law is also shown in Figure 15.7a and b.

We may see that plotting these formulas depending on v or λ makes a difference in their appearance. The exponent of v and λ, with which Planck's law falls "off" to longer wavelength or smaller frequency, respectively, is different.

C. Total emitted energy, Stefan-Boltzmann Law, Kirchhoff's Law, and Wien's Law

If we integrate the energy density per frequency interval du/dv over the frequency range from 0 to ∞, we obtain the energy density u as

$$u = \frac{8}{15}\pi^5\frac{k^4 T^4}{c^3 h^3} \tag{15.31}$$

(see problem 5). The energy density is proportional to the fourth power of the temperature. In Figure 15.6, the area under the curves having T as a parameter is proportional to u.

If we want to measure u, we can bring a radiation-measuring instrument before the hole at the blackbody (see Figure 15.1). According to Chapter 11, the power leaving the hole into the solid angle $d\Omega$ is

$$dW = L_B\,da\cos\theta\,d\Omega \tag{15.32}$$

where L_B is the radiance of the blackbody, da is the area of the emitter, and θ is the angle that the center line of the solid angle makes with the normal of the emitter. For the total solid angle 2π we may calculate with the result of problem 4 in Chapter 11,

$$W = L_B\,da\,\pi \tag{15.33}$$

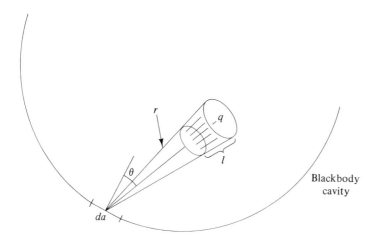

Fig. 15.8
The power emitted from the surface element da traverses the volume element $dV = q\ell$ in time ℓ/c.

The quantity of interest is the radiant exitance, that is, the power per unit area

$$\frac{W}{da} = L_B \pi \tag{15.34}$$

In the cavity of a blackbody, the absorbed power per area, W_a/da is equal to the emitted power per area W_e/da. The ratio depends only on the frequency and temperature, that is,

$$\frac{W_a/da}{W_e/da} = \frac{W_a}{W_e} = g(v, T) \tag{15.35}$$

where $g(v, T)$ is related to Planck's formula. Equation 15.35 is called Kirchhoff's law for the blackbody.

To relate the energy density u to L_B, we consider a small volume in the blackbody (Figure 15.8). The power passing through the volume $dV = \ell \cdot q$ in the time dt is

$$W\,dt = da(\cos\theta)L_B\,d\Omega\,dt \tag{15.36}$$

We have $d\Omega = q/r^2$ and $dt = \ell/c$. For $W\,dt$, we then have $u\,dV$, and using $dV = q\ell$, we have

$$\begin{aligned} W\,dt = u\,dV &= \frac{L_B\,da(\cos\theta)q\ell}{r^2 c} \\ &= \frac{L_B\,da\cos\theta}{r^2 c}\,dV \end{aligned} \tag{15.37}$$

Integration over the volume of the blackbody results in

$$uV = \frac{L_B}{c}V\int\frac{da\cos\theta}{r^2} \tag{15.38}$$

In problem 5, we find the value 4π for the integral and have

$$u = L_B \frac{4\pi}{c} \qquad (15.39)$$

Using equation 15.31, we have

$$L_B = \frac{2}{15}\left(\frac{\pi^4 k^4 T^4}{c^2 h^3}\right) \qquad (15.40)$$

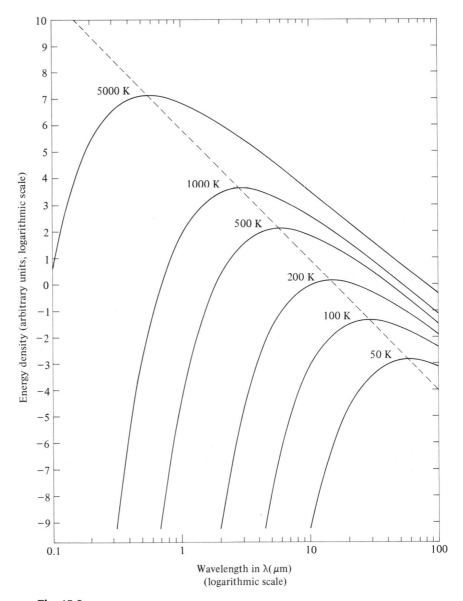

Fig. 15.9
Planck's law plotted in logarithmic form, showing Wien's law of displacement of maxima with temperature.

15. Light Sources

From equation 15.34, the radiant exitance is

$$\frac{W}{da} = L_B \pi = \frac{2}{15} \left(\frac{\pi^5 k^4 T^4}{c^2 h^3} \right) \qquad (15.41)$$

Equation 15.41 is called the **Stefan-Boltzmann law**. The constant

$$\sigma = \frac{2}{15} \left(\frac{\pi^5 k^4}{c^2 h^3} \right)$$

has the value $\sigma = 5.6703 \times 10^{-8}\,W/m^2 \cdot K^4$.

As already mentioned, a wire heated to increased temperatures changes color from red to white. The maximum of the emitted radiation shifts its peak from lower to higher frequencies. This is expressed as

$$\nu_{max} \propto T \qquad (15.42)$$

and is shown in Figure 15.9. It is called **Wien's displacement law** and will be derived in problem 6. Our eye is most sensitive at a frequency corresponding to about 555 nm. Close to that frequency (480 nm), a blackbody of temperature $T = 6000$ K has maximum emission. The sun has just this temperature.

3. ATOMIC EMISSION

In contrast to blackbody radiation, which is continuous over a large frequency region, atomic emission is confined to certain well-defined frequency intervals. The emission spectrum has the appearance of a resonance structure. Only at certain frequencies, characteristic for the type of atom, light will be emitted. This fact is used in chemistry and astrophysics to identify atoms. See also the color table at the beginning of this book.

The atomic emission is usually obtained by using a gas discharge. The atoms are brought into a vapor state of appropriate pressure, and a certain voltage is applied through electrodes. The electrons freed by the collision of atoms excite the electrons of other atoms. When the electrons fall back into lower states, light is emitted. This works well for relatively light atoms, but for heavier atoms, we must make special provisions to get the atoms into a gaslike state. However, we are well familiar with the mercury and sodium lamps used for highway and parking illumination.

The atoms are described as having certain well-defined energy levels. The mechanical analog is any vibrating system having resonance frequencies, such as a vibrating string on a violin.

A. Bohr's Model

We call the energy levels of the atom E. To get a mathematical description, we assume that electrons are moving around the nucleus and that the nucleus has a positive charge and is at rest. For the hydrogen atom, we have only one electron, and the nucleus is about 2000 times heavier.

We want to calculate the possible energy levels the electron can occupy. We

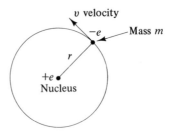

Fig. 15.10
Bohr's model of hydrogen. The electron orbits around the nucleus.

will not apply quantum mechanics but will use Bohr's model in connection with de Broglie's relations. In so doing, we will be able to obtain, with simple mathematics, the correct result for the energy levels of the hydrogen atom. In analogy we will also be able to understand the energy schematic of other atoms and ions, which will help us understand lasers. Finer details are, of course, lost. We start out with the kinetic and potential energy of a point mass m with charge $-e$ moving around a fixed charge $+e$ in a circle (Figure 15.10). The total energy is

$$E = \frac{1}{2}mv^2 - \frac{Ke^2}{r} \tag{15.43}$$

where K is the constant of Coulomb's law. The kinetic energy and potential energy are related through the fact that the centripetal and attractive forces are equal:

$$\frac{Ke^2}{r^2} = m\frac{v^2}{r} \tag{15.44}$$

From equations 15.43 and 15.44, we have

$$E = -\frac{1}{2}\left(\frac{Ke^2}{r}\right) \tag{15.45}$$

The electrons are restricted to certain orbits. We introduce this into our model by the use of de Broglie's relation, associating with each particle a wave of a certain wavelength given by

$$\lambda = \frac{h}{mv} \tag{15.46}$$

where h is Planck's constant. Considering a wave of this wavelength, we apply the standing-wave concept, that is, the concept that the phase of the wave after one turn should be the same as at the beginning. In other words, we want the circular orbits of radius r to be a multiple of the wavelength of the particle, that is,

$$2\pi r = n\lambda = \frac{nh}{mv} \tag{15.47}$$

where n is an integer.

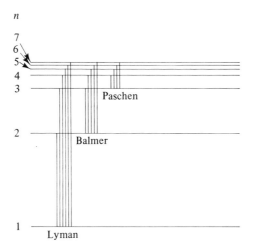

Fig. 15.11
Diagram of energy levels and transitions for Bohr's model. The electrons change from
n_i to n_f and emit light.
Lyman series: $n_f = 1, n_i = 2, 3, 4, \ldots$
Balmer series: $n_f = 3, n_i = 3, 4, 5, \ldots$
Paschen series: $n_f = 3, n_i = 4, 5, 6, \ldots$
The energy difference between the state $n = 1$ and $n = \infty$ is the dissociation energy; for
hydrogen, it is 13.6 eV.

For v^2, we obtain

$$v^2 = \frac{n^2 h^2}{(2\pi r m)^2} \tag{15.48}$$

and using equation 15.44, we get

$$\frac{1}{r} = \frac{K e^2 m 4\pi^2}{n^2 h^2} \tag{15.49}$$

For the energy, equation 15.45 gives us

$$E = -\frac{2\pi^2 K^2 e^4 m}{n^2 h^2} \tag{15.50}$$

The only variable in the expression for the energy is the integer n, introduced
through the standing-wave condition. All other quantities are fundamental constants. It is a strong point for this theory that no empirical constants are used
and that equation 15.50 agrees surprisingly well with the experiment. However,
this agreement does not hold for any atom other than the hydrogen atom. The
Schrödinger equation of quantum mechanics gives the details for these atoms.

Figure 15.11 shows the energy schematic of the hydrogen atom. The transition frequencies are obtained from the differences between the energy levels as

$$E_{n_f} - E_{n_i} = v_{n_i, n_f} h = \frac{2\pi^2 K^2 e^4 m}{h^2}\left(\frac{1}{n_i^2} - \frac{1}{n_f^2}\right) \tag{15.51}$$

where i and f stand for the initial and final states. Light of energy $E = hv$ is
emitted.

3. Atomic Emission   515

B. Summary of Energy Levels and Electronic Transitions for Atoms Other than Hydrogen

The Schrödinger equation and the Pauli principle supply us with what we need to understand the atomic energy schematics and transitions. We will now consider the notation needed to understand the energy schematics of atoms other than hydrogen. In Figure 15.12, we show the schematic of an atom having up to Z electrons with a nucleus having a positive charge of Ze.

Principal quantum number and angular momentum quantum number. We start with the schematic of the hydrogen atom. The quantum number n is called the **principal quantum number**. The question is, How many electrons can be placed in each level $n = 1, 2, 3, 4$? First, the Schrödinger equation tells us that to each n there belong $n - 1$ values corresponding to the angular momentum quantum number ℓ of the electrons.

Magnetic quantum number and degeneracy. Second, there are $2\ell + 1$ substates to each state labeled by ℓ. These substates have different energy values

Number of different m $(2\ell + 1)$:	1	3	5	7	Number of electrons in filled state n	Maximum number of electrons with spin up	Maximum number of electrons with spin down	Shell
n | ℓ:	0	1	2	3				
4	4s	4p	4d	4f	32	16	16	N
3	3s	3p	3d		18	9	9	M
2	2s	2p			8	4	4	L
1	1s				2	1	1	K

Fig. 15.12
Energy levels and quantum numbers for a few states of a hydrogen-like atom. The notation 1s or 4f refers to individual electrons at the various levels.

if the atom is in a magnetic field. The corresponding quantum number is called the **magnetic quantum number** m. There are $2\ell + 1$ different orientations of ℓ in the magnetic field. If energy levels are characterized by different quantum numbers but they all have the same energy, they are called **degenerate energy levels**. In the hydrogen atom, all levels numbered by ℓ and m belong to the same state, n, and are degenerate.

Spin states. Third, there is a quantum number characterization for each electron related to an angular momentum around their "axis," called the **spin**. In a magnetic field, the spin has two possible positions, "up" and "down," each corresponding to a spin quantum number s. The projection along the magnetic field direction has quantum numbers $+\frac{1}{2}$ and $-\frac{1}{2}$.

Pauli's principle and occupation rule. The occupation of the energy levels by the electrons is governed by Pauli's principle that each electron must have a different set of quantum numbers, including the spin quantum numbers. For each n, there are $n - 1$ values for ℓ; for each ℓ, there are $2\ell + 1$ values for m; and each electron has two values for its spin. Table 15.2 presents the electron configurations for the first 25 elements in the periodic system.

Build-up principle of atoms. The quantum number n is labeled by K, L, M, and so on, and the quantum number ℓ by s, p, d, and f (lowercase letters give the state of an individual electron, and capital letters specify compound atomic states). The electrons are placed according to the lowest available value of ℓ for each n and the two possibilities for the spin. For a particular value of ℓ, we have $2\ell + 1$ values of m, the total number being $2(2\ell + 1)$. For $Z = 2$ and $Z = 10$, all possibilities for $n = 1$ and $n = 2$ are used up. The K and L shells are filled.

We see from Table 15.2 that this beautiful building system of atoms comes to an end with $Z = 19$, where the $4s$ electron apparently has lower energy than the $3d$ electron. Quantum mechanics explains this with the interaction of the electrons, which we have so far ignored.

Figure 15.12 shows the correspondence of the different quantum numbers n, ℓ, m and spin to the few lowest energy states of a hydrogen-like atom. Also indicated are the total number of electrons for the states labeled $n = 1, 2, 3$, and 4 and the notation for the individual electrons as $1s$ to $4f$.

C. Atomic Transitions: The Sodium Atom

After briefly discussing the characterization of the electrons in an atom, we will turn to the question of transitions of electrons between atomic energy states, leading to absorption or emission of light. Let us consider the case of sodium, an atom with many electrons (Figure 15.13).

From Table 15.2, we see that the "outermost" or "valence" electron for sodium is the $3s$ electron. Figure 15.13 starts with this level on the bottom. As in the case of hydrogen, this electron makes up most of the emission of the sodium atom spectrum. For $4s$, $5s$, and so on, we have a series of lines similar to those for the hydrogen atom. For the angular momentum equal to 1, that is, P states, we have two series of almost equal levels, and similarly for D and F, and so on. These

Table 15.2
Electron configurations of the first 25 elements

		K	L	M	N
		n: 1	2	3	4
Z	Element	ℓ: s	s p	s p d	s p d f
1	H Hydrogen	1			
2	He Helium	2			
3	Li Lithium	2	1		
4	Be Beryllium	2	2		
5	B Boron	2	2 1		
6	C Carbon	2	2 2		
7	N Nitrogen	2	2 3		
8	O Oxygen	2	2 4		
9	F Fluorine	2	2 5		
10	Ne Neon	2	2 6		
11	Na Sodium	2	2 6	1	
12	Mg Magnesium	2	2 6	2	
13	Al Aluminum	2	2 6	2 1	
14	Si Silicon	2	2 6	2 2	
15	P Phosphorus	2	2 6	2 3	
16	S Sulfur	2	2 6	2 4	
17	Cl Chlorine	2	2 6	2 5	
18	Ar Argon	2	2 6	2 6	
19	K Potassium	2	2 6	2 6	1
20	Ca Calcium	2	2 6	2 6	2
21	Sc Scandium	2	2 6	2 6 1	2
22	Ti Titanium	2	2 6	2 6 2	2
23	V Vanadium	2	2 6	2 6 3	2
24	Cr Chromium	2	2 6	2 6 5	1
25	Mn Manganese	2	2 6	2 6 5	2

Electrons having the same n value are in states called "shells" and named by K, L, M, and so on. For each n, we have $n - 1$ substates of angular momentum quantum number ℓ, named $s, p, d, f, \ldots$ There are $2\ell + 1$ possible values of m and two spin states. Therefore, the maximum occupation of K and L are

$$K: n = 1 \quad 2[(2 \cdot 0 + 1)] = 2$$
$$L: n = 2 \quad 2[(2 \cdot 0 + 1) + (2 \cdot 1 + 1)] = 8$$

(The M shell is more complicated.)

states are compound states in which the angular momentum and the spin appear together as the characteristic quantum number j. For $P = 1$ and spin quantum number $s = \frac{1}{2}$, we have the two combinations $j = \frac{1}{2}$ and $j = \frac{3}{2}$, corresponding to the orientation of the spin antiparallel to the angular momentum and parallel, respectively. These j values are written as a subscript on the P, D, and F characterizations. Since the energy levels corresponding to the two j states are very close, the P, D, and F lines appear as doublets. This is indicated by the superscript 2 preceding each letter relating to the total spin as well as to the multiplicity $2S + 1$, where S is the total spin. This is even done for the S state, which is not a doublet. We have to remember that the spectroscopic notation comes from the days of empirical investigations of the atomic spectra, long before Niels Bohr, Werner Heisenberg, Erwin Schrödinger, and Wolfgang Pauli brought some

15. Light Sources

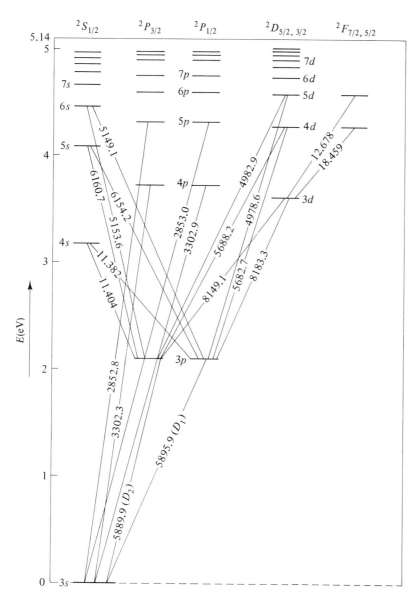

Fig. 15.13
Energy-level scheme for sodium. The energy level of 3s, that is, the ground state of the outermost electron, has been chosen as 0. (From H. G. Kuhn, *Atomic Spectra*, figure III, 17. By permission of Longman Group Limited.)

order into it. The letters *S*, *P*, and *D* have their origin in the labeling of lines by spectroscopists as "sharp," "principal," or "diffuse."

We now want to discuss why the hydrogen atom has transitions connecting every level *n* to every other level *n*, whereas in Figure 15.13, we see a lot of transitions, but not between all levels. A transition leading to absorption or emission of radiation is associated with a change in the dipole moment. This is similar to the case for electromagnetic radiation. For dipole transitions, we have

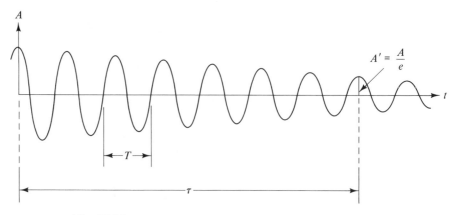

Fig. 15.14
Wave train decreasing over the time τ to the value $A/e = A/2.71$.

a corresponding condition for allowing only certain transitions. This is called the **selection rule**, and it states that the quantum number ℓ must change by 1, either up or down. This explains why in Figure 15.13 we only have transitions between columns of levels labeled ℓ. For the hydrogen atom, where the ℓ levels are all degenerate with the n levels, we find simple transition series, as shown in Figure 15.11.

4. BANDWIDTH

A. Natural Line Width

We have calculated the emitted energy of blackbody radiation and shown the emission curves depending on temperature in Figure 15.6. For the hydrogen atom, we showed the energy schematic leading to line spectra in Figure 15.11. We can see that the blackbody spectrum extends over a broad spectral range. The hydrogen spectrum consists of lines. Around the frequency of the lines there is emission of radiation in a narrow range.

In Chapter 2, we mentioned these two types of light sources and also the laser. The laser has a much smaller band width than the lines of the hydrogen atom. In Chapter 2, we characterized these three light sources by their coherence length. We will start at that point, but will now consider the time of the emission process rather than the length of the emitted wave train (Figure 15.14). Such an emitted wave may be described by (space part suppressed)

$$A = A_0 e^{i\omega_0 t} e^{-(t/\tau)} \tag{15.52}$$

where ω_0 is the angular frequency corresponding to the difference between the two states of the oscillator involved in the transition,

$$E_2 - E_1 = h\nu \tag{15.53}$$

and τ (lifetime) is a constant, that is, the time until the amplitude has decreased to a value of $A/e = A/2.71$.

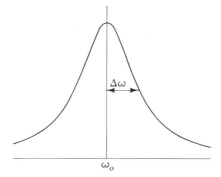

Fig. 15.15
Plot of the frequency spectrum of the wave schematically shown in Figure 15.14.

We have already discussed the fact that a nonmonochromatic wave can be represented by the superimposition of many monochromatic waves having certain well-defined frequencies and amplitudes. The sizes of these amplitudes are determined by a Fourier transformation integral and are called the frequency distribution. In our case, we have ($A_0 = 1$)

$$y(\omega) = \frac{1}{2\pi} \int_0^\infty e^{i\omega_0 t} e^{-(t/\tau)} e^{-i\omega t}\, dt \tag{15.54}$$

which yields

$$y(\omega) = \frac{1}{2\pi} \left(\frac{-1}{i(\omega_0 - \omega) - (1/\tau)} \right) \tag{15.55}$$

Taking the intensity as the square of the amplitude, we have

$$I(\omega) = \frac{1}{4\pi^2} \left(\frac{\tau^2}{1 + (\omega_0 - \omega)^2 \tau^2} \right) \tag{15.56}$$

This function is plotted in Figure 15.15 (also see problem 8). If $(\omega_0 - \omega)^2$ times τ^2 equals 1, then we have half the intensity. Therefore, the half-width (at half-height) is given by

$$\Delta\omega = \frac{1}{\tau} \tag{15.57}$$

From equation 15.57, we have the result that the length of the emitted wave, or the time for the emission, is inversely proportional to the line width.

For any oscillator with losses, the product of ω_0 and τ is called the quality factor Q of the oscillator. The ratio $Q/2\pi$ is equal to τ/T, that is, the number of periods in the wave until the $1/e$ attenuation point. If we divide ω_0 by the full width at half height $\Delta\omega$, we have $\omega_0/\Delta\omega = Q$. The fringes of interference and emerging spectral lines from oscillations in cavities may be characterized by Q. A large value of Q presents a very narrow line, associated in some cases with high resolution of the device.

We have used the concept of an electron circling an atom for the time τ during which the radiation is emitted and has at the end of the emission process

4. Bandwidth   **521**

the amplitude $A/2.71$. We can also say that the oscillator in the upper state has the "life time" τ. If the lifetime is large, the emitted lines are narrow, and vice versa. The Einstein coefficient A_{21} of spontaneous emission can be related to τ as follows.

We consider the number N of emitting atoms. Their change, that is, the number of atoms going from the higher to the lower state, is described by the "rate equation," which we know from nuclear decay problems

$$dN = -A_{21} N \, dt \tag{15.58}$$

Integration yields

$$N = N_0 e^{-A_{21}t} = N_0 e^{-(t/\tau)} \tag{15.59}$$

from which we read off the relation

$$\tau = \frac{1}{A_{21}} \tag{15.60}$$

We will see in the next chapter that there are energy states in atoms and molecules that have long lifetimes. An average atomic state has a lifetime of 10^{-8} s; long lifetimes are of the order 10^{-3} s. These states are important to laser action and are called **metastable states**.

B. Collision Broadening

The interpretation of τ in the classical sense is that the electron circulates around the nucleus and emits the wave until by the time τ, the original amplitude A is only

$$A \frac{1}{e} = A \frac{1}{2.71}$$

If we consider that the emission process of the atoms may be interrupted by collisions, the time τ is shorter and a broader line width will result (see equation 15.56). The interruptions depend on the free path the oscillator can travel between collisions. The length between collisions is called the free path length l. It is shorter for higher temperatures. If the emitted light has the wavelength λ, the number of wavelengths fitting into the free path length is $n = l/\lambda$. Comparing lines for shorter and longer wavelengths, we find that for longer wavelengths, n becomes smaller and the emission process is interrupted after a smaller number of wavelengths. Consequently, the lines are broader.

C. Doppler Broadening

From special relativity, we know that the frequency of the observed light is different depending on whether the emitting atom is traveling to the observer or away from the observer. This is called the **Doppler effect**.

We have a mechanical analog that qualitatively gives the same result. A car with a blowing horn seems to emit sound with a higher frequency when traveling toward us, a lower frequency when traveling away from us.

The frequency shift is given as

$$v - v_0 = \frac{v}{c} v_0 \tag{15.61}$$

The dependence of the intensity on the frequency interval follows from the Boltzmann distribution as

$$I(v) = I_{v_0} \exp^{(-E_K/kT)} = I_{v_0} \exp^{[-1/2(mv^2/kT)]} = I_{v_0} \exp^{[-1/2(mc^2/kT)(v-v_0/v_0)^2]} \tag{15.62}$$

where I_{v_0} is the intensity at the center of the line, m is the mass of the oscillator, c is the speed of light, k is the Boltzmann constant, and T is the absolute temperature.

In problem 9, we will calculate the half-width at half-height for the Doppler effect; the result is

$$\Delta v = 2v_0 \sqrt{\frac{2kT}{mc^2} \ln 2} \tag{15.63}$$

In the visible spectral region, the Doppler width is in general larger than the collision width; for longer waves, the reverse is true.

D. Length Standard

The fact that in the visible spectral region the Doppler width of a spectral line is larger than the collision broadening has been used for the selection of an atomic emission in the visible spectral region as a length standard. The Doppler width can be minimized by selecting an atom with a large mass that may emit light at low temperatures. Krypton 86 has been introduced as a length standard. Light is emitted at $T = 63$ K at wavelength $\lambda_{air} = 605.616$ nm, and the line width is

$$\frac{\Delta v}{v} = 3 \times 10^{-7} \tag{15.64}$$

In comparison, for blackbody radiation at $T = 6000$ K, the center frequency corresponds to 0.4 μm. The width at half-height is approximately at 0.2 μm and 0.9 μm, giving a width of $\Delta\lambda = 0.7$ μm, and for

$$\left| \frac{\Delta v}{v} \right| = \left| \frac{\Delta\lambda}{\lambda} \right|$$

we have $0.7/0.4 \simeq 2$.

5. LASERS

We know that light waves and radio waves are both electromagnetic radiation. We discussed two types of light waves in the two preceding sections: blackbody radiation, which is emitted over a broad frequency range and has a very broad line width, and atomic emission, which has line spectra and consequently has a narrow frequency range for each line. If we could make a very narrow filter, we could filter out of both types of light a very narrow frequency interval. Making the filter narrower would result in longer and longer wave trains. Taking this

5. Lasers

Fig. 15.16
Oscillators (dots) and waves (W) in a cavity formed by two mirrors (M).

process to the limit, we could obtain monochromatic light, but with zero intensity. Actually, such a filtering process would come to an end much earlier, since the energy of radiation for wave trains of finite length would already drop to a level not observable with any existing detector. Radio waves, on the other hand, carry very small amounts of energy and have very long wave trains. They are emitted by oscillations of an antenna and not by the type of oscillators discussed in the previous sections.

To produce light waves with long wave trains, we might first try to imitate the production of radio waves or microwaves. In fact, this has been tried. Making the mechanical size of the oscillators smaller and smaller could produce millimeter waves. But to get to the visible spectral region, a new principle had to be used—the **microwave** (or light) **amplification by stimulated emission of radiation (MASER or LASER)**.

The MASER principle was developed in 1950. Microwaves were amplified by stimulated emission of radiation, and the microwaves obtained had very narrow bandwidths with appreciable intensities. An outgrowth of this work has been the atomic clocks with a bandwidth so small that they are one of the best time standards today.

The question of how this principle could be applied in the visible spectral region was discussed in a famous paper by Charles H. Townes and Arthur L. Schalow in 1958.* The principles laid out in this paper led to the first solid-state laser by T. H. Maiman and the first gas laser by A. Javan, D. R. Herriott, and W. R. Bennett, Jr. Townes received the Nobel Prize for his work in 1964.

A. Population Inversion

In our discussion of blackbody radiation, we saw how running waves in the cavity get absorbed and reemitted from the oscillators on the walls. The energy for the emission by the oscillators was supplied by the thermal reservoir of temperature T. We now assume that the oscillators are in the volume of the cavity and that the walls are highly reflecting, so that the waves may travel forward and backward very often and may form standing waves (Figure 15.16).

If we can bring energy into the oscillators and have it picked up by the running waves, we will get an amplification of the waves in the cavity. We then

---

*A. L. Schalow and C. H. Townes, Infrared and optical masers, *Phys. Rev.* **112**, 1940 (1958).

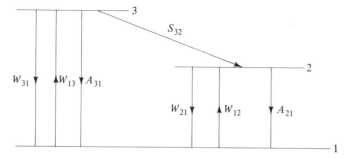

Fig. 15.17
Schematic of a three-level laser. S_{32} is a radiationless transition. The transition probability W_{iK} and A_{iK} spontaneous emission probability have been explained in Section 2.

can insert a small hole into one of the mirrors and "couple" out some of the radiation. If we heat up a gas consisting of atoms, we find, according to the Boltzmann distribution, that there are always more atoms in the lower energy states than in the higher ones. What we need is just the opposite, called **population inversion**. There are other methods besides pumping heat energy into the states of atoms. One way is irradiation with a high-intensity lamp whose specific frequency will lift electrons from the lower state m to the higher state n. Another method is to have the atoms in a gas discharge. Collisions with electrons and other atoms may bring a larger number of oscillators into a higher state than can be obtained from the Boltzmann distribution. A third method uses current flow in a solid-state device (to be explained later).

Rate equation. Let us consider a system of energy levels consisting of a ground state, an excited state, and below the excited state a third level (Figure 15.17). For transitions between level 3 and the ground state, we have induced absorption, induced emission, and spotaneous emission. The transition probabilities are described by W_{31}, W_{13} ($W_{iK} = B_{iK}$ times the radiation density per frequency interval), and A_{31}, respectively. A similar discription holds for level 2. Between levels 2 and 3, we assume only "radiationless transitions" in the down direction, and the probability is described by S_{32}. The number of all oscillators involved is assumed to be fixed, and we have

$$N_0 = N_1 + N_2 + N_3 \tag{15.65}$$

where N_1, N_2, and N_3 are the number of oscillators in the states 1, 2, and 3, respectively.

The change, in time, of the number of oscillators in states 2 and 3 is proportional to the number of oscillators in these states. For level 3, we have absorption from level 1 and spontaneous and induced emission. In addition, we have the radiationless transition to level 2, that is,

$$\frac{dN_3}{dt} = W_{13}N_1 - (W_{31} + A_{31} + S_{32})N_3 \tag{15.66}$$

and for level 2 we get

$$\frac{dN_2}{dt} = W_{12}N_1 + S_{32}N_3 - (W_{21} + A_{21})N_2 \tag{15.67}$$

For the steady state, the time derivatives are zero and we obtain the ratio N_2/N_1 if we assume that A_{31} is small compared to W_{31}:

$$\frac{N_2}{N_1} = \left(\frac{W_{13}S_{32}}{W_{31} + S_{32}} + W_{12} \right)(A_{21} + W_{21})^{-1} \tag{15.68}$$

(see problem 10). From Section 1B, we can set $W_{13} = W_{31}$ and $W_{12} = W_{21}$. We assume further that in the process being considered, the radiationless transitions from 3 to 2 are fast, that is, $S_{32} \gg W_{13}$, and from equation 15.68, we have

$$\frac{N_2}{N_1} = \frac{W_{13} + W_{12}}{W_{12} + A_{21}} \tag{15.69}$$

(see problem 10). From equations 15.65 and 15.68, we can also write

$$\frac{N_2 - N_1}{N_0} = \frac{W_{13} - A_{21}}{W_{13} + A_{21} + 2W_{12}} \tag{15.70}$$

(see problem 10).

Threshold condition.   For a system for which we have $W_{13} = A_{21}$, we will have $N_2 = N_1$ in the steady state. When there are as many oscillators in the upper state as in the lower state, we have the beginning of population inversion, and oscillations in the cavity can start.

Minimum power.   To get an idea how much power is necessary to get to laser oscillations, we calculate the induced absorption of level 3. $P_3 = nh\nu$, with $n = NW_{13}$. This must be larger than $P' = NA_{21}h\nu$ because of the condition $W_{13} > A_{21}$. For $\lambda = 0.5 \times 10^{-4}$ cm (i.e., the middle of the visible spectrum), $A_{21} = 1/\tau_{sp} = \frac{1}{3} \times 10^3$, and with $N = 10^{19}$ cm^{-3}, we have

$$NA_{21}h\nu = NA_{21}h\frac{c}{\lambda} = \frac{10^{19}\text{ cm}^{-3} \cdot 0.33 \times 10^3(1/\text{s}) \cdot 7 \times 10^{-34}\text{ W} \cdot \text{s}^2 \cdot 3 \times 10^{10}\text{ (cm/s)}}{0.5 \times 10^{-4}\text{ cm}}$$

$$\simeq 1400 \text{ W/cm}^3 \tag{15.71}$$

The coefficient of spontaneous emission of level 2 has been taken as $A_{21} = \frac{1}{3} \times 10^3$; that is, the lifetime (see equation 15.60) of the state in the oscillator is large compared to what we normally observe, which is of the order of 10^{-8}. Having such a large lifetime, state 2 is called a metastable state. If we did not use such a metastable state but assumed for the coefficient of spontaneous emission $A_{21} = \frac{1}{3} \times 10^8$, we would get for the minimum power

$$P = 10^5 \cdot 1400 \text{ W/cm}^3 \tag{15.72}$$

B. Pumping Schemes

The helium-neon laser.   The helium-neon (He-Ne) laser was the first gas laser. It was built at Bell Telephone Laboratories in 1961. The mixture of helium

15. Light Sources

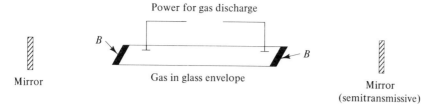

Fig. 15.18
Schematic of a gas laser. *B* are Brewster windows.

and neon gas (five parts He to one part Ne) is enclosed in a gas tube, and a discharge is operated from the two ends of the tube (Figure 15.18). The energy schematic of He and Ne is shown in Figure 15.19. The schematic is for the compound state. The electron configuration is also indicated.

We have a four-level scheme for the laser action. Through the gas discharge, the He atom is exited into the 2^3S and 2^1S states. These states have a long lifetime; they can hold the energy longer than other He states. Collision with the Ne atoms transfers the energy to the $2S$ and $3S$ states of Ne. Small differences in the energy levels of He and Ne are compensated for by the kinetic energy in the collision process. Energy from the 2^1S level transfers to the $3S$ level of Ne by resonance. As we know from the mechanical coupled oscillator, energy transfer is easily possible if the two oscillators have almost the same energy levels. The Ne $(2S)$ level becomes the upper level with the population inversion; the lower level is the Ne $(2P)$. The Ne $(2P)$ level decays with radiation to the $1S$ level in 0.01 μs, whereas the spontaneous emission from the Ne $(2S)$ to Ne $(2P)$ takes much longer, 0.1 μs. Therefore, population inversion can be established. The corresponding laser frequency is 1.15 μm.

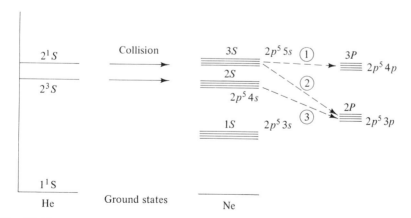

Fig. 15.19
Energy schematic of the He-Ne laser. Capital letters refer to the (compound) atomic state, lowercase letters to electron configuration. Frequencies of laser action: (1) $\lambda = 3.3912$ μm; (2) $\lambda = 0.6328$ μm; (3) $\lambda = 1.1523$ μm.

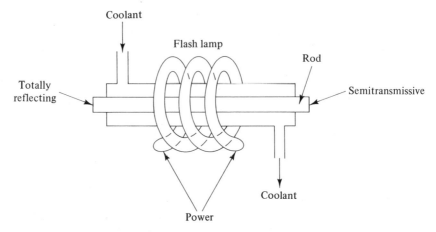

Fig. 15.20
Schematic of a solid-state laser.

The Ne (3S) level is the upper state for the laser action at 0.6328 μm (6328 Å), with the lower state the Ne (2P). This is the red laser line characteristic for the He-Ne laser. There is a third transition, Ne (3S) to Ne (3P) with a wavelength of 3.39 μm, appearing in the infrared, but it is less frequently used.

Solid-state lasers. The ruby laser was the very first laser, developed by Maiman in 1960. The active medium consists of Cr^{+3} ions, 0.05 percent by weight in an Al_2O_3 crystal. The pumping was done optically by a flash lamp (Figure 15.20). The light travels in a rod, one end of which is completely silvered, the other end only partially. The energy schematic is shown in Figure 15.21. The light from the flash lamp brings the ions into the states $^4F_1$ and $^4F_2$. Nonradiative transitions transfer energy to the two states $2\bar{A}$ and $\bar{E}$. (Some of the spectroscopic

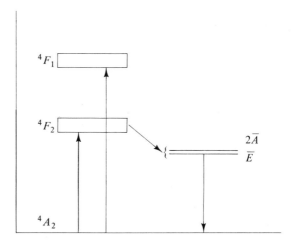

Fig. 15.21
Three-level energy schematic of a ruby laser.

notation used here has not been covered; for more detail, see Yariv.[*] Laser emission takes place by transitions from $2\bar{A}$ and $\bar{E}$ to the $^4A_2$ state. The transition $\bar{E} \rightarrow {}^4A_2$ has frequencies of 0.6929 and 0.6943 μm corresponding to the red output of the ruby laser. Spontaneous transitions from $^4F_1$ and $^4F_2$ to $^4A_2$ are about 100 times less probable than the transfer to $2\bar{A}$ and $\bar{E}$. About 99 percent of all ions get transferred from the F states into these states.

Semiconductor lasers. Semiconductor lasers are physically small and in general have low power output but are of great importance for many applications, such as in integrated optics (see Chapter 17). For the following, also see the discussion in Chapter 14, Sections 12 and 13. Solids have their electrons in bands instead of levels (as is the case for single atoms). Let us consider a model crystal made up of atoms having only one electron to contribute to the band. The band is made of all states of all atoms. In the band, each state can be occupied by two electrons, a result of Pauli's principle. Therefore, such a band is only half-filled. If the atoms had two electrons, instead of one, as we assumed, the band would be filled. Just as energy levels are separated from one another in the atom, the bands are separated from one another. The separation energy is called gap energy E_G.

Depending on the material, we can have large gaps or small gaps, filled bands and unfilled bands. An insulator is a material with a filled band, above the band a large gap, and the next band empty. In a conductor the uppermost band is not completely filled. There are a lot of electrons available for conduction. A semiconductor has a small gap, and through thermal excitation, electrons are excited from the lower band into the upper band, where they can then participate in the conduction. In the lower band they leave holes. Impurities may have electronic states located in the gap region. If an impurity atom tends to give up an electron, which goes to the upper band, it is called a donor. If it attracts an electron from the lower band, it is called an acceptor. An acceptor leaves a hole in the lower band. A p-type semiconductor has the lowest band almost filled. What is not filled is called a hole. Electrons have negative charges, holes have positive charges, and the material is called p-type. For n-type semiconductors, the lowest band is just a little bit filled (Figure 15.22a), and if the two materials are joined without a current going through, the electrons in both parts have the same highest energy state, called the Fermi level.

A semiconductor laser is shown in Figure 15.23. When a current is applied in the "forward" direction, the energy schematic changes (see Figure 15.22b). There is a region where electrons in the n-type material are located above holes of the p-type region. Stored energy is located in the upper state, and running electromagnetic waves in the crystal can pick up the energy and start laser action.

---

[*] A. Yariv, *Quantum Electronics* (New York: Wiley, 1967) and P. A. Tipler, *Modern Physics*, cited on page 505.

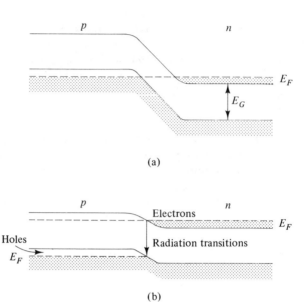

(a)

(b)

Fig. 15.22
Schematic of a p-n junction. E_G is gap energy; E_F is energy of Fermi level. (a) Zero-applied field. (b) Forward bias voltage. Radiation is emitted from the region containing electrons and holes.

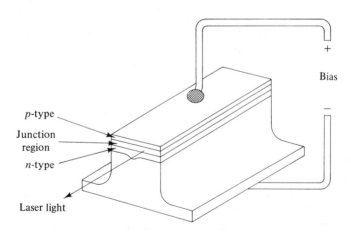

Fig. 15.23
Schematic of a gallium-arsenide (Ga-As) p-n junction semiconductor laser.

C. Cavity Resonators and Modes

In Chapter 1, we discussed the arrangement of two mirrors to form a resonator cavity. After one "round trip," the geometrical optical ray points in the same direction as it pointed at the beginning. We can use flat and spherical mirrors in various combinations to form resonators. Figure 15.24 shows a number of such

15. Light Sources

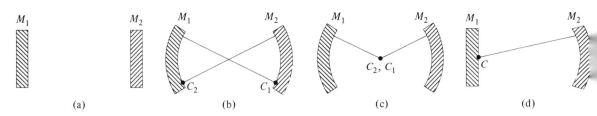

Fig. 15.24
Two-mirror arrangement for cavity formation. (a) Plane parallel mirrors. (b) Confocal configuration. (c) Concentric configuration. (d) Hemispherical configuration. Cases (a), (b) and (d) were discussed in Chapter 1.

combinations. The configurations in Figure 15.24a, b, and d were discussed in Chapter 1, Section 9.

Besides the geometrical optical point of view, we consider light as waves in the cavity. If we assume two parallel mirrors, the simplest condition for standing waves is

$$d = m\frac{\lambda}{2} \tag{15.73}$$

where m is an integer. Into the same distance d also fits a slightly different wavelength λ', with $m + 1$ half-wavelengths,

$$d = (m + 1)\frac{\lambda'}{2} \tag{15.74}$$

The difference in wavelength, $\Delta\lambda$, is expressed in the difference of the frequencies as

$$\Delta v = \frac{c}{2d} \tag{15.75}$$

where c is the speed of light. This Δv is called the mode separation (longitudinal).

If the laser is designed so that only one longitudinal mode can exist, we have a single-frequency laser. We now assume that we have only one wavelength. We remember that the wave has a lateral extension, and by traveling between the two mirrors, diffraction will occur at each mirror. While the diffraction losses may be very small because the ratio of the wavelength to the radius of the mirror, λ/a, is small, the speed of light is very large, and we have many reflections by the mirrors at distance d of the cavity. Using diffraction theory such as was formulated by Kirchhoff's formula (see Chapter 3), we can find modes in the cavity that theoretically have least diffraction losses. The principal idea is the same as for the traveling ray in the cavity. The wave must "travel" into itself. The ratio of the distance between the mirrors to the radius of the mirror, d/a, is an important parameter from the point of view of diffraction. The product of a/λ and a/d is called the Fresnel number N:

$$\frac{a}{\lambda}\left(\frac{a}{d}\right) = N \tag{15.76}$$

Diffraction losses will be small when N is much greater than 1.

For the description of the modes in the resonator, we start with the standing-wave condition in three dimensions, as developed for the blackbody. We assume here that the walls of the box reflected the running waves, and we have for the standing-wave condition

$$\frac{M_X^2}{\ell_X^2} + \frac{M_Y^2}{\ell_Y^2} + \frac{M_Z^2}{\ell_Z^2} = \frac{1}{(\lambda/2)^2} \tag{15.77}$$

where M_X, M_Y, and M_Z had their origin in the standing-wave condition in each direction, where each one was connected to a set of traveling waves

$$Ae^{i(k_X X \pm \omega t)}, \quad A'e^{i(k_Y Y \pm \omega t)}, \quad A''e^{i(k_Z Z \pm \omega t)} \tag{15.78}$$

In the one-dimensional case, when $M_X = 0$, the corresponding k_X value in the wave is also zero, and no "running wave" exists in that direction. In the three-dimensional case, we know that some M values may be zero and the corresponding k values also, but still there will be a standing wave. For the laser, one dimension of the box is much larger than the other two. Calling this large number $M_Z = q$, and setting $\ell_X = \ell_Z = \ell$, we have

$$\frac{M_X^2 + M_Y^2}{\ell^2} + \frac{q^2}{d^2} = \frac{1}{(\lambda/2)^2} \qquad q \gg M_X, M_Y \tag{15.79}$$

As in the standing-wave pattern in one dimension, the number $M + 1$ gives the number of nodes or nodal lines in the pattern. In three dimensions, we assume

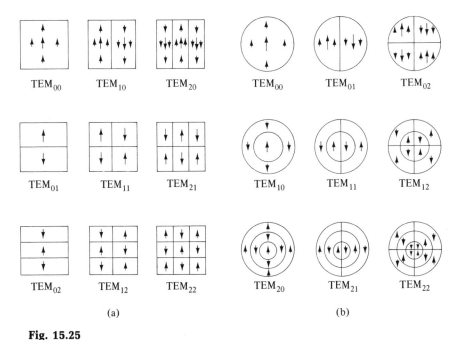

(a)   (b)

Fig. 15.25
Fabry-Perot resonators. (a) With plane parallel square-shaped reflectors. (b) With round, spherical mirrors. (See Chapter 17, Section 2C, for an explanation of TEM.) [From H. Kogelnik and T. Li, *Applied Optics* **5**, 1550 (1966).]

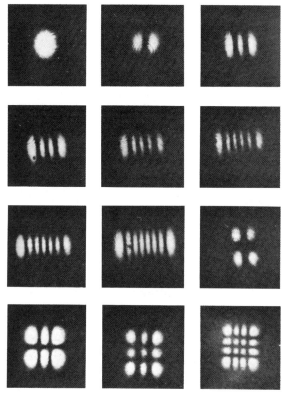

Fig. 15.26
Observations of some of the patterns presented in Figure 15.25. [From H. Kogelnik
and W. W. Rigrod, *Proc. IRE* (Correspondence) **50** (February 1962), p. 220. Copyright
© 1972 IEEE.]

a large number for q, and for the first few modes, we have

$$
\begin{array}{lll}
M_X = 0 & M_Y = 0 & q \\
M_X = 1 & M_Y = 0 & q' \\
M_X = 0 & M_Y = 1 & q'' \\
M_X = 1 & M_Y = 1 & q'''
\end{array}
\qquad (15.80)
$$

Unlike the standing-wave pattern in one dimension (where our only condition
is that the waves are reflected at the walls), we do not have a small E field close
to the walls over the entire surface. Figure 15.25 shows the modes for rectangular
and round mirrors, and Figure 15.26 shows a photograph of the experimental
observations. The approach used in the calculation of the modes in the rectangular
box is different from that using diffraction theory. We solve Maxwell's equations
using the appropriate boundary conditions. The modes are described by products
of Hermite polynomials, similar to the ones found for the harmonic oscillator in
quantum mechanics. These polynomials are denoted by a characteristic number
starting with 0, and the numbering of the modes in the X and Y directions is
done accordingly. In this way, the characteristic numbers also give the number

of nodes or nodal lines. Characterization of the modes by nodes or nodal lines is independent of the particular functions (geometry) describing the modes. In Figure 15.25, we observe that the nodal lines for the rectangular and circular modes of the rectangular and circular geometry of the mirrors are analogous and equal in their characterization by the indices i and j in the TEM_{ij} notation.

So far, we have only discussed flat mirrors. For spherical mirrors, we proceed in the same way. For the resonance wavelength for a system of two confocal mirrors of equal size, we have the relation

$$\frac{4d}{\lambda} = 2q + (1 + M_X + M_Y) \tag{15.81}$$

where, as before, d is the length of the cavity and q, M_X, and M_Y are the mode numbers in the three directions. Although q is a large number, different q choices for the same wavelength λ give us different $M_X + M_Y$ choices. For example, if we decrease q by 1 and increase $M_X + M_Y$ by 2, we have a condition applying to the same wavelength. Increasing $M_X + M_Y$ by 2 can be done in three different ways. As a result, we have a number of modes characterized by q, M_X, and M_Y for the same wavelength, and we speak of degenerate modes of the same wavelength. Figure 15.27 shows lines of equal phase of a mode in a confocal resonator. We can see that this line is conforming at the mirrors with the curvature of the mirrors. Also, the cross section of the Gaussian beam diameter is shown in Figure 15.28. In Figure 15.29, the Fresnel number given in equation 15.76 is plotted against the diffraction loss per reflection for confocal and circular plane reflectors for a number of modes. The Fresnel number is almost the same for a large region of losses per reflection for the confocal reflectors, almost independent of the particular mode. For circular plane mirrors, the Fresnel number drops rapidly for high diffraction losses per reflection. This shows that the confocal arrangement is advantageous.

The quality value Q of an oscillator is defined as $Q = \omega_0/\Delta\omega$. For a simple R-C-L circuit, it is $Q = L\omega_0/R$, where ω_0 is the resonance frequency $\omega_0^2 = 1/LC$. The bandwidth of the resonance line is $\Delta\omega = R/L$. For a laser, Q can be expressed as $Q = \omega t_{photon}$, where t_{photon} is the decay time constant of the mode.*

D. Threshold, Gain Curve, and Oscillations

In Chapter 7, we discussed how a light wave traversing a medium of thickness d is attenuated exponentially, that is,

$$I = I_0 e^{\alpha d} \tag{15.82}$$

where α is the extinction coefficient and is negative. For the wave traveling forward and backward in the cavity, the coefficient α is positive. The cavity is filled with the "active" medium having a population inversion; that is, there are more atoms in the upper state N_2 than in the lower state N_1. An initial very small wave will pick up energy, and for one full round trip between the two mirrors a distance d apart, we have an increase in intensity of

$$I = I_0 e^{\alpha 2d}$$

---

*See A. Yariv, p. 235.

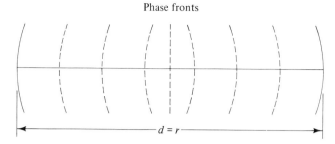

Fig. 15.27
Schematic of phase fronts in a confocal cavity. Distance between mirrors d; radius r. [From G. D. Boyd and H. Kogelnik, *BSTJ* **41**, 1347 (1962).]

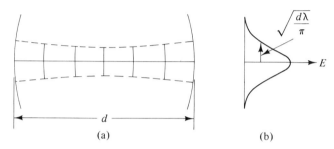

(a) (b)

Fig. 15.28
TEM$_{00}$ mode in a confocal cavity. (a) Surfaces of constant phases are indicated. (b) Cross section of the E field in the cavity. [From G. D. Boyd and J. P. Gordon, *BSTJ* **40**, 489 (1961).]

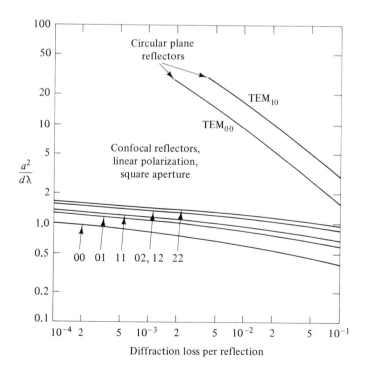

Fig. 15.29
Fresnel number indicating losses for several modes. [From G. D. Boyd and J. P. Gordon, *BSTJ* **40**, 489 (1961).]

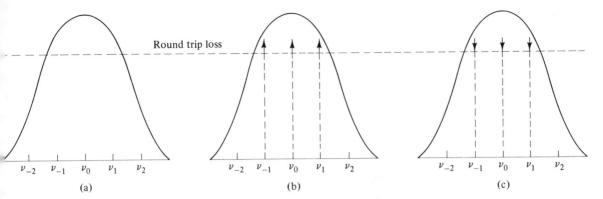

Fig. 15.30
Amplification around the loss level for different modes.

where α is proportional to the spontaneous emission A_{21} of the higher level 2, to the difference $\Delta N = N_2 - N_1$, that is, the population difference, and the line shape of the emitted line $g(\omega - \omega_0)$, that is,

$$\alpha \propto A_{21} \cdot \Delta N \cdot g(\omega - \omega_0) \tag{15.83}$$

We describe the losses by reflection, scattering, and diffraction by the factor δ (for one round trip) and have

$$I - I_0 = \delta I \tag{15.84}$$

For the condition in which the initial small wave increases, we have

$$I - I_0 \geq \delta I_0 \tag{15.85}$$

or

$$e^{\alpha 2d} - 1 \geq \delta \tag{15.86}$$

For one round trip, $\alpha 2d$ is small compared to 1, and we have to consider

$$1 + \alpha 2d - 1 \geq \delta \quad \text{or} \quad \alpha 2d \geq \delta \tag{15.87}$$

The threshold condition, that is, the situation in which the wave is increasing so much that the losses are offset, is

$$\alpha_T 2d = \delta \tag{15.88}$$

To have the initial wave grow, we must have $\alpha 2d \geq \delta$. With optical pumping, we keep $N_2 - N_1 > 0$ and oscillation starts at the threshold condition. As the oscillation increases above the threshold condition, the upper state depletes, and since α depends on the difference $N_2 - N_1$, the value of α drops. It decreases to the value around the threshold condition. There the condition for starting oscillations is fulfilled, and the value of α rises again. After some initial deviations from this level, it will stay there for the equilibrium. Figure 15.30 shows the sequence of events. Since the gain curve without oscillations will have a reduced height at the frequencies of oscillation, we speak of "hole burning" (the gain curve is inhomogeneously broadened) (Figure 15.31). Depending on the particular shape and height of the gain curve, only certain modes at certain frequencies can

15. Light Sources

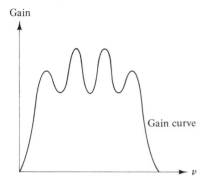

Fig. 15.31
Gain curve with "hole burning."

oscillate. The losses are influenced partly by the cavity design, such as reflection and diffraction losses. Different cavity designs will prefer the oscillation of some modes over others.

We have seen that the gain decreases after oscillation starts. We can prevent this decrease by stopping the oscillations momentarily while pumping continues and bringing $N_2 - N_1$ to a higher value. Then, if oscillations are again allowed, all the light quanta in level 2 are available for stimulated emission (Figure 15.32). While a "shutter" is closed, the pumping brings the oscillators into the higher

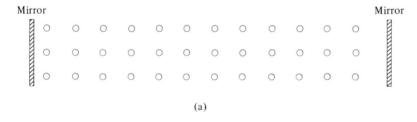

(a)

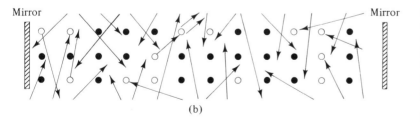

(b)

Fig. 15.32
(a) Atom in the cavity in the lower energy state (open circles). The shutter is closed. (b) The "pump" light excites the atoms into the excited state (black circles). The shutter is closed. (c) The shutter is open and the light traveling between the mirrors can pick up more energy from the excited atoms.

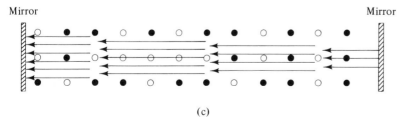

(c)

5. Lasers   **537**

state 2, shown by dark circles. When the shutter is open, the light waves can run back and forth and pick up light so that the amplitude increases very fast, thereby depleting the population of state 2.

E. Brewster's Angles

If the wave travels between the mirrors M, as shown in Figure 15.33, the wave will be partly reflected and partly transmitted at the interface at the active section of the laser. We can bring the reflection mirrors into the cavity for a gas laser or make the ends of the rod reflecting, as has been done in the case of the solid-state laser. In the case of the semiconductor laser, we are doing just that, because there is no choice from the point of view of space.

For gas lasers, we have the choice of sealing off the active medium with Brewster windows; these are flat surfaces under a specific angle to the axis (Figure 15.33). (This is explained in detail in Chapter 4.) The angle depends on the refractive index of the material inside and outside of the window. Brewster windows have the property that only one component of polarization of light is not reflected but passes through without losses. This component is used for the laser action, and laser light coming from cavities with Brewster windows is predominantly plane-polarized.

F. Q-Switching and Laser Pulses

If we rotate one mirror of the cavity around an axis perpendicular to the laser beam with a certain velocity, we would have only momentarily the two mirrors parallel. Only during that short moment could the cavity be used for "laser action." The pumping action, however, goes on while the cavity cannot function. In the moment the cavity functions, all the laser energy pumped into the excited states is ready for laser action. A high energy, short pulse of laser light is emitted. Today, mirrors are not used for Q-switching the cavity. The name Q-switching

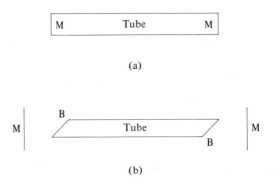

(a)

(b)

Fig. 15.33
The use of Brewster windows B with a gas laser. Mirrors, M. (a) Mirrors in the tube of a gas laser. (b) Brewster windows will let pass only one direction of polarization, but without reflection losses. The mirrors are outside of the tube and may easily be adjusted.

15. Light Sources

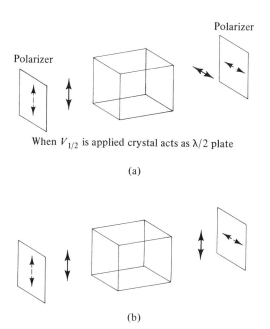

When $V_{1/2}$ is applied crystal acts as $\lambda/2$ plate

(a)

(b)

Fig. 15.34
Electrooptical shutter. (a) When the voltage is applied, the linearly polarized light incident on the crystal will be rotated by 90° and can pass the second polarizer. (b) If no voltage is applied, no light can pass.

comes from the quality factor Q, discussed in Section 4C. The value of Q changes drastically when the cavity is functioning.

Acoustooptical or electrooptical devices are used for Q-switching. In the acoustooptical device, discussed in Chapter 18, acoustic waves are generated in a crystal, and the light of the laser is periodically scattered by these waves, depleting the quality of the cavity and producing pulses only during the time interval when there is no scattering. This happens when the acoustic waves produced have no density changes within the period of vibration. The waves have frequencies of a few hertz to about 50 kHz, and the pulses have a duration of $100 - 500 \times 10^{-9}$ seconds.

Electrooptical devices use the linear or quadratic Pockels effect, as discussed in Chapter 5. In both cases, a voltage induces a uniaxial crystal. When the voltage is periodically applied, the crystal is periodically a homogeneous medium and a uniaxial crystal. The condition for the applied voltage and the length of the active medium may be chosen in such a way that the crystal changes between a homogeneous medium and an effective $\lambda/2$ plate. Such a plate rotates the direction of incident linear polarized light by 90°. The shutter is placed between crossed polarizers (see Chapter 6). When the voltage is applied, the full intensity can pass the second polarizer (see Figure 15.34). No light can pass when the voltage is zero. One of the polarizers can be replaced by a Brewster window. KDP is frequently used employing the linear Pockels effect. A calculation of the half voltage, that is, the voltage necessary for a phase shift of π, is presented in

Chapter 6. KDP may be used for modulation into the GHz range. For the quadratic Pockels effect, for example, for KTN, a much smaller half-voltage of about 75 V is necessary, compared to KDP.

In the foregoing sections, we have discussed the various principles involved in operating gas lasers, solid-state lasers, and semiconductor lasers. By their operating principles and materials, gas lasers are applicable for the visible spectrum to the far infrared; solid-state lasers and semiconductor lasers for the visible and near infrared.

6. THREE EXAMPLES OF COMMERCIALLY AVAILABLE LASERS

We will now discuss three types of lasers that are commercially available: a gas laser, a solid-state laser, and a semiconductor laser. The three lasers chosen for discussion have important applications for laboratory research, and for material processing.

A. The CO_2 Laser

The CO_2 laser is a high-power gas laser that is commercially available in a large range of output powers from a few watts to tens of kilowatts. Commercially, its applications are primarily in the processing of dielectric materials. Although high-power lasers are also applied to metals, most dielectrics absorb the output wavelength of the CO_2 laser better than metals.

A gas discharge is used for optical pumping. The tube is water-cooled and filled with a mixture of CO_2, N_2, and He gas. The output wavelength is around 10 μm in the infrared. A schematic of a CO_2 laser is shown in Figure 15.35. On one side of the cavity, the tube has an uncoated flat plate that acts as an output coupler. An antireflection coated window allows the beam to exit the outer tube.

The curved mirror is mounted on a piezoelectric transducer. Changing the

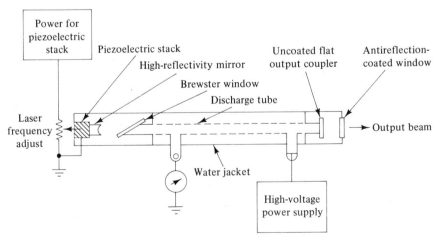

Fig. 15.35
Schematic of a tunable CO_2 laser. (Courtesy of Line Lite Laser Corporation, Mountain View, Calif.)

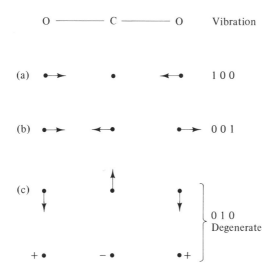

Fig. 15.36
Vibrations of the CO_2 molecule. (a) Symmetric stretching vibration. (b) Asymmetric stretching vibration. (c) Vibration perpendicular to the line connecting the atoms. The signs $+$ and $-$ indicate vibration in and out of the plane of the paper, respectively. The two vibrations have the same energy and are called degenerate vibrations.

voltage on the piezoelectric crystal will change the length of the crystal, and by changing the length of the cavity, we can tune the cavity to particular frequencies of the CO_2 laser spectrum.

The CO_2 laser uses molecular transitions, in contrast to the atomic transitions discussed for the He-Ne laser. The CO_2 molecule vibrates in four normal vibrations, as shown in Figure 15.36. The energy schematic and laser transitions are shown in Figure 15.37. The N_2 and CO_2 molecules collide in the gas discharge and excite the CO_2 molecule into its (001) state. Collisions with electrons also participate in these excitations. From the (001) state, oscillations are possible involving the (100) state and the (020) state. Depopulation of these states to lower states involving the (010) state are indicated in Figure 15.37a. There is a further complication. The rotation of the CO_2 molecule around the axis perpendicular to the line of the atoms corresponds to additional energy levels. These levels have much lower energy and are superimposed on the vibrational states, shown for the (001) and (100) states in Figure 15.37b. The rotational levels are numbered by the quantum number J from zero on. For transitions between the rotational states, we must have a change by 1 of the rotational quantum numbers J. This is similar to what we found for the atomic states discussed for sodium, where the quantum number $\ell (S, P, D,)$ must change by 1 in the case of sodium (see Figure 15.13). If the J number of the upper state is larger than the J number of the lower state, the transitions belong to the R branch; if smaller, they belong to the P branch. Figure 15.38 shows the R and P branches for the transitions (001)–(020) and (001)–(100).

All these rotation-vibration transitions have different frequencies. The length of the cavity can be tuned with the piezoelectric transducer to have oscillations only at one frequency.

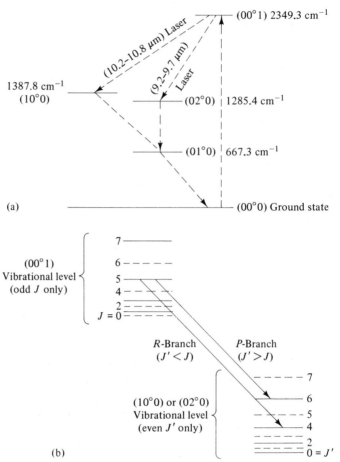

Fig. 15.37
Transitions of the CO_2 molecule used in laser operation. (a) Vibrational levels involved in laser action near 10 μm. (b) Rotational splitting of vibrational transitions. (Courtesy of Line Lite Laser Corporation, Mountain View, Calif.)

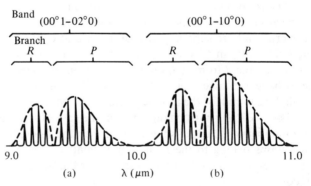

Fig. 15.38
CO_2 molecule, P and R branches of the rotation-vibration transitions for (a) $(00°1–02°0)$ and (b) $(00°1–10°0)$ transitions. (Courtesy of Line Lite Laser Corporation, Mountain View, Calif.)

(a)

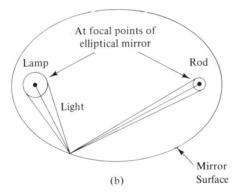

(b)

Fig. 15.39
(a) Photograph of a YAG laser. (Courtesy of U.S. Laser Corp.) (b) Elliptical mirror cavity for pumping the rod with light from the lamp.

B. The YAG Laser

The YAG laser is a solid-state laser that is used in the laboratory and has industrial applications in cutting, drilling, welding, and soldering of metals. The YAG laser output is strongly absorbed by metals and many semiconductors. By removing conductive material from a resistor in a microcircuit, we can trim the resistivity to a desired value. We can also scribe on absorbing materials such as semiconductors.

For the active medium, an yttrium-aluminum-garnet rod is used, containing neodymium ions (Nd^{3+}) (1 percent doping level is normally used). The optical pumping is done by high-power krypton lamps mounted at the focal line of an elliptical reflecting cavity (Figure 15.39). The rod and lamps are immersed in a

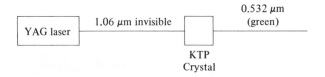

Fig. 15.40

Frequency-doubling experiment. The light incident on the KTP crystal has wavelength 1.06 μm. The second harmonic has wavelength 0.532 μm.

cooling-water system, and output power of up to several hundred watts in continuous-wave operation are obtainable. The ends of the rod are generally flat, and the mirrors for the cavity are separated from the rod.

The laser operates in the infrared at 1.06 μm and the single-mode power may be as high as 10–20 W. Between the mirrors and the rod an acoustooptical Q switch can be placed for pulsed operation. Pulse repetition rates may be varied from a single shot to 50 kHz, with pulse width down to below 100 ns. The diameter of the rods is typically 3–6 mm, and length is in the 50–100 mm range. By placing a special crystal into the beam, a nonlinear effect called frequency doubling can be achieved (see also Chapter 18). Figure 15.40 shows schematically the experimental setup of the frequency-doubling experiment. A photograph of a frequency-doubling experiment is shown at the end of this book. The incident beam has a wavelength of 1.06 μm, which is in the near infrared (invisible). The frequency-doubled beam is produced by a KTP crystal (see Chapter 5), and the output beam has a wavelength of 0.532 μm with a green color in the visible.

C. The Diode Laser

Semiconductor lasers with very small dimensions have been developed for use in integrated and fiber optics. Figure 15.41 shows a schematic of the cross section of such a diode.

The n-type substrate has a Gaussian-shaped groove. On the top is a p-type electrode; on the bottom is an n-type electrode; in between are various n- and p-type layers. The ends of the 250 μm cavity are coated with Al_2O_3 to improve the reflectivity. Figure 15.42 shows the output for various currents. Below threshold, with a current of 15.68 mA (Figure 15.42a, b), only spontaneous emission takes place. Above threshold (Figure 15.42c, d), with increasing current, only a single mode is supported. The output area for the diode is very small. To couple it into a fiber, we can use a taper, as well as a small lens with a graded index of refraction (see Chapter 17).

7. LASER SPECTROSCOPY

A. Single and Multiple Laser Lines for Spectroscopy

The lasers discussed so far were presented as narrow-band light oscillators. Spectroscopy is usually involved in the measurement of the absorption or emission

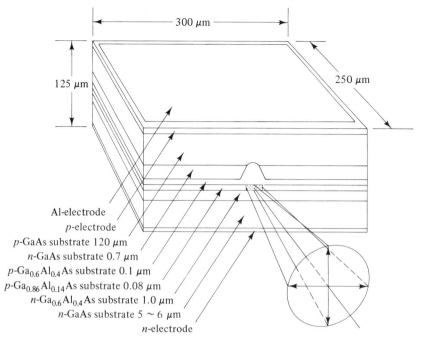

Fig. 15.41
Schematic of layer structure of a laser diode. The cavity and the emerging light are indicated. (Courtesy of Sharp Electronics Corporation.)

properties of a sample over a certain frequency range. An oscillator that cannot be tuned over a frequency range has only limited spectroscopic applications. In microwave spectroscopy, where the oscillators can only be tuned over a narrow band of frequencies, we use the available narrow band for high resolution of some spectroscopic problems.

The hyperfine structure of spectral lines is positioned in such a narrow frequency range that a laser can be used to investigate it. The high intensity over the narrow range of interest is advantageous for the spectroscopic resolution. Also, the cavity can be used to tune the laser over a narrow range.

A laser may also be used to assist certain spectroscopic procedures, such as changing the population of the Boltzmann distribution by continuously exciting certain higher states or supplying enough energy for a chemical reaction, thereby assisting in the observation of certain spectroscopic processes. We can also use many narrow-appearing laser lines to investigate the absorption properties of a sample and obtain noncontinuous spectral information.

We have seen in the discussion of the CO_2 laser that the vibration-rotation lines of the 10 μm band are numerous (see Figure 15.38). We also saw how a transducer in the laser is used to change the laser cavity and therefore supply the condition whereby only one line of the rotation-vibration transitions can be used for "laser action." If we investigate a sample around the 10 μm range with many different lines of the available P and R branches of the CO_2 bands (see Figure 15.38), we get a noncontinuous spectrum of the sample.

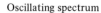

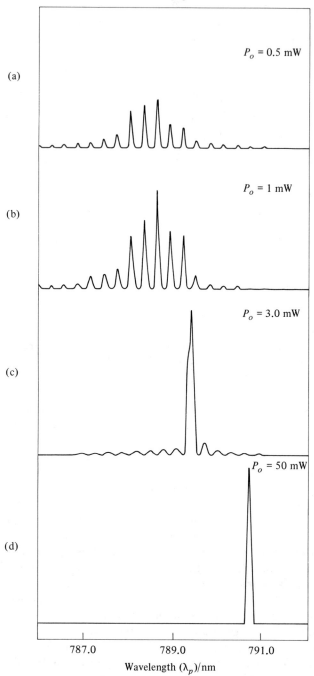

Fig. 15.42
Wavelength-dependent output oscillations of a diode laser. With increasing current.
(a)–(d), the output can be changed to single-mode oscillation. (Courtesy of Sharp
Electronics Corporation.)

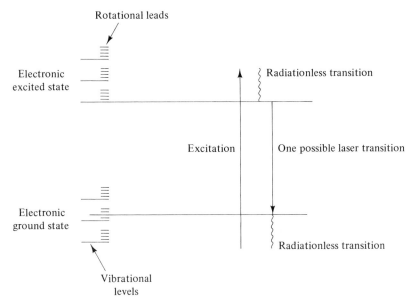

Fig. 15.43
Electronic ground state and first excited state for dye laser operation. Both states are composed of vibrational and rotational levels. The excitation transitions, laser transitions, and radiationless transitions are indicated.

A similar principle has been used in the far infrared. There we use a CO_2 laser as the pump laser and produce population inversion in different gases placed in a second laser cavity. These gases produce laser action on a number of lines. Through tuning and filtering we can come up with a set of laser lines covering a large spectral region in a noncontinuous way.

B. Dye Lasers

The principle of using many laser lines that are narrowly spaced for spectroscopic purposes is used in the dye laser. Dyes are organic chemicals that fluoresce under the incident light. The incident light is absorbed and reemitted into a lower frequency range. The dye is in a solvent, and pump light from the outside excites electrons from the ground state into the first excited state. The ground state and the first excited state are made up of a large and densely spaced number of vibrational and rotational sublevels, as shown in Figure 15.43. The electrons move down in the excited state with radiationless transitions until they are at the bottom of the ladder. There, a population inversion is obtained. The large number of states in the lower electronic state gives the possibility of laser transitions to various lower states, depending on the tuning of the cavity. From the lower laser state, even lower states may be occupied by radiationless transitions (see Figure 15.43).

In Figure 15.44, we see a schematic arrangement of a dye laser. The grating performs the tuning of the cavity for different angles of tilting. In Figure 15.45,

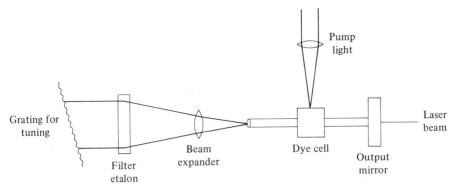

Fig. 15.44
Schematic of a dye laser. The grating and the output mirror form the cavity. The etalon is used as a filter, and the dye cell is the active medium pumped in from the outside.

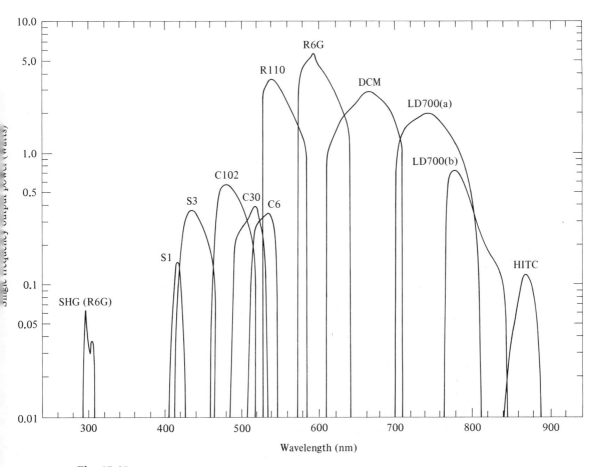

Fig. 15.45
The frequency range covered with 11 different dyes in a single frequency ring dye laser. The power output is indicated. Good overlapping is achieved from 400 to 900 nm.
[From T. F. Johnston, Jr., R. H. Brady, and W. Proffitt, *Applied Optics* **21**, 2307 (1982).]

we see the power output of a single-frequency ring-dye laser corresponding to 11 dyes covering the wavelength range from 400 to 900 nm. See also the color table at the end of this book.

Dye lasers can produce very short pulses, in the picosecond and femtosecond range, that is, in the range of 10^{-12} and 10^{-15} s per pulse, by mode locking. Such pulses can be used to test the dynamic processes of molecules in a liquid, such as how long it takes for two molecules oriented at time $t = 0$ in a certain way to be oriented in a different way. A similar question can be investigated for molecules adsorbed on surfaces or for electronic states in solids.

Problems

1. *Modes equation.* Derive equation (15.8),

$$k^2 = \pi^2 \left[\left(\frac{M_X}{\ell_X} \right)^2 + \left(\frac{M_Y}{\ell_Y} \right)^2 + \left(\frac{M_Z}{\ell_Z} \right)^2 \right] = \frac{\omega^2}{c^2}$$

similar to the two-dimensional derivation leading to equation 15.7.

2. *Radiation density equation.* Derive equation (15.12),

$$dZ = \tfrac{1}{8}(4\pi)R^2\, dR = 4\pi \frac{\ell^3}{c^3} v^2\, dv$$

similar to the two-dimensional case leading to equation 15.11.

3. *Elimination of ultraviolet catastrophe.* The Rayleigh-Jeans law diverges for small wavelengths. Consider Planck's formula,

$$\frac{du}{dv} = \frac{8\pi h v^3}{c^3} \frac{1}{e^{(hv/kT)} - 1}$$

Show that for Planck's formula, $\lim\limits_{v \to \infty} \dfrac{du}{dv}$ approaches 0, whereas in the Rayleigh-Jeans law it approaches ∞.

4. *Rayleigh-Jeans law depending on wavelength.* The Rayleigh-Jeans law was derived as

$$\frac{du}{dv} = 8\pi kT \frac{v^2}{c^3}$$

Show that for the radiation density per wavelength interval we get

$$\frac{du}{d\lambda} = \frac{8\pi kT}{\lambda^4}$$

5. *Stefan-Boltzmann law.*
   a. Integrate du/dv for Planck's law from 0 to ∞, and use the integral formula

$$\int_0^\infty \frac{x^3\, dx}{e^x - 1} = \frac{\pi^4}{15}$$

   With

$$\frac{W}{da} = \frac{c}{4} u$$

and show that for the radiant exitance we have

$$\frac{W}{da} = \frac{2}{15}\left(\frac{\pi^5 k^4 T^4}{c^2 h^3}\right)$$

b. Calculate the constant

$$\sigma = \frac{2}{15}\left(\frac{\pi^5 k^4}{c^2 h^3}\right)$$

to be $5.66 \times 10^{-8} (W/m^2 \cdot K^4)$.

c. Show that

$$\int_{\substack{\text{over} \\ \text{sphere}}} \frac{da \cos\theta}{r^2} = 4\pi$$

6. *Wien's law.* Rewrite du/dv as $du/d\lambda$, and show that for the maximum of $du/d\lambda$, we have

$$\lambda_{\max} \cdot T = 2.899 \times 10^{-3} \text{ m} \cdot \text{K}$$

7. *Hydrogen atom, Rydberg constant, and transition frequencies.*
   a. The transition frequencies are given as

$$v_{n_i, n_f} = \frac{2\pi^2 K^2 e^4 m}{h^3}\left(\frac{1}{n_i^2} - \frac{1}{n_f^2}\right)$$

Using the definition $\hbar = h/2\pi$, where h is Planck's constant, we get

$$\frac{v_{n_i, n_f}}{c} = \left(\frac{1}{\lambda}\right)_{n_i, n_f} = \frac{K^2 e^4 m}{4\pi c t r^3}\left(\frac{1}{n_i^2} - \frac{1}{n_f^2}\right)$$

where m is the electron mass and c is the speed of light. Show that for

$$R = \frac{K^2 e^4 m}{4\pi c \hbar^3}$$

we have the value close to 109,720 cm^{-1} as originally given by Rydberg. Take $K = 8.988 \times 10$ N $\cdot$ m^2/C.
   b. The energy differences for the hydrogen atom can also be expressed as

$$\Delta E = E_{n_f} - E_{n_i} = E_0\left(\frac{1}{n_i} - \frac{1}{n_f}\right)$$

where

$$E_0 = \frac{2\pi^2 m K^2 e^4}{h^2} = 13.6 \text{ eV}$$

First calculate a table; then plot ΔE as a function of n_i for the Balmer series $n_f = 2$ for $n_i = 2$ to 10. The value of ΔE for $n_i \to \infty$ is the ionization energy for the hydrogen atoms with the electron originally in the state E_2.

8. *Lorentz profile.* Calculate the width at half-height of an oscillator line (Lorentz). The intensity as a function of frequency is given as

$$I(\omega) = \frac{E_0^2}{4\pi^2((\omega - \omega_0)^2 + (1/\tau)^2)}$$

For the calculation, start from

$$\tfrac{1}{2} I(\omega_0) = I(\omega_{1/2})$$

where $\omega_{1/2}$ is the angular frequency for which the intensity has half its maximum.

9. *Doppler broadening.* Calculate the half-width for Doppler broadening. The intensity $I(v)$ is given as

$$I_{v_0} \exp^{[-1/2(mc^2/kT)(v-v_0/v_0)^2]}$$

as in problem 8, start with

$$\tfrac{1}{2}I(v_0) = I(v_{1/2})$$

10. *Rate equation for three-level lasers*

a. Consider the steady state

$$dN_3/dt = 0 \quad \text{and} \quad dN_2/dt = 0 \text{ (for equation 15.67)}$$

   derive

$$\frac{N_2}{N_1} = \left(\frac{W_{13}S_{32}}{W_{31} + S_{32}} + W_{12}\right)(A_{21} + W_{21})^{-1}$$

   Note: A_{31} is small compared to W_{31}.

b. We know that $W_{12} = W_{21}$, $W_{31} = W_{13}$, and $S_{32} \gg W_{13}$. Derive

$$\frac{N_2}{N_1} = \frac{W_{13} + W_{12}}{W_{12} + A_{21}}$$

c. Consider

$$\frac{N_2 - N_1}{N_1} = \frac{W_{13} + W_{12}}{A_{21} + W_{12}} - 1$$

   Use

$$\frac{N_0}{N_1} = \frac{N_1 + N_2 + N_3}{N_1}$$

   and derive

$$\frac{N_2 - N_1}{N_0} = \frac{W_{13} - A_{21}}{W_{13} + A_{21} + 2W_{12}}$$

11. *Minimum power for laser oscillation.* Calculate the minimum power

$$NA_{21}hv[\text{W} \cdot \text{cm}^{-3}]$$

to obtain laser oscillation for the unfavorable case where $A_{21} = \tfrac{1}{3} \times 10^8$, corresponding to equation 15.71. Assume that $h = 7 \times 10^{-34}$ W·s^2, $c = 3 \times 10^{10}$ cm/s, $N = 10^{19}$ cm^{-3}, and $\lambda = 0.5 \times 10^{-4}$ cm.
Ans. 14×10^7 W/cm^3

12. *Pulsed laser.* Consider a pulsed laser having pulse durations of 10^{-4} s and a peak power of 1 kW. If the repetition rate is 100 pulses per second, what is the comparable power of a continuous-wave laser?
Ans. 10 W

Scattering of Light | **16**

1. INTRODUCTION

In Chapter 3, we saw how incident light is diffracted at an aperture. The important criterion in observing diffraction is that the dimensions of the aperture have the same order of magnitude as the wavelength of the incident light. If the aperture is several orders of magnitude larger or smaller, diffraction patterns are not observed.

In discussing diffraction, we assumed that the amplitude of the electromagnetic wave was zero at the obstacle and equal to the incident wave at the open area. We used this formulation for the boundary condition in the electromagnetic approximation leading to the Kirchhoff-Fresnel integral. In the formulation of Huygens' principle, we assumed that the secondary waves were originating only from the open area of the aperture. The transmission, reflection, and absorption properties of the obstacle were ignored.

In Chapter 8, we studied how an oscillator is driven by an incident wave. We saw how the properties of the oscillator showed up in the amplitude of the resulting oscillation and how we can use such a model to describe the optical properties of a material. We interpreted the oscillation of the forced damped oscillator as the reemitted light of the incident light and obtained a relationship between the incident and transmitted light depending on parameters of the oscillator representing the material.

The scattering of light involves both aspects—diffraction at an aperture and absorption and reemission by an oscillator. The electromagnetic wave is diffracted by the object, but now the material properties must be taken into account. Electromagnetic theory is used to describe the scattering of light. The material properties enter the calculation through the boundary conditions.

A. The Driven Oscillator Model and Dipole Radiation

In this section, we will discuss the radiation emitted by a dipole. This will serve as a useful model for discussing a number of scattering phenomena.

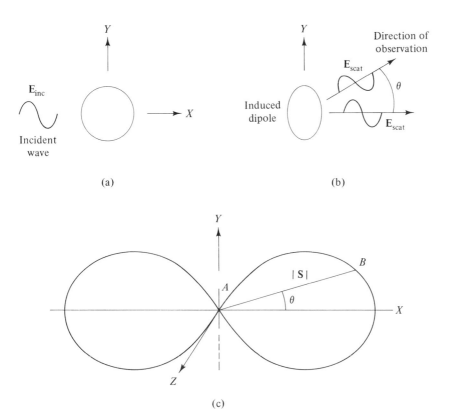

Fig. 16.1
Scattering on a spherical charge distribution by induced dipole emission. (a) Incident wave on a spherical charge distribution. (b) Induced dipole reemission. (c) Intensity distribution of the emission of a dipole. The length from A to B indicates the intensity in direction θ.

In Chapter 7, we studied the forced oscillator with eigenfrequency ω_0. The incident wave with frequency ω was driving the oscillator to vibrate with this same frequency. Depending on the parameters of the oscillator and the frequency of the incident light, the amplitude and phase of the oscillator change with respect to the incident light. At resonance, that is, when $\omega = \omega_0$, the amplitude was maximum and the phase difference $90°$. The direction of the incident, transmitted, and reradiated light was the same.

Now we want to study the emission of electromagnetic radiation by an oscillating dipole. Of interest is the intensity of the emitted light as a function of the various angles the dipole makes with the direction of emission. Experimentally, we find a distribution as shown in Figure 16.1. The intensity distribution of a vibrating dipole is mathematically presented as

$$I = (\text{const})\frac{p^2}{\lambda^4}\cos^2\theta \tag{16.1}$$

where p is the magnitude of the dipole moment, θ is the angle between the direction of the emitted light and the forward direction, and λ corresponds to

Fig. 16.2
Vibrating dipole. The charge $q(t)$ depends on the time as $q_0 \cos \omega t$,
where q_0 is a constant.

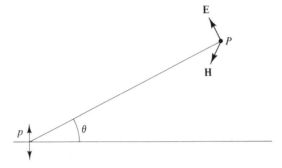

Fig. 16.3
Electric and magnetic field vectors **E** and **H** of the emitted light at angle θ. **E** is in the
plane of the paper; **H** is perpendicular to it.

the eigenfrequency ω of the dipole. The intensity is maximum in the X direction,
minimum in the Y direction, and symmetric around the X axis. Equation 16.1
can be derived using electromagnetic theory for the emitted intensity of a vibrating charge along the distance s. See Figure 16.2. The charge q has a time
dependence of

$$q(t) = q_0 \cos \omega t \tag{16.2}$$

and the change of the dipole moment is given as

$$p(t) = s\,q(t) = p_0 e^{i\omega t} \tag{16.3}$$

The **E** and **H** vector components of the emitted amplitude are derived from
Maxwell's equations, which are reformulated using the vector potential.* With
the appropriate initial conditions, we obtain ($r \gg s, \lambda$)

$$E = \frac{\mu_0 \ddot{p}}{4\pi} \left(\frac{\cos \theta}{r} \right) e^{i2\pi[(t/T)-(r/\lambda)]} \tag{16.4}$$

and

$$H = \frac{\mu_0 \ddot{p}}{4\pi c} \left(\frac{\cos \theta}{r} \right) e^{i2\pi[(t/T)-(r/\lambda)]} \tag{16.5}$$

Both components, E and H, are perpendicular to the direction of propagation
of the emitted light, (Figure 16.3), and $\ddot{p}$ is the second time derivative of the dipole
moment (see equation 16.3). From the definition of the Poynting vector **S** and

---

*See, for example, D. J. Griffiths, *Introductions to Electrodynamics* (Englewood Cliffs, N.J.:
Prentice-Hall, 1981).

equation 16.3, we obtain the intensity as

$$I = |\mathbf{S}| = \frac{\mu_0 \omega^4 p_0^2}{16\pi^2 c} \left(\frac{\cos^2 \theta}{r^2} \right) \tag{16.6}$$

We see that the factor $1/r$ in E and H is responsible for the $1/r^2$ law of attenuation of the intensity. The second time derivative of the dipole moment in the expression of E and H each contributes a factor ω^2, resulting in the $\omega^4 \propto 1/\lambda^4$ dependence for the intensity. The angle dependence of E and H is the same, resulting in the factor $\cos^2 \theta$.

2. SCATTERING WHEN THE WAVELENGTH IS LARGE COMPARED TO THE DIAMETER OF THE PARTICLE AND THE PARTICLES ARE RANDOMLY ARRANGED

A. Blue Sky, Red Sky

The light coming from the sun seems to be white light. The superposition of different wavelengths with their respective intensities creates the impression (to our eyes and brain) of "no color." According to the theory of complementary colors, if one wavelength of the spectrum of white light is either suppressed or enhanced, we see color. The blue appearance of the day sky has been explained by physicists as preferred scattering of some wavelength of the light coming from the sun. Lord Rayleigh presented a simple argument showing that the wavelength dependence of the scattered light is $1/\lambda^4$. The scattered intensity depends on the volume V of the particle, the distance r from the particle, the wavelength λ of the scattered light, and refractive indices n_1 of the particle and n_2 of its environment. The particle is assumed to be spherical, and for the scattered intensity I depending on the incident intensity I_0, we have

$$I = f(V, r, \lambda, n_1, n_2) I_0 \tag{16.7}$$

The function f is dimensionless. Since n_1 and n_2 are also dimensionless, the dimensions of V, r, and λ must cancel one another. The field of the dipole is proportional to the volume of the particle. Since we must square the field of the dipole to get the intensity, we must square the volume and get the dimension $[\text{meter}]^6$. The intensity falls off with $1/r^2$. That leaves the dimension $[\text{meter}]^{-4}$ for λ, and consequently, we expect the $1/\lambda^4$ dependence, as was found for the emitted radiation of a dipole.

Lord Rayleigh could show with simple experiments that sunlight is scattered according to these considerations. The color of the sky is explained by the short or long distance the light travels at day and in the evening. The ratio of the intensity of scattered light for red light (7000 Å) to the insensity of scattered light for blue light (4000 Å) is, according to the $1/\lambda^4$ dependence,

$$\frac{\text{Red}}{\text{Blue}} = \left(\frac{7}{4} \right)^4 \simeq 2^4 = 16 \tag{16.8}$$

While both kinds of light are scattered, the blue light is scattered much more. Thus, the sky appears blue. At sunset, the light from the sun must travel a much

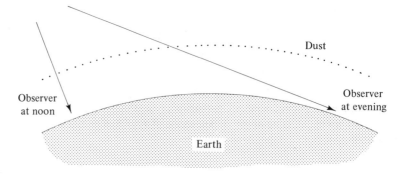

Fig. 16.4
Light from the sun must travel a great path through the atmosphere at sunset. The blue light is scattered away, and only the red light scatters for our observation.

longer path through the atmosphere. The scattered light getting to our eye is red light, since the blue light has been scattered so much that only the red light remains to be scattered (Figure 16.4). The scatterers in the atmosphere are the air molecules, which are small compared to the wavelength of the incident light.

B. Polarization

The light emitted from a dipole is polarized in the plane containing the axis of the dipole. In the case of scattering by sunlight, the incident light is unpolarized and can induce dipoles in all directions perpendicular to the direction of propagation. The only restriction is that the dipoles cannot emit light in the direction of their axes. If we look at light scattered to us at sunset and use a polarizer, we will find that the light is partially polarized.

Let us consider the model case shown in Figure 16.5. The incident light has amplitudes in directions Y and Z and propagates in direction X. The observer looks along the Z direction. At the scatter center, light can only be scattered in directions Y and Z, and not in direction X. Since Z is the direction of propagation of the scattered light, the scattered light can only vibrate in direction Y. For this case, we have not used the fact that the dipoles cannot radiate light along their axes. Also, the dipoles were not oriented in any particularly direction. However, we assumed a very special orientation of the incident light and the observed light.

C. Rayleigh Scattering

The scattering of light by particles or collections of particles will depend on whether the wavelength of the light is large or small compared to the diameter of the particle. The scattering of light also depends on the shape of the particle as well as on its material, for example, whether the particle is metallic or dielectric. It also depends on the arrangement of the particles, on whether the distances between them are random or periodic. Another important distinction is whether or not the frequency of the scattered light is different from the incident light, as is the case for the Raman effect.

We consider the case when the wavelength is large compared to the diameter

16. Scattering of Light

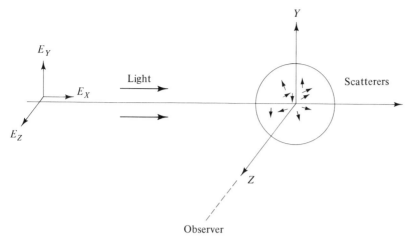

Fig. 16.5
Scattering of unpolarized light from induced dipoles. The direction of observation is perpendicular to the direction of incident light.

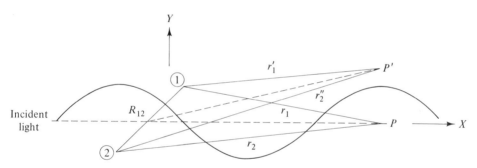

Fig. 16.6
Scattering from two particles. The incident wave has a large wavelength compared to the diameter of the particles.

of the particle. In this case, it is difficult to perform scattering experiments on single particles, and we must consider the case of collections of particles that are randomly spaced.

Let us consider the scattering from two particles (Figure 16.6). The particles are separated by the distance R_{12} and have the distances r_1 and r_2 from the point of observation P, which is located in the direction of the incident light. The particles act as dipoles; that is, their dipole moment is induced by the incoming light, and they each reradiate a wave at slightly different times. These waves must travel slightly different optical paths to the observation point. As a result, they arrive at P with different phase constants.

We can describe the superposition of the two waves as

$$E_{\text{scat}} = E_1 + E_2 = Ae^{i(\omega t + \phi_1)} + Ae^{i(\omega t + \phi_2)} \tag{16.9}$$

where E_{scat} is the scattered electric field. The intensity is obtained the same way

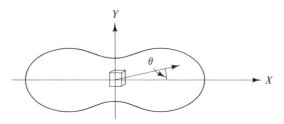

Fig. 16.7
Scattering from a collection of randomly oriented particles.

as in Chapter 2, and we have

$$I(\theta) = |E_{\text{scat}}|^2 = 4A^2 \cos^2\left(\frac{\phi_1 - \phi_2}{2}\right)$$

$$= 2I^{(1)}[1 + \cos(\phi_1 - \phi_2)]$$

(16.10)

where $I^{(1)}$ is the single-particle scattering intensity. As discussed for interference effects, the intensity $I(\theta)$ can be larger than $2I^{(1)}$ for certain values of $\phi_1 - \phi_2$. If the location of the two particles varies in a random way, we obtain for $\cos(\phi_1 - \phi_2)$ all possible values between $+1$ and -1. Taking the average, we obtain, similar to the "random" case discussed in Chapter 2,

$$\langle \cos(\phi_1 - \phi_2) \rangle = 0$$

(16.11)

For a collection of particles, the intensity depends on the angle θ is found to be

$$I(\theta) = \tfrac{1}{2} I_0 (1 + \cos^2 \theta)$$

(16.12)

(see Figure 16.7). The particles are assumed to be randomly oriented and the dipoles illuminated by unpolarized light. The constant I_0 is the forward-scattered intensity for $\cos \theta = 0$. To understand Figure 16.7, we regard the unpolarized incident light as being composed of a superposition of polarized light with all possible orientations. The scattering dipoles each get excited and re-radiate a pattern as shown in Figure 16.1. For three orientations of the dipoles with respect to two polarization directions of the incident light, the scattered light components are as shown in Figure 16.8. We can view the pattern in Figure 16.7 as being generated by a superposition of lobe pattern resulting from the three cases shown in Figure 16.8.

D. Extinction Constant for Rayleigh Scattering

The incident radiation of power S is scattered by the particles. The amount of scattered radiation is related to the incident radiation by the extinction constant K:

$$K = \frac{\Delta S}{S}\left(\frac{1}{\Delta x}\right)$$

(16.13)

where ΔS is the power of scattered radiation per unit solid angle. The scatter

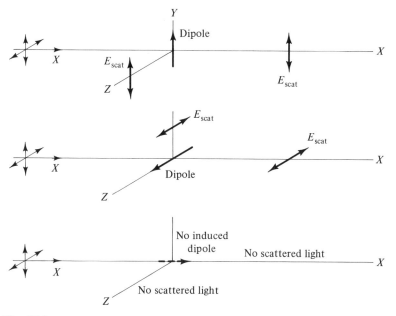

Fig. 16.8
Scattering of unpolarized light on a dipole oriented in three different directions.
Direction X is the direction of the incident light, direction Z the direction of
observation.

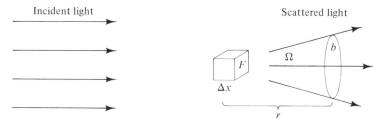

Fig. 16.9
Incident radiation, scattered radiation, scatter volume, solid angle Ω, and cross
section b, distance r.

volume is equal to $\Delta x \cdot F$, where F is the cross section and Δx the length (Figure
16.9).

The incident power S per unit area depends on the electric field as

$$S = \tfrac{1}{2}\varepsilon_0 c E^2 \tag{16.14}$$

and for the cross section F, we have

$$S_F = \tfrac{1}{2}\varepsilon_0 c E^2 F \tag{16.15}$$

The power scattered by one dipole is, according to equation 16.6,

$$\Delta S_S = \frac{\mu_0 \pi^2 c^3 p_0^2}{\lambda^4}\left(\frac{\cos^2\theta}{r^2}\right) \tag{16.16}$$

The solid angle is defined as b/r^2. Dividing equation 16.16 by the solid angle yields

$$\Delta S = \frac{\mu_0 \pi^2 c^3 p_0^2 \cos^2 \theta}{b\lambda^4} \qquad (16.17)$$

Averaging over the angle θ gives the factor $\frac{4}{3}\pi$ (see problem 3) and we have

$$\overline{\Delta S} = \frac{4}{3}\left(\frac{\mu_0 \pi^3 c^3 p_0^2}{b\lambda^4}\right) \qquad (16.18)$$

For N particles, we have

$$\overline{\Delta S} = N\frac{4}{3}\mu_0 \left(\frac{\pi^3 c^3 p_0^2}{b\lambda^4}\right) \qquad (16.19)$$

The reemitted amplitude of a particle's light is described by the amplitude of a forced oscillator (see Chapter 7):

$$p_0 = eu = \frac{e^2}{m}\left(\frac{1}{(\omega_0^2 - \omega^2) - i\gamma\omega}\right)E \qquad (16.20)$$

We can simplify equation 16.20 for our case of dipole scattering where the particle size is small compared to the wavelength. As a consequence, the eigenfrequency of the dipoles is high compared to the frequency of the incident light. Also, we neglect damping, and equation 16.20 gives us

$$p_0 = \frac{e^2}{m\omega_0^2}E \qquad (16.21)$$

Taking equation 16.13 as

$$K = \frac{\overline{\Delta S}}{S_F}\left(\frac{1}{\Delta x}\right)$$

and introducing $v = F\,\Delta x$, we have, with equations 16.19 and 16.21,

$$K = \left(\frac{e^4 N}{6\varepsilon_0^2 m^2 \pi v_0^4 vb}\right)\frac{1}{\lambda^4} \qquad (16.22)$$

where $\omega_0 = 2\pi v_0$. We call the number of particles per unit volume N_v and get

$$K = \left(N_v \frac{e^4}{6\pi\varepsilon_0^2 m^2 v_0^4 b}\right)\frac{1}{\lambda^4} \qquad (16.23)$$

All the quantities within parentheses are constants, and for illustration surposes, let us use

$$K = (\text{const})\frac{1}{\lambda^4} \qquad (16.24)$$

R. Pohl presents a demonstration of equation 16.24 with a sodium chloride (NaCl) crystal containing strontium ions in a ratio of 1 Sr : 1000 Na$^+$. The strontium ions cause defects in the crystal lattice. Measurements of the extinction constant K from $\lambda = 0.2\ \mu m$ to $\lambda = 1\ \mu m$ are shown in Figure 16.10a. These same

16. Scattering of Light

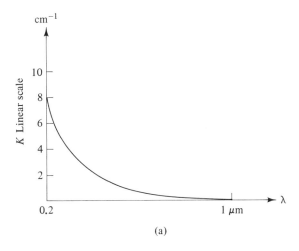

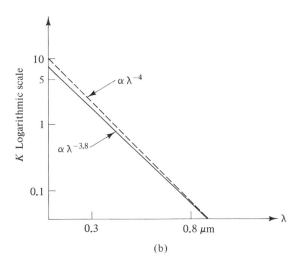

Fig. 16.10
Extinction coefficient plotted over the wavelength. (a) Linear plot. (b) Logarithmic plot.

data are plotted on a logarithmic scale in Figure 16.10b. We obtain a dependence of K with $1/\lambda^{3.8}$, which is close to the theoretical value $1/\lambda^4$.

3. SCATTERING OF LIGHT WHEN THE WAVELENGTH IS EQUAL TO OR SMALLER THAN THE DIMENSION OF THE PARTICLE

If we take the wavelength of the light to be very short (e.g., X rays), then X-ray diffraction can be considered. We can also choose λ to be the wavelength of visible or infrared light, in which case single-particle scattering experiments of micron-size particles may be discussed.

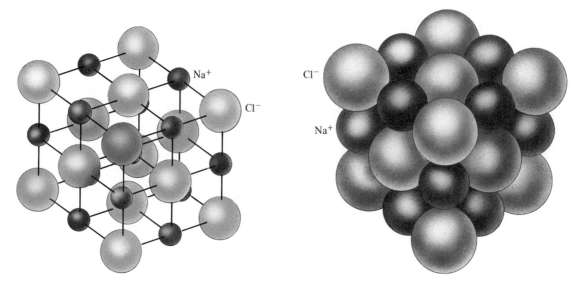

Fig. 16.11
Face-centered cubic crystal of Na and Cl ions.

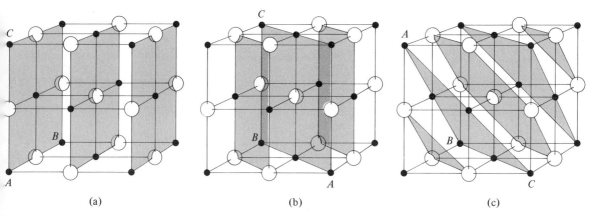

(a)   (b)   (c)

Fig. 16.12
All ions of the crystal may be placed in different sets of planes. (a) One plane of the set is the
plane *A*, *B*, *C*. (b) One plane of the set is the plane through the diagonal of the top *C*, *B*, *A*.
(c) One plane of the set is the plane through the diagonal of the cube *A*, *B*, *C*. The distance
between planes for each set is constant. This distance is different for different sets.

A. Bragg Scattering

Consider the regular positions of ions in a simple crystal. Figure 16.11 shows the
arrangement of Na and Cl ions in the face-centered cubic crystal lattice. The
distance between Na and Cl ions is $a = 2.82$ Å.

In Figure 16.12, we see how all atoms can be considered as being positioned
on a set of planes. Different sets of planes can be chosen containing all ions.

The X-ray wavelength may be chosen from a molybdenum target bombarded
with electrons of 20 kV. The wavelength is $\lambda = 0.6$ Å. The ions are considered

         16. Scattering of Light

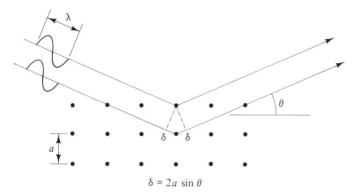

$$\delta = 2a \sin \theta$$

Fig. 16.13
Phase difference between two waves reflected at two planes.

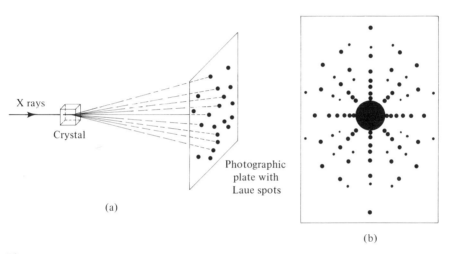

(a)

(b)

Fig. 16.14
Schematic of the diffraction experiments with X rays. (a) The crystal is made of a regular arrangement of atoms in three dimensions, just like a three-dimensional grating. (b) The two-dimensional photographic plate showing the diffraction maxima. The diffraction pattern is reminiscent of an aperture of regularly spaced openings in two dimensions.

spherical particles; remembering that the diameter of the hydrogen atom in the Bohr model is of the order of 1 Å, we see that the wavelength of the X rays is of the same order of magnitude as the dimension of the ions.

The X rays are considered as plane waves incident at the planes of atoms separated by the distance d (Figure 16.13). Each plane reflects the incident wave, and interference is due to the light reflected by all planes and traveling in the same direction. For interference maxima, we have

$$2a \sin \theta = m\lambda \qquad m, \text{ integer} \qquad (16.25)$$

This is called the **Bragg condition**. Figure 16.14a shows the experimental realization as first proposed by Max v. Laue, and a **Laue pattern** is shown in Figure

3. Scattering of Light and Dimension of Particle

16.14b. The photograph reminds us of a diffraction pattern obtained from regularly spaced openings in two dimensions.

The calculation of the diffraction pattern assumes that the incident wave creates a secondary wave at each atom. These secondary waves become plane waves after traveling a great distance. The waves are created at regular distances, resulting in regular phase differences. The summation of all the waves at the observation point will result in maxima and minima depending on the phase difference. This is similar to what we discussed in Chapter 3. There we saw that the Fourier transformation is involved when the incident and reemitted waves are considered plane waves. In this three-dimensional case, similar formulations can be attained.

What is different from our discussion of diffraction is the fact that now there are no Huygens waves to consider. The secondary wave created at each atom is the induced dipole emission of the incident wave and not a fictitious wave according to Huygens' principle. Because the observed pattern and the theory are so similar to the pattern and theory of diffraction, this method is called **X-ray diffraction**. It is one of the most powerful methods to determine the structure of solid-state materials. Using computers, we can draw conclusions about the positions of the atoms in the solid from the observed scattering pattern. This can be done for complicated structures such as biological molecules.

B. Scattering by a Long Dielectric Cylinder

Light whose wavelength is approximately 500 mμ is scattered by a dielectric fiber of about the same diameter. The light is incident perpendicular to the axis of the cylinder. It is in the plane of observation, that is, the scatter plane, which is also perpendicular to the axis of the cylinder (Figure 16.15). The incident light is polarized. The intensity of the light polarized parallel to the axis is $I_{\parallel}$; the intensity

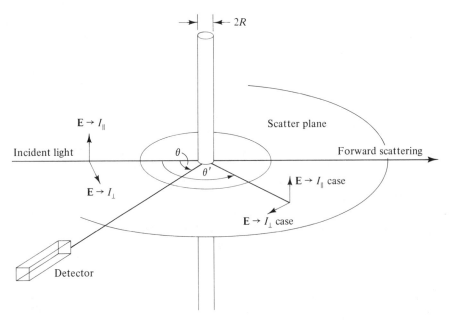

Fig. 16.15
Schematic of the scatter experiment on an infinite dielectric cylinder.

16. **Scattering of Light**

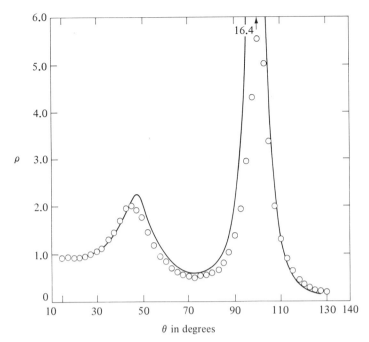

Fig. 16.16
Ratio $\rho = I_\perp/I_\parallel$ of scattered light. Wavelength $\lambda = 546$ mμ. Experimental results are shown as ○○○○○. Theoretical calculations are shown as a solid curve, using $n = 1.46$ and $\alpha = 3.70$. From the value of $\alpha = 2\pi R/\lambda$ follows the diameter of the fiber as 320 mμ. [From W. A. Farone and M. Kerker, *J. Opt. Soc. Am.* **56**, 481 (1966).]

of the light polarized perpendicular to the axis is $I_\perp$. For the mathematical analysis, we use the size parameter $\alpha = 2\pi R/\lambda$, where R is the radius of the cylinder and λ is the wavelength of the incident light.

The E components of the incident and scattered light for both directions of polarization are also shown in Figure 16.15. The scattered light is measured as the ratio $\rho = I_\perp/I_\parallel$. This is convenient because we do not have to measure the absolute intensity of light in a second scattering experiment. For the same angle setting, one has only to change the polarization of the incident light and record the two readings. In Figures 16.16 and 16.17, we see the results for a Pyrex fiber investigated using light of wavelength $\lambda = 546$ mμ and $\lambda = 436$ mμ. From the theoretical analysis, we obtain both the refractive index of the fiber and the diameter. In this particular case, there was no difference between the refractive indices of the two different wavelengths.

Although the theoretical approach to the calculation of the scattered intensity is in general simple, the mathematical details are very difficult. The scattered intensities $I_\parallel$ and $I_\perp$ in Fourier series are given as

$$I_\parallel = \frac{2}{\pi k r}\left| b_0 + 2\sum_{m=1}^{\infty} b_n \cos m\theta \right|^2$$

$$I_\perp = \frac{2}{\pi k r}\left| a_0 + 2\sum_{m=1}^{\infty} a_n \cos m\theta \right|^2$$

(16.26)

3. Scattering of Light and Dimension of Particle

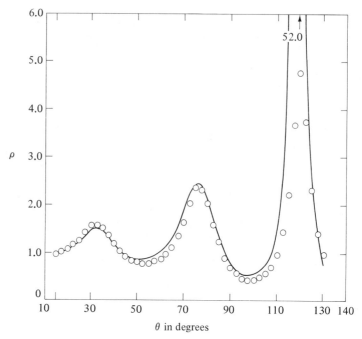

Fig. 16.17
Ratio $\rho = I_\perp/I_\parallel$ of scattered light. Wavelength $\lambda = 436$ mμ. Experimental results are shown as ○○○○○. Theoretical calculations are shown as a solid curve, using $n = 1.46$ and $\alpha = 4.62$. From the value of $\alpha = 2\pi R/\lambda$ follows the diameter of the fiber as 320 mμ, the same value obtained for the same fiber from the experimental results of Figure 16.16. [From W. A. Farone and M. Kerker, *J. Opt. Soc. Am.* **56**, 481 (1966).]

where r is the distance from the cylinder axis and k is equal to $2\pi/\lambda$, where λ is the wavelength outside the cylinder. The coefficients a_m and b_m must be determined from the boundary conditions and depend on n and α. For the dielectric cylinder, they are composed of Bessel and Hankel functions.

This method is very accurate in determining the refractive index and the radius of the fiber. The determination of the radius with an electron microscope has been less accurate.

C. Scattering from a Metal Cylinder

Incident light of wavelength 10.6 μm from a CO_2 laser is scattered by a copper wire of radius 121 μm. The incident light is polarized and parallel to the axis of the cylinder. The scatter plan is perpendicular to the axis of the cylinder (Figure 16.18). The incident light is chopped for amplification purposes. The intensity of the scattered light is measured with a detector driven by a motor at a rate of 2°/min over a range of 15°–50°. The detector is a liquid-nitrogen-cooled PbSnTe detector. The experimental results for copper and brass wires are shown in Figure 16.19a and b, where the size parameter is indicated. We see that the fringes for the smaller-size brass wire are separated by a larger distance than for the larger-size copper wire. This can be understood from what we discussed for diffraction. As

16. Scattering of Light

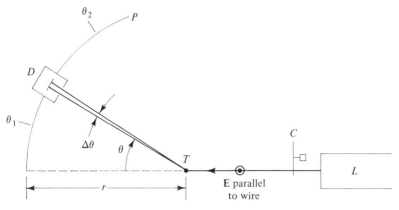

Fig. 16.18
Experimental setup for scattering of CO_2 laser light at a wire. D is the detector, P is the point of measurement on the scale of angles θ, r is the distance from target T, $\mathbf{E}$ is the electric field vector polarized parallel to the wire, C is the chopper, and L is the laser.

a first approximation of scattering, we can use diffraction theory. For scattering, however, we must satisfy the boundary conditions of the electromagnetic theory for the interface of the "inner" and "outer" media. The results of theoretical calculations for both wires are shown in Figure 16.19c and d.

We have seen that reflection from metals can be described by a complex refraction index with large values of $\bar{n}$ and $\bar{K}$, compared to the case of transmission through a lossless dielectric. Analysis of the results of the scattering experiment gives us $\bar{n} = 10$ and $\bar{K} = 12$ for copper, compared to literature values at 10.6 μm of $\bar{n} = 12$ and $\bar{K} = 60$. To obtain the optical constants of metals, ellipsometry is used for more accurate results, as discussed in Chapter 7.

D. Single-Particle Scattering from a Water Droplet

A water droplet with radius of 3.5 μm is taken as a spherical particle. YAG laser light of $\lambda = 1.06$ μm polarized perpendicular to the scattering plane is used for the experiment. Since the incident intensity of the laser light is quite high, an array of detectors has been built to record the scattered light simultaneously and eliminate the need for scanning (Figure 16.20). The fringe pattern shown in Figure 16.21 is similar to what we saw for the cylinders (see Figure 16.19). However, the mathematical theory for the analysis is different. The expansion in Fourier series uses spherical harmonics to determine the constants a_m and b_m (see equation 16.26).

4. SCATTERING WITH CHANGES IN THE FREQUENCY

The types of scattering considered so far could all be classified as "elastic." That is, the wavelength of the scattered waves were not changed by the interaction. If we view the problem in terms of the photon picture, an elastically scattered photon is a photon that has the same energy ($h\nu$), and thus the same frequency,

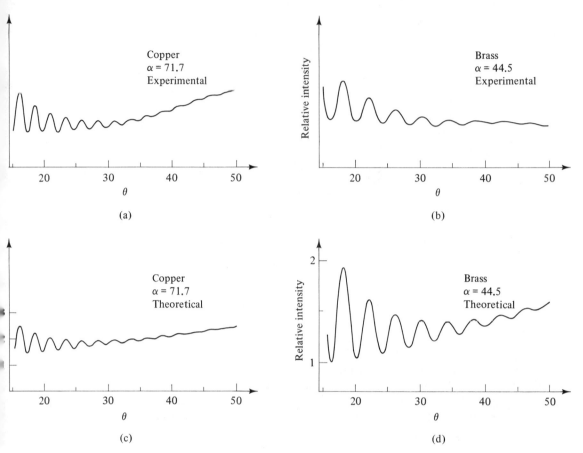

Fig. 16.19
Experimental and theoretical results of scattering on copper and brass wires. Relative scattering intensity is plotted as a function of angle θ. (a, b) Experimental results obtained with CO_2 laser light for $\alpha = 2\pi R/\lambda$, as indicated. (c, d) Theoretical calculations. [After A. Cohen, L. D. Cohen, R. D. Haracz, V. Tomaselli, J. Colosi, and K. D. Möller, *J. Appl. Phys.* **56**, 1331 (1984).]

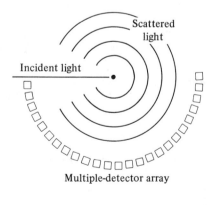

Fig. 16.20
The scattering of a water droplet can be observed simultaneously at different angles by arranging a number of detectors on a circle between the forward-scattering direction and the backward-scattering direction.

16. Scattering of Light

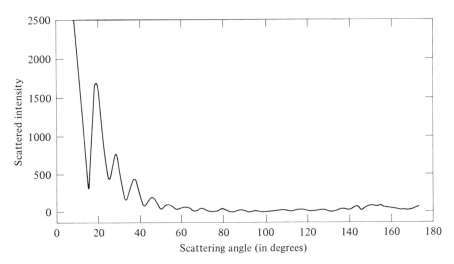

Fig. 16.21
Result of the scattering on a water droplet depending on angles from 0° (forward-scattering direction) to 180° (backward).

as it did before it was scattered; it changes only direction. An inelastically scattered photon changes both direction and frequency.

A. Compton Scattering

As an extreme example of inelastic scattering of light, we can consider the scattering of X rays by electrons and ions. Let us assume that the X rays have a wavelength of 0.711 Å and that they are incident on carbon atoms in graphite. An observation of a scattered X ray is shown in Figure 16.22a. The electron and the scattered X ray are observed at various angles ϕ and θ, respectively, with reference to the incident light. Figure 16.22b shows the scattered X rays for angles $\theta = 45°, 90°$, and $135°$. The dependence of the wavelength difference on the angle θ is given by **Compton's formula**

$$\lambda_2 - \lambda_1 = \frac{h}{mc}(1 - \cos\theta) \tag{16.27}$$

where m is the mass of the scatterer (problem 5).

If the X ray is scattered by a loosely bounded electron, m is the mass m_e of the electron. For scattering by electrons that are more strongly bonded, m is the mass of the ion, which is about the same as the atom. The more strongly bounded electron forms, for scattering purposes, an entity with the nucleus. The mass of the atom is about 10^4 times larger than m_e. If the mass of the electron is used in Compton's formula, the angle and wavelength shift refers to inelastic scattering. If the mass of the atom is used, we obtain almost no wavelength shift. The corresponding peak is called the unmodified scattered X ray. These two peaks are shown in Figure 16.22b. The wavelength shift is calculated by using the theorems of conservation of momentum and conservation of (relativistic) energy. This will be discussed in problem 5.

4. Scattering with Changes in the Frequency

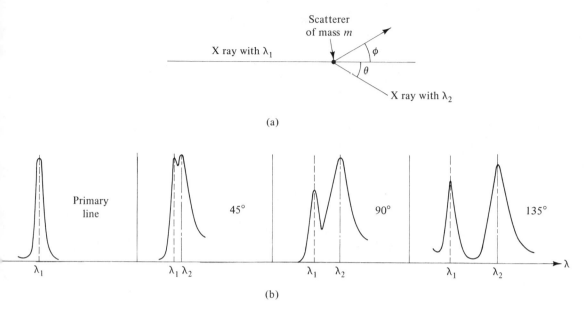

(a)

(b)

Fig. 16.22
(a) Scattering of the incident X ray with wavelength λ_1 on the scatterer with mass m. The scattered X ray has wavelength λ_2. (b) Observation in direction $\theta = 0°, 45°, 90°$, and $135°$. The first peak is due to an unmodified X ray of wavelength λ_1. The second is scattered on a loosely bound electron; it has wavelength λ_2.

B. Scattering and the Model of the Linear Chain of Ions

The light scattered by a solid shows three closely positioned frequencies. There is a spectral line with the same frequency as the incident light, and this line is called the Rayleigh line. At higher and lower frequencies, two other lines appear. These are produced by the Raman effect and are much weaker than the Rayleigh line. The frequency difference of both lines from the Rayleigh line is the same. From the frequency shift, we can obtain important information about the vibrational properties of the solid. To describe the frequency shifts, we represent the solid by a linear chain made of positive and negative charged ions (Figure 16.23a).

There are two different types of vibrations possible. In one type, the positive charged ions move opposite to the negative charged ions. This is called the **optical mode** (see Figure 16.23a). The oscillation keeps the center of mass in each "unit cell" at rest (see Figure 16.23b). A cell is a repetitive unit of one positive ion and one negative ion. The oscillations are associated with a change of a dipole moment. In the other type of vibration, called **the acoustical mode**, neighboring ions move in the same direction (see Figure 16.23a).

In Figure 16.23b, we see the modes for a very long wavelength. For the acoustical mode, the oscillation frequency becomes smaller and smaller with longer and longer wavelengths. For the optical mode, all the positive ions move as one line against the line of the negative ions, and the frequency is

$$\Omega_0 = \left[2f \left(\frac{1}{m_+} + \frac{1}{m_-} \right) \right]^{1/2} \tag{16.28}$$

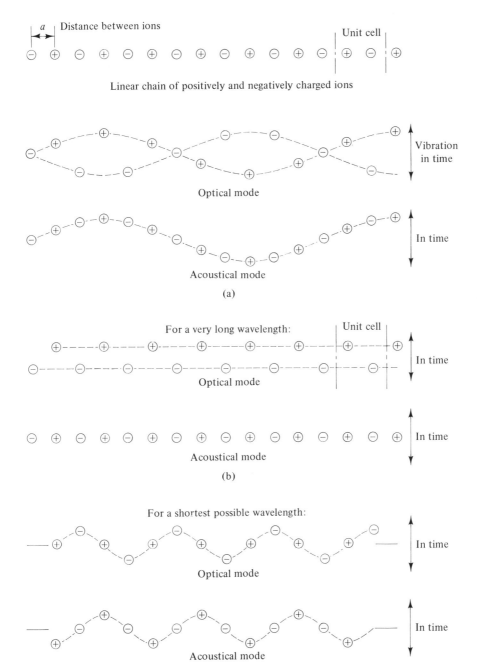

Fig. 16.23
Linear chain of positive and negative ions, with a distance a between ions. (a) Optical mode and acoustical mode of the linear chain. (b) Optical mode and acoustical mode at a very long wavelength. (c) Optical mode and acoustical mode at the shortest possible wavelength.

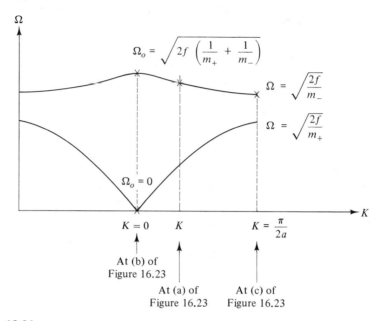

Fig. 16.24
Dispersion curve for the linear chain. Crosses indicate modes discussed for the optical and acoustical modes for three frequencies [see Figure 16.23 for (a), (b), and (c)].

where m_+ and m_- are the masses of the ions (we assume that $m_+ > m_-$) and f is the force constant describing the force between them. The fundamental mode corresponds to the mode where the ions in each cell oscillate against each other and the phase relation is the same for all cells.

We call each arrangement of the positive and negative ions at equilibrium a "sublattice." The fundamental mode is the vibration of the two sublattices against each other. Equation 16.28 describes the same frequency obtained for the vibration of a free diatomic molecule having masses m_+ and m_- and force constant f.

What happens for a very short wavelength? There is a limit of wavelike motion because of the regular distance a of the ions. This limit is at $\lambda = 4a$. In the optical mode, the heavy masses m_+ are all at rest. In the acoustical mode, the lighter masses m_- are all at rest (see Figure 16.23c). The corresponding frequencies are indicated in Figure 16.24. There we also indicate by crosses the frequencies corresponding to the wavelengths of the three cases discussed earlier. The solid line is a plot of the calculated frequencies versus the wavelengths, which is called a dispersion curve. For convenience, this curve is plotted as Ω versus $K = 2\pi/\lambda$. There are frequencies for which no wavelike motion exists. This is called the gap.

Scattering can be compared to collisions in mechanics. There we use the fact that energy and momentum are conserved before and after the collisions (also see the Compton effect, described earlier). For a "collision of waves," we must employ the corresponding expressions.

For light waves, energy is given as $E = \hbar\omega$ and momentum is

16. Scattering of Light

$$p = \frac{\hbar\omega}{c} = \frac{h\nu}{c} = \frac{h}{\lambda} = \hbar k \qquad (16.29)$$

We know that both k and p can be vectors.

The vibrational modes in the solid are described in a similar way using the angular frequency Ω and the K vector of the modes. We have

$$E = \hbar\Omega \quad \text{and} \quad \mathbf{P} = \hbar\mathbf{K}$$

The use of $\hbar$ indicates that the modes of the crystal are considered quantized. In analogy to light waves, we speak of a **phonon** of energy E and momentum $\mathbf{P}$. For conservation of energy and momentum, we have

$$\hbar\omega = \hbar\omega' + \hbar\Omega$$
$$\hbar\mathbf{k} = \hbar\mathbf{k}' + \hbar\mathbf{K} \qquad (16.30)$$

where the primed quantities belong to the scattered light. The first equation just tells us that the difference in frequency of incident and scattered light is the frequency of the mode in the crystal. The second equation is the conservation of $\mathbf{k}$ vectors, or momentum.

C. Scattering by the Acoustical Modes: Brillouin Scattering

The scattering by light on the acoustical mode is referred to as Brillouin scattering. In the crystal, the light has the frequency ω that is equal to $(c/n)k$; the frequency of the vibration in the crystal is similarly $\Omega = v_S K$. The phonon velocity in the crystal v_S is of the order of 10^5 cm/s, which is small compared to the speed of light in the crystal; that is, $v_S \ll c/n$. We assume the frequency of the incident light to be much larger than the phonon frequency in the solid, that is, $\Omega \ll \omega$. It follows, that $v_S K \ll (c/n)k$, and we can assume that $k \simeq k'$. From Figure 16.25, we have

$$K = 2k \sin\frac{\psi}{2} \qquad (16.31)$$

Introducing the preceding relations yields

$$\Omega = 2v_S \frac{2\pi}{\lambda} n \sin\frac{\psi}{2} \qquad (16.32)$$

For example, consider the case where the incident wavelength λ is 400 Å, that is 0.4×10^{-5} cm, and $v_S = 5 \times 10^5$ cm/s ($\omega = 5 \times 10^{15}$ 1/s) and we assume $n = 1.5$. We take $\sin(\psi/2) = 1$ and get

$$\Omega = \frac{(2 \cdot 2 \cdot 3.14) \cdot 5 \times 10^5 \text{ cm/s} \,(1.5) \cdot 1}{0.4 \times 10^{-5} \text{ cm}} = 2 \times 10^{12} \text{ 1/s}$$

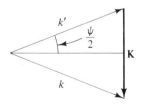

Fig. 16.25
K vector diagram for scattering.

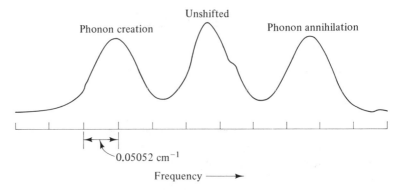

Fig. 16.26
Light with wavelength 6328 Å scattered on water. [From G. B. Benedek, J. B. Lastovka, K. Fritsch, and T. Greytak, *J. Opt. Soc. Am.* **54**, 1284 (1964).]

For the frequency shift, we have $\omega - \omega' = \Omega$, and for the fractional change,

$$\frac{\Omega}{\omega} = \frac{2 \times 10^{12}}{5 \times 10^{15}} = 0.4 \times 10^{-3} = 4 \times 10^{-4} \tag{16.33}$$

which is very small.

Figure 16.26 shows the scattering of He-Ne laser light in water at right angles (to the incident light), where $\lambda = 6328$ Å. The frequency Ω was determined to be 4×10^9 cps, and the velocity is calculated to be 1.5×10^5 cm/s.*

In the scattering process, the wave in the crystal may take away energy from the incident light, lowering the frequency of the scattered light. But the wave in the crystal may also add energy to the incident light, raising the frequency of the scattered light. Both types of scattered light are observed in the experiment shown in Figure 16.26.

Using the phonon picture we can say that a phonon is "created" by the incident light and that the scattered light has a smaller frequency. The process is described by

$$\hbar\omega = \hbar\omega' + \hbar\Omega$$
$$\hbar\mathbf{k} = \hbar\mathbf{k}' + \hbar K \tag{16.34}$$

The reverse process is "annihilation" of a phonon and results in a higher frequency of the scattered light.

D. Scattering of Light by the Optical Modes: Raman Scattering

The scattering of light on the optical modes is called **Raman scattering**. As in the discussion of the acoustical modes, we observe scattered light with the same frequency (Rayleigh lines) and weak scattered light at higher and lower frequencies. In a spectral observation, the weak lines are so close to the exiting strong line that they are difficult to observe. Historically, Raman scattering was first observed with molecules. Molecules vibrate, and each molecular vibration

---

*After G. B. Benedek, J. B. Lastovka, K. Fritsch, and T. Greytak, *J. Opt. Soc. Am.* **54**, 1284 (1964).

corresponds to a certain amount of energy. In the scattering process, this energy is added or subtracted from the incident light. The vibration frequency of the molecular vibrations is much higher than the frequency of the modes in solids; therefore, the weak scattered lines are further removed from the exiting line.

E. Raman Scattering on Molecules in the Gas or Liquid State

Molecules undergo mechanical vibrations. The number of normal vibrations depends on the number of atoms in the molecule. A system of N atoms has $3N$ degrees of freedom. The molecule has three translatory motions of its center of mass. A nonlinear molecule has three degrees of freedom of overall rotation. Therefore, there are $3N - 6$ vibrations. Some of these vibrations carry with them a change of the dipole moment of the molecule. They can be observed with infrared light in absorption. For most vibrations, the incident light changes the polarizability of the molcule. This also happens if there is no change of the dipole moment while the molecule vibrates. The incident light induces a dipole moment in the molecule. Reradiation from the induced dipole results in the scattered light, and the vibration may be observed in Raman scattering.

There are molecules with a certain type of symmetry in which some vibrations can be observed in infrared absorption, others in Raman scattering. There are also molecules for which certain vibrations cannot be observed with either or both methods.

For a simple theory of Raman scattering, we assume light incident on molecules contained in a certain volume as gas or liquid. In the classical theory, the incident light is presented as

$$E = E_0 \sin 2\pi v_0 t \qquad (16.35)$$

The incident light induces an electric dipole moment

$$p = \alpha E = \alpha E_0 \sin 2\pi v_0 t \qquad (16.36)$$

where α is the polarizability. If the excited molecule oscillates at a frequency v_m, the polarizability changes to

$$\alpha = \alpha_0 + \alpha_1 \sin 2\pi v_m t \qquad (16.37)$$

Introducing equation 16.37 into equation 16.36 gives us

$$p = \alpha E = (\alpha_0 + \alpha_1 \sin 2\pi v_m t) E_0 \sin 2\pi v_0 t$$
$$= \alpha_0 E_0 \sin 2\pi v_0 t + \alpha_1 E_0 (\sin 2\pi v_m t)(\sin 2\pi v_0 t) \qquad (16.38)$$

The second term is the product of two time-dependent sine terms. This term contains the nonlinearity of the process. We can write

$$p = \alpha_0 E_0 \sin 2\pi v_0 t + \tfrac{1}{2}\alpha_1 E_0 [\cos 2\pi (v_0 - v_m)t + \cos 2\pi (v_0 + v_m)t] \quad (16.39)$$

The reemitted light has the frequency v_0, that is, the Rayleigh line. The frequencies $v_0 \pm v_m$ are the frequencies of the scattered light. A schematic of the Raman spectrum of carbon tetrachloride (CCl_4) is shown in Figure 16.27. We see that the intensities of the lines associated with a particular vibrational mode are not equal. The so-called **Stokes lines** ($v_0 - v_m$) have a higher intensity than the anti-

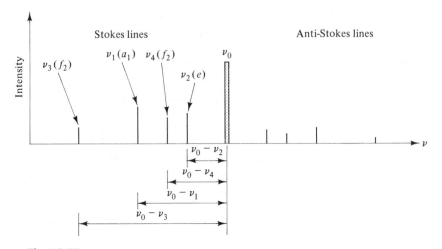

Fig. 16.27
Schematic of the Raman spectrum of CCl_4. The four vibrational frequencies are indicated with their symmetry notation according to the T_d group.

Stokes lines ($v_0 + v_m$). We know that in thermal equilibrium, the fraction of all molecules having the energy E is given by the Boltzmann factor $e^{-E/kT}$. There are fewer molecules in the states with higher values of E. The anti-Stokes lines have higher energy than the Stokes lines, and accordingly, their intensity is lower. The Stokes and anti-Stokes lines are produced in the scattering process; the equivalent energy of a molecular vibration is added or subtracted from the energy of the incident light quantum.

5. SCATTERING AND THE INDEX OF REFRACTION

In Chapter 7, we used the forced damped oscillator to derive an expression for the refractive index. In this chapter, we have used the same model to describe scattering from small particles with induced dipoles. In both cases, we discussed the excitation of the dipole by the incident light and the reemission by the dipoles. In the simple case discussed for the index of refraction, the amplitude of the forced oscillator was found to be

$$u = \frac{eE_0/m}{(\omega_0^2 - \omega^2) - i\gamma\omega} \tag{16.40}$$

For the complex amplitude u, we may write $u = ae^{-i\phi}$ and get

$$a^2 = \frac{e^2(E_0/m)^2}{(\omega_0^2 - \omega^2)^2 + \gamma^2\omega^2} \tag{16.41}$$

and

$$\tan\phi = \frac{\gamma\omega}{\omega_0^2 - \omega^2} \tag{16.42}$$

　　　　16. Scattering of Light

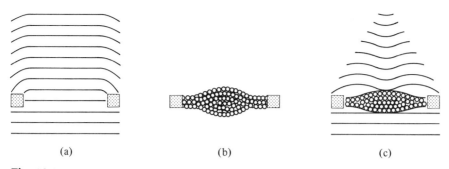

Fig. 16.28
(a) Aperture through which plane waves are traveling. (b) Collection of small spheres forming a lens. (c) The collection of spheres at the aperture, resulting in lenslike action. (From Pohl, *Einführung in die Optik.*)

where ω is the frequency of the incident light and ω_0 is the resonance frequency of the induced oscillator. If these two frequencies are almost equal, we obtain $\phi = 90°$, that is, the reemitted wave has a phase retardation of $90°$ with respect to the incident wave. Not all of the incident wave excites the dipoles. The light that passes the sample and the phase-retarded reemitted light interfere in the forward direction. As a result, the superposition of these two waves propagate at a velocity v that is smaller than the "outside" velocity c. We know that $v = c/n$ and may see the connection between scattering and the refractive index. A mechanical demonstration using water waves is given by Pohl*. Figure 16.28a shows an aperture through which plane waves are traveling. Diffraction occurs at the edges, as expected. Figure 16.28b shows a collection of small spheres assembled into a lens-shaped arrangement. In Figure 16.28c, the lens is mounted at the aperture and focusing is observed. In a similar way, we can expect focusing of light waves. The scattering of water waves off the spheres has produced a refractive medium that converges the incident plane waves. Using the formula for the focal length of a biconvex lens

$$\frac{1}{f} = (n - 1)\frac{2}{R} \tag{16.43}$$

with radius R of curvature and inserting $f = 8.75$ cm and $R = 7$ cm (Pohl's values) gives $n = 1.4$, a very reasonable number.

Problems

1. *Randomly oriented particles.* For a collection of randomly oriented particles, compare the intensity of light scattering into the direction $X-Y$ ($\theta = 45°$) and Y ($\theta = 90°$) to what is scattered into the X ($\theta = 0°$) direction.
   Ans. $X-Y$ to X: ratio 0.75; Y to X: ratio 0.5

*R. Pohl, *Introduction to Optics* (New York, Springer-Verlag, 1948), page 185 (in German).

2. *Calculation of K.* Calculate equation 16.22 from equations 16.14, 16.19, and 16.21 by using

$$K = \frac{\Delta S}{S_F}\left(\frac{1}{\Delta x}\right) \quad \text{and} \quad v = F\,\Delta x$$

3. *Average of* $\cos^2\theta$. For the dependence on angle θ for dipole emission, we have

$$|\mathbf{S}| \propto \cos^2\theta$$

where θ is given as

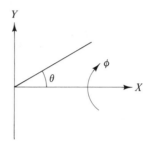

The angle ϕ is defined around the X axis. Calculate the average of $\cos^2\theta$ over all directions in three dimensions.
Ans. $4\pi/3$

4. *Bragg condition.* Consider a cubic crystal. The incident radiation has wavelength $\lambda = 1.6$ Å. The lattice constant is $a = 4.2$ Å. At what angle will we observe diffraction in first order ($n = 1$)?
Ans. $\theta = 10.98$ degrees

5. *Compton scattering.* For the scattering of an X ray by an electron, the energy of the photon is $E = h\nu$, the momentum of the photon is $p = h/\lambda$, the energy of the electron before scattering is $E_0 = mc^2$, where m is the rest mass of the electron, and the energy of the electron after scattering is $E = \sqrt{E_0^2 + p_e^2 c^2}$, where c is the speed of light and p_e is the momentum of the electron after collision. We do not have to specify p_e because it will be eliminated in the derivation.
   a. *Conservation of momentum.* We have the vector equation

$$\mathbf{p}_1 + \mathbf{p}_e = \mathbf{p}_2 + \mathbf{p}_e$$
$$(\text{before} = \text{after})$$
$$\mathbf{p}_e(\text{before}) = 0$$

and subscript 1 and 2 refer to "before" and "after" for the X ray, respectively. Show that we have

$$p_e^2 = p_1^2 + p_2^2 - 2p_1 p_2 \cos\theta$$

where θ is the angle between $\mathbf{p}_1$ and $\mathbf{p}_2$.
   b. *Conservation of energy.*

$$p_1 c + E_0 = p_2 c + \sqrt{E_0^2 + p_e^2 c^2}$$

Eliminate p_e^2 and obtain

$$\lambda_2 - \lambda_1 = \frac{h}{mc}(1 - \cos\theta)$$

16. Scattering of Light

6. *Linear chain.* Consider a free molecule having the masses

$$m_+ = m_\mathrm{H} = 1.7 \times 10^{-27} \text{ kg} \quad \text{and} \quad m_- = m_\mathrm{Li} = 11.4 \times 10^{-27} \text{ kg}$$

Assume that the force constant is $f = 7 \times 10^{-5}$ dyne/cm, corresponding to 7 N/m.
a. Calculate the vibrational frequency of this hypothetical molecule with

$$v = \frac{1}{2\pi} \sqrt{2f\left(\frac{1}{m_\mathrm{H}} + \frac{1}{m_\mathrm{Li}}\right)}$$

and give v in cm^{-1} and the corresponding value λ_R in micrometers.
b. Using the same parameters, calculate the frequencies in cm^{-1} for $k = \pi/2a$, that is,

$$v_- = \frac{1}{2\pi c} \sqrt{\frac{2f}{m_-}} \quad \text{and} \quad v_+ = \frac{1}{2\pi c} \sqrt{\frac{2f}{m_+}}$$

Check the result for v_+ by multiplying $\sqrt{m_-/m_+}$ by v_-.
Ans. a. $v = 517 \text{ cm}^{-1}$, $\lambda_R = 19 \ \mu\text{m}$; b. $v_- = 186 \text{ cm}^{-1}$, $v_+ = 481 \text{ cm}^{-1}$

7. *Brillouin scattering.*
   a. Calculate the shifted frequency ω' from $\omega - \omega' = \Omega$. First calculate

$$\Omega = 2v_s \frac{2\pi}{\lambda} n \sin\frac{\phi}{2}$$

Using $v_s = 5.5 \times 10^5$ cm/s, $\lambda = 4358.3$ Å (mercury lamp line), $n = 1.55$, and $\sin(\phi/2) = 1$.
   b. Calculate $\omega = 2\pi v$ from $\lambda v = c$ for $\lambda = 4358.3$ Å; use $c = 2.9979 \times 10^{10}$ cm/s.
   c. Calculate $\omega' = \omega - \Omega$.
   d. Convert Ω into λ_s (in angstroms) and ω' into λ' (in angstroms) and compare the difference corresponding to $\omega - \omega'$ to λ_s.

8. *Raman effect.* The exciting line for the Raman effect is the 4358.3 Å line of mercury. The vibrational frequencies of CCl_4 are $v_2 = 218 \text{ cm}^{-1}$, $v_4 = 314 \text{ cm}^{-1}$, $v_1 = 458 \text{ cm}^{-1}$, and $v_3 = 775 \text{ cm}^{-1}$. Assume that we want to give an interpretation of a spectrograph in an angstrom scale. Calculate the four Raman lines for CCl_4 in angstroms.

Integrated Optics | **17**

1. INTRODUCTION

Invention of the laser created a great deal of excitement among communication engineers. Until that time, frequencies of the electromagnetic spectrum up to the microwave region had been used for communication purposes. The amount of information carried by an electromagnetic wave is directly proportional to the frequency of the wave carrying it. In a simplified way, this can be seen by considering the transmission of pulses per second. The presence or absence of a pulse received at the end of a light guide constitutes one bit of information. A wave that has a maximum only 100 times per second can only transmit a maximum of 100 pulses per second. Since the laser uses light frequencies that are so much higher than the highest frequency used at that time, the amount of information that could be transmitted could be so much higher. For example, light transmitted by a single fiber can carry 44.7 million pulses per second. Why did we have to wait for the laser? Laser light is monochromatic, with a very small frequency bandwidth and high intensity. The wave trains of laser light are very long, and when we wish to transport pulses, we cannot use wave trains that break up during transmission.

What devices are needed for light wave communication? The telephone has a microphone that acts as an emitter of information. The electromagnetic waves get modulated and then pass through the wires, which are transmitters. On the other end, we have the receiver, which reproduces the information in our language. When this system is used over long distances, we must reamplify the electromagnetic waves because they become attenuated. Also, we must consider distortion of the original information, which has to be restructured to make the voice acceptable at the receiving end.

A light communication system that was used by the Navy at least until World War II was operated in the following way. A sailor encoded the Morse code, and a blinker was used to emit the light pulses; the atmosphere was the transmitting medium; and the eye and brain of another sailor acted as receiver and decoder of the message. In this system, distortion of the message was minimized, since pulses could also be detected if they were distorted.

Light communication systems presently follow the same basic idea. To eliminate the distortion of the amplitude modulation system, we code and decode the messages. The elements of a light communication system are therefore the light source, which is the laser; the coding device, which is a modulator; the transmitter, which is the fiber; the demodulator; and the receiver, which is a photon detector. Lasers and modulators are solid-state electrooptical devices. For lasers see Chapter 15; modulators, Chapters 5 and 15; detectors, Chapter 14.

In the following sections, we will discuss the transmission of light through rectangular strips, films, and fibers and the necessary manipulation of the light to get it in and out of the light guides. We will also consider filters and mode converters.

2. COMPONENTS AND MODES

Let us first discuss light propagation through thin films. The principles can be readily understood, and the results can be applied later, with some modification, to the propagation of light through fibers.

Light can be propagated between two metal plates by reflection on the plates. For dielectric films, the reflection can be accomplished by total internal reflection or through the use of metal films attached to the dielectric film on both sides. To bring the light into the dielectric plate using total internal reflection, we can make use of a prism, as shown in Figure 17.1. Such devices can also be used for laser light, of course, but they turn out to be very sensitive to ambient temperature changes and mechanical vibrations. For example, if the thickness of the plate changes, we find that over a long distance, the direction of the outcoming beam is displaced. Such disturbances can be avoided if we employ mode propagation of a monochromatic light in thin films. As an added advantage, we can make all the devices much smaller and adapt them to the small diameters of transmitting fibers or output apertures of solid-state lasers.

A. Modes and Mode Propagation

Let us consider a box with reflecting walls and waves propagating between the walls. In the derivation of the formula for blackbody radiation, we use such a box to find the number of possible standing waves depending on the frequency interval under consideration (see Chapter 15). There we wished to find the

Fig. 17.1
A prism is used to make incident light totally internally reflected in a plane parallel plate. The prism may be an integral part of the plate; it acts as a "light funnel."

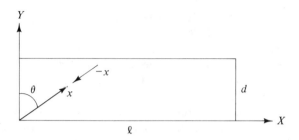

Fig. 17.2
Two waves traveling in the x and $-x$ direction in a two-dimensional box of length ℓ and width d.

radiation intensity depending on the frequency and temperature. Here we are interested in a more restricted aspect of the waves in the box. We want to consider laser light of one single frequency traveling back and forth in the box, and we want to describe the propagation of the electric field in the box as it relates to the dimensions of the box.

We restrict this consideration to a two-dimensional box of width d and length ℓ. We assume two waves traveling in opposite directions along the x direction and are interested in possible standing-wave conditions. The wave makes the angle θ with the normal, which is the Y direction (Figure 17.2). The wave moving in the x direction may be expressed as

$$e^{i[(\mathbf{kx})-\omega t]} = e^{i(k_X X + k_Y Y - \omega t)} = e^{-i\omega t}e^{ik_X X}e^{ik_Y Y} \tag{17.1}$$

where the vector $\mathbf{x} = \mathbf{X}_0 X + \mathbf{Y}_0 Y$ and $\mathbf{X}_0$ and $\mathbf{Y}_0$ are unit vectors and X and Y are the coordinates. We have

$$\mathbf{k} = \mathbf{X}_0 k_X + \mathbf{Y}_0 k_Y \quad \text{and} \quad \mathbf{x} = \mathbf{X}_0 X + \mathbf{Y}_0 Y$$

We can also write

$$\mathbf{k} = (k_X, k_Y), \quad \mathbf{x} = (X, Y) \tag{17.2}$$

with the components

$$k_X = k \sin \theta, \quad k_Y = k \cos \theta, \quad (\mathbf{k} \cdot \mathbf{k}) = k_X^2 + k_Y^2 = k^2 = \left(\frac{2\pi}{\lambda}\right)^2 = \frac{\omega^2}{c^2}, \quad |\mathbf{k}| = k$$

The standing-wave condition for the X and Y directions separately are

$$\frac{\ell}{\lambda/2} = \bar{m}_X \quad \text{and} \quad \frac{d}{\lambda/2} = \bar{m}_Y \tag{17.3}$$

From the requirement that the standing wave pattern be periodic, we have from $e^{ik_X X}$ for $X = \ell$ that $k_Y = \pi\bar{m}_X/\ell$, and from $e^{ik_Y Y}$ for $Y = d$ that $k_Y = \pi\bar{m}_Y/d$, where $\bar{m}_X$ and $\bar{m}_Y$ are integers.

Using the dot product

$$(\mathbf{k} \cdot \mathbf{k}) = \frac{\omega^2}{c^2} = \left(\frac{2\pi}{\lambda}\right)^2 \tag{17.4}$$

and introducing the standing-wave condition formulated for k_X and k_Y results in

$$k_X^2 + k_Y^2 = \frac{\pi^2 \bar{m}_X^2}{\ell^2} + \frac{\pi^2 \bar{m}_Y^2}{d^2} = \frac{\omega^2}{c^2} = \frac{4\pi^2 v^2}{c^2} \qquad (17.5)$$

or

$$\frac{\bar{m}_X^2}{\ell^2} + \frac{\bar{m}_Y^2}{d^2} = \frac{1}{(\lambda/2)^2} \qquad (17.6)$$

Equation 17.6 shows that $\bar{m}_X$ and $\bar{m}_Y$ are not independent; that is, we have to look for pairs of $\bar{m}_X$ and $\bar{m}_Y$ after ℓ and d have been chosen. See also Chapter 15, Section 2.

If $\ell \to \infty$, we are back at the standing-wave condition for the Y direction alone, that is, considering waves only for $\theta = 0°$. Similarly, for $d \to \infty$, we have the standing-wave condition for the X direction, or for $\theta = 90°$.

A standing wave characterized by $\bar{m}_X$ and $\bar{m}_Y$ is called a **mode**. We will now discuss the possible values of $\bar{m}_X$ and $\bar{m}_Y$ for two specific examples, $\bar{m}_Y = 1$ and $\bar{m}_Y = 2$. The first case, $\bar{m}_Y = 1$, is the lowest mode in the Y direction, and we have

$$\frac{\bar{m}_X^2}{\ell^2} + \frac{1}{d^2} = \frac{1}{(\lambda/2)^2} \qquad (17.7)$$

If d has the length of the standing-wave pattern in the Y direction, that is, $d = \lambda/2$, then ℓ must be infinite ($\bar{m}_X$ finite). If $\ell \neq \infty$, then d must be chosen larger than $\lambda/2$.

We can make d so large that the $\bar{m}_Y = 2$ mode is possible:

$$\frac{\bar{m}_X^2}{\ell^2} + \frac{4}{(d')^2} = \frac{1}{(\lambda/2)^2} \qquad (17.8)$$

Now d' must be larger than λ so that $4/(d')^2 < 4/\lambda^2$ and there is still "room" for $\bar{m}_X^2/\ell^2$ to fulfill equation 17.8.

A standing wave can be represented by the superposition of two running waves. The choice of sine waves ensures that the standing wave has nodes at the metal walls. For the X direction, the superposition of

$$u_{\text{to right}} = A \sin(k_X X - \omega t)$$

and

$$u_{\text{to left}} = A \sin(k_X X + \omega t)$$

results in

$$u = 2A \cos(\omega t) \sin(k_X X)$$

The space part is proportional to

$$\sin k_X X \qquad (17.9)$$

and it has intensity maxima and nodes along the X direction. In the Y direction, for $\bar{m}_Y = 1$, we have one intensity maximum between $Y = 0$ and $Y = d$; and for $\bar{m}_Y = 2$, we have two intensity maxima (Figure 17.3).

Each of the two running waves making up the standing wave in the x

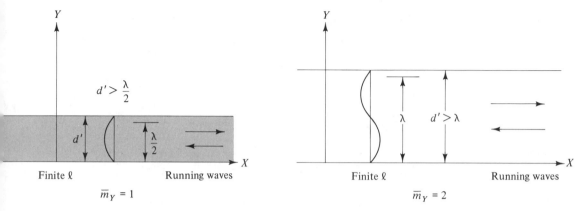

Fig. 17.3
Running waves in the X direction and standing waves in the Y direction for $\bar{m}_Y = 1$ and $\bar{m}_Y = 2$ and finite ℓ.

direction will show the nodes in the Y direction because we can separate the waves in the X and Y directions while equation 17.6 is satisfied. We can interpret this as a wave traveling through a thin film of thickness d and having a node pattern in the Y direction. The choice of d will determine the mode to be supported (e.g., $\bar{m}_Y = 1$, $\bar{m}_Y = 2$).

The possible angles θ at which the waves may travel in the x direction are now restricted by the choice of $\bar{m}_X$ and $\bar{m}_Y$ and may be calculated by observing that $k_X = k \sin \theta$ and $k_Y = k \cos \theta$, so that we get

$$\tan \theta = \frac{k_X}{k_Y} \tag{17.10}$$

If we choose $\bar{m}_Y = \infty$, we have $\theta = 0°$, that is, standing waves in the X direction. If $\bar{m}_X = \infty$, there is a standing wave in the Y direction ($\theta = 90°$). If we use the k_X and k_Y components to characterize the waves, we may indicate a standing wave in two dimensions as shown in Figures 17.4, 17.5, and 17.6.

Figure 17.6 demonstrates that the wave is represented by a standing-wave pattern in the Y direction and propagation in the X direction. The running waves described by $\mathbf{k} = (-k_Y, k_X)$ and $\mathbf{k}' = (k_{Y'} k_X)$ each fold forward and backward in an opposite way, producing a standing-wave pattern in the Y direction and a moving wave in the X direction.

Electromagnetic waves propagating in a plane are called TE waves (mode) if the E vector is perpendicular to the plane and TM waves (mode) if the B vector is perpendicular to the plane. (Figure 17.7). The mode number is often added to this notation. We have used $\bar{m}$ to represent the number of maxima of the mode in the X or Y direction. Often, m is used to represent the number of nodes. There is always one more maxima than nodes, and consequently, $m = \bar{m} - 1$.

B. Modes in Dielectrics

We will now consider a thin dielectric film with refractive index n_2 embedded in a medium with lower index n_1 (on both sides). If a wave's angle of incidence θ is

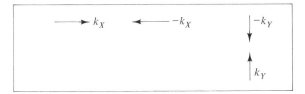

Fig. 17.4
Wave vector **k** of standing waves in the X and Y directions.

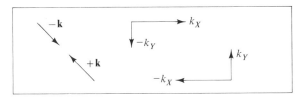

Fig. 17.5
Wave vector **k** of standing waves in the X direction can be decomposed into standing waves in the X and Y directions.

Fig. 17.6
Interpretation of the superimposed waves described by opposite k_Y but same k_X components.

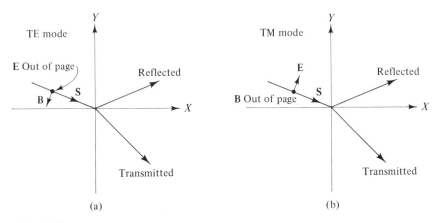

Fig. 17.7
(a) TE mode. The plane of the paper is the plane of incidence. The **E** vector is perpendicular to the plane; the **B** vector is in the plane. (b) TM mode. For the same plane of incidence, the **B** vector is perpendicular to the plane, and the **E** vector is in the plane.

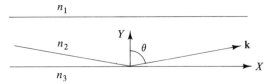

Fig. 17.8
Total internal reflection of a wave incident from the inside on the boundary n_2-n_1 where $n_2 > n_1$ and $\theta > \theta_c$ (critical angle).

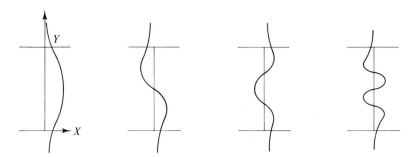

Fig. 17.9
Mode structure in the Y direction for an infinitely long film (in the X direction).

large enough, we have total internal reflection, as shown in Figure 17.8. As discussed for the metal plate sandwich, we have internal reflection restricted to certain modes with mode angle $\theta_{\overline{m}}$. We know that at total internal reflection, we have a decreasing electric field outside the film. We expect that a mode calculation for this case results in a somewhat modified mode picture, but with some similarity to what we found for the metal plate sandwich. Inside the film, we have similar wave forms for the amplitude with the same number of nodes. Outside, we have exponentially decreasing tails, as shown in Figure 17.9. The outside wave is called the evanescent field.

The propagation of the modes with $\overline{m}$ and $\theta_{\overline{m}}$ is now described by the wave vector $k_X = k \cos \theta$ with $\lambda = \lambda_0/n_2$, where n_2 is the refractive index of the dielectric. In vacuum, we have $k_0 = 2\pi/\lambda_0$, assuming $n_1 = 1$.

$$k_X = \frac{2\pi}{\lambda_0/n_2} \sin \theta_{\overline{m}} = k_0 n_2 \sin \theta_{\overline{m}} \tag{17.11}$$

$$k_Y = \frac{2\pi}{\lambda_0/n_2} \cos \theta_{\overline{m}} = k_0 n_2 \cos \theta_{\overline{m}} \tag{17.12}$$

with

$$k_X^2 + k_Y^2 = k_0^2 n_2^2 \tag{17.13}$$

where the factor n_2 enters the formula from what we learned in Chapter 2. We characterize the direction of the wave resulting from the superposition of the waves with $\mathbf{k}$ and $\mathbf{k}'$ by a superposition of waves with k_X and k_Y and k_X and $-k_Y$, respectively (Figure 17.10).

17. Integrated Optics

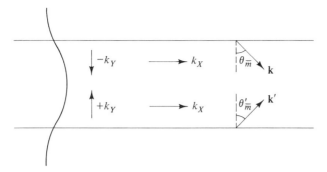

Fig. 17.10
The mode with $\bar{m}_Y = 1$ propagating in the X direction in a dielectric film.

The phase velocities in the X and Y directions are obtained from

$$v_X = \frac{\omega}{k_X} = \frac{\omega}{k_0 n_2 \cos \theta_{\bar{m}}} \quad \text{or} \quad v_X = \frac{2\pi v}{(2\pi/\lambda_0) n_2 \sin \theta_{\bar{m}}} = \frac{c}{n_2 \sin \theta_{\bar{m}}} \quad (17.14a)$$

and

$$v_Y = \frac{c}{n_2 \cos \theta_{\bar{m}}} \quad (17.14b)$$

where $c = \lambda_0 v$. This tells us (as was also true for the metal plate sandwich) that different modes have different phase velocities.

Although total reflection has been known for centuries, some 30 years ago, the phase shift of a total reflected wave was measured and found not to be zero. This is called the **Goos-Haenchen shift** (Figure 17.11). In such a shift, the ray behaves as if it were reflected from a plane in the evanescent field region some distance b away, depending on the penetration depth in the less dense medium. This phase shift is important for the case where we have two different media above and below. The penetration depth at the two interfaces is different. Thus, the shift is different for the two boundaries, and the mode pattern is influenced. We can see this in the wave picture, since many reflections take place over the length of the traveled space, and the accumulation of the additional path length is not symmetric.

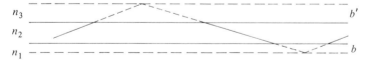

Fig. 17.11
Goos-Haenchen shift at total reflection. The wave behaves as if it were reflected at a plane displaced into the outer medium. Note that b depends on n_1, and b' depends on n_3.

Let us now consider the simple case where $n_1 = n_3$. With respect to the possible values of θ, we found restrictions on the possible values of $\bar{m}_X$ and $\bar{m}_Y$ for the metal plate sandwich. This is also true here, but only for angles larger than the critical angle θ_c. We know from Chapter 4 that this angle is determined by $n_2 \sin \theta_c = n_1$. The critical angle θ_c restricts the region of total internal reflec-

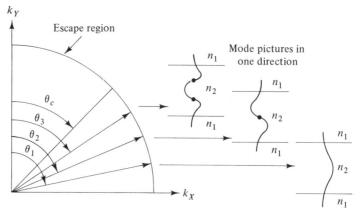

Fig. 17.12
Simplified schematic of supported modes in a dielectric film on a substrate. For angles θ_1, θ_2, and θ_3 modes are indicated with $\bar{m}_Y = 1, 2, 3$ propagating in the film.

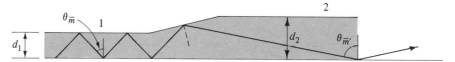

Fig. 17.13
A wave reflected inside a dielectric film through a tapered section, from a section with thickness d_1 to a section with d_2, changes the reflection angle from $\theta_{\bar{m}}$ to $\theta_{\bar{m}'}$, respectively.

tion. For angles smaller than θ_c, the light will escape and travel in the medium outside the film. This is the situation depicted in Figure 17.12. For three angles, the picture of the supported modes is shown.

C. Thin Films with Different Thicknesses: "Refraction at the Interface"

Having discussed the mode pattern in some detail, we now turn to the question of mode propagation. Let us assume that we have excited a certain mode characterized by the angle $\theta_{\bar{m}}$. Can we force the light into a different mode, one with the angle $\theta_{\bar{m}'}$? Since the phase velocity v_X in direction X depends on the angle θ, the phase velocity would change. To study this, we take two films with different thicknesses and connect them with a taper, that is, a linearly increasing connection, as shown in Figure 17.13. The light will go from section 1 to section 2, changing the angle θ from $\theta_{\bar{m}}$ to $\theta_{\bar{m}'}$. From a simple ray tracing, we see that $\theta_{\bar{m}}$ is smaller than $\theta_{\bar{m}'}$, and from

$$v_X = \frac{c}{N_1 \sin \theta_{\bar{m}}} \quad \text{and} \quad v'_X = \frac{c}{N_2 \sin \theta_{\bar{m}'}}.$$

it follows that the phase velocity in medium 2 can be smaller. We can experimentally demonstrate such a behavior if we produce two films of different thickness, connected by a taper section, as shown in Figure 17.14. There we go from a two-dimensional light path to a three-dimensional path.

17. Integrated Optics

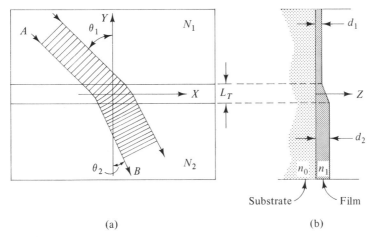

Fig. 17.14
A light beam guided in a planar dielectric guide is refracted at a step of the film of thickness d. The difference in film thickness between regions 1 and 2 causes their effective indices of refraction N_1 and N_2 to be different. (a) Top view. (b) Side view. [From R. Ulrich and R. J. Martin, *Applied Optics* **10**, 2077 (1971).]

In Figure 17.14, we see the light traveling up and down in the film along a direction θ_1 with respect to the normal of the taper section. The light travels from A to B. According to Fermat's principle, the path from A and B is chosen so that the time is a minimum. The light has different phase velocities before and after passing the taper section. We can apply Snell's law and consider the region of the taper section the boundary. Here we observe, in the same way, a bending of the direction of travel after the light has passed the taper region. The bending is, as in Snell's law, toward the medium in which the speed is smaller. This bending is described mathematically by Snell's law,

$$N_1 \sin \theta_1 = N_2 \sin \theta_2 \tag{17.15}$$

where N_1 and N_2 are the *effective refraction indices* in media 1 and 2. They are given by the ratio of the speed of light in a vacuum to the phase velocity:

$$N = \frac{c}{v_{\text{ph}}} \tag{17.16}$$

This is similar to what was found in geometrical optics. An experimental demonstration is shown in Figure 17.15 for two ZnS films with thicknesses $d_1 = 700$ Å and $d_2 = 2400$ Å and the use of He-Ne light.

In Figure 17.16, we see a calculation of N_m for a ZnS film ($n_1 = 2.35$) on a glass substrate ($n_0 = 1.51$) and for $\lambda = 6328$ Å (red laser light). We see that the effective refractive index N_m is increasing for each mode with the film thickness d and that the values are between n_1 and n_2. The solid line is for TE modes, the broken line for TM modes.

We can also demonstrate total internal reflection if the light is incident from the thicker film at a large angle $\theta > \theta_c$, where θ_c is the corresponding critical

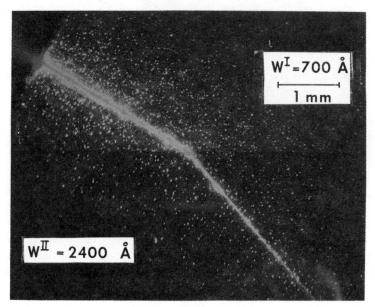

Fig. 17.15
Demonstration of the refraction of a guided laser beam. The light guide is formed by
ZnS films vacuum deposited on a glass substrate. The thicknesses $d_1 = W^I$ and
$d_2 = W^{II}$ of the ZnS films are indicated. The He-Ne laser beam propagates as a TE
($m = 0$) mode. [From R. Ulrich and R. J. Martin, *Applied Optics* **10**, 2077 (1971).]

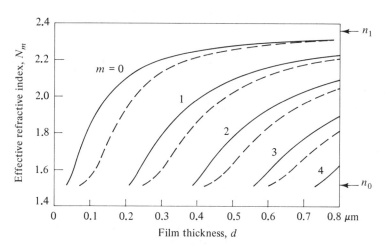

Fig. 17.16
The effective index of refraction N_m of planar dielectric waveguide as a function of the
film thickness d, given here for the example of a ZnS film on a glass substrate ($n_0 = 1.51$;
$n_1 = 2.35$; $\lambda = 6328$ Å). The parameter at the curves is the mode number $m = \bar{m} - 1$.
Solid lines indicate TE modes; broken lines, for TM modes. [From R. Ulrich and
R. J. Martin, *Applied Optics* **10**, 2077 (1971).]

17. Integrated Optics

angle given by

$$\sin \theta_c = \frac{N_1}{N_2} \qquad (17.17)$$

This is schematically shown in Figure 17.17 and demonstrated in Figure 17.18.

We can continue to produce geometrical film arrangements similar to the devices used in geometrical optics. Figure 17.19 shows schematically the formation of a prismlike film structure, and Figure 17.20 shows the formation of a prismlike and a lenslike structure.

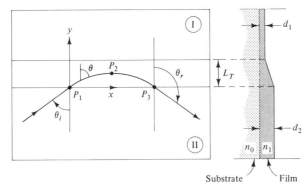

Fig. 17.17
The total reflection of a guided beam at a step of film thickness may be understood from the curved path of the wave inside the transition region L_T of nonuniform film thickness. [From R. Ulrich and R. J. Martin, *Applied Optics* **10**, 2077 (1971).]

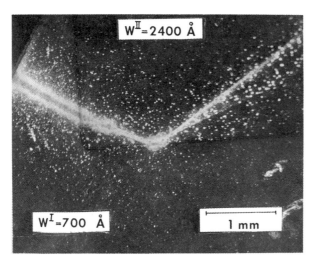

Fig. 17.18
Demonstration of the total reflection of a guided laser beam at a step of film thickness. The beam is launched in the thin region $d_1 = W^I$, refracted when entering the thick region $d_2 = W^{II}$, and totally reflected at the boundary to the thin region. (ZnS on glass, TE_0 mode, $\lambda = 6328$ Å.) [From R. Ulrich and R. J. Martin, *Applied Optics* **10**, 2077 (1971).]

2. Components and Modes   59

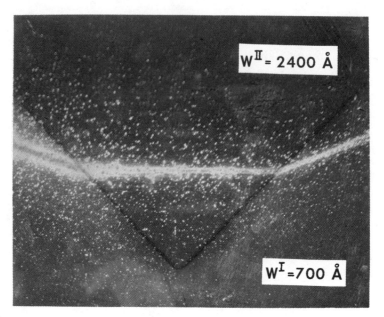

Fig. 17.19
Thin-film optical element of a prismlike structure; thicknesses $d_1 = W^I$, $d_2 = W^{II}$.
[From R. Ulrich and R. J. Martin, *Applied Optics* **10**, 2077 (1971).]

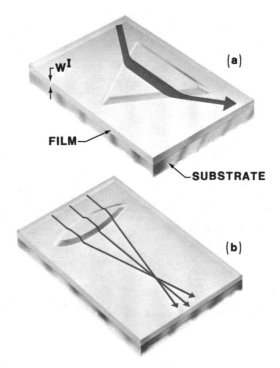

Fig. 17.20
Thin-film optical elements of a prismlike and lenslike structure (W^I is thickness).
[From R. Ulrich and R. J. Martin, *Applied Optics* **10**, 2077 (1971).]

D. Coupling Devices

To use a thin film or a transmitting fiber, we must bring enough light into it, since attenuation occurs at some distances. Even if large amounts of light enter the film, amplification may be necessary after long distances. The power density in a light guide may become quite large. If power of 150 mW traverses the area of 3 μm times 5 μm, we have a power density of 10^{10} W/m^2. For comparison, the solar constant is about 10^3 W/m^2.

One device that may bring in more light is the prism coupler used for the thick plate mentioned at the beginning of this chapter. From the point of view of the materials, it is difficult to make such a large prism on such a thin film. However, we can use a prism made of a material with a comparable index of refraction and put it as close as possible to the film. There will be an air gap, and it must be determined how significant it is. We know that total internal reflection produces an evanescent field outside the medium having the larger index of refraction. If the exponential decrease of this field is interrupted by the boundary of a material with a comparable refractive index, light is transmitted through the gap and can travel in the thin film, as shown in Figure 17.21. A smaller gap will couple more light into the waveguide. This phenomenon is sometimes called optical tunneling. To excite a mode in the film, the components of the wave vectors in the film must be equal (Figure 17.22). By changing the angle θ_2 of the total internal reflected light in the prism, different modes can be excited.

The way the prism is used in Figure 17.22 is not very efficient, since the

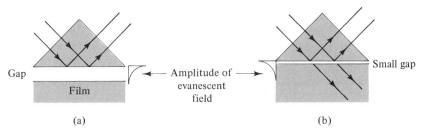

(a)   (b)

Fig. 17.21
The light internally reflected in a prism will (a) not get into the underlying film if the gap is large but (b) will get into the underlying film if the gap is small enough.

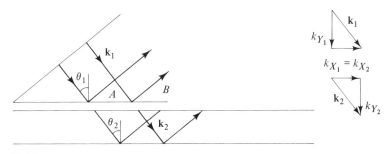

Fig. 17.22
Equal components of the wave vector in the film excite a mode. θ_1 is the angle of total reflection of the prism; θ_2 is the angle of reflection of the excited mode in the film; $\mathbf{k}_1 = (k_{X_1}, k_{Y_1})$ and $\mathbf{k}_2 = (k_{X_2}, k_{Y_2})$ are the k vectors of the light in the prism and in the film, respectively.

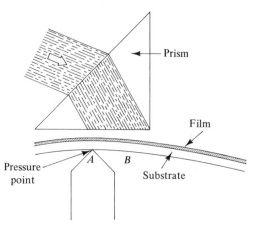

Fig. 17.23
Light is coupled into the film by pressing the film against the prism at point A, where the light is incident in the prism. The light cannot couple back out in the area B, because the bent film has a larger gap. [From P. K. Tien, *Applied Optics* **10**, 2395 (1971).]

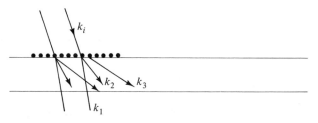

Fig. 17.24
Diffraction at a grating results in light propagating in different directions, and these directions correspond to different k vectors.

situation for coupling in at area A is the same as at area B, and at B we will find that light is coupled out as well. To avoid this, a rectangular prism is used, and by pressing on the substrate of the film, close contact is obtained at the spot A where the light is coupled in. The film bends, and larger separation from the prism makes it more difficult for the light to enter the film at B. In the region where the prism no longer exists, the light cannot couple back out (Figure 17.23).

Matching the prism coupler to the film requires continuity of k_x, which is done by choosing the correct angle of incidence at the prism. The matching of the k_x components can also be achieved by using a grating. The incoming light is diffracted in the different orders that propagate in different directions. Again, by choosing the right angle of incidence, the k_x components of the incoming light can be matched to those of the mode to be excited in the film (Figure 17.24).

Another waveguide coupler is the tapered coupler. It uses the principle that a mode in a film can propagate only if the film has a certain thickness. Below that thickness, the mode cannot move. This can be seen from the calculation of the effective refractive indices (see Figure 17.16), where the curves for $m = 0, 1,$ 2 are displaced to the right, that is, with increasing thickness of the film. From

17. Integrated Optics

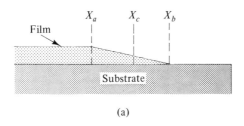

(a)

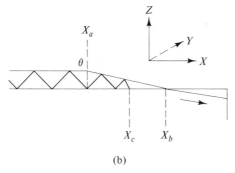

(b)

Fig. 17.25
Coupling of light into the substrate film by a tapered section. At X_c the coupling takes place. [From P. K. Tien, *Applied Optics* **10**, 2395 (1971).]

that figure, a thickness $d = 0.7$ μm allows modes with $m = 0, 1, 2, 3$ to propagate, whereas a thickness of 0.1 μm can only support the $m = 0$ mode. Consider light traveling along a taper on top of a film, as shown in Figure 17.25. The k_X component of the k vector is made smaller and smaller, reducing the angle θ with the normal. At the cutoff point X_c at an angle θ close to but less than θ_c, the mode cannot travel in the taper any longer, and the light enters the film.

E. Losses and Filters

Having seen that a considerable amount of energy density is present in a light guide, we must discuss the losses of such waveguides as well as the effects occurring when large energy densities exist. We saw how light propagates in thin-film guides and how it can be coupled in and out. In integrated optics, very small devices are considered in a similar manner to the chips used in integrated electronics. If the devices are made small, we must consider the possibility that light will have to go around a corner. We would like to determine how this can be done with minimal losses.

To understand the problems involved, we must consider **Cerenkov radiation**. If a charged particle moves through a vacuum with a constant speed v, it carries the electromagnetic field with it without radiating away energy. If the particle moves at 0.9 c through the vacuum, the charge also moves at that speed. The situation changes when the charged particle moves through a medium having a refractive index of, for example, $n = 2$. For the electromagnetic field, the highest speed for transporting energy is, in our case, $c/n = c/2$, but we are trying to force

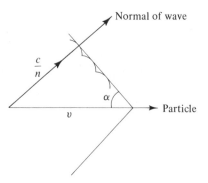

Fig. 17.26
Cerenkov radiation. The particle moves with velocity v through a medium of refractive index n. The radiation propagates in direction $(90° - \alpha)$.

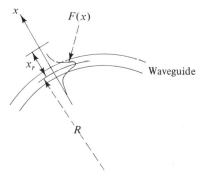

Fig. 17.27
"Inside" and "outside" electric field intensity propagating in a curved waveguide (fiber). R is bend radius; x_r is the coordinate of a point in the "radiating" outside field.

it through at 0.9 c. The field resists this attempt by radiating light with speed c/n at the angle α in the direction of the charged particle. The angle α is given by $v \sin \alpha = c/n$, as shown in Figure 17.26. This is called Cerenkov radiation.

Now, let us consider the bent film shown in Figure 17.27. If the field in the film moves, then the outside evanescent field, which in principle extends to infinity, moves with it, and far away from the film, it has to move faster than c (for a vacuum). Since it cannot do so, losses will occur through radiation. This simplified picture has been confirmed by elaborate calculations of this complex situation. A two-dimensional dielectric waveguide having a 1 percent smaller refractive index on the outside shows very small losses for bent radii of curvature longer than 1 mm and about 100 times more for radii of 0.6 mm. For shorter radii, the losses increase to a prohibitive level, as shown in Figure 17.28. A bend radius of about 1 mm is large for integrated optics, and its impact can be seen in a ring waveguide, which acts as a filter.

A ring waveguide can act as a filter when, for example, it is placed between two light guides. If frequencies f_1, f_2, and f_3 travel in light guide 2, f_2 could be filtered out by a ring waveguide of resonance frequency f_2 (Figure 17.29). The

17. Integrated Optics

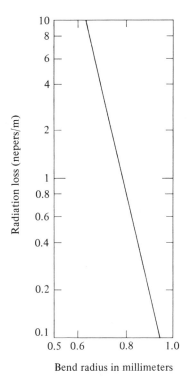

Fig. 17.28
Radiation loss versus bend radius for a two-dimensional dielectric waveguide.

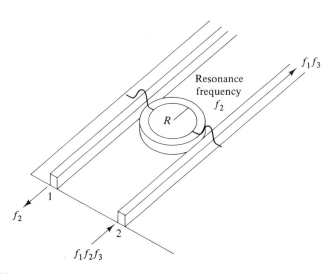

Fig. 17.29
Ring waveguide of resonance frequency f_2 acting as a filter between light guides 1 and 2.

evanescent field traveling in guide 1 can excite a mode in the ring guide. The light travels in the ring, and if it meets a cyclical boundary condition, it will be a mode. The resonance frequency f_2 of the mode depends on the travel time once around and therefore on the diameter R of the ring. With the same mechanism, the light in guide 2 will be excited with the resonance frequency f_2 of the ring, and the component with frequency f_2 is filtered out of guide 1 into guide 2. We see that the condition to have the bend radii not too small (i.e., about 1 mm) influences the size of the integrated optical device considerably.

3. LIGHT TRANSMISSION THROUGH FIBERS

A. Communication and Fibers

Most of us have seen fiber-optic trees sold in stores as a novelty. They are made with a number of fibers emerging from a common point, the end of the stem. Light is conducted through the fibers from the lamp positioned at the stem. The conduction through the fibers (with diameters of about 10 μm) is totally due to internal reflection. The conditions for total internal reflection are different for skew and nonskew rays, but in general are similar to what we discussed for thin films. Some of the light escapes, as shown in Figure 17.30. The fiber is protected on the outside by a coating, called cladding, having a refractive index slightly smaller than the core and also serving the purpose of preventing scratches from occurring that add to light leakage.

A bundle of these fibers can be used for image transformation, since if the eye is used, a 10 μm diameter fiber gives sufficient resolution. Such flexible imaging devices are used in medicine (e.g., to look into the stomach or bladder).

For optical communication purposes, there is a problem with such fibers, as demonstrated in Figure 17.31. In Figure 17.31a, we see that the input pulse is internally reflected. Since the rays are reflected at different angles, they do not all meet at the same point on the axis, resulting in a distorted output pulse. In Figure 17.31b, however, the index of refraction profile is such that all rays are reflected at angles that allow them to meet at the same point on the axis. As seen in the figure, as a consequence, modes propagate through the fiber, and the resulting output pulse is much less distorted than in (a). In our discussion of modes in thin films, we saw that different angles with the normal may be used for the characterization of different modes with different phase velocities. This is also true for fibers. The fact that different modes travel with different velocites is called **mode dispersion**.

For communication purposes, we must transmit light over large distances. Since laser light is used, which has a very narrow frequency bandwidth, frequency dispersion is very small. But both frequency dispersion and mode dispersion must be very small, since dispersion will give a distortion of the pulses in the communication process. This can be understood by considering the Fourier decomposition of the input pulses. All components must travel with the same speed to the output point in order to reconstruct the pulse without distortion. The mode dispersion can be eliminated if we give up the idea of a fiber whose core has a constant refractive index and consider fibers having a graded index of refraction.

17. Integrated Optics

(a)

Fig. 17.30
Light transmission by fibe[
(a) Fiber-optic lamp as
advertised by Edmund
Scientific Company, Bar-
rington, N.J. (b) Fiber-opt[
light guides used in an
experimental setup. (This
picture first appeared on t[
cover of the February 198[
issue of *Lasers & Applicatio*[
Torrance, CA. Reprinted
with permission.)

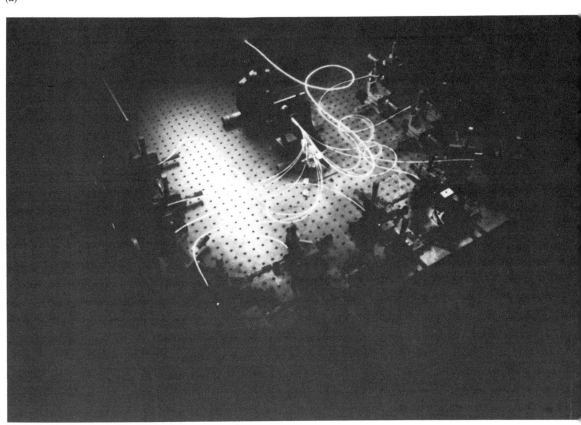

(b)

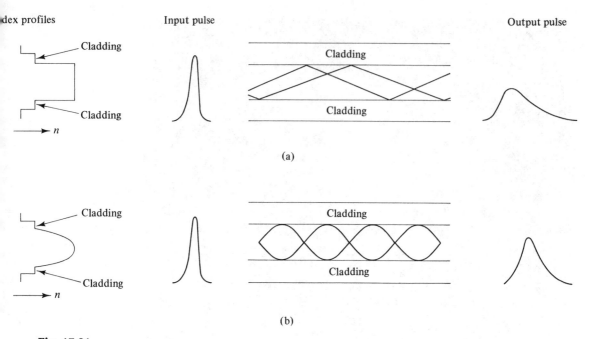

Fig. 17.31
Schematic of propagation of modes in fibers. The output pulse is disturbed. (a) Stepped index profile. (b) Graded index profile. The output pulse is less disturbed.

Figure 17.31b shows a graded index of refraction. The index has a parabolic dependence with respect to the radius of the fiber. All rays under different angles with the axis at the beginning point arrive at the same point some distance further on; by continuing to travel in such a way, the rays arrive at the output point at the same time. Such a fiber is called a **self-focusing fiber**.

Single-mode fiber-optic cables can be manufactured using, for example, light from a GaAsInP diode laser. Mode and frequency dispersion is very small, and transmission of 100 km can be obtained before an amplifier is needed for reamplification.

B. Fibers and Modes

In the section on thin films, we discussed modes propagating in the films for a thickness of several wavelengths. The same discussion may be applied to fibers. We can demonstrate the problem of the graded index of refraction using a stack of dielectric films having decreasing indices of refraction depending on their distance from the bottom, where the light enters (Figure 17.32). Rays starting with angles θ_1, θ_2, and θ_3 at a point A undergo refraction and total internal reflection in such a way that all rays arrive at the same point B at the same angle that they had at point A.

To understand the final result—that in a continuous medium, such behavior of the rays can be obtained by a dependence of the refractive index in a parabolic way—we must study the deviation of the ray as depending on the change in the

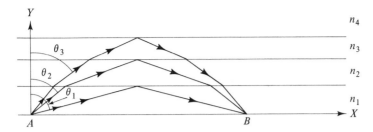

Fig. 17.32
Light internally reflected in a stack of films with decreasing indices of refraction n_1 to n_4. Depending on the angle of incidence, the light is totally reflected by one of the interfaces.

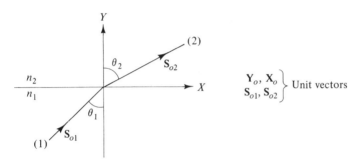

Fig. 17.33
Coordinates and vector notation for the ray equation.

refractive index. The equation describing this is called the **ray equation for an inhomogeneous medium**. We will derive this equation only for the two-dimensional case.

Let us consider a ray refracted at an interface, as shown in Figure 17.33, and the expression

$$S_{02}n_2 - S_{01}n_1 = \alpha Y_0 + \beta X_0 \qquad (17.18)$$

where α and β are parameters to be determined. Equation 17.18 states just that a vector of length n_2 in the direction of ray 2 minus a vector of length n_1 in the direction of ray 1 must be in the X, Y plane. The parameters α and β must now be determined. If we take the cross product with the vector Y_0, we obtain

$$[S_{02} \times Y_0]n_2 - [S_{01} \times Y_0]n_1 = \beta[X_0 \times Y_0] \qquad (17.19)$$

and if Z_0 is the vector out of the plane of the paper, we have

$$(\sin \theta_1)n_1 Z_0 - (\sin \theta_2)n_2 Z_0 = \beta Z_0 \qquad (17.20)$$

Application of Snell's law shows that $\beta = 0$. This result tells us that the difference of the vectors $S_{02}n_2 - S_{01}n_1$ (see equation 17.18) is parallel to a vector normal to the interface of the two media with different refractive indices ($\alpha \neq 0$).

If we study the continuous case and consider a differential of the preceding

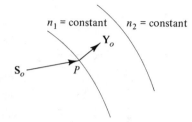

Fig. 17.34
Surfaces of constant n in a medium with
a continuously variable refractive index.

situation, we have, from equation 17.18 with $\beta = 0$,

$$\frac{\mathbf{S}_{02}n_2 - \mathbf{S}_{01}n_1}{\Delta S} = \frac{\alpha \mathbf{Y}_0}{\Delta S} \tag{17.21}$$

(Figure 17.34). For the left side of equation 17.21, we can write $d(\mathbf{S}_0 n)/ds$, where $\mathbf{S}_0$ is the unit vector in the direction of the ray at point P (see Figure 17.34). In Figure 17.33, $\mathbf{Y}_0$ was perpendicular to the interface of n_1 and n_2. In Figure 17.34, $\mathbf{Y}_0$ is perpendicular to the surfaces $n = $ constant; therefore, the right side of equation 17.21 is proportional to grad n. Let us call $f(y)$ the function that might be necessary to write equation 17.21 as

$$\frac{d}{ds}(\mathbf{S}_0 n) = f(y) \text{ grad } n \tag{17.22}$$

To get a quantitative expression for $f(y)$, we multiply equation 17.22 by $n\mathbf{S}_0$ as a dot product:

$$(n\mathbf{S}_0)\frac{d}{ds}(\mathbf{S}_0 n) = f(y)n(\mathbf{S}_0 \cdot \text{grad } n) \tag{17.23}$$

The left side is easily obtained as

$$\frac{1}{2}\left(\frac{d}{ds}\right)(n\mathbf{S}_0)^2 = \frac{1}{2}\left(\frac{d}{ds}\right)n^2 = n\frac{dn}{ds} \tag{17.24}$$

For the right side, since the dot product, $(\mathbf{S}_0 \cdot \text{grad } n)$, equals dn/ds, we have $f(y)n(dn/ds)$, which is the same as the left side if $f(y) = 1$, and we have from equation 17.22

$$\frac{d}{ds}(\mathbf{S}_0 n) = \text{grad } n \tag{17.25}$$

While this equation has been derived for the special case defined here, it is true in general for the three-dimensional case.

For our two-dimensional case, the unit vector $\mathbf{S}_0$ can be presented with its components in the X and Y directions as

$$\mathbf{S}_0 = \frac{dX}{ds}\mathbf{X}_0 + \frac{dY}{ds}\mathbf{Y}_0 \tag{17.26}$$

Its length is

$$\sqrt{\left(\frac{dX}{ds}\right)^2 + \left(\frac{dY}{ds}\right)^2} = 1 \tag{17.27}$$

If we write the components of equation 17.25 in the X and Y directions and observe that $(S_0)_Y = dY/ds$ (see equation 17.26), we have

For Y: $\quad \dfrac{d}{ds}\left(n(Y)\dfrac{dY}{ds}\right) = \dfrac{dn(Y)}{dY} \qquad$ For X: $\quad \dfrac{d}{ds}\left(n(Y)\dfrac{dX}{ds}\right) = \dfrac{dn(Y)}{dX} \qquad$ (17.28)

where we have emphasized that n depends on Y only. We will apply these equations to our problem and assume that the fiber is very thin, as it is in reality. That is, we will apply the equation to the paraxial case, where θ is close to $90°$ (measured with respect to the normal). Since n is a function of Y only, we have in this approximation $dX \simeq ds$, and the equation for X in equation 17.28 is identically fulfilled. For the equation for Y in equation 17.28, we have

$$\frac{d}{dX}\left(n(Y)\frac{dY}{dX}\right) = \frac{dn(Y)}{dY} \quad \text{or} \quad \frac{d^2Y}{dX^2} = \frac{1}{n(Y)}\left(\frac{dn(Y)}{dY}\right) \qquad (17.29)$$

We introduce the parabolic profile

$$n(Y) = n_0(1 - 2\eta^2 Y^2)$$

where $2\eta^2$ determines the change in the refractive index with Y, and $2\eta^2 Y^2 \ll 1$. For η, we have the dimension $1/\text{cm}$.

We want to show that all rays starting from one point for various angles θ arrive at the same point a little bit further on at the axis. Introducing $n(Y)$ into equation 17.29 yields

$$\frac{d^2Y}{dX^2} - \frac{n_0(-2\eta^2 2Y)}{n_0(1 - 2\eta^2 Y^2)} = 0 \qquad (17.30)$$

We can neglect the term $2\eta^2 Y^2$ in the denominator and finally arrive at an equation of the harmonic oscillator type.

$$\frac{d^2Y}{dX^2} + 4\eta^2 Y = 0 \qquad (17.31)$$

A solution is

$$Y = A\sin 2\eta X \qquad (17.32)$$

The maximum for $Y(X)$ occurs at the distance $X = L/4$, where L is the length of the period. At the first maximum, $\sin 2\eta X = 1$, giving us

$$\frac{\pi}{2} = 2\eta\frac{L}{4} \quad \text{or} \quad \eta L = \pi \qquad (17.33)$$

The product of L and η is a constant, where η is the constant characterizing the change in the refractive index.

Introducing this result into equation 17.32 yields

$$Y = A\sin 2\pi\frac{X}{L}$$

For the assumed refractive index profile $n_0(1 - 2\eta^2 Y^2)$, the period is independent of η. All rays having different angles θ at the point $X = 0$ will meet at $X = L/2$

3. Light Transmission Through Fibers

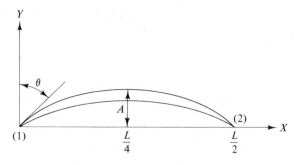

Fig. 17.35
Rays starting at point 1 at different angles all arrive at point 2.

and also at $X = L$. Consequently, we have the family of curves depicted in Figure 17.35. All rays coming from point 1 at different angles will arrive at point 2. Together, they correspond to a mode. We have used geometrical optical considerations. Electromagnetic calculations can also be used to describe the modes and mode propagation, similar to the modes in the plane parallel plate or in resonators (see also Chapter 15).

C. Fabrication of Fibers

How can we fabricate fibers that have such a special graded refractive index? We must find low-loss materials that can be manipulated in such a way that a continuous change in the refractive index can be obtained. Such a material is silica. If we dope the silica with germanium, the refractive index increases; if we dope the silica with boron, it decreases. The process is schematically illustrated in Figure 17.36. A tube of fused quartz is heated, and different amounts of $SiCl_4$ and $GeCl_4$ can flow through the tube, depending on the particular composition desired at that moment. A stream of O_2 carries the mixture through, and a torch heating the tube uniformly frees the Cl_2. The desired mixture of Si and Ge is deposited on the inside walls. After the refractive index of silica has been reached, the process is continued by mixing more and more boron into the silica, using the same pocedure. This is called a preform. At the end, the tube is heated and pulled in length. Extremely long fibers can be made having an outside diameter in the range of 100 μm. The refractive index as a function of the radius R of such a fiber is shown in Figure 17.37.

What is the final attenuation of such a fiber? Mode dispersion and frequency

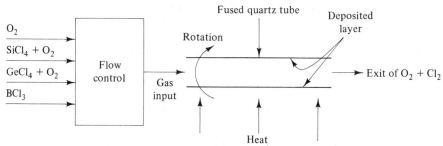

Fig. 17.36
The modified chemical-vapor deposition process for preparing the preforms from which fibers are drawn. In this process, the first layers deposited become the cladding and the last layers the core.

17. Integrated Optics

dispersion have already been minimized and low-loss materials have been selected. An absorption spectrum of a fiber in the spectral region around 1 μm is shown in Figure 17.38. We see that an absorption peak is present at about 0.95 μm, which originates from impurities (OH) and their absorption spectrum. The minimum of the absorption curve is obtained between 1.0 and 1.1 μm. Here we remember that the YAG : Nd laser light has a wavelength of 1.064 μm. The attenuation at that point is less than 2 dB/km. How small is this? Consider the fact that the numerical value of $10 \log_{10} S/N$ is called dB, where S is the signal and N the noise. If the ratio S/N is 10, we speak of having an attenuation of 10 dB equal to $10 \log_{10} 10$; for an S/N equal to 100, we get 20 dB. If we have an attenuation value of 2 dB/km, then after 5 km we have an attenuation of 10 dB. Fibers such as the one described here can be used for optical communication.

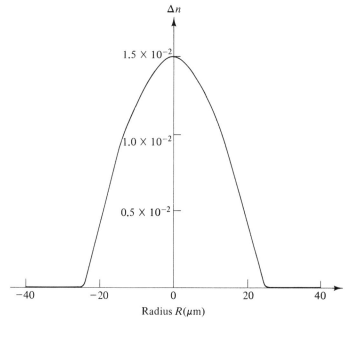

Fig. 17.37
The refractive index profile of a fiber. The nearly parabolic radial distribution minimizes mode dispersion.

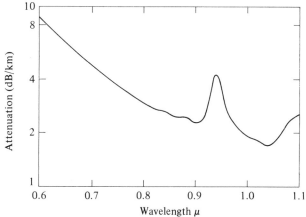

Fig. 17.38
An absorption spectrum typical of glass fibers drawn from preforms; the peak is due to the vibration of contaminants.

3. Light Transmission Through Fibers

We have seen how the parabolic profile of the index of refraction makes it possible for modes to propagate. Propagation of modes will maintain the shape of the pulse, and that is important for proper communication. We have assumed that the modes are single-frequency modes. In reality, more than one wavelength is involved. The pulses are composed of modes of different wavelengths propagated differently, even if the refractive index is the same for each mode. Moreover, if the refractive index is the same for each mode, additional pulse changes may occur. To compensate for dispersion in the fibers and keep the attenuation low, complicated index of refraction profiles are used (Figure 17.39).

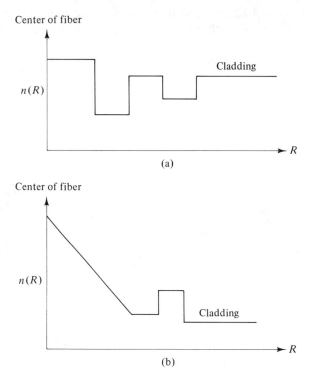

Fig. 17.39
Refraction index for fibers. (a) The refractive index has different segments between the center of the fiber and the cladding. (b) The refractive index has a linearly changing section and a segmented section between the center of the fiber and the cladding.

D. Lenses with Graded Refractive Index

For the parabolic profile of the index of refraction, we saw how all the rays starting at point $X = 0$ under different angles θ arrive at the point $X = L/2$ (see Figure 17.35). If we make a section of a rod having this profile of refractive index, we have a lens. For the parabolic profile, we found that $\eta L = \pi$, and by choosing η, we can make L a desirable length. Making the length of the lens equal to multiples of $L/2$, we have a focus at the end of the lens. If we choose $L/4$, $3/4L$, and so on, we have the light leaving the lens parallel. In Figure 17.40, we show these two cases and some other combinations.

17. Integrated Optics

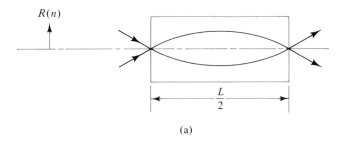

(a)

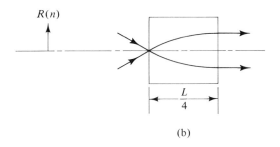

(b)

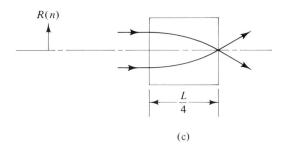

(c)

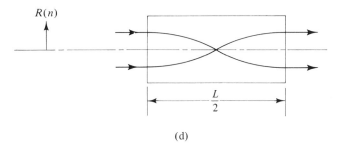

(d)

Fig. 17.40
Lenses made from a material having a graded refractive index. (a) Transfer of focal point to focal point. (b) Transfer of light from focal point to parallel light. (c) Reverse case of (b). (d) Transfer of parallel light to parallel light.

Problems

1. *Angle of k vector.* The angle between the k vector (direction of propagation of the wave) and the normal of the plates of a cavity was called θ. We may approximate

$$\tan \theta = \frac{\overline{m_X}/\ell}{\overline{m_Y}/d}$$

Calculate the angle θ for

$$\overline{m_X} = 50; \qquad \overline{m_Y} = 1$$
$$\overline{m_X} = 50; \qquad \overline{m_Y} = 2$$
$$\overline{m_X} = 25; \qquad \overline{m_Y} = 1$$
$$\overline{m_X} = 25; \qquad \overline{m_Y} = 2$$

Use $d = 1$, $\ell = 100$. Two of those modes have the same angle θ.
Ans. $26.6°$, $14°$, $14°$, $7.1°$

2. *Phase velocity.* For $\theta = 26°$, $14°$, and $7°$, calculate the phase velocity in a medium of refractive index $n_2 = 1.7$. Express v_X and v_Y as multiples of $c = 3 \times 10^8$ m/s in vacuum.

3. *Tapered section of a thin-film waveguide.* Consider a tapered section having an angle α with the horizontal.

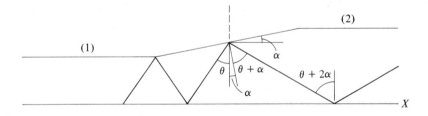

The angle θ is changing from θ to $\theta + 2\alpha$. How is the phase velocity v_X changing if $\alpha = 2°$ and $\theta = 14°$? Give the change as a multiple of $c = 3 \times 10^8$ m/s and observe that $v_1 - v_2 = \Delta v$ is negative, that is, the phase velocity decreases going from section 1 to section 2.
Ans. $\Delta v = -0.53c$

4. *Conical light pipe.* Consider a conically shaped tube with a large opening of radius r_1, small opening of radius r_2, and length x. The inner surfaces are perfectly reflecting. Consider only rays in a plane containing the axis. We want to investigate the relationship between the angle at which an extreme ray enters the cone, α_1, and the angle at which it emerges, α_2. To determine α_2, we do not have to trace the ray through the cone, but can "fold" the cone and consider a straight-line path as shown in the figure [see (b)]. If the ray is tangent to the dotted circle, $\alpha_2 = 90°$ and all rays with incident angles less than the corresponding α_1 will pass through the cone. If $\alpha_2 > 90°$, some rays will be reflected backward and not pass through.

17. Integrated Optics

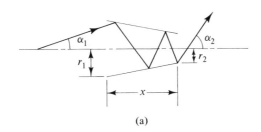

(a)

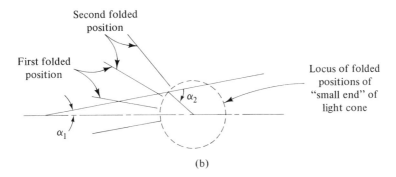

Second folded
position

First folded
position

Locus of folded
positions of
"small end" of
light cone

α_2

α_1

(b)

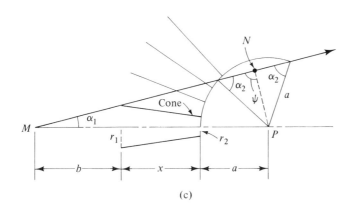

(c)

a. Derive the following expression relating x, r_1, r_2, α_1, and α_2 by referring to the figure above.

$$x = \left(1 - \frac{r_2}{r_1}\right)\frac{r_1 \cos \alpha_1}{(r_2/r_1)\sin \alpha_2 - \sin \alpha_1}$$

One may first deduce the following relations:

$$a = \frac{x}{(r_1/r_2) - 1}$$

$$\tan \alpha_1 = \frac{r_1}{b}$$

$$\frac{\sin \alpha_1}{a \sin \alpha_2} = \frac{1}{b + x + a}$$

Problems

b. Show that for the case where $x \rightarrow \infty$, we obtain

$$r_2 \sin \alpha_2 = r_1 \sin \alpha_1$$

This corresponds to Abbe's sine condition and tells us that for a long tube (e.g., optical fibers), the image quality is conserved by transmission.

5. *Grating coupler.* Consider the diffraction on an amplitude grating as discussed in problem 6 of Chapter 3. There we found $a(\sin \theta + \sin \phi) = m\lambda$ for the condition of constructive interference.

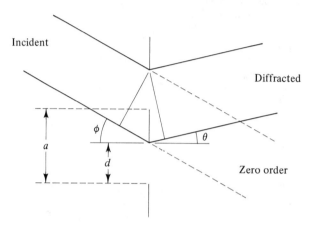

Now consider light from the left under angle $\phi = 0°$. The right side has the refractive index n_2.

a. Show that for the condition of constructive interference at normal incidence ($\phi = 0°$), we get

$$a(n_2 \sin \theta) = m\lambda$$

b. Assume now that the wave on the left is incident under the angle ϕ and that the refractive index on the left is n_1. Show that we now have, for the maximum condition,

$$a(n_2 \sin \theta + n_1 \sin \phi) = m\lambda$$

We can obtain the k values for the different orders of diffraction by considering $k = (2\pi/\lambda)n$ and calculate λ from the above conditions.

c. Calculate k for $m = 0, 1, 2$.

6. *Decibel problem.* The attenuation is given in decibels, and we have $1\,dB = 10 \log_{10} S/N$.

a. Give the ratio S/N for $20\,dB$.

b. In optics, we present the attenuation in experiments as $I = I_0 e^{-\alpha d}$, where α is in $1/cm$ when the path length is given in centimeters. How many decibels correspond to the attenuation if $\alpha = 0.1\,cm^{-1}$ and $d = 10$ cm?

Ans. (a) 100 (b) 4.34

7. *Ray equation.* Write out the formulas for the derivation of equations 17.22 and 17.23.

17. Integrated Optics

8. *Index of refraction at Y_{max} for a fiber.* Consider the index of refraction $n = n_0(1 - 2\eta^2 Y^2)$.

a. Show that the value for Y_{max} depending on θ (angle with the normal) may be expressed as

$$Y_{max} = \frac{(\cot \theta) L}{2\pi}$$

b. Show that the variation of the refraction index depending on Y_{max} is

$$n_{opt} = n_0(1 - \tfrac{1}{2}\cot^2 \theta)$$

c. Calculate the refractive index ratio n_{opt}/n_0 if we launch the ray with angles $2°$, $4°$, and $6°$ to the horizontal.

Acoustooptics and Nonlinear Effects | 18

1. INTRODUCTION

We want to discuss how local changes of material density or electrical polarization can be studied by interaction with light. The generation of acoustic waves in the material and the scattering of light on running acoustic waves or acoustically generated interference patterns leads to applications of laser light modulation, radio frequency detection, and holography. Extremely large changes in electronic polarizability, as can be obtained by laser light through materials, leads to nonlinear dielectric constants. Similarly, as discussed for Raman scattering, frequency changes of the incident light are observed that can lead to second harmonic generation.

2. SCHLIEREN METHOD

If the density in a material is changing from one point to another, the refractive index also changes. We will discuss two related methods of observation.

A. Geometrical Optics

A burning candle heats the air above it. The density of the air (and burned wax) above the candle is different for different locations, depending on local temperatures. Different densities result in different refractive indices, as we know from our discussion on the graded refractive index in Chapter 17. Figure 18.1 shows how a laser beam is bent passing through a solution with a decreasing concentration of material and consequently with a decreasing refractive index.

When light passes above a burning candle, we see on the wall bright and not-so-bright areas, moving in accordance with the movement of the flame. The brighter spots are formed by light bent into the area; at the not-too-bright spots, light is lost. This phenomenon can be used for the analysis of the density of flow

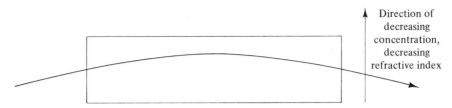

Fig. 18.1
Traversal of a laser beam through a medium with a gradient of the refractive index.

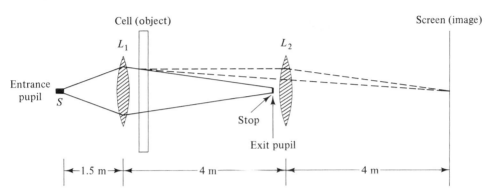

Fig. 18.2
Observation of light bent by inhomogeneous density areas. The bent light "misses" the stop at the position of the image (exit pupil) of the light-emitting object (entrance pupil). It is brought to the observation screen by lens L_2. (Dimensions for demonstration by Pohl: L_1: $f = 1$ m, diam. 12 cm; L_2: $f = 4$ m.)

of gases. It is called the **Schlieren method**. The analysis can be made much more sensitive if we use only the light bent by the areas with inhomogeneous refractive indices and if we eliminate the light of the direct illumination. Figure 18.2 depicts an experimental setup. The light from the source is passing through a condenser lens L_1, illuminating the cell that contains the material with locations of inhomogeneous refractive indices. The direct light is eliminated by the stop, and the bent light is brought into the focal plane S by lens L_2. For the geometrical optical arrangement, see the discussions on the projector, entrance pupil, exit pupil, and field of view in Chapter 12 (Figure 12.4).

B. Wave Optics and the Knife Edge Method

There is a "wave" aspect to this Schlieren method experiment that makes the observation of small refractive index changes more sensitive. We can consider the area of changing refractive indices as a phase grating. Looking at it, we use the stop to block out the zero-order light in the focal plane of the image-forming lens. We know from our discussion of the phase grating in Chapter 10 that the grating becomes visible like an amplitude grating. Often, the stop is used as a knife edge to block off the zero order and all other orders on one side. This allows observation of the change in certain directions.

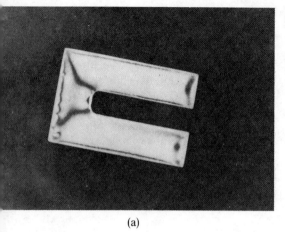
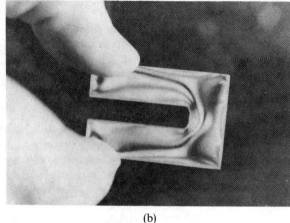

(a)   (b)

Fig. 18.3
(a) U-shaped piece of clear plastic between crossed polarizers. (b) The U-shaped piece is under stress from outside forces. (From Hecht and Zajac, *Optics*, p. 261.)

3. POLARIZED LIGHT AND CHANGES IN ELECTRICAL POLARIZABILITY

The polarizability of the electrons in a material may change from point to point. This can sometimes be accompanied by a change in density.

A. Stretched Mylar and Stress in Plastic

In Chapter 5, we saw how different refractive indices can be produced if we stretch a sheet of Mylar. The refractive index along the direction of the stress is different from the refractive index perpendicular to it. With polarized light, we can demonstrate this difference as in a uniaxial crystal.

Figure 18.3 again shows a U-shaped plastic piece photographed between crossed polarizers. The internal stress of the plastic, which is sometimes produced in the hardening process or if stress is applied from the outside (e.g., by bending), produces density changes in the material. The density changes in turn create changes in the refractive index. Such changes are also related to different orientations of the plastic's molecular chains. Therefore, differences occur in the refractive indices parallel and perpendicular to the local stress field. They show up in polarized light between "crossed polarizers." (This has been discussed in detail in Chapter 5.)

B. Uniaxial Crystals Produced by Electric and Magnetic Fields

We have also seen how different refractive indices can be produced by application of an electric or magnetic field (see the Kerr and Pockels effects in Chapter 5 and the Faraday effect in Chapter 7). From the point of view of the effect on the incident light, all these phenomena are described by the model of the uniaxial crystal and the use of polarization of light.

4. SCATTERING OF LIGHT ON ACOUSTIC WAVES

A. Brillouin Scattering

In Chapter 16, we saw how light is scattered by oscillations of a crystal. The oscillations were described by the model of the simple linear chain of atoms, and the excitation of the oscillations were of a thermal nature. In the discussion we used the eigenfrequencies of the linear chain.

A laser beam of angular frequency ω_2 traversing a quartz crystal can produce a traveling acoustic wave of angular frequency ω_S in the crystal. The wave is longitudinal and periodically changes the density of the material; therefore, it is called an acoustic wave. Scattering of the laser frequency ω_2 on the acoustic wave results in a scattered beam of frequency $\omega_2 - \omega_S$ (Figure 18.4a).

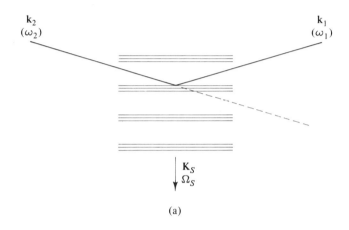

(a)

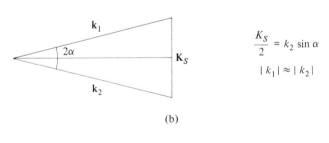

(b)

$$\frac{K_S}{2} = k_2 \sin \alpha$$

$$|k_1| \approx |k_2|$$

Fig. 18.4
Scattering of a laser beam on an acoustic wave. (a) Scattering schematic. (b) Wave vector diagram (c) Bragg scattering analog. [The notation is according to A. Yariv, *Quantum Electronics* (New York: Wiley, 1967), p. 432.]

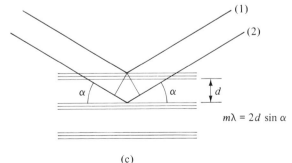

$$m\lambda = 2d \sin \alpha$$

(c)

Let us assume that the acoustic wave is generated separately from the incident light beam. The scattering process is schematically shown in Figure 18.4a, and the wave vector diagram is shown in Figure 18.4b. The scattering of light on a set of planes resembles Bragg scattering, as shown in Figure 18.4c. However, the acoustic wave moves, and the planes of denser and less dense material are not stationary.

B. Bragg Cell Approach and Doppler Effect

First we assume that the planes of the materials of different densities are stationary, as in a standing-wave pattern, and obtain the "Bragg condition" for the "diffracted light" in direction α:

$$2\lambda_S \sin \alpha = m\lambda_2 \qquad m = 1, 2, 3 \tag{18.1}$$

where λ_S is equal to the spacing of the "planes." Equation 18.1 is also valid if the planes move. Now we assume that the wave moves; that is, we have scattering of the light wave by the acoustic wave. If the light wave in the crystal with refractive index n moves with velocity c/n, and if the acoustic wave has the velocity v_S, we have, from Doppler's formula

$$\omega_2 - \omega_1 = \frac{2\omega_2 v_S n \sin \alpha}{c} \tag{18.2}$$

If we introduce equation 18.1 with $m = 1$ into equation 18.2 with

$$\omega_2 = 2\pi \frac{c/n}{\lambda_2}$$

and

$$\lambda_S \frac{\Omega_S}{2\pi} = v_S$$

we have

$$\omega_2 - \omega_1 = \frac{2 \cdot 2\pi(c/n\lambda_2)\lambda_S(\Omega_S/2\pi)n(\lambda_2/2\lambda_S)}{c} = \Omega_S \tag{18.3}$$

The difference in angular frequency between the incident wave and the scattered wave is equal to the angular frequency of the acoustic wave.

C. Energy and Wave Vector Conservation

In Chapter 16, we considered Brillouin scattering, which is similar to light scattering on acoustic waves. For our description, we used the laws of conservation of energy and momentum. In our case, we can express the two conservation laws as

$$\hbar\omega_2 = \hbar\omega_1 + \hbar\Omega_S \tag{18.4}$$

$$\hbar\mathbf{k}_2 = \hbar\mathbf{k}_1 + \hbar\mathbf{K}_S \tag{18.5}$$

(see Figure 18.4a and b), where ω, $\mathbf{k}$ refers to the optical waves and Ω_S, $\mathbf{K}_S$ to the "acoustic" wave. Equation 18.4 states the same result obtained in equation 18.3.

18. Acoustooptics and Nonlinear Effects

From Figure 18.4b, we have

$$K_S = 2k_2 \sin \alpha \qquad (18.6)$$

or

$$\lambda_2 = 2\lambda_S \sin \alpha \qquad (18.7)$$

which is the same result as stated in equation 18.1 We see that the two descriptions in Sections B and C are equivalent.

D. Acoustooptical Modulator

In Chapter 15, we mentioned how modulation or Q-switching of the laser beam converts continuous-wave laser output into pulsed output. The pulses can be very strong during a very short time.

We also discussed electrooptical modulators, or shutters. Uniaxial crystals are obtained from a homogeneous material by application of an electric or magnetic field. The shutter is placed between crossed polarizers. If no field is applied, no light can pass. Application of a field changes the crystal to a uniaxial crystal. The plane of the incident polarized light is turned so that it can pass the analyzer.

The acoustooptical modulator produces acoustic waves. Suppose we want laser light to pass the modulator when no acoustic waves are produced. To interrupt the light, acoustic waves are produced, and according to Figure 18.4, diffraction occurs. During the time these waves are produced, the incident laser light is scattered into a certain direction. It is lost for the direct pass of the laser light between the mirrors of the cavity. About ninety percent of the incident light can be deflected. To obtain pulses, we must switch the acoustic waves on and off. Devices operating in the 100 KHz to 50 MHz range are available from industrial manufacturers.

We can reverse the order of the "working condition" to the "interruption condition." It is possible to have the deflected light participate in the laser action cycle and have the "no acoustic wave" condition for the loss cycle. We know that there is a slight frequency change in the scattered light, which must be tolerated in the operation of the laser.

E. Acoustooptical Deflector

It is often desirable that a laser beam be deflected and for a short time scan over a certain angle. An example is bar-code reading at the supermarket checkout to determine the price of a product.

An acoustooptical deflector can be constructed by changing the frequency of the radio-frequency transducer that drives (or creates) the running acoustic waves in the cell. Changing the frequency of the driver changes the wavelength of the acoustic wave. In the Bragg cell (see Figure 18.4), the laser beam of wavelength λ is diffracted into the angle α with an acoustic wave of wavelength λ_S. This is described for the first order in the symmetric case of Figure 18.4c as

$$\lambda = 2\lambda_S \sin \alpha \qquad (18.8)$$

where α is the same angle the incident and diffracted light makes with the plane

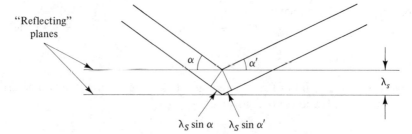

Fig. 18.5
Bragg condition for the "asymmetric case" where $\alpha \neq \alpha'$.

of the acoustic wave. For a change of λ_S, we must again account for the path length differences between the first and second beams, as we did in Figure 18.4. This is shown in Figure 18.5. The Bragg (constructive interference) condition that is obtained is

$$\lambda = \lambda_S(\sin \alpha + \sin \alpha') \tag{18.9}$$

The dependence of the angle α' of the diffracted light beam on the frequency of the acoustic wave ($v_S = v_S/\lambda_S$) for laser light of frequency $v = c/\lambda$ is

$$\sin \alpha' = -\sin \alpha + \left(\frac{c}{v v_S}\right) v_S \tag{18.10}$$

A variation of v_S results in a range of α'.

F. Acoustooptical Receiver

The scattering of light by an acoustic wave in a crystal is also used for transforming radio-frequency signals into "light" signals. The schematic of such a device is shown in Figure 18.6.

The radio-frequency signal makes a transducer vibrate with the same frequency. The transducer is of piezoelectric material, such as zinc oxide or lithium niobate. The vibrations of the transducer generate an acoustic wave of the same frequency in the Bragg cell. The frequency of the acoustic wave can be as high as 1–2 GHz.

The light source may be a semiconductor laser, and a detector is placed where the scattered light is expected to arrive. Up to this point, we have assumed that the acoustic wave has one wavlength (frequency). If we have two frequencies contained in the incident radio-frequency signal, we have two "sets" of planes of the acoustic waves and will have two different light beams "diffracted" at two different angles and detected at two separate spots by two different detectors. As a result, the two superimposed input radio-frequency signals are now separated. The acoustooptical receiver can "disentangle" the frequency components of the input signal. This is similar to Fourier transform spectroscopy (see Chapter 8, where we used a Fourier transform to find the spectrum from the interferogram). In applications, lens L_2 in Figure 18.6 is called the Fourier transform lens.

18. Acoustooptics and Nonlinear Effects

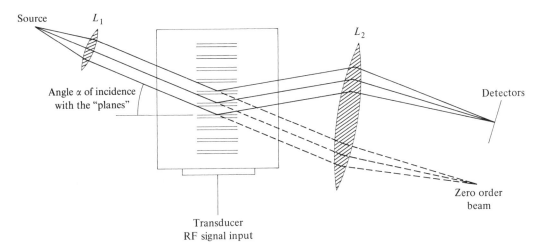

Fig. 18.6
Schematic of an acoustooptical receiver. L_1 is the collimating lens, and L_2 is the Fourier transform lens.

5. ACOUSTIC HOLOGRAPHY

In Chapter 10, we discussed holography and how laser light can be used for the reconstruction of an image. We can make the hologram with an argon laser (green light) and do the reconstruction with a He-Ne laser (red light). How much difference in wavelength can be used for the two processes? We have seen that we can make a hologram with X rays of $\lambda = 60$ Å and reconstruct it with He-Ne laser light of 6328 Å—that is, a factor of 100 for the two wavelengths. This experiment is shown again in Figure 18.7a.

An acoustic hologram is produced on the water surface by two sources under water (see Figure 18.7b). The object is placed in one sound beam, and the other beam serves as a reference beam. A He-Ne laser is again used for image reconstruction. The wavelength of the laser light is about 100,000 times smaller than the wavelength of the sound wave. Whether or not the image can be reconstructed with light of such a short wavelength depends on the highest order under which diffraction can be observed. For example, we consider in Figure 18.8 diffraction on Michelson's echelon grating, where large path differences of the order of 20,000 wavelengths (in the glass) were used to obtain high-resolution spectra. Holograms are made by interference, and not by mechanical procedures as in the production of echelon gratings. They are accurate enough so that a factor of 100,000 for the path difference of interfering beams is possible and reasonable for reconstruction of the image.

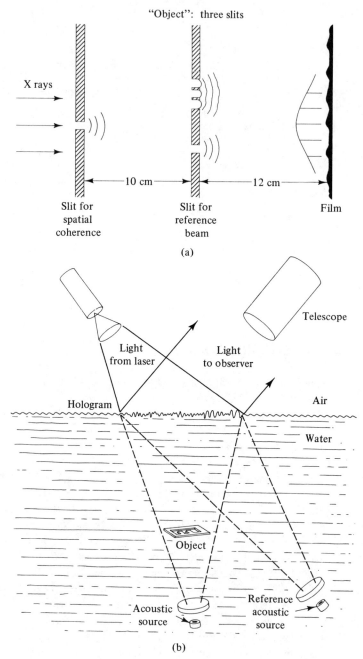

"Object": three slits

X rays

|←———— 10 cm ————→|←———— 12 cm ————→|

Slit for
spatial
coherence

Slit for
reference
beam

Film

(a)

Telescope

Light
from laser

Light
to observer

Hologram

Air

Water

Object

Acoustic
source

Reference
acoustic
source

(b)

Fig. 18.7
Holography using X rays and acoustic waves. (a) Hologram produced by X rays. The
first slit is for obtaining coherent illumination at the object consisting of three 3 μm
wide slits and the reference beam consisting of a 2.5 μm wide opening. The hologram is
produced on the film. The object is reconstructed by illuminating the film with laser
light. (b) Two acoustic sources produce waves of equal wavelength and produce a
hologram on the water surface. One beam contains the object. Laser light reflected
from the water surface reconstructs the object seen through the telescope.

18. Acoustooptics and Nonlinear Effects

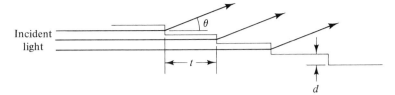

Fig. 18.8
Michelson's echelon grating. The dielectric plates have thickness $t = 18$ mm, steps separation of $d = 1$ mm, and refractive index $n = 1.5$. An interference pattern is observed at angle θ. The path difference of interfering beams is of the order of 20,000.

6. RAMAN SCATTERING

In Chapter 16, we discussed Brillouin and Raman scattering. They were discussed separately from other scattering effects because the scattered light changes its frequency. We presented Raman scattering with a simple calculation showing that the scattered light contains higher harmonics of the incident light. The observation of Raman scattering with nonlaser sources takes a great deal of time, since the ratio of Raman-scattered light to incident light is very small. With the application of laser light, this situation is drastically changed. The strong electric field of about 10^7 V/cm produced by the laser in the material also produced new effects.

A. Stimulated Raman Effect

In the normal Raman effect, we saw that the molecule absorbs the incident light of frequency f_0 and reemits light of higher and lower frequencies. The difference in frequency is equal to the frequencies of molecular vibrations. The energy of a molecular mode is added or subtracted from the energy of the incident light. Spectral lines with frequencies smaller than f_0 are called Stokes lines; lines with frequencies larger than f_0 are called anti-Stokes lines.

Consider now the molecules in a cavity, similar to the atoms of the He-Ne laser. Laser light is absorbed by the molecules, and the Raman-scattered light can now travel back and forth in the cavity. If enough light is pumped into the Raman frequency emission, laser action occurs.

It is possible, however, that amplification of the Raman frequency occurs without the cavity arrangement. The laser light excites a large number of molecules in a certain volume. The emitted wave can pick up so much stimulated light by traveling through this volume that it forms a laserlike beam after traversal.

7. SELF-FOCUSING

The strong electric field established by the incident laser light is further enlarged by self-focusing. For strong fields, the dielectric constant is expressed as

$$\varepsilon_{\text{tot}} = \varepsilon_1 + \varepsilon_2 \overline{E^2} \tag{18.11}$$

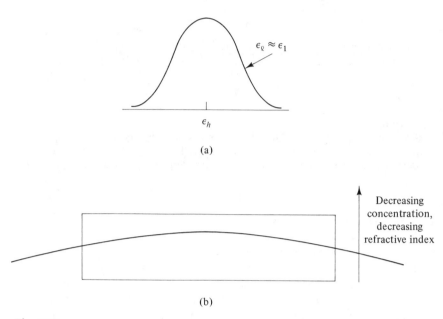

Fig. 18.9
(a) Gaussian-type cross section of a laser beam. The high field (h) is in the middle, the low field (ℓ) on the margins. (b) As a model case, we show a laser beam traversing a medium with a graded refractive index. The index is decreasing to the top.

where $\overline{E^2}$ is the average of the squared time-dependent electric field. As a simple model, consider a laser beam with a Gaussian distribution of its cross section traversing such a medium. The effective dielectric constant is larger in the middle of the beam than on the outer parts (Figure 18.9). We know from the experiment with the graded refractive index that the light beam is bent toward the area of the larger refractive index. We can also argue that the velocity of the rays on the low-field side has a value of $v_\ell = c/n_\ell$, which is larger than the velocity on the high-field side ($v_h = c/n_h$), because $n_h > n_\ell$. The outer rays run faster, and an initial plane wave gets bent toward the middle.

8. SECOND HARMONIC GENERATION

A. Theoretical Aspects

In Chapter 7, we studied the electrical polarization of electrons in a solid. The incident wave displaces the electrons of the atoms or ions and induces dipole moments. The dipoles reradiate the light. Depending on the resonance and damping properties of the polarizability, we characterized the medium by n and K. While the frequency of the incident light is the same inside and outside, the wavelength is decreased to λ/n on the inside. The velocity is $v = c/n$, and the damping is described by K. The energy corresponding to electrical polarization by the electric field is assumed to be small compared to the binding energy of the electrons.

18. Acoustooptics and Nonlinear Effects

Intense laser light can have an electric field strength of the order of 10^7 V/m. The corresponding energy is similar to the binding energy of the electrons.

Let us consider a strong oscillating electric field introduced by laser light in a crystal that has no center of symmetry. The positions of the atoms in such a crystal cannot be brought into equivalent positions by the symmetry operation called "inversion." As an example of this symmetry operation, recall that in Chapter 6 we saw that left-hand and right-hand helical molecules cannot be brought into equivalent "atomic positions" by rotation, but only by a mirror imaging. If we reflect all the points in space at the origin, we have inversion [point$(x, y, z) \to$ point$(-x, -y, -z)$]. Examples of crystals having no center of symmetry are piezoelectrical crystals. We discussed "artificial" uniaxial crystals produced by an electric field in Chapter 5.

B. Nonlinear Polarizability

We consider now a strong electrical field in a crystal. The electric field of the laser now polarizes the electrons in the arrangement of the atoms in an asymmetric way. The polarization is described by

$$P = \alpha_1 E + \alpha_2 E^2 \tag{18.12}$$

where α_1 and α_2 are material constants describing the electric susceptibility.

The first term is the only term we considered in Chapter 7. The additional appearance of the term in E^2 is analogous to what was discussed for the Raman effect and self-focusing. It is called the nonlinear term. The electric field of the incident laser light is described by $E = E_0 \sin \omega t$. Introduction into equation 18.12 gives us for the polarization

$$P = \alpha_1 E_0 (\sin \omega t) + \alpha_2 E_0^2 (\sin \omega t)^2 \tag{18.13}$$

We can write for

$$\sin^2 \omega t = \tfrac{1}{2}(1 - \cos 2\omega t)$$

and have

$$P = \alpha_1 E_0 \sin \omega t + \frac{1}{2}\alpha_2 E_0^2 - \frac{\alpha_2 E_0^2}{2} \cos 2\omega t \tag{18.14}$$

Once again, as discussed for the Raman effect, the frequency components of the polarization are related to the frequency components of the reemitted wave. Besides the incident light of frequency f_o, one observes a second harmonic of frequency $2f_o$. A photograph of such an experiment is shown in the color table appearing at the end of the book.

C. Conditions for Observations

We see that by using simple theoretical arguments, such a second harmonic wave should be produced. For the actual observation, we must choose suitable crystals and the direction of observation with respect to the crystal axis. The reason is that the wave with frequency $2f_o$ has a different velocity in the crystal than the original wave of frequency f_o. This is a result of the frequency dependence of the refractive index.

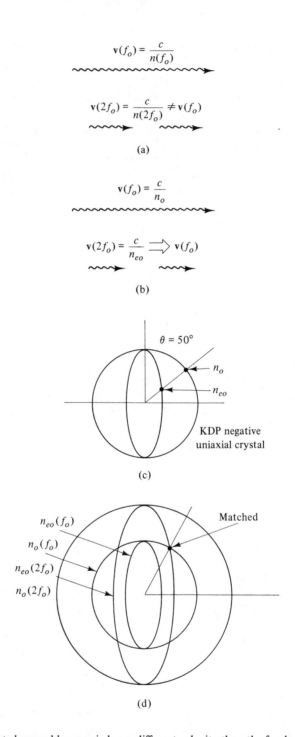

Fig. 18.10
(a) The generated second harmonic has a different velocity than the fundamental.
(b) Matching condition. The index n_o has the same value for f_o as n_{eo} has for $2f_o$ in the direction of $\theta = 50°$. (c) Surfaces of the refractive indices for f_o for KDP. (d) Super-position of the surfaces of refraction indices for f_o and $2f_o$ for KDP. The matching indices at $\theta = 50°$ for $n_o(f_o)$ and $n_{eo}(2f_o)$ are indicated by a dot.

18. Acoustooptics and Nonlinear Effects

The fundamental wave produces secondary harmonics in traversing the crystal. The secondary waves produced at different times are not in phase with one another, since the secondary waves travel with a different velocity. Therefore, the secondary waves are not added "in phase" and undergo attenuation by random averaging. In a uniaxial crystal, if the fundamental wave travels as an ordinary ray and the second harmonics travel as an extraordinary ray, the problem can be solved. We can choose a direction where the ordinary ray (with f_o) and the extraordinary ray (with $2f_o$) have just the desired refractive indices so that they all travel with the same velocity (Figure 18.10). To use this scheme, it is important that the original wave in the crystal produce the second harmonic in the perpendicularly polarized direction. Whether this can be done or not depends on the type of crystal used.

In the mathematical description of the polarizability of the crystal, a polarization tensor appears. If certain coefficients of that tensor are not zero, "coupling" from the direction of the ordinary ray to the direction of the extra-ordinary ray is possible. For KDP, which is a uniaxial crystal, the matching condition is fulfilled for $\theta = 50°$ (see Figure 18.10).

Second harmonic generation of 532.3 nm can be accomplished with barium sodium niobate crystals from an Nd : YAG laser of 1.06 μm. (See the color photograph at the end of this book.) Matching can also be accomplished through temperature change, and a variety of crystals can be used.

Problems

1. *Continuously changing refractive index.* A light ray enters a medium of refractive index n_Δ from a medium of refractive index n.

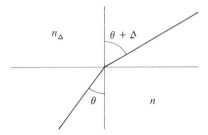

We want to calculate the refractive index n_Δ if θ is between 50° and 80° and the change in θ is $\Delta = 1°$. Calculate n_Δ/n for $\theta = 50°, 60°, 70°, 80°$.
(*Note:* You will find that n_Δ changes less for larger angles but more for smaller angles for 1° derivation. A continuously decreasing refractive index would bend the ray.)
Ans. $50°: n_\Delta = .9857n$ $80°: n_\Delta = .9971n$

2. *Scattering on an acoustic wave.*
   a. Check the calculation for $\omega_2 - \omega_1 = \Omega_s$ (equation 18.3).
   b. Check the orders of magnitude of ω and Ω_s for

$$\lambda = 0.5 \times 10^{-4} \text{ cm}, \quad c = 3 \times 10^{10} \text{ cm/s}$$

and

$$\lambda_s = 1.6 \times 10^{10} \text{ cm}, \quad v_s = 5 \times 10^5 \text{ cm/s}$$

3. *Bragg condition.*
   a. Go through the derivation of equation 18.10.
   b. Calculate α' from $\alpha = 25°$, $v_S = 10^{10}$ 1/s, $v_S = 5 \times 10^5$ m/s, $v = 3 \times 10^{15}$ 1/s, and $c = 3 \times 10^{10}$ m/s.
   c. As a check, calculate from $\lambda/\lambda_S = 0.2$ the value of λ_S for λ corresponding to $v = 3 \times 10^{15}$ 1/s, and (as a check) then calculate v_S from λ_S.

4. *Michelson echelon grating.* Consider the echelon grating used in the "forward" direction (i.e., $\theta = 0°$).

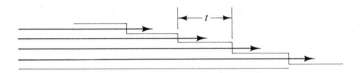

   The glass has refractive index $n = 1.5$, and $t = 18$ mm. The light used has wavelength $\lambda = 0.5 \times 10^{-4}$ cm. Show that we have constructive interference for an order number of 18,000.

5. *Raman effect and Stokes and anti-Stokes lines.* The Boltzmann distribution gives us the fraction of molecules in the state E at temperature T as

$$N = N_0 e^{-(E/kT)}$$

   In the Raman effect, the molecules absorb the incident photon of energy $h\nu$. In the emission process, the energy of a molecular vibration $h\nu_v$ is either added or subtracted. If the emitted energy of the photon is lower, we have $E_S = h(v - v_v)$; if it is higher, we have $E_{AS} = h(v + v_v)$, where S stands for Stokes and AS for anti-Stokes emission. According to the Boltzmann distribution, we have

$$N_S > N_{AS}$$

   Calculate N_S/N_{AS} using $\overline{v}_v = 500$ cm^{-1}, $\overline{v} = 22938$ cm^{-1}, $k = 1.4 \times 10^{-23}$ J/K, $T = 300$ K, and $h = 6.6 \times 10^{-34}$ J$\cdot$s, and show that for this case we have

$$\frac{N_S}{N_{AS}} \simeq 100$$

   Note:

$$1 \text{ cm}^{-1} = \frac{v \text{ 1/s}}{c \text{ cm/s}}$$

   where $c = 2.9979 \times 10^{10}$ cm/s.

6. *Power for self-focusing.* Yariv gives the power level for self-focusing as

$$P_C = \frac{\varepsilon_0 n^3 c \lambda^2}{4\pi n_2'}$$

   where the total dielectric constant is given as

$$\varepsilon_{\text{total}} = \varepsilon + \varepsilon_2 E^2$$

   E being the electric field. The following values are given:

and

$$\lambda_s = 1.6 \times 10^{10} \text{ cm}, \quad v_s = 5 \times 10^5 \text{ cm/s}$$

3. *Bragg condition.*
   a. Go through the derivation of equation 18.10.
   b. Calculate α' from $\alpha = 25°$, $v_S = 10^{10}$ 1/s, $v_S = 5 \times 10^5$ m/s, $v = 3 \times 10^{15}$ 1/s, and $c = 3 \times 10^{10}$ m/s.
   c. As a check, calculate from $\lambda/\lambda_S = 0.2$ the value of λ_S for λ corresponding to $v = 3 \times 10^{15}$ 1/s, and (as a check) then calculate v_S from λ_S.

4. *Michelson echelon grating.* Consider the echelon grating used in the "forward" direction (i.e., $\theta = 0°$).

   The glass has refractive index $n = 1.5$, and $t = 18$ mm. The light used has wavelength $\lambda = 0.5 \times 10^{-4}$ cm. Show that we have constructive interference for an order number of 18,000.

5. *Raman effect and Stokes and anti-Stokes lines.* The Boltzmann distribution gives us the fraction of molecules in the state E at temperature T as

$$N = N_0 e^{-(E/kT)}$$

   In the Raman effect, the molecules absorb the incident photon of energy hv. In the emission process, the energy of a molecular vibration hv_v is either added or subtracted. If the emitted energy of the photon is lower, we have $E_S = h(v - v_v)$; if it is higher, we have $E_{AS} = h(v + v_v)$, where S stands for Stokes and AS for anti-Stokes emission. According to the Boltzmann distribution, we have

$$N_S > N_{AS}$$

   Calculate N_S/N_{AS} using $\overline{v_v} = 500$ cm^{-1}, $\overline{v} = 22938$ cm^{-1}, $k = 1.4 \times 10^{-23}$ J/K, $T = 300$ K, and $h = 6.6 \times 10^{-34}$ J·s, and show that for this case we have

$$\frac{N_S}{N_{AS}} \simeq 100$$

   Note:

$$1 \text{ cm}^{-1} = \frac{v \text{ 1/s}}{c \text{ cm/s}}$$

   where $c = 2.9979 \times 10^{10}$ cm/s.

6. *Power for self-focusing.* Yariv gives the power level for self-focusing as

$$P_C = \frac{\varepsilon_0 n^3 c \lambda^2}{4\pi n_2'}$$

   where the total dielectric constant is given as

$$\varepsilon_{\text{total}} = \varepsilon + \varepsilon_2 E^2$$

   E being the electric field. The following values are given:

The fundamental wave produces secondary harmonics in traversing the crystal. The secondary waves produced at different times are not in phase with one another, since the secondary waves travel with a different velocity. Therefore, the secondary waves are not added "in phase" and undergo attenuation by random averaging. In a uniaxial crystal, if the fundamental wave travels as an ordinary ray and the second harmonics travel as an extraordinary ray, the problem can be solved. We can choose a direction where the ordinary ray (with f_o) and the extraordinary ray (with $2f_o$) have just the desired refractive indices so that they all travel with the same velocity (Figure 18.10). To use this scheme, it is important that the original wave in the crystal produce the second harmonic in the perpendicularly polarized direction. Whether this can be done or not depends on the type of crystal used.

In the mathematical description of the polarizability of the crystal, a polarization tensor appears. If certain coefficients of that tensor are not zero, "coupling" from the direction of the ordinary ray to the direction of the extraordinary ray is possible. For KDP, which is a uniaxial crystal, the matching condition is fulfilled for $\theta = 50°$ (see Figure 18.10).

Second harmonic generation of 532.3 nm can be accomplished with barium sodium niobate crystals from an Nd : YAG laser of 1.06 μm. (See the color photograph at the end of this book.) Matching can also be accomplished through temperature change, and a variety of crystals can be used.

Problems

1. *Continuously changing refractive index.* A light ray enters a medium of refractive index n_Δ from a medium of refractive index n.

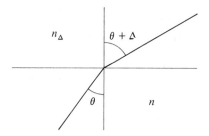

We want to calculate the refractive index n_Δ if θ is between 50° and 80° and the change in θ is $\Delta = 1°$. Calculate n_Δ/n for $\theta = 50°, 60°, 70°, 80°$.
(*Note:* You will find that n_Δ changes less for larger angles but more for smaller angles for 1° derivation. A continuously decreasing refractive index would bend the ray.)
Ans. 50°: $n_\Delta = .9857n$ 80°: $n_\Delta = .9971n$

2. *Scattering on an acoustic wave.*
   a. Check the calculation for $\omega_2 - \omega_1 = \Omega_s$ (equation 18.3).
   b. Check the orders of magnitude of ω and Ω_s for

$$\lambda = 0.5 \times 10^{-4} \text{ cm}, \quad c = 3 \times 10^{10} \text{ cm/s}$$

$$n = \sqrt{\frac{\varepsilon}{\varepsilon_0}}$$

$$(n_2')_{esu} = 10^{-11}, \quad \lambda = 10^{-6} \text{ m}$$

$$\varepsilon \simeq \varepsilon_0, \quad (n_2')_{esu} = 9 \times 10^8 (n_2')_{mks}$$

$$\varepsilon_0 = 8.25 \times 10^{-12} \frac{C^2}{N \cdot m^2}$$

Yariv arrives at $P_c \simeq 2 \times 10^4$ W. Show that the value is correct.

7. *Nonlinear polarizability.* Consider the polarizability of a hydrogen-like atom. We have
$$P_a = \alpha E$$
For hydrogen, we have $\alpha = 1.7 \times 10^{-41}$ mC²/N. We take $E = 5 \times 10^5$ N/C (mks units).
a. Calculate P_a.

We have expressed the polarizability in one dimension as
$$P = \alpha_1 E + \alpha_2 E^2$$

In three dimensions, P and E are vectors and α_1 and α_2 are tensors. In crystals of point group symmetry $\bar{4}2$ m, we have for the three components of the nonlinear polarizability
$$P_X = 2d_{14} E_Z E_Y$$
$$P_Y = 2d_{14} E_Z E_X$$
$$P_Z = 2d_{36} E_X E_Y$$

where E_X, E_Y, and E_Z are the components of the electric field and d_{14} and d_{36} are constants. In particular, for KDP we have $d_{14} = d_{36}$. For the conversion, we have
$$d_{esu} = 3.7 \times 10^{-15} d_{mks}$$

where ε_0 is included in the coefficient d.
b. Calculate in mks units
$$P_{ne} = 2d_{14} E^2 \quad \text{for} \quad E = 10^8 \text{ N/C}$$

Ans. a. 8.5×10^{-36} ml; b. 11.8×10^{-8} ml

8. *Review of the "pancake model" and second harmonic generation.*

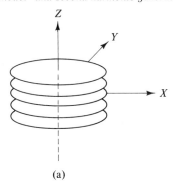

(a)

For waves vibrating in the X-Y plane, the ordinary refractive index applies; for waves vibrating parallel to the Z axis, the extraordinary index applies. For the propagation vectors, we have

k_X (polarization of wave ∥ to Y) $k_X \rightarrow$ ordinary $\rightarrow k_{Xo}$

k_X (polarization of wave ∥ to Z) $k_X \rightarrow$ extraordinary $\rightarrow k_{Xeo}$

k_Z (polarization of wave ∥ to Y) $k_Z \rightarrow$ ordinary $\rightarrow k_{Zo}$

k_Z (polarization of wave ∥ to X) $k_Z \rightarrow$ ordinary $\rightarrow k_{Zo}$

For a negative uniaxial crystal for the k-ellipsoid ($n_e < n_o, k_e < k_0$) we have the following k surfaces:

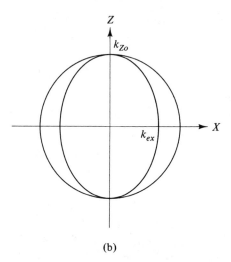

(b)

We may draw the index ellipsoid for $n_{ex}(2f_o)$ and $n_o(f_o)$:

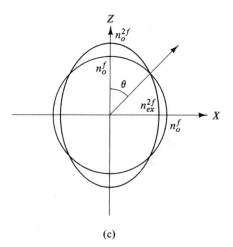

(c)

We have superimposed the index ellipsoid for the second harmonic, which has values $n_0(2f_o)$ in the Z direction and $n_{ex}(2f_o)$ in the X direction. Index matching occurs at

18. Acoustooptics and Nonlinear Effects

angles θ_m, which may be calculated from

$$\frac{\cos^2 \theta_m}{[n_o(2f_o)]^2} + \frac{\sin^2 \theta_m}{[n_{ex}(2f_o)]^2} = \frac{1}{[n_o(f_o)]^2}$$

a. Show that we have

$$\sin^2 \theta_m = \frac{1/[n_o(f_o)]^2 - 1/[n_o(2f_o)]^2}{1/[n_{ex}(2f_o)]^2 - 1/[n_o(2f_o)]^2}$$

b. For KDP, we have for $\lambda = 6940\text{Å}$

$$n_{ex}(f_o) = 1.466$$
$$n_o(f_o) = 1.506$$

and for the second harmonic

$$n_{ex}(2f_o) = 1.487$$
$$n_o(2f_o) = 1.534$$

Show that $\theta_m = 50.4°$.

Constants

Speed of light	c	2.99793×10^8 m/s
Electron charge	e	1.60292×10^{-10} C
Electron mass	m	9.10956×10^{-31} kg
Proton mass	M_p	1.67261×10^{-27} kg
Planck's constant	h	6.62620×10^{-34} Js
	$\dfrac{h}{2\pi}$	1.05459×10^{-34} Js
	$\dfrac{e}{m}$	1.75880×10^{11} C/kg
Bohr radius	r_0	5.29177×10^{-11} m
Compton wavelength	$\dfrac{h}{m_0 c}$	2.42631×10^{-12} m
Stefan-Boltzmann constant	k	1.38066×10^{-23} J/K
Avogadro's number	N	6.02217×10^{23} particles/mol

$$100 \ \mu\text{m} \Rightarrow 3000 \text{ GHz}$$
$$100 \ \mu\text{m} \Rightarrow 100 \text{ cm}^{-1}$$
$$10 \ \mu\text{m} \Rightarrow 1000 \text{ cm}^{-1}$$
$$1 \text{ meV} = 10^{-3} \text{ eV} = 1.6 \times 10^{-16} \text{ joule} \Rightarrow 8.07 \text{ cm}^{-1}$$
$$100 \text{ nm} = 1000 \text{ Å}$$
$$10000 \text{ Å} = 1 \ \mu$$
$$1 \text{ Å} = 10^{-8} \text{ cm} = 10^{-10} \text{ m}$$

Formulas

$$\sqrt{-1} = i \qquad i^2 = -1$$

$$z = a + ib = r(\cos\phi + i\sin\phi) = re^{i\phi}$$

$$\cos x = \frac{e^{ix} + e^{-ix}}{2} \qquad \sin x = \frac{e^{ix} - e^{-ix}}{2i}$$

$$e^x = 1 + x + \frac{x^2}{2!} + \frac{x^3}{3!} + \cdots, \quad \sin x = x - \frac{x^3}{3!} + \cdots, \quad \cos x = 1 - \frac{x^2}{2!} + \cdots$$

$$s_n = a + aq + aq^2 + \cdots + aq^{n-1} = a\frac{q^n - 1}{q - 1}$$

$$if\ |q| < 1, \quad n \to \infty \qquad s_\infty = \frac{a}{1 - q}$$

$$x^2 + ax + b = 0 \qquad x_{1,2} = -\frac{a}{2} \pm \sqrt{\left(\frac{a}{2}\right)^2 - b}$$

$$(1 \pm x)^n \cong 1 \pm nx \qquad |x| \ll 1$$

$$\begin{pmatrix} a_1 & b_1 \\ a_2 & b_2 \end{pmatrix} \times \begin{pmatrix} c_1 & d_1 \\ c_2 & d_2 \end{pmatrix} = \begin{pmatrix} a_1c_1 + b_1c_2 & a_1d_1 + b_1d_2 \\ a_2c_1 + b_2c_2 & a_2d_1 + b_2d_2 \end{pmatrix}$$

$$\begin{vmatrix} a_1b_1c_1 \\ a_2b_2c_2 \\ a_3b_3c_3 \end{vmatrix} = a_1\begin{vmatrix} b_2c_2 \\ b_3c_3 \end{vmatrix} - a_2\begin{vmatrix} b_1c_1 \\ b_3c_3 \end{vmatrix} + a_3\begin{vmatrix} b_1c_1 \\ b_2c_2 \end{vmatrix}$$

$$\begin{vmatrix} b_1c_1 \\ b_2c_2 \end{vmatrix} = b_1c_2 - b_2c_1$$

	0	30°	45°	60°	90°	180°	270°	360°
sin	0	$\dfrac{1}{2}$	$\dfrac{1}{2}\sqrt{2}$	$\dfrac{1}{2}\sqrt{3}$	1	0	-1	0
cos	1	$\dfrac{1}{2}\sqrt{3}$	$\dfrac{1}{2}\sqrt{2}$	$\dfrac{1}{2}$	0	-1	0	1
tan	0	$\dfrac{1}{3}\sqrt{3}$	1	$\sqrt{3}$	∞	0	∞	0
cot	∞	$\sqrt{3}$	1	$\dfrac{1}{3}\sqrt{3}$	0	∞	0	∞

$$\sin^2\alpha + \cos^2\alpha = 1 \qquad \frac{\sin \alpha}{\cos \alpha} = \tan \alpha \qquad \frac{\cos \alpha}{\sin \alpha} = \cot \alpha$$

$$\tan \alpha = \frac{1}{\cot \alpha} \qquad \frac{1}{\cos^2\alpha} = 1 + \tan^2\alpha \qquad \sin \alpha = \frac{\tan \alpha}{\sqrt{1 + \tan^2\alpha}}$$

$$\cos \alpha = \frac{1}{\sqrt{1 + \tan^2\alpha}}$$

$$\sin(90° \pm \alpha) = +\cos \alpha \qquad \sin(180° \pm \alpha) = \mp\sin \alpha$$

$$\cos(90° \pm \alpha) = \mp\sin \alpha \qquad \cos(180° \pm \alpha) = -\cos \alpha$$

$$\tan(90° \pm \alpha) = \mp\cot \alpha \qquad \tan(180° \pm \alpha) = \pm\tan \alpha$$

$$\cot(90° \pm \alpha) = \mp\tan \alpha \qquad \cot(180° \pm \alpha) = \pm\cot \alpha$$

$$\sin(-\alpha) = -\sin \alpha$$

$$\cos(-\alpha) = +\cos \alpha$$

$$\tan(-\alpha) = -\tan \alpha$$

$$\cot(-\alpha) = -\cot \alpha$$

$$\sin(\alpha \pm \beta) = \sin \alpha \cos \beta \pm \cos \alpha \sin \beta$$

$$\cos(\alpha \pm \beta) = \cos \alpha \cos \beta \mp \sin \alpha \sin \beta$$

$$\tan(\alpha \pm \beta) = \frac{\tan \alpha \pm \tan \beta}{1 \mp \tan \alpha \cdot \tan \beta}$$

$$\cot(\alpha \pm \beta) = \frac{\cot \alpha \cdot \cot \beta \mp 1}{\cot \beta \pm \cot \alpha}$$

$$\sin 2\alpha = 2 \sin \alpha \cos \alpha$$

$$\cos 2\alpha = \cos^2\alpha - \sin^2\alpha = 1 - 2 \sin^2\alpha = 2 \cos^2\alpha - 1$$

$$\sin 2\alpha = \frac{2 \tan \alpha}{1 + \tan^2\alpha} \qquad\qquad \cos 2\alpha = \frac{1 - \tan^2\alpha}{1 + \tan^2\alpha}$$

$$\tan 2\alpha = \frac{2 \tan \alpha}{1 - \tan^2\alpha} = \frac{2}{\cot \alpha - \tan \alpha}, \qquad \cot 2\alpha = \frac{\cot^2\alpha - 1}{2 \cot \alpha} = \frac{1}{2}(\cot \alpha - \tan \alpha)$$

Formulas

$$1 + \cos \alpha = 2 \cos^2 \frac{\alpha}{2}, \quad 1 - \cos \alpha = 2 \sin^2 \frac{\alpha}{2}$$

$$\tan \alpha = \sqrt{\frac{1 - \cos 2\alpha}{1 + \cos 2\alpha}} = \frac{\sin 2}{1 + \cos 2\alpha} = \frac{1 - \cos 2\alpha}{\sin 2\alpha} = \frac{2 \tan \frac{\alpha}{2}}{1 - \tan^2 \frac{\alpha}{2}}$$

$$\sin \alpha + \sin \beta = 2 \sin \frac{\alpha + \beta}{2} \cdot \cos \frac{\alpha - \beta}{2}$$

$$\sin \alpha - \sin \beta = 2 \cos \frac{\alpha + \beta}{2} \cdot \sin \frac{\alpha - \beta}{2}$$

$$\frac{\sin \alpha + \sin \beta}{\cos \alpha + \cos \beta} = \tan \frac{\alpha + \beta}{2}$$

$$\cos \alpha + \cos \beta = 2 \cos \frac{\alpha + \beta}{2} \cdot \cos \frac{\alpha - \beta}{2}$$

$$\cos \alpha - \cos \beta = -2 \sin \frac{\alpha + \beta}{2} \cdot \sin \frac{\alpha - \beta}{2}$$

$$\frac{\sin \alpha - \sin \beta}{\cos \alpha + \cos \beta} = \tan \frac{\alpha - \beta}{2}$$

$$\frac{\tan \alpha + \tan \beta}{\cot \alpha + \cot \beta} = \tan \alpha \cdot \tan \beta$$

$$\tan \alpha \pm \tan \beta = \frac{\sin(\alpha \pm \beta)}{\cos \alpha \cos \beta} \qquad \frac{1 + \tan \alpha}{1 - \tan \alpha} = (\tan 45° + \alpha)$$

$$\cot \alpha \pm \cot \beta = \frac{\pm \sin(\alpha \pm \beta)}{\sin \alpha \sin \beta} \qquad \frac{\cot \alpha + 1}{\cot \alpha - 1} = \cot(45° - \alpha)$$

$$\cos \alpha + \sin \alpha = \sqrt{2} \sin(45° + \alpha) = \sqrt{2} \cos(45° - \alpha)$$

$$\cos \alpha - \sin \alpha = \sqrt{2} \cos(45° + \alpha) = \sqrt{2} \sin(45° - \alpha)$$

$$\cot \alpha + \tan \alpha = \frac{2}{\sin 2\alpha} \qquad \cot \alpha - \tan \alpha = 2 \cot 2\alpha$$

$$(u \cdot v)' = uv' + u'v$$

$$\left(\frac{u}{v}\right)' = \frac{u'v - v'u}{v^2}$$

$$(\sin x)' = \cos x \qquad (e^x)' = e^x$$

$$(\cos x)' = -\sin x \qquad (\ln x)' = \frac{1}{x}$$

$$(\tan x)' = \frac{1}{\cos^2 x} \qquad (\arcsin x)' = \frac{1}{\sqrt{1 - x^2}}$$

$$(\cot x)' = \frac{-1}{\sin^2 x} \qquad (\arccos x)' = -\frac{1}{\sqrt{1 - x^2}}$$

Formulas

$$\int u \, dv = uv - \int v \, du$$

$$\int dx = x \qquad \int x^n \, dx = \frac{x^{n+1}}{n+1} \quad (n \neq -1) \qquad \int \frac{dx}{x} = \ln x$$

$$\int \sin x \, dx = -\cos x \qquad\qquad\qquad \int \cos x \, dx = \sin x$$

$$\int \cot x \, dx = \ln \sin x \qquad\qquad\qquad \int \frac{dx}{\sin^2 x} = -\cot x$$

$$\int \frac{dx}{\cos^2 x} = \tan x$$

$$\int \frac{dx}{1 - x^2} = \ln \sqrt{\frac{1+x}{1-x}}$$

$$\int \frac{dx}{\sqrt{1 - x^2}} = \arcsin x \qquad\qquad\qquad \int \frac{dx}{1 + x^2} = +\arctan x$$

$$\int e^x \, dx = e^x \qquad\qquad \int a^x \, dx = \frac{a^x}{\ln a} \qquad \int \frac{dx}{x \pm a} = \ln(x \pm a)$$

Formulas

Index

Index

Index